Numerical Analysis
for Computer Science

Numerical Analysis
for Computer Science

Irving Allen Dodes

NORTH HOLLAND
NEW YORK · OXFORD

Elsevier North Holland, Inc.
52 Vanderbilt Avenue, New York, New York 10017

Distributors outside the United States and Canada:

Thomond Books
(A Division of Elsevier/North-Holland Scientific Publishers, Ltd.)
P.O. Box 85
Limerick, Ireland

Library of Congress Cataloging in Publication Data

Dodes, Irving Allen.
    Numerical analysis for computer science.

    Bibliography: p.
    Includes index.
    1. Numerical analysis–Data processing.
I. Title.
QA297.D59   519.4   77-24547
ISBN 0-444-00238-3

Manufactured in the United States of America

How do I love thee? Let me count the ways.
Sonnets from the Portuguese
ELIZABETH BARRET BROWNING

*Dedicated to my lovely wife, Dorothy, whose
only computing skill, an important one,
is that of counting the ways.*

Contents

Computer Programs

Preface

I have worked and taught in the field of computing since 1952. This book is the result of my experience with students ranging from very superior seniors at the Bronx High School of Science, to sophomores in a community college, to graduate students in mathematics, to teachers of mathematics and the sciences.

The entire book is intended to serve as a textbook and reference book for a one-year course which has as its aim a first serious exploration of computer programming and its mathematical background. Selected sections, as indicated at the end of the preface, can be used to serve a one-semester course.

The main content of this book involves the illumination and solution of nine major elementary problems of computing:

1. Making a table of values from a formula
2. Finding a formula from a table of values
3. Solving an equation, linear or non-linear
4. Solving a system of equations, linear or non-linear
5. Solving matrix problems, e.g. inversion and eigenvalues
6. Finding a sum efficiently
7. Integrating numerically
8. Finding the numerical solution of an ordinary differential equation
9. Finding the solution of a system of ordinary differential equations

Although these topics are more or less common to all books in the field, the emphasis in this book is quite different from any other. *First*, the mathematical background of the problem is explored as thoroughly as possible. If background theorems in algebra, calculus or finite differences are needed, they are (usually) displayed and proved in the book. In very few cases, this would have led too far afield and a reference had to be given. However, the book is almost self-contained in this respect. *Second*, an attempt is made to push the computer to its utmost accuracy. Computers operate in millionths or even billionths of a second per operation. Because the input-output mechanisms cannot operate at a speed comparable with that of computation, there is virtually no difference in turnaround time between a sloppy program and a precise program. Almost all the programs in this book are done precisely, in double precision. *Third*, the book starts with error analysis. The analysis of errors is carried throughout the book, and the concepts developed in Chapter 1 are used in every illustrative program. This first chapter is an obvious logical beginning for a serious course, but may not be psychologically satisfactory for everyone. For this reason, a summary of programming hints, based upon

Chapter 1, has been placed at the end of the chapter for those who prefer to postpone the study of the first chapter. This should be read before starting Chapter 2. So far as I know, no other book at this level has so detailed a discussion of error analysis. This reflects my deep feeling that error analysis is essential for expert programming. _Fourth_, the book does not bother to discuss algorithms which are merely of historical interest, e.g. Euler's method for differential equations, unless there is a special point to make, e.g. the _caveat_ in the use of Simpson's rule. A smaller number of methods were chosen for good error characteristics, and these are delved into thoroughly. For this reason, a great deal of time and space is devoted to Chebyshev-Hastings methods. These are easy; they are the ones usually used in computer software; and they are very good. I think the book is unique in this respect.

I have included one sample of every important program as worked out in class. A list of these programs can be found following the Table of Contents. I think that I should explain that I usually teach, or reteach, **FORTRAN** (or **WATFIV**) along with the numerical-analysis course. (The usual programming course does not meet my needs.) Therefore, you will find that the beginning programs are done with simple techniques, and the programs increase in sophistication throughout the course. For example, along with one program, I reviewed the computed **GO TO**, and with another the variable **FORMAT**. No claim is made that the programs, as displayed, are optimized for space or speed. They do the job and do it accurately. That is all I wanted as a first trial.

In the main, the problem sections are designed to afford practice. I usually assign different problems to different students. Enough information is provided in the answer section at the end of the book, or with the problems, to enable the student to check his results. There is ample opportunity for research by varying the problems and comparing results. In fact, the comparison of class results is one of the most exciting outcomes of this approach.

We come to the problem, ever present, of typographical and other mistakes. I have tried very hard to eliminate these. The material has been taught in its present form for at least ten years, with different sections omitted each time. (There was not enough time to do it all.) It would be foolish to believe that there are no mistakes. And mistakes are very irritating. However, e^{x^2} and e^{x2} look alike to a non-mathematician, and so do 3.141592 and 3.141952. A note about mistakes will be most appreciated.

Part of my teaching technique is to "step" through a problem by a sample calculation before writing the program. In order to prepare the many presentations for this book, I needed a very good calculator or computer at home. I am very grateful to Mr. Stan Rose, of the Wang Laboratories, for lending me a calculator and, afterwards, a small computer, for this purpose. I am also grateful to Dr. George Grossman, Director of Mathematics for New York City, for his help in certain derivations, and to my colleagues at Kingsborough for their various acts of mercy—in particular my good friend and associate, Louis Lampert, who has been of inestimable assistance.

IRVING ALLEN DODES
Forest Hills, New York

A Suggested Time Schedule
for Lectures

It would be both impractical and impertinent to write a time schedule for another teacher, another class or another college. The following is then merely a suggestion for (1) the one-year course and (2) the one-semester course. In both cases, it is assumed that students are assigned reading in advance of each lecture, and that lab time (for preparation and discussion of programs) and time for tests is to be added. Each semester should be a minimum of four semester-hours, preferably five, depending upon the previous preparation of the students.

SECTION	HOURS (YEAR COURSE)	HOURS (TERM COURSE)
1.1	2	—
1.2	2	—
1.3	1	—
1.4	2	—
1.5	1	—
2.1	2	1
2.2	1	1
2.3	1	1
2.4	1	1
3.1	2	2
3.2	1	—
3.3	2	2
3.4	2	2
3.5	2	—
4.1	1	—
4.2	1	1
4.3	2	2
5.1	1	—
5.2	2	1
5.3	1	1
5.4	1	—
6.1	2	2
6.2	1	—
6.3	2	—
6.4	2	—
6.5	2	—

SECTION	HOURS (YEAR COURSE)	HOURS (TERM COURSE)
7.1	1	1
7.2	2	1
8.1	1	—
8.2	2	2
9.1	2	—
9.2	2	2
10.1	2	2
10.2	1	—
10.3	1	—
11.1	2	2
11.2	2	2
11.3	2	2
12.1	1	1
12.2	1	1
12.3	1	1
12.4	2	2
12.5	2	2
13.1	1	—
13.2	2	2
13.3	2	2
14.1	2	—
14.2	2	—
14.3	2	—
14.4	1	—
14.5	2	—
Total	81	42

Numerical Analysis for Computer Science

An Introductory Comment

Part of the purpose of this book is to explore those methods of computation (*algorithms*) which guarantee maximum accuracy. Would you like to test your present intuition concerning procedure? In the following, the literal symbols stand for real numbers like 123.45678. It is assumed that the computer is of the third generation or later, with fixed-word and binary format internally, like the IBM 360. Which of the following statements are *true*? Assume that there are no overflows or other gross mistakes on the part of the programmer.

1. $a+(b+c)=(a+b)+c$.
2. $a(bc)=(ab)c$.
3. Some Taylor series are alternating and some are monotonic. If a series is alternating, the error of truncation is no greater than the first term neglected. The error for a monotonic series can be estimated with somewhat more difficulty, e.g. the Lagrange error term. From the standpoint of accuracy, if there is a choice, it is better to use the alternating series.
4. If a function is analytic in the neighborhood of (x_0, y_0), a Taylor series is one of the best methods for approximating the function.
5. A series used to evaluate a function must be convergent.
6. To solve a linear system $AX = B$, where A is a matrix and both X and B are vectors, the Gauss-Jordan method using $a_{11}, a_{22}, \ldots, a_{nn}$ as pivots in that order is an excellent method.
7. To check an equation or a system of equations, $f(X) = 0$, substitute in the original. If the result of the substitution, $f_k(X)$, satisfies $|f_k(X)| \leqslant \epsilon$, where ϵ is sufficiently small, the result is satisfactory.
8. It is a good general rule to use a short formula in preference to a long one, if there is a choice. For example, the statistical definition of the *standard deviation* is

$$\sigma = \sqrt{\frac{\Sigma(X-M)^2}{N}}$$

where X is a raw score and M is the mean of a column of N raw scores. By algebraic manipulation, the following equivalent formula can be derived:

$$\sigma = \frac{\sqrt{N\Sigma(X^2)-(\Sigma X)^2}}{N}$$

It is more accurate to use the first formula.

9. Simpson's rule is an excellent method for numerical integration (quadrature).
10. In a numerical integration, accuracy can be improved by making the intervals smaller.

It would be surprising to find many readers who, at this point, would know that only statement 2 is true. The others are false for various reasons. We shall start with some of the reasons in the first chapter.

Chapter 1

Introduction to Error Analysis

This chapter deals with the unavoidable errors involved in computer operations with floating-point numbers. Happily, errors often cancel each other—but this is unpredictable. There is always that time when the errors add instead of canceling. It is important for the computer scientist to write his program in such a way that the piling up of errors is minimized. For those who wish to postpone the thorough treatment of error analysis in this chapter, a short summary of programming hints is given at the end (Section 1.6). All the programs in this book were written with error minimization in mind.

You have just signed a six-figure check and obtained a gleaming computer. It seems reasonable to expect that if you read a decimal number like 0.13579246 into the machine and print it out (without doing anything with the number), then it will print out correctly. We shall show, however, that the correct *echo* occurs only rarely. Hence, if you add two decimal numbers like these, or subtract them, or do anything at all with them, there is almost no chance that the answer will print exactly right to the same number of places.

In Figure 1-1(a), a short FORTRAN program is displayed. This program enters ten decimal numbers $a_1, a_2, \ldots, a_{10}$ into the computer. The result of the programs shows how the computer converted these numbers (we shall explain) to the form shown in Figure 1-1(b). Figure 1-1(c) shows the output, which is to be compared with the original input in Figure 1-1(a).

The crimes committed by the computer or by the computer procedure are called *errors*. (This is different from *mistakes*, which we never make.)

We define error as follows:

ERROR, ϵ

$$\epsilon_x = x_T - x \tag{1.1}$$

where ϵ_x is the error in x, x_T is the true value of x, and x is the computed value of x.

RELATIVE ERROR, i

$$i_x = \epsilon_x / x \qquad (1.2)$$

where i_x is the *relative error* or *inherent error* in x.

It is natural to wonder why (1.2) has the computer value of x in the denominator instead of the true value. The reason is, of course, that we don't ordinarily know the true value. If we did, we wouldn't be doing the computation in the first place. In spite of this, we can often estimate the error and, therefore, the relative error.

This chapter is intended to furnish a background for an understanding of the errors involved in a computed result, especially in a hexadecimal, fixed-word machine like the IBM 360 or 370, the most popular type at this time. We shall discuss two types of errors at input time, and two types of procedural errors.

Before launching into the discussion, it is necessary to define three types of input numbers:

EXACT NUMBERS   Certain constants and any number obtained by a count. *Examples*: π, 7 (girls), \$23.14 (i.e., 2314 pennies).

APPROXIMATE NUMBERS   Numbers obtained by an estimate or measurement. *Examples*: 3.1415926 as an estimate for π, 28 in the phrase "about 28 people", 23.1° centigrade, 5 grams.

STATISTICAL NUMBERS   Numbers obtained by a process of statistical analysis, sometimes recognizable by a form such as 2.852 ± 0.009 volts, in which a standard measure of error is included. This standard measure is usually the standard deviation (σ) in modern works; the probable error (P.E.) in older ones.

We must also define and explain *accuracy* and *precision*.

ACCURACY   The number of significant figures *(sf)* in a datum.

PRECISION   The order of magnitude of the last figure of a datum.

ILLUSTRATIVE PROBLEM 1-1   Find the accuracy and precision of (a) 54.273, (b) 0.00178921, (c) 0.048730, (d) 2700, (e) 2700.0.

Solution   (a) This has 5 sf (its accuracy). Its precision is 0.001, or 10^{-3}. (b) This has 6 sf. The *leading* or *high-order* zeros are merely placeholders. Its precision is 10^{-8}. (c) This has 5 sf. Its precision is 10^{-6}. (d) There is no way to tell whether the two rightmost (*low-order*) zeros are placeholders or significant figures. The person

```
        DIMENSION A(10)
C INPUT
        A(1)=.12345678
        A(2)=.23456789
        A(3)=.34567890
        A(4)=.45678901
        A(5)=.56789012
        A(6)=.67890123
        A(7)=.78901234
        A(8)=.89012345
        A(9)=.90123456
        A(10)=.13579246
C THE FOLLOWING SKIPS TO A NEW PAGE AND SHOWS HOW THE MACHINE
C CONVERTS THE INPUT FOR ITS OWN USE.
        WRITE(3,101)
101     FORMAT(1H1,T10,'INTERNAL FORM'/)
        DO 1 I=1,10
1       CALL PDUMP(A(I),A(I),0)
C OUTPUT.
        WRITE(3,103)
103     FORMAT(1H0,T10,'OUTPUT'/)
        DO 2 I=1,10
2       WRITE(3,102)I,A(I)
102     FORMAT(1H0,T2,'A(',I2,')=',F9.8)
        END
```

(a)

INTERNAL FORM

OUTPUT

002898	401F9ADD

A(1)=.12345678

00289C	403C0CA4

A(2)=.23456788

0028A0	40587E69

A(3)=.34567887

A(4)=.45678896

0028A4	4074F01F

A(5)=.56789011

0028A8	4091613F

A(6)=.67890120

A(7)=.78901231

0028AC	40ADCC78

A(8)=.89012343

0028B0	40C9FCB6

A(9)=.90123451

A(10)=.13579243

0028B4	40E3DF21

(c)

0028B8	40E6B74E
0028BC	4022C34B

(b)

Fig. 1-1 (a) An input-output program. (b)Internal form. (c)Output.

who supplies the input should either tell how many sf there are or else rewrite the number in scientific notation. (In scientific notation, all the digits are significant.) For example, suppose only one of the two low-order zeros is significant. Then it could be written as 2700 (3 sf) or else as 2.70×10^3. In either case, the accuracy is 3 sf and the precision is 10^1, or 10. (e) This has 5 sf. The precision is 0.1 or 10^{-1}. If the zeros were not significant, there would be no sense in writing .0 after the 2700.

1.1 INPUT ERRORS

1.1.1 The Internal Form of a Number

Most modern computers operate with a fixed length of number. In the IBM 360, the decimal number may be 8, 16 or 32 digits long; these are called, respectively, single-precision, double-precision and quadruple-precision. In most modern computers, floating-point numbers are converted from decimal into binary for internal use. These binary numbers are expressed in the hexadecimal system (base 16) or in the octal system (base 8). For output, the numbers are reconverted to decimal.

1.1.2 The Hexadecimal System

In the *decimal* system, there are ten distinct digits: $0, 1, 2, \ldots, 9$. The value of a number depends upon the placement of each digit. For example, 623.45 represents

$$6 \times 10^2$$
$$+2 \times 10^1$$
$$+3 \times 10^0$$
$$+4 \times 10^{-1}$$
$$+5 \times 10^{-2}$$

In the *hexadecimal* system (professionals call it *hex* for short), there are 16 distinct digits: $0, 1, 2, \ldots, 9$, A, B, C, D, E, F. Subscripts x or *10* identify the base being used if there is a question. The digits of the number $2A7.F_x$ represent

2×16^2	or	$2 \times 10_x^2$
10×16^1	or	$A \times 10_x^1$
7×16^0	or	$7 \times 10_x^0$
15×16^{-1}	or	$F \times 10_x^{-1}$

a total of (decimal) 784.9375_{10}. We mention in passing that $427_8 = 4 \times 8^2 + 2 \times 8^1 + 7 \times 8^0 = 279_{10}$.

Although we will seldom need to do the conversion from decimal to hex or hex to decimal (the computer does this internally), we must demonstrate a method in order to show the source of the error which is introduced.

ILLUSTRATIVE PROBLEM 1.1-1   Convert 387.62000_{10} into hex.

Solution   The whole-number part is obtained by repeated division:

$$16\,\overline{)387}\quad 3$$
$$16\,\overline{)24}\quad 8$$
$$1$$

This indicates that $387_{10} = 183_x$. In effect, the first division finds how many 16s there are (there are *24*) and how many 1s (there are *3*). The second division is really a division of the original number by 256, since *two* successive divisions were done. It tells how many 16^2s there are (there is *1*) and how many 16s (there are *8*). Note that the answer is read upwards.

It would be possible to convert the fractional part of the number using repeated divisions by 16^{-1}, 16^{-2}, and so on. An easier but equivalent procedure is to use repeated multiplication by 16, 16^2,... as follows:

$0.62000 \times 16 = 9.92000$	first hex digit is *9*
$0.92000 \times 16 = 14.72000$	second hex digit is *E*
$0.72000 \times 16 = 11.52000$	third hex digit is *B*
$0.52000 \times 16 = 8.32000$	fourth hex digit is *8*
$0.32000 \times 16 = 5.12000$	fifth hex digit is *5*
$0.12000 \times 16 = 1.92000$	sixth hex digit is *1*
$0.92000 \times 16 = 14.72000$	(beginning of repetition)

The result is an unending hexadecimal:

$$387.62000_{10} + 183.9\,EB851\,EB851\,EB851 \cdots_x$$

As you will see, this kind of result causes a *conversion error*. To lead into this, we illustrate the reverse conversion.

ILLUSTRATIVE PROBLEM 1.1-2   Convert $183.9EB_x$ to decimal.

Solution

$$1 \times 16^2 = 256$$
$$8 \times 16^1 = 128$$
$$3 \times 16^0 = 3$$
$$9 \times 16^{-1} = 0.5625$$
$$14 \times 16^{-2} = 0.0546875 \qquad (E = 14)$$
$$\underline{11 \times 16^{-3} = 0.0026855\ldots \qquad (B = 11)}$$
$$387.6198730\ldots_{10}$$

1.1.3   Normalized Form

We mentioned that an 8-digit (single-precision) decimal number is converted into binary internally and expressed in hex. This hex number has a 6-digit *mantissa* and a 2-digit *characteristic*. For example, we showed that

$$387.62000_{10} = 183.9EB851\,EB851\ldots_x \qquad (1.1.1)$$

The hex number is *normalized* by shifting the hexadecimal point to the left of the first non-zero digit. Thus, remembering that the symbol 10 in hex is the same as 16 in decimal,

$$183.9\,EB851\,EB851\ldots_x \rightarrow 0.1839\,EB851\,EB851\ldots_x \times 10_x^3 \qquad (1.1.2)$$

In (1.1.2), the right member is multiplied by 10_x^3 to compensate for the three-digit leftward shift of the hexadecimal point.

The *mantissa* is now chopped to 6 digits for single precision (14 digits for double precision), and 40_x is added to the exponent to obtain the characteristic. The result is put together with characteristic first:

$$43\,1839EB \qquad (1.1.3)$$

(A space has been inserted to facilitate reading. There is no space left in the computer.) The addition of $40_x = 64_{10}$ to the exponent is done to avoid working with negative exponents. It does not concern us except to explain the internal form shown in Fig. 1-1(b).

1.1.4　The Error of Conversion

From Illustrative Problems 1 and 2, you see that if 387.62000_{10} is read into a computer and promptly read out again, the *error of conversion* introduced is

$$
\begin{array}{r}
387.62000 \\
-\ 387.61987 \\
\hline
0.00013
\end{array}
$$

which represents a relative error of $0.00013/387.61987 = 3.35 \times 10^{-7}$. This may not seem like much, but when further calculations are done it is quite possible to multiply the error by a large factor, as you will see.

ILLUSTRATIVE PROBLEM 1.1-3　Display the internal single-precision form of $0.00413\,6AF_x$.

Solution　This is $0.4136AF_x \times 10_x^{-2}$. Using $-2_x + 40_x = 3E_x$, the required form is 3E4136AF. The characteristic is 3E, the exponent is -2, and the mantissa is 4136AF.

ILLUSTRATIVE PROBLEM 1.1-4　Find the decimal value (8 sf) of a floating-point number stored internally as 3F ABC000.

Solution　The characteristic is 3F. Therefore the exponent is $3F_x - 40_x = -1$. The mantissa is ABC000. The *unnormalized* hexadecimal is therefore $0.0ABC_x$. The decimal value is $0.04193\,1152_{10}$, found as follows:

$$0 \times 16^{-1} = 0.$$
$$+ A \times 16^{-2} = 0.0390625 \qquad (A = 10)$$
$$+ B \times 16^{-3} = 0.0026855466 \qquad (B = 11)$$
$$+ C \times 16^{-4} = 0.001831056 \qquad (C = 12)$$

ILLUSTRATIVE PROBLEM 1.1-5   Find the decimal value to 8 sf of a floating-point number stored internally as 40 FFFFFF.

Solution

$$F \times 16^{-1} = 0.9375 \qquad [F = 15]$$
$$+ F \times 16^{-2} = 0.05859375$$
$$+ F \times 16^{-3} = 0.0036621094$$
$$+ F \times 16^{-4} = 0.0002288818$$
$$+ F \times 16^{-5} = 0.0000143051$$
$$+ F \times 16^{-6} = 0.0000008941$$
$$\overline{\qquad\qquad 0.9999999404}$$

Chopping to 8 sf, the result is 0.99999994.

1.1.5 Chopping and Symmetric Roundoff

We shall investigate the results of two methods of roundoff. In *chopping roundoff* to hundredths, a decimal number like 23.469 is rounded to 23.46, dropping the 9. In *symmetric roundoff*, the same number becomes 23.47. The chopping method is characteristic of computers, the second of some calculators.

For the moment, in order to clarify the problem, we shall assume *unrealistically* that we are working with a decimal computer which has a fixed word length of 4 digits, everything else being chopped. Consider the decimal number 23.4673, which we shall write in normalized form as 0.234673×10^2. We shall separate the parts to be kept from the parts to be chopped by writing the number as

$$(0.2346 + 0.000073) \times 10^2$$

which can be rewritten, for our purposes, as

$$(0.2346 + [0.73 \times 10^{-4}]) \times 10^{-2}$$

When the computer chops, the result is 0.2346×10^{-2}.

More generally, any floating-point number can be written, before rounding, in the form

$$(p + [q \times b^{-t}]) \times 10^n$$

where p is the mantissa, t is the length of the mantissa permitted by the computer, b is the base, and n is the exponent which becomes the characteristic.

ILLUSTRATIVE PROBLEM 1.1-6   Express (a) 1234.5678, (b) 0.0012345678 in normalized unrounded form for a hypothetical decimal computer with fixed word length of 5 digits.

Solution   (a) $(0.12345 + 0.678 \times 10^{-5}) \times 10^{4}$, (b) $(0.12345 + 0.678 \times 10^{-5}) \times 10^{-2}$.

For later use, note that in decimal

$$0.1000 \leqslant |p| \leqslant 0.9999 < 1 \qquad (1.1.5)$$

$$0 \leqslant |q| \leqslant 0.9999 < 1 \qquad (1.1.6)$$

whereas in hexadecimal

$$0.1000 \leqslant |p| \leqslant 0.FFFF < 1 \qquad (1.1.7)$$

$$0 \leqslant |q| \leqslant 0.FFFF < 1 \qquad (1.1.8)$$

For the discussion which follows, we shall also need the fact that in decimal a number like $0.19999\ldots$ is mathematically identical to 0.2; in hex, a number like $0.A3D7F\,FFF\ldots$ is mathematically identical to 0.A3D8.

1.1.6   The Chopping Theorem

We are going to develop a theorem which expresses, for any computer that chops, the maximum possible roundoff error and relative error. Before doing so, we pause to take some arithmetic examples. In these examples, we assume an unrealistic hypothetical decimal computer with fixed word length of 4.

Case I. Given: $a_T = 0.123499\ldots \times 10^{n}$.

$$a_T = 0.1235 \times 10^{n}$$

$$\epsilon_a = a_T - a = (0.1235 - 0.1234) \times 10^{n}$$

$$= 10 \times 10^{-5} \times 10^{n}$$

$$|\epsilon_a| = 10 \times 10^{-5} \times 10^{n}$$

$$i_a = \epsilon_a / a = (1 \times 10^{-4}) \times 10^{n} / (0.1234 \times 10^{n})$$

$$= 8.10 \times 10^{-4}$$

$$|i_a| = 8.10 \times 10^{-4}$$

Case II. Given: $b_T = 0.12342 \times 10^n$.

$$\epsilon_b = b_T - b = (0.12342 - 0.1234) \times 10^n$$
$$= 2 \times 10^{-5} \times 10^n$$
$$|\epsilon_b| = 2 \times 10^{-5} \times 10^n$$
$$i_b = \epsilon_b / b = (2 \times 10^{-5} \times 10^n)/(0.1234)$$
$$= 1.62 \times 10^{-4}$$
$$|i_b| = 1.62 \times 10^{-4}$$

Case III. Given: $c_T = -0.12347 \times 10^n$.

$$\epsilon_c = c_T - c = [(-0.12347) - (-0.1234)] \times 10^n$$
$$= -7 \times 10^{-5} \times 10^n$$
$$|\epsilon_c| = 7 \times 10^{-5} \times 10^n$$
$$i_c = \epsilon_c / c = 5.67 \times 10^{-4}$$
$$|i_c| = 5.67 \times 10^{-4}$$

Case IV. Given $d_T = -0.1234999\ldots \times 10^n$.

$$\epsilon_d = d_T - d = -0.1235 \times 10^n$$
$$|\epsilon_d| = 0.1235 \times 10^n$$
$$i_d = \epsilon_d / d = 8.10 \times 10^{-4}$$
$$|i_d| = 8.10 \times 10^{-4}$$

From these four examples, you might guess that the *worst* absolute error, for a 4-digit decimal machine, is $10 \times 10^{-4} \times 10^n$, and the worst absolute relative error is 10×10^{-4}.

THE CHOPPING THEOREM   If a computer uses base b for conversion and chops to a t-digit mantisssa, then

$$|i_a| \leqslant b \times b^{-t} \qquad (1.1.9)$$

where i_a is the relative error of a.

Proof   Consider the unrounded normalized number

$$a_T = [p + q \times b^{-t}] \times b^n$$

where $0.1 \leqslant |p| < 1$ and $0 \leqslant |q| < 1$. After chopping,

$$a = p \times b^n \tag{1.1.10}$$

$$\epsilon_a = a_T - a = q \times b^{-t} \times b^n \tag{1.1.11}$$

$$i_a = \epsilon_a / a = \frac{q}{p} \times b^{-t} \tag{1.1.12}$$

$$|i_a| = \frac{|q|}{|p|} \times b^{-t} \tag{1.1.13}$$

The *maximum* value of $|q|$ is 1, and the *minimum* value of $|p|$ is 0.1. If q and p are decimal (i.e., $b = 10$), then the maximum for the fraction is 10_{10}. If q and p are hexadecimal (i.e., $b = 16$), then the maximum for the fraction is 10_x, which is the same as 16_{10}. In general, the maximum is the base, b. This completes the proof. $\square$

ILLUSTRATIVE PROBLEM 1.1-7   An octal (base 8) machine which chops has a word length of 8. (a) What is the maximum absolute relative error? (b) What is the maximum absolute error of a number N?

Solution   (a) 8×8^{-8}, which is about 4.77×10^{-7}. (b) $4.77 \times |N| \times 10^{-7}$.

ILLUSTRATIVE PROBLEM 1.1-8   A hexadecimal machine which chops has a mantissa of 6. Find the maximum relative error.

Solution 16×16^{-6}, approximately 9.54×10^{-7}.

1.1.7 The Symmetric-Roundoff Theorem

Because many calculators use symmetric roundoff, we shall consider it briefly. We start with the same examples as those in the previous section, showing only the results.

Case I.   *Case III.*

$$\epsilon_a = 0$$ $$\epsilon_c = (-3 \times 10^{-5}) \times 10^n$$
$$i_a = 0$$ $$i_c = -2.4 \times 10^{-4}$$

Case II.   *Case IV.*

$$\epsilon_b = (2 \times 10^{-5}) \times 10^n$$ $$\epsilon_d = (3 \times 10^{-5}) \times 10^n$$
$$i_b = 1.6 \times 10^{-4}$$ $$i_d = 2.4 \times 10^{-4}$$

From these few examples, you might guess that the absolute relative error for symmetric rounding is much less than that for chopping. We omit the proof of the following.

THE SYMMETRIC-ROUNDOFF THEOREM   If a computer uses base b for conversion and rounds symmetrically to a t-digit mantissa, then

$$|i_a| \leqslant \frac{b}{2} \times b^{-t} \qquad (1.1.14)$$

Note that the symmetric-rounding error is, at worst, one-half the chopping error.

ILLUSTRATIVE PROBLEM 1.1-9   A Wang electronic calculator displays 10 digits and rounds symmetrically. It is a decimal machine. (a) What is its maximum relative error? (b) What is the maximum absolute error of a number N? (c) Compare the result with a hexadecimal computer which chops and uses a 6-digit mantissa for single precision.

Solution   (a) $b = 10$ and $t = 10$. Therefore $|i_a| \leqslant 5 \times 10^{-10}$. (b) $|e_a| \leqslant 5 \times |N| \times 10^{-10}$. (c) For the computer, $b = 16$ and $t = 6$; $|i_a| \leqslant 16 \times 16^{-6} \sim 9.54 \times 10^{-7}$. The calculator is $(9.54 \times 10^{-7})/(5 \times 10^{-10})$ or about 1900 times as accurate, so far as roundoff is concerned.

1.1.8   Practical Considerations

As professionals, we naturally wish our computers and our programs to avoid the introduction of errors. However, as we have shown, the introduction of errors is inevitable even if no computations are done.

For pure mathematical work, this can be quite serious if the input is supposed to be extremely precise numbers. Unless multiple precision is used, the precision of the input can be destroyed. For scientific work, the situation is not so bad. It is true that in a typical hexadecimal machine, the relative error of conversion can be as large as 9.54×10^{-7} before calculations. However, the relative error of the input datum is usually more than that, anyhow. A scientific observation of 27.813 has a possible error as large as 0.0005 and, consequently, a relative error of 1.80×10^{-5}. When the conversion error is added, the input error becomes about 1.90×10^{-5}, not a drastic change. For the computer error in a scientific calculation to be very important, the input datum would have to be something like 27.813728 so that its relative error would be 1.80×10^{-8} before conversion. Few scientific observations are that accurate. In such a case, we would naturally use double or quadruple precision to obtain accurate results.

The true importance of this section is that every time a datum is entered into a machine, a conversion error must be added to the *approximation error* inherent in the datum, and every time a calculation is done, the inherent error is inextricably involved. We shall investigate this further in the sections which follow.

PROBLEM SECTION 1.1

A. In each of the following, (a) convert to hex, chop the mantissa to 6 hex digits, express in normalized form; (b) convert this back to decimal (8 sf); (c) find the relative error of conversion in the form $x.xx \times 10^{-n}$. In part (a), do not round off symmetrically.

(1) 15.275   (4) 409.826

(2) 21.418   (5) 5233.618

(3) 358.937   (6) 6927.411

B. Assume that the computer is a decimal machine with fixed word length of 4. Show the internal form after chopping:

(7) 26.543   (10) 0.00043 329

(8) 81.005   (11) 2.71828 1828

(9) π (12) 0.00061 180

C. Express in unrounded form for a decimal computer with a fixed word length of 4 decimal digits; find the absolute relative error under chopping; find the absolute relative error under symmetric rounding:

(13) 8.124877   (16) 581.63572

(14) 9.023506   (17) 0.001588946

(15) 158.26228   (18) 0.001973926

1.2   ERRORS IN ADDITION AND SUBTRACTION

1.2.1   Addition of Floating-Point Numbers

We have discussed two kinds of errors at input time: the approximation error of the original datum, and the additional error of conversion. Together, they are often called the *inherent error*. This error is attached to the number and interferes with every calculation.

Now we are also going to discuss two kinds of *procedural errors*: those caused by roundoff in arithmetic operations, and those caused by the truncation of infinite series. To understand the roundoff problem, it is necessary to delve, very briefly, into the mechanics of the procedure used by the computer to accomplish addition. To clarify the method, we once again use the hypothetical decimal computer with word length of 4.

ILLUSTRATIVE PROBLEM 1.2-1   Add 1234 and 357.

Solution $1234 = 0.1234 \times 10^4$, $357 = 0.3570 \times 10^3$. The machine subtracts exponents (here, $4 - 3 = 1$) and inserts placeholder zeros for the datum with the smaller exponent. For the purpose of addition, 357 becomes *temporarily* 0.0357×10^4.

$$0.1234 \times 10^4$$
$$+ \ 0.0357 \times 10^4$$
$$\overline{0.1591 \times 10^4}$$

In this case, there is no error in the procedure.

Now we shall consider $1234 + 357.9$. Different computers handle this problem differently, and the same computer, using different kinds of instructions, may handle the same problem differently. In a hypothetical computer with fixed word length 4, we have $1234 = 0.1234 \times 10^4$ and $357.9 = 0.3579 \times 10^3$. The latter must be shifted to align the decimal points and should temporarily become 0.03579×10^4. However, in a 4-digit decimal machine, (1) the 9 may be chopped, (2) the number may be rounded to 0.0358, or (3) the extra digit may be kept as a so-called *guard digit*. We omit discussion of the third case, because it seems to be rather difficult to find out when the guard digit is kept and when it is not.

ILLUSTRATIVE PROBLEM 1.2-2   Given: $a = 1234$ and $b = 357.9$. Find the sum under (a) chopping, (b) symmetric rounding.

Solution

(a)
$$a = 0.1234 \times 10^4$$
$$\underline{b = 0.357 \times 10^4} \quad \text{[temporarily shifted]}$$
$$a + b = 0.1591 \times 10^4$$

(b)
$$a = 0.1234 \times 10^4$$
$$\underline{b = 0.0358 \times 10^4} \quad \text{[shifted and rounded]}$$
$$a + b = 0.1592 \times 10^4$$

At this point, someone usually asks for the "true answer." In fact, there is no way to answer this question unless more is known about the original data. If a and b are *exact*, then 1591.9 is the true answer. If a and b are measurements, then

1234 is between 1233.5 and 1234.5
357.9 is between 357.85 and 357.95

so that the true value is somewhere between 1591.35 and 1592.45.

From a mathematical standpoint, we can regard subtraction as an operation in which a sign is changed and then algebraic addition takes place. This is not exactly what happens inside a computer, but the result is the same.

1.2.3 Roundoff Errors in Addition

The computation is quite straightforward.

ILLUSTRATIVE PROBLEM 1.2-3   Find $|i_{a+b}|$ if $a = 1234$ and $b = 357.9$ assuming (a) chopping, (b) symmetric roundoff. Assume a fixed-word computer using 4 decimal digits.

Solution   (a) From Illustrative Problem 1.2-2(a), $a + b = 0.1591 \times 10^4$ and the error is 0.00009×10^4. Therefore, $|i_{a+b}| = 5.66 \times 10^{-4}$. (b) From Illustrative Problem 1.2-2(b), $a + b = 0.1592 \times 10^4$ and the error is 0.00001×10^4. Then $|i_{a+b}| = 0.63 \times 10^{-4}$.

In Illustrative Problems 1.2-4 and 1.2-5, we assume a decimal computer of word length 4.

ILLUSTRATIVE PROBLEM 1.2-4   Find $|i_{a+b}|$ if $a = 9934$, $b = 87.99$, assuming (a) chopping, (b) symmetric roundoff.

Solution

(a)
$$a = 0.9934 \times 10^4$$
$$b = 0.0087 \times 10^4 \quad \text{[temporarily shifted]}$$
$$a + b = 1.0021 \times 10^4 \quad \text{[temporary]}$$
$$= 0.1002 \times 10^5 \quad \text{[the chopped answer]}$$

The true value is 0.1002199×10^5. The error is $(0.100 \times 10^{-4}) \times 10^5$. Then $|i_{a+b}| = 1.99 \times 10^{-4}$.

(b) The result is the same as (a) in this case, even though the intermediate results are different.

ILLUSTRATIVE PROBLEM 1.2-5   Find $|i_{a+b}|$ if $a = 1234$, $b = 0.9999$, assuming (a) chopping, (b) symmetric roundoff.

Solution

(a)
$$a = 0.1234 \times 10^4$$
$$b = 0.0000 \times 10^4$$
$$a + b = 0.1234 \times 10^4$$

The true value is 0.12349999×10^4. Therefore, the error is $(0.9999 \times 10^{-4}) \times 10^4$ and $|i_{a+b}| = 8.10 \times 10^{-4}$. (b) $a + b = 0.1235 \times 10^4$, the error is $(0.0001 \times 10^{-4}) \times 10^4$, and $|i_{a+b}| = 8.10 \times 10^{-8}$.

1.2.4 Error Theorem for Sums

Let us assume that we wish to find

$$S = a + b + c + \cdots \tag{1.2.1}$$

where a, b, c... have errors due to rounding, or conversion, or other factors. Then the error of the sum can be estimated by forming the differentials and assuming that $\Delta S \sim dS$.

$$\Delta S = da + db + dc + \cdots \qquad (1.2.2)$$

This is equivalent to

$$\epsilon_S^I = \epsilon_a^I + \epsilon_b^I + \epsilon_c^I + \cdots \qquad (1.2.3)$$

where ϵ^I is the input error. To find the relative errors i^I, we use the relationship

$$i_x^I = \epsilon_x^I / x \qquad (1.2.4)$$

Substituting for ϵ in (1.2.3), we obtain

$$S i_S^I = a i_a^I + b i_b^I + c_c^I + \cdots \qquad (1.2.5)$$

leading to the following theorem.

ERROR THEOREM FOR SUMS   If $S = a + b + c + \cdots$, then the input error ϵ_S^I is given by

$$\epsilon_S^I = \epsilon_a^I + \epsilon_b^I + \epsilon_c^I + \cdots \qquad (1.2.6)$$

and the relative input error i_S^I is given by

$$i_S^I = \left(\frac{a}{S} \right) i_a^I + \left(\frac{b}{S} \right) i_b^I + \left(\frac{c}{S} \right) i_c^I + \cdots \qquad (1.2.7)$$

Since a, b, c... may be positive or negative, the theorem holds equally well for differences.

In this chapter, we are interested in *maximum* error, i.e., the situation where all errors add and none cancel. For this reason, we are going to work with the absolute-value relationships derivable from (1.2.6) and (1.2.7). Using the extended triangle-inequality theorem,[1] we have

$$|\epsilon_S^I| \leqslant |\epsilon_a^I| + |\epsilon_b^I| + |\epsilon_c^I| + \cdots \qquad (1.2.8)$$

$$|i_S^I| \leqslant \left| \frac{a}{S} \right| |i_a^I| + \left| \frac{b}{S} \right| |i_b^I| + \left| \frac{c}{S} \right| |i_c^I| + \cdots \qquad (1.2.9)$$

It is often possible to place an upper limit on $|i|$. (We shall explain.) Let R^I be the largest of $|i_a^I|$, $|i_b^I|$,.... We write this requirement as

$$R^I = \max(|i_a^I|, |i_b^I|, \ldots) \qquad (1.2.10)$$

---

[1]See Apostol, T. M., *Calculus*, Vol. I, Blaisell, 1961, p. 32 for a proof of the extended triangle-inequality theorem.

Then, from 1.2.9,

$$|i_s^I| \le \frac{1}{|S|}(|a|+|b|+|c|+\cdots)R^I \tag{1.2.11}$$

For example, if the input data are scientific and of the form *xx.xxx*, then we assume the maximum input error is 5×10^{-4}, i.e., half the precision of the datum. Now, note that the *smallest* number of the form *xx.xxx* is 10.000, so that the *largest* relative error is $(5\times10^{-4})/10.000=5\times10^{-5}$. We would add the conversion error to this and use the result as R^I. On the other hand, if the input data have negligible errors and we are using single precision in a fixed-word computer like the IBM 360 or 370, we would use $R^I=9.54\times10^{-7}$, as previously explained.

1.2.5   The Effect of Roundoff Error

We shall use the symbols ϵ^O, i^O, and R^O for the procedural errors (read: *other* error). Then, the total errors are given by

$$\epsilon = \epsilon^I + \epsilon^O \tag{1.2.12}$$

$$i = i^I + i^O \tag{1.2.13}$$

We start with some numerical examples to clarify the situation. As usual, for explanatory purposes, we are assuming a decimal computer with a fixed word length of 4. To simplify further, we are assuming that there is no inherent error in the input, i.e., $\epsilon^I=0$.

ILLUSTRATIVE PROBLEM 1.2-6   Add 9999, 999, 99 and 9 in that order. The true sum is 11,106.

Solution   We remind you that addition is a binary operation so that we must find $9999+999$, then add 99, then add 9.

	CHOPPING	SYMMETRIC
	0.9999×10^4	0.9999×10^4
	$+0.0999\times10^4$	$+0.0999\times10^4$
	0.1099×10^5	0.1100×10^5
	$+0.0009\times10^5$	$+0.0010\times10^5$
	0.1108×10^5	0.1110×10^5
	$+0.0000\times10^5$	$+0.0001\times10^5$
	0.1108×10^5	0.1111×10^5
Answer	11,080	11,110
$|\epsilon|$	26	4
$|i|$	23×10^{-4}	3.60×10^{-4}

ILLUSTRATIVE PROBLEM 1.2-7   Add 9, 99, 999, and 9999 in that order. The true sum is 11,106.

Solution

	CHOPPING	SYMMETRIC
	0.0900×10^2	0.0900×10^2
	$+0.9900 \times 10^2$	$+0.9900 \times 10^2$
	0.1080×10^3	0.1080×10^3
	$+0.9990 \times 10^3$	$+0.9990 \times 10^3$
	0.1107×10^4	0.1107×10^4
	$+0.9999 \times 10^4$	$+0.9999 \times 10^4$
	0.1110×10^5	0.1111×10^5
Answer:	11,100	11,110
$\mid \epsilon \mid$	6	4
$\mid i \mid$	5.41×10^{-4}	3.60×10^{-4}

Illustrative Problems 1.2.6 and 1.2.7 reveal the astonishing fact that addition is not necessarily associative on a computer and that one order of adding may lead to less procedural error than another. We shall come to a definite recommendation at the end of the section which follows.

1.2.6   Total Error of a Sum

We can diagram the situation for the addition of two numbers in a *process chart* (Figure 1.2-1).

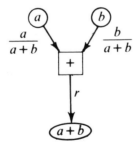

Fig. 1.2-1   Process chart for $(a+b)$.

Read this starting at the top. The input consists of two numbers, a and b. They may be original data, in which case they have inherent errors consisting of (1) the approximation error of the data, and (2) the error of conversion; or they may be the result of other calculations. In either case, they come to the top of the chart with inherent errors ϵ_a^I and ϵ_b^I, and inherent relative errors i_a^I and i_b^I. These errors are multiplied by the error factors shown at the sides of the errors; the error factors were taken from

the error theorem for sums. After addition, the result is rounded, usually
by chopping. The *total error* is then

$$i_{a+b} = i_{a+b}^I + i_{a+b}^O = \frac{a}{a+b}i_a^I + \frac{b}{a+b}i_b^I + r \tag{1.2.14}$$

ILLUSTRATIVE PROBLEM 1.2-8 (a) Make a process chart for $u = (a+b)+c$. (b) Find
the total error. (c) If $a > b > c > 0$ and the numbers have approximation relative
error $\leqslant 5 \times 10^{-6}$ and conversion relative error $\leqslant 9.54 \times 10^{-7}$, estimate the maximum total error. (d) Under the same circumstances, if the input has approximation
relative error $\leqslant 3 \times 10^{-9}$, estimate the maximum total error.

Solution

(a) The process chart is shown in Fig. 1.2-2.

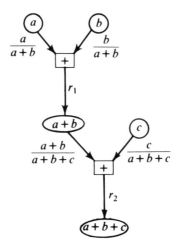

Fig. 1.2-2   Process chart for $u = (a+b)+c$.

(b)
$$i_u = \frac{a+b}{a+b+c}i_{a+b} + \frac{c}{a+b+c}i_c^I + r_2 \tag{1.2.15}$$

$$i_{a+b} = \frac{a}{a+b}i_a^I + \frac{b}{a+b}i_b^I + r_1 \tag{1.2.16}$$

$$i_u = \frac{1}{a+b+c}\left[ai_a^I + bi_b^I + (a+b)r_1 + ci_c^I + (a+b+c)r_2 \right] \tag{1.2.17}$$

$$i_u = \frac{1}{a+b+c}\left[(ai_a^I + bi_b^I + ci_c^I) + (a+b)r_1 + (a+b+c)r_2 \right] \tag{1.2.18}$$

We now use

$$R^I = \max(|i_a^I|, |i_b^I|, |i_c^I|) \tag{1.2.19}$$

$$R^0 = \max(|r_1|, |r_2|) \tag{1.2.20}$$

Because we took a, b and c as positive, we can write

$$|i_u| \leqslant \frac{a+b+c}{a+b+c} R^I + \frac{2a+2b+c}{a+b+c} R^O \qquad (1.2.21)$$

$$|i_u| \leqslant R^I + \frac{2a+2b+c}{a+b+c} R^O \qquad (1.2.22)$$

(c) The inherent error of the input must include the error of conversion:

$$R^I = 5 \times 10^{-6} + 9.54 \times 10^{-7} = 5.95 \times 10^{-6}$$
$$R^O = 9.54 \times 10^{-7}$$

These values are substituted in (1.2.22) along with the values of a, b and c.

(d)
$$R^I = 3 \times 10^{-9} + 9.54 \times 10^{-7} = 9.57 \times 10^{-7}$$
$$R^O = 9.54 \times 10^{-7}$$

These values are substituted in (1.2.22).

ILLUSTRATIVE PROBLEM 1.2-9 What happens, in the preceding problem, if a, b and c are not all of the same sign?

Discussion We simplified the problem by assuming a, b and c were all positive. It would have come out the same if they were all negative. However, if they do not agree in sign, we have the following in place of (1.2.21):

$$|i_u| \leqslant \frac{|a|+|b|+|c|}{|a+b+c|} R^I + \frac{|2a|+|2b|+|c|}{|a+b+c|} R^O \qquad (1.2.23)$$

Now, if $a+b+c$ is close to zero, the resulting error of the computation can be immense. For example, suppose there is no input error apart from conversion. Then $R^I = R^O = R$ and from (1.2.23) we can write

$$|i_u| \leqslant \frac{|3a|+|3b|+|2c|}{|a+b+c|} R \qquad (1.2.24)$$

Now, let $a=1$, $b=1.5$ and $c=-2.495$. We take the usual single-precision value for R, 9.54×10^{-7}. Then $|i_u|$ becomes 4.79×10^{-4}, a huge error for a modern computer. For the same values of a, b and R but different values of c, here are some other values of $|i_u|$:

c	$|i_u|$
-2.4995	4.77×10^{-3}
-2.49995	4.77×10^{-2}
-2.499995	4.77×10^{-1}
-2.4999995	4.77
-2.49999995	47.7

In the last case, the maximum possible error will be almost *50* times the correct answer.

We pause to emphasize an important rule for computing.

PRACTICAL RULE FOR DIFFERENCES   If possible, avoid computation using differences close to zero.

ILLUSTRATIVE PROBLEM 1.2-10   Discuss the use of the quadratic formula on a computer.

Discussion   The formula is

$$x = \frac{-b \pm \sqrt{b^2 - 4ac}}{2a}$$

If b is positive, it is likely that $\sqrt{b^2 - 4ac}$ will be close to b in value and the numerator close to zero when the $+$ sign is used. Before writing the program, change the formula for computational use:

$$x = \frac{-b + \sqrt{b^2 - 4ac}}{2a} \cdot \frac{-b - \sqrt{b^2 - 4ac}}{-b - \sqrt{b^2 - 4ac}}$$

$$x = \frac{-2c}{b + \sqrt{b^2 - 4ac}} \tag{1.2.25}$$

If b is negative, perform the same operation for the minus case.

1.2.7   Partial Errors

Because Illustrative Problem 1.2-8 was simple, the algebra was not too devastating. We shall show an alternative method which leads to the same results and which is far easier when the problem is difficult.

Let ϵ^O and i^O be the errors except for input, and let ϵ^I and i^I be the input errors. Then, as we have shown,

$$\epsilon = \epsilon^I + \epsilon^O \tag{1.2.26}$$

$$i = i^I + i^O \tag{1.2.27}$$

It is easy to calculate the partial errors separately. In Illustrative Problem 1.2-9, if we assume that all the input errors (i_a^I, i_b^I and i_c^I) are zero, then Figure 1.2-2 leads to

$$i_{a+b+c}^O = \frac{a+b}{a+b+c} i_{a+b}^O + r_2 \tag{1.2.28}$$

$$i_{a+b}^O = r_1 \tag{1.2.29}$$

$$\therefore \quad i_{a+b+c}^O = \frac{a+b}{a+b+c} r_1 + r_2 \tag{1.2.30}$$

$$i_{a+b+c}^O = \frac{1}{a+b+c} \left[(a+b)r_1 + (a+b+c)r_2 \right] \tag{1.2.31}$$

We can find i_{a+b+c}^I by applying the theorem (1.2.7) or else directly by differentiating:

$$S = a + b + c \qquad (1.2.32)$$

$$\Delta S = da + db + dc \qquad (1.2.33)$$

$$\frac{1}{S} = \frac{1}{S}\left[a\frac{da}{a} + b\frac{db}{b} + c\frac{dc}{c} \right] \qquad (1.2.34)$$

$$i_S^I = \frac{1}{S}\left[ai_a^I + bi_b^I + ci_c^I \right] \qquad (1.2.35)$$

We shall frequently use this alternative procedure when there is no theorem to use, or when it is more convenient to differentiate than to find the theorem. We can generalize the method of estimating the result of input error as follows:

$$y = f(x) \qquad (1.2.36)$$

$$\Delta y \sim f'(x)\,dx \qquad (1.2.37)$$

where $\Delta y = \epsilon_y^I$.

ILLUSTRATIVE PROBLEM 1.2-11   Compare the two procedures $u = (a+b)+c$ and $v = a+(b+c)$ if $|a| > |b| > |c|$.

Solution   The input errors are independent of the procedure. We need only compare the other errors—either ϵ_u^O and ϵ_v^O, or i_u^O and i_v^O, whichever is easier. The process chart in Figure 1.2-2 leads to

$$|i_u^O| \le \frac{|2a| + |2b| + |c|}{|a+b+c|} R^O \qquad (1.2.38)$$

where $R^O = \max(|r_1|, |r_2|)$. To find the corresponding expression for v, permute the symbols as follows:

$$\begin{pmatrix} a & b & c \\ b & c & a \end{pmatrix}$$

Then immediately from (1.2.38),

$$|i_v^O| \le \frac{|2b| + |2c| + |a|}{|a+b+c|} R^O \qquad (1.2.39)$$

The question is: which is larger, (1.2.38) or (1.2.39)? We compare as follows:

	$2|a| + 2|b| + |c|$	$2|b| + 2|c| + |a|$
Subtract $2|b|$:	$2|a| + |c|$	$2|c| + |a|$
Subtract $|a| + |c|$:	$|a|$	$|c|$

But we were given that $|a| > |c|$. Therefore (2.1.38) is larger and the v procedure is better, because the maximum error in that procedure is less than that in the u procedure.

This leads to another practical rule.

PRACTICAL RULE FOR SUMS   In finding the sum of a set of numbers, arrange the set so that they are in ascending order of absolute value.

1.2.8 The Error of a Series

We shall frequently have to deal with the error of a finite series. For the series

$$a_1 + a_2 + a_3 + a_4$$

we shall define *forwards addition* as

$$((a_1 + a_2) + a_3) + a_4$$

and *backwards addition* as

$$a_1 + (a_2 + (a_3 + a_4))$$

The effect of Illustrative Problem 1.2-11 and the practical rule was to show that if the original series is arranged in descending order of absolute value, the addition should be done backwards to reduce roundoff error.

Repeated application of the process chart, Figure 1.2-2, leads to the following theorems:

FORWARDS-SUM ERROR THEOREM If a series of n terms, $a_1, a_2, \ldots, a_n$, is added forwards, the total roundoff error is given by

$$|\epsilon_S| \leqslant |\epsilon_1^I + \epsilon_2^I + \ldots + \epsilon_n^I| + R^O|a_n + 2a_{n-1} + 3a_{n-2} + \ldots + (n-1)(a_2 + a_1)| \tag{1.2.40}$$

where R^O is the absolute value of the largest roundoff error, and $\epsilon_1^I, \epsilon_2^I, \ldots, \epsilon_n^I$ are inherent errors. In summation notation this becomes

$$|\epsilon_S| \leqslant \left| \sum_1^n \epsilon_k^I \right| + R^O \left| \sum_1^{n-1} k a_{n-k+1} + (n-1)a_1 \right| \tag{1.2.41}$$

BACKWARDS-SUM ERROR THEOREM If a series of terms, $a_1, a_2, \ldots, a_n$, is added backwards, the total roundoff error is given by

$$|\epsilon_S| \leqslant |\epsilon_1^I + \epsilon_2^I + \ldots + \epsilon_n^I| + R^O |a_1 + 2a_2 + 3a_3 + \ldots + (n-1)(a_{n-1} + a_n)|$$

(1.2.42)

In summation notation, this becomes

$$|\epsilon_S| \leqslant \left| \sum_1^n \epsilon_k^I \right| + R^O \left| \sum_1^{n-1} k a_k + (n-1)a_n \right|$$

(1.2.43)

ILLUSTRATIVE PROBLEM 1.2-12   (a) What is the best way to add 5, 10, 15, and 20 in a computer with a fixed word length of 4? (b) What is the best way to add 5678, 901.2, 12.34, 34.56 and 7.890 in the same computer?

Solution   (a) It doesn't matter. Integers have no roundoff error. (b) $7.890 + 12.34 + 34.56 + 901.2 + 5678$.

PROBLEM SECTION 1.2

A. Add the numbers as if in a decimal machine with fixed word length of 4. Calculate ϵ and i.

(1) $83.72 + 1.529$ (4) $3.583 + 0.001856$

(2) $75.46 + 8.223$ (5) $0.02455 + 0.0005123$

(3) $6.258 + 0.0022281$ (6) $0.09345 + 0.0008244$

B. (7) (a) Make a process chart for $S = ((a+b)+c)+d$.
   (b) Show by differentiation that

$$|i_S^I| \leqslant \frac{1}{|S|}(|a|+|b|+|c|+|d|)R^I$$

   (c) From the process chart, show that

$$i_S^O = \frac{1}{S}[(a+b)r_1 + (a+b+c)r_2 + (a+b+c+d)r_3]$$

$$|i_S^O| \leqslant \frac{1}{|S|}[3|a|+3|b|+2|c|+|d|]R^O$$

   (d) What is the best way to add if $|c| > |a| > |d| > |b|$?
(8) (a) Form a process chart for $(a+b)+(c+d)$.
   (b) Show that i_S^I is the same as in 7(b).
   (c) From the process chart, show that

$$i_S^O = \frac{1}{S}[(a+b)r_1 + (c+d)r_2 + (a+b+c+d)r_3]$$

$$|i_S^O| \leqslant \frac{2}{|S|}[|a|+|b|+|c|+|d|]R^O.$$

   (d) Show that $((a+b)+c)+d$ is a better procedure whenever $|a| + |b| < |d|$.

1.3 FLOATING-POINT MULTIPLICATION AND DIVISION

1.3.1 Floating-Point Multiplication

This should have no surprises. For example, in a hypothetical decimal machine with a mantissa length of 4,

$$23.58 \times 0.15628 = 0.2358 \times 10^2 \times 0.1562 \quad (chopped)$$
$$= 0.03683 \times 10^2 \quad (temporarily)$$
$$= 0.3683 \times 10^1$$

If we assume the true value is 3.6850824, the errors are as follows:

$$\epsilon_{a \times b} = 3.6850824 - 3.683 = 2.08 \times 10^{-3}$$
$$i_{a \times b} = (2.08 \times 10^{-3})/3.683 = 5.65 \times 10^{-4}$$

1.3.2 Error Theorem for Products

Let $da, db, dc, \ldots$, be the input errors in $a, b, c, \ldots$, and let $P = abc \ldots$ be the product. Using the approximation formula (1.2.37),

$$\Delta P = d(abc\ldots) = \frac{P}{a}da + \frac{P}{b}db + \frac{P}{c}dc + \ldots \tag{1.3.1}$$

$$\epsilon_P^I = P[\epsilon_a/a + \epsilon_b/b + \epsilon_c/c + \ldots] \tag{1.3.2}$$

$$i_P^I = i_a^I + i_b^I + i_c^I + \ldots \tag{1.3.3}$$

ERROR THEOREM FOR PRODUCTS If $P = abc\ldots$, where $a, b, c, \ldots$, have inherent errors $\epsilon_a, \epsilon_b, \epsilon_c, \ldots$, then the errors are given by (1.3.2) and (1.3.3).

ILLUSTRATIVE PROBLEM 1.3-1 We wish to compare two procedures: $u = [(a \times b) \times c] \times d$, and $v = [(a \times b) \times (c \times d)]$. (a) Draw the process charts for the two procedures. (b) Which is the better procedure? (c) Find the maximum total error for the better procedure, assuming that a, b, c, d enter with error $\leqslant 5 \times 10^{-6}$ and that the roundoff error $\leqslant 9.54 \times 10^{-7}$. (d) Find the maximum total error for the better procedure, assuming inherent errors of $a, b, c, d \leqslant 5 \times 10^{-9}$ and roundoff as in part (c).

Solution (a) The process charts are shown in Figure 1.3-1.

(b) For all procedures, the input errors are the same. Using the theorem

$$i_P^I = i_a^I + i_b^I + i_c^I + i_d^I \tag{1.3.4}$$

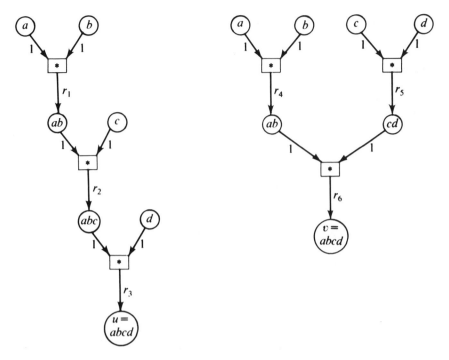

Fig. 1.3-1   Process charts for Problem 1.

we have

for the u procedure:

$$i_u^O = i_{abc}^O + r_3.$$

$$i_{abc}^O = i_{ab}^O + r_2$$

$$i_{ab}^O = r_1$$

$$i_u^O = r_1 + r_2 + r_3$$

for the v procedure:

$$i_v^O = i_{ab}^O + i_{cd}^O + r_6 \qquad (1.3.5)$$

$$i_{ab}^O = r_4 \qquad (1.3.6)$$

$$i_{cd}^O = r_5 \qquad (1.3.7)$$

$$i_v^O = r_4 + r_5 + r_6 \qquad (1.3.8)$$

Let $R^O = \max(|r_1|, |r_2|, |r_3|, |r_4|, |r_5|, |r_6|)$

$$|i_u^O| \leqslant 3R^O \qquad\qquad |i_v^O| \leqslant 3R^O \qquad (1.3.9)$$

Answer   The u and v procedures are equally accurate.

(c) From (1.3.4) and (1.3.9), using

$$R^I = \max(|i_a^I|, |i_b^I|, |i_c^I|, |i_d^I|) \qquad (1.3.10)$$

$$|i_P| \leqslant 4R^I + 3R^O \qquad (1.3.11)$$

and the given values, we have

Answer $\qquad\qquad\qquad |i_P| \leqslant 2.66 \times 10^{-5} \qquad (1.3.12)$

(d) Using (1.3.11) with the given values, we have

Answer $|i_P| \leqslant 6.70 \times 10^{-6}$ (1.3.13)

1.3.3 Floating-Point Division

Here is an example. We use the hypothetical computer once more.

$$13.594/278.2 = (0.1359 \times 10^2)/(0.2782 \times 10^3)$$
$$= 0.04884 \times 10^{-1}$$
$$= 0.4884 \times 10^{-2}$$

We take 0.4886412×10^{-2} as an approximation to the true answer. Then the errors are as follows:

$$\epsilon = 2.41 \times 10^{-6}$$
$$i = 4.93 \times 10^{-4}$$

1.3.4 Error Theorem for Division

Let da and db be the errors in a and b, and let $q = a/b$. Then, for $b \neq 0$,

$$\Delta q = \frac{b\,da - a\,db}{b^2} \tag{1.3.14}$$

$$\epsilon_q^I = \frac{b\epsilon_a^I - a\epsilon_b^I}{b^2} \tag{1.3.15}$$

$$i_q^I = i_a^I - i_b^I \tag{1.3.16}$$

We should not be misled by the minus sign in (1.3.16). To find the maximum error of the quotient, we will have to assume that the relative errors are *opposite* in sign:

$$|i_q^I| \leqslant |i_a^I| + |i_b^I| \tag{1.3.17}$$

ERROR THEOREM FOR QUOTIENTS If $q = a/b$, then the errors are given by (1.3.15) and (1.3.16).

ILLUSTRATIVE PROBLEM 1.3-2 (a) Make process charts for $u = (a \times b)/c$ and $v = a \times (b/c)$. (b) Which is a better procedure? (c) Find the maximum total error for the better procedure assuming that a, b, c enter with maximum relative error $\leqslant 5 \times 10^{-5}$

and that the maximum relative roundoff error $\leqslant 9.54 \times 10^{-7}$. (d) Find the maximum total error for the better procedure assuming that a, b, c enter with error $\leqslant 5 \times 10^{-10}$.

Solution   (a) The process charts are shown in Figure 1.3-2.

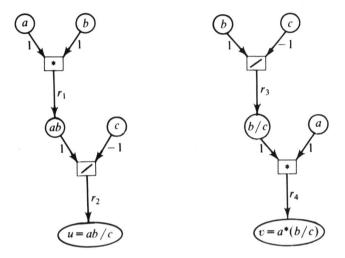

Fig. 1.3-2   Process charts for Illustrative Problem 1.3-2.

(b) For all procedures, the input error is the same. We differentiate to find the input error: Let

$$y = \frac{ab}{c}. \tag{1.3.18}$$

Then

$$\Delta y = \frac{c(a\,db + b\,da) - ab\,dc}{c^2} \tag{1.3.19}$$

$$\epsilon_y^I = \frac{a}{c}\epsilon_b^I + \frac{b}{c}\epsilon_a^I - \frac{ab}{c^2}\epsilon_c^I \tag{1.3.20}$$

Dividing the left member by y and the right member by ab/c,

$$\frac{\epsilon_y^I}{y} = \frac{a}{c}\epsilon_b^I \frac{c}{ab} + \frac{b}{c}\epsilon_a^I \frac{c}{ab} - \frac{ab}{c^2}\epsilon_c^I \frac{c}{ab} \tag{1.3.21}$$

$$i_y^I = i_b^I + i_a^I - i_c^I \tag{1.3.22}$$

Of course, the result (1.3.22) could have been obtained from a combination of the product theorem and quotient theorem without going to all this trouble.

For the u procedure:	For the v procedure:	
$i_u^O = i_{ab}^O + r_2$	$i_v^O = i_{b/c}^O + r_4$	(1.3.23)
$i_{ab}^O = r_1$	$i_{b/c}^O = r_3$	(1.3.24)
$i_u^O = r_1 + r_2$	$i_v^O = r_3 + r_4$	(1.3.25)

$$\text{Let} \quad R^O = \max(|r_1|, |r_2|, |r_3|, |r_4|)$$

$$|i_u^O| \leqslant 2R^O \qquad |i_v^O| \leqslant 2R^O \tag{1.3.26}$$

Answer   The u and v procedures are equally accurate.

(c) From (1.3.22) and (1.3.26),

$$|i| \leqslant 3R^I + 2R^O \tag{1.3.27}$$

Using $R^I = 5 \times 10^{-5} + 9.54 \times 10^{-7} = 5.10 \times 10^{-5}$ and $R^O = 9.54 \times 10^{-7}$, we obtain

$$|i| \leqslant 1.55 \times 10^{-4} \tag{1.3.28}$$

(d) Using the given quantities,

$$|i| \leqslant 4.77 \times 10^{-6} \tag{1.3.29}$$

1.3.5   Mixed Elementary Operations

ILLUSTRATIVE PROBLEM 1.3-3   (a) Draw process charts for $u = a(b + c)$ and $v = ab + ac$. (b) Which is the better procedure?

Solution   (a) The proocess charts are displayed in Figure 1.3-3.

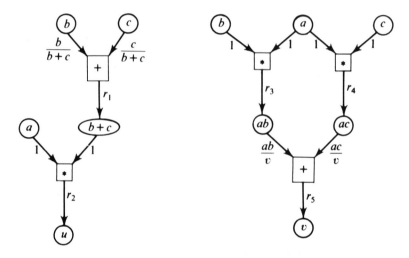

Fig. 1.3-3   Process charts for Illustrative Problem 1.3-3.

(b) For all procedures, the input error is the same. Since the total error was not required in this problem, we need not calculate the input error at all. The other error is calculated as follows:

For the u procedure:

$$i_u^O = i_{b+c}^O + r_2$$

$$i_{b+c}^O = r_1$$

$$i_u^O = r_1 + r_2$$

For the v procedure:

$$i_v^O = \frac{ab}{ab + ac} i_{ab}^O + \frac{ac}{ab + ac} i_{ac}^O + r_5 \tag{1.3.30}$$

$$i_{ab}^O = r_3, \qquad i_{ac}^O = r_4 \tag{1.3.31}$$

$$i_v^O = \frac{abr_3 + acr_4 + (ab + ac)r_5}{ab + ac} \tag{1.3.32}$$

Let $R^O = \max(|r_1|, |r_2|, |r_3|, |r_4|, |r_5|)$

$$|i_u^O| \leqslant 2R^O \qquad |i_v^O| \leqslant 2R^O \tag{1.3.33}$$

Answer   The two procedures are equally accurate.

PROBLEM SECTION 1.3

A. In (1)–(6), assume a decimal computer chopping after four digits in the mantissa; (a) find the true answer, (b) find the chopped answer, (c) calculate the absolute error, (d) calculate the absolute relative error.

(1) 153.4×0.38699	(4) $371.58/0.23386$
(2) 512.7×0.11576	(5) $0.058229/0.0051444$
(3) $0.071893 \times 0.0023888$	(6) $0.073186/0.0092281$

B. In (7)–(10), proceed as in part A.

(7) $35.623 \times 18.115/6.22481$
(8) $72.819 \times 27.433/8.15526$
(9) $(62.583 \times 17.114)/(8.0457 \times 0.31448)$
(10) $(31.842 \times 61.890)/(9.1568 \times 0.15729)$

C. In (11)–(14), show that the u and v procedures are equally accurate.

(11) $u = (x \times x) \times (x \times x)$, $v = x \times (x \times (x \times x))$
(12) $u = (ab)/c$, $v = (a/c) \times b$
(13) $u = a/(b \times c)$, $v = (a/b)/c$
(14) $u = (ab)/(cd)$, $v = (a/c) \times (b/d)$

D. In (15) and (16), show that u is better than v.

(15) $u = 4a$, $v = a + a + a + a$
(16) $u = 3ab$, $v = ab + ab + ab$

E. (a) Make process charts. (b) Find the input error for each. (c) Find the roundoff error for each. (d) Which is better?

(17) $u = (x - y) + y$, $v = x$
(18) $u = x \times x + 2x$, $v = x(x + 2)$
(19) $u = 1 + x(1 + x)$, $v = x \times x + 2x + 1$
(20) $u = x + y$, $v = (x \times x - y \times y)/(x - y)$

1.4   TRUNCATION ERRORS

1.4.1   Introduction

We have discussed and described the inherent error (including the conversion error) and one kind of procedural error, that due to roundoff. The other kind of procedural error which concerns us is the *truncation error*. This is caused by the fact that the computer cannot add an infinite set of terms. It must stop somewhere. The difference between the infinite series and the finite summation is defined as the *theoretical truncation error*.

The situation in computing is that we need to use logarithms, exponentials, trigonometric functions, hyperbolic functions and so on. These are *built-in subroutines* supplied by the manufacturer. In most cases, a series is used to compute these functions. The research team of the manufacturer compares a sample of the true values with the corresponding computed values using the built-in subroutine. The results are published in a manual supplied by the manufacturer. We shall call the published errors *actual truncation errors.* They differ from the theoretical truncation errors in that they include roundoff.

The following table is excerpted from IBM Manual GC 28-6596-4. The numbers under T give the relative truncation errors. We shall use these numbers for illustrative problems and for the exercises. However, they apply only to the IBM 360 machine at the time of publication. All manufacturers improve the subroutines from time to time. For actual problems on the job, it is necessary to refer to the current manual for the computer being used.

> *Explanation*   (1) To find the maximum relative truncation error in the machine calculation of ln 0.85527, look at the first line, because 0.85527 is in the proper region. The maximum truncation relative error is

$$\frac{1.64 \times 10^{-7}}{\ln 0.85527}$$

about 1.05×10^{-6} in absolute value. We shall be using the symbols D and E to represent the characteristic. For example, 1.2E5 and 1.2D5 both mean 1.2×10^5. The last column of the table shows the method used by the programmers to arrive at the computer value of ln 0.85527. (We shall devote two chapters to this important method.) (2) To find the maximum relative truncation error in the machine calculation of $\sqrt{56.2289}$ in double precision, look at the last line. According to the table, the maximum relative truncation error is 1.08×10^{-16} for any double-precision square root, and the method used is the Newton-Raphson method (which we shall take up in the next chapter). This means that the error would be $56.2289 \times 1.08 \times 10^{-16}$, which is about 6.07×10^{-15}.

1.4.2   The Total-Error Theorem for Truncation

We shall prove that the truncation error adds on to all the other errors.

TOTAL-ERROR THEOREM FOR TRUNCATION   The total relative error of a computation is found by adding the input error to the truncation error as given in the manufacturer's manual.

> *Discussion*   Imagine that the manufacturer has invented a subroutine to multiply a number by 2. Unfortunately, the number reads in with an error of conversion. You enter a 10 and (we are exaggerating for effect) it becomes an 11. Then the subroutine is called, which (let us say) has a truncation relative error of 10%, as reported in the manufacturer's manual. Instead of the correct answer, 20, we obtain $22 + 2.2 = 24.2$. In effect, we have found the wrong function of the wrong number. This, in brief, is the situation.

Table 1.4-1
Truncation Relative Errors for Selected Subprograms

MATHEMATICAL FUNCTION	PRECISION	REGION	T	METHOD
ln	single	[0.5, 1.5]	1.64D-7$/\ln x$	Chebyshev
		other	1.05D-6	
	double	[0.5, 1.5]	1.85D-16$/\ln x$	Chebyshev
		other	3.31D-16	
$\cos^{-1}$	single	all	1.85D-7	Chebyshev
	double	all	2.72D-16	
$\sin^{-1}$	single	all	8.56D-7	Chebyshev
	double	all	2.40D-16	
$\tan^{-1}$	single	all	9.75D-7	Continued
	double	all	2.08D-16	fraction
cos	single	$[0, \pi]$	1.47D-7$/\cos x$	Chebyshev
		$[-10, 0)$	1.42D-7$/\cos x$	
		$(\pi, 100]$	1.35D-7$/\cos x$	
	double	$[0, \pi]$	1.79D-16$/\cos x$	Chebyshev
		$[-10, 0)$	1.76D-16$/\cos x$	
		$(\pi, 100]$	2.65D-15$/\cos x$	
sin	single	$[0, \pi/2]$	1.59D-6	Chebyshev
		$(\pi/2, 10]$	1.41D-7$/\sin x$	
		$(10, 100]$	1.46D-7$/\sin x$	
	double	$[0, \pi/2]$	4.08D-6	Chebyshev
		$(\pi/2, 10]$	1.64D-16$/\sin x$	
		$(10, 100]$	2.69D-15$/\sin x$	
tan	single	$[0, \pi/4]$	1.56D-6	Continued
		$(\pi/4, \pi/2]$	6.58D-5	fraction
		$(\pi/2, 10]$	4.92D-5	
		$(10, 100]$	3.35D-5	
	double	$[0, \pi/4]$	5.25D-16	Continued
		$(\pi/4, \pi/2]$	1.67D-12	fraction
		$(\pi/2, 10]$	1.57D-13	
		$(10, 100]$	3.75D-12	
cosh	single	[0, 5]	1.31D-6	Continued fraction
	double	[0, 5]	4.81D-16	Chebyshev
sinh	single	[0, 5]	1.20D-6	Chebyshev
	double	[0, 0.34657]	2.10D-16	Chebyshev
		(0.34657, 5]	3.59D-16	
exp	single	[0, 1]	4.65D-7	Continued
		(1, 170]	4.69D-7	fraction
	double	[0, 1]	2.27D-16	Chebyshev
		(1, 20]	2.31D-15	
		(20, 170]	2.33D-15	
sq. root	single	all	8.70D-7	Newton-
	double	all	1.08D-16	Raphson

Proof of the Theorem   We wish to find

$$y = f(x) \tag{1.4.1}$$

where f may be ln, sin, exp, etc. Let us suppose, for a moment, that the f subroutine is perfect, but that x has an inherent error, so that we actually find

$$y + \epsilon^I = f(x + \Delta x) \tag{1.4.2}$$

We now recall the mean-value theorem, the proof of which can be found in any book on elementary calculus:

MEAN-VALUE THEOREM   If a function $f(x)$ is continuous on the interval $[a, b]$ and possesses a derivative at every interior point of (a, b), then there exists a point θ such that $f(b) = f(a) + (b - a)f'(\theta)$, where θ is between a and b.

For our purposes, we let $a = x$ and $b = x + \Delta x$. Then

$$f(x + \Delta x) = f(x) + \Delta x f'(\theta) \tag{1.4.3}$$

Substituting in (1.4.2) and using (1.4.1), we have

$$\epsilon^I = \Delta x f' \tag{1.4.4}$$

We return to the truncation problem. Instead of finding $f(x)$ we actually find $(f + \Delta f)(x + \Delta x)$. We write the latter as $(f + g)(x + \Delta x)$. Instead of (1.4.2), we have

$$y + \Delta y = (f + g)(x + \Delta x) \tag{1.4.5}$$

$$y + \Delta y = f(x + \Delta x) + g(x + \Delta x) \tag{1.4.6}$$

The transition from (1.4.5) to (1.4.6) is not a trivial one. It is explained and justified in books dealing with the properties of transformations.[2] In (1.4.6), the quantity $g(x + \Delta x)$ is the truncation error as given in the manufacturer's manual. Then, using the mean-value theorem and (1.4.4),

$$y + \Delta y = f(x) + \Delta x f'(\theta) + g(x + \Delta x) \tag{1.4.7}$$

$$\Delta y = \epsilon^I + \epsilon^T \tag{1.4.8}$$

completing the proof. □

Comment   In specific instances, we do know the approximating function, and it might even be possible to make a stab at the function causing the truncation error. However, the general case is another

---

[2]Birkhoff, G. and Maclane, S., *A Survey of Modern Algebra*, Macmillan, 1965, p. 195.

matter. In practice, the maximum relative truncation error and, in fact, the maximum total error are many times the actual errors because the assumption that none of the errors cancel is an unlikely one. We shall have more to say about this in problems throughout the book. There have been some attempts at a statistical treatment of errors, but to date the results have not been applicable to practical problems.

1.4.3 Errors in Logarithmic Functions

First we calculate the input error:

$$y = \ln x \tag{1.4.9}$$

$$\Delta y = \frac{1}{x} dx \tag{1.4.10}$$

$$\epsilon_{\ln x}^I = \frac{1}{x} \epsilon_x^I \tag{1.4.11}$$

Divide the left member by y and the right member by $\ln x$:

$$\frac{\epsilon_{\ln x}^I}{y} = \frac{\epsilon_x^I}{x \ln x} \tag{1.4.12}$$

But $\epsilon_{\ln y}^I / y$ is $i_{\ln y}^I$, and ϵ_x^I / x is i_x^I. Therefore,

$$i_{\ln x}^I = i_x^I / \ln x \tag{1.4.13}$$

The procedural error is read directly from Table 1.4-1. If x is between 0.5 and 1.5 and we are working in single precision, the truncation relative error is $1.64 \times 10^{-7} / \ln x$, so that the total error is

$$i_{\ln x} = i_{\ln x}^I + T_{\ln x} \tag{1.4.14}$$

$$i_{\ln x} = \frac{i_x^I + 1.64 \times 10^{-7}}{\ln x} \tag{1.4.15}$$

TOTAL-ERROR THEOREM FOR $\ln x$

$$i_{\ln x} = i^I r_x / \ln x + T_{\ln x} \tag{1.4.16}$$

ILLUSTRATIVE PROBLEM 1.4-1 (a) Draw a process chart for $y = \ln x$. (b) If x is about 2 and enters with a relative error $\leqslant 5 \times 10^{-9}$, find the maximum total error. (c) If x is about 50 with the same input error, find the maximum total error.

Solution

(a) See Figure 1.4-1.

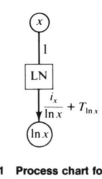

Fig. 1.4-1   Process chart for Problem 1.

(b) From Table 1.4-1, $T_{\ln} \leqslant 1.05 \times 10^{-6}$ in single precision. We also take $|i_x| \leqslant 5 \times 10^{-9} + 9.54 \times 10^{-7}$. Using (1.4.16),

$$|i_{\ln x}| \leqslant \left| \frac{9.59 \times 10^{-7}}{\ln 2} + 1.05 \times 10^{-6} \right| = 2.43 \times 10^{-6}$$

$$|\epsilon_{\ln x}| = |i_{\ln x}| \times |\ln 2| = 1.69 \times 10^{-6}$$

(c)

$$|i_{\ln x}| \leqslant \left| \frac{9.59 \times 10^{-7}}{\ln 50} + 1.05 \times 10^{-6} \right| = 1.30 \times 10^{-6}$$

$$|\epsilon_{\ln x}| = |i_{\ln x}| \times |\ln 50| = 5.07 \times 10^{-6}$$

ILLUSTRATIVE PROBLEM 1.4-2   Do the same calculations for double precision assuming the input error $\leqslant 5 \times 10^{-17}$.

Solution

$$|i_x| \leqslant 5 \times 10^{-17} + 2.22 \times 10^{-16} = 2.72 \times 10^{-16}$$

$$|T_{\ln x}| \leqslant 3.31 \times 10^{-16}$$

The calculations result in the following.

$$|\epsilon_{\ln 2}| \leqslant 5.01 \times 10^{-16}$$

$$|\epsilon_{\ln 50}| \leqslant 1.57 \times 10^{-15}$$

1.4.4   Errors in Exponential Functions

$$y = e^x \tag{1.4.17}$$

$$\Delta y = e^x \, dx \tag{1.4.18}$$

$$\epsilon_{\exp x}^I = e^x \epsilon_x^I$$

We divide the left member by y and the right member by e^x:

$$i_{\exp x}^I = \epsilon_x^I \tag{1.4.19}$$

But

$$\epsilon_x = x i_x \tag{1.4.20}$$

$$\therefore \quad i_{\exp x}^I = x i_x^I \tag{1.4.21}$$

TOTAL-ERROR THEOREM FOR $\exp x$

$$i_{\exp x} = x i_x^I + T_{\exp x} \tag{1.4.22}$$

ILLUSTRATIVE PROBLEM 1.4-3 (a) Make a process chart for $y = e^x$. (b) Assuming that x enters with conversion error only and that the computation is in single precision, calculate the maximum total error when x is about 5.7.

Solution

(a) The process chart is shown in Figure 1.4-2.

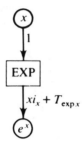

Fig. 1.4-2   Process chart for Problem 3.

(b)
$$i_{\exp x} = x i_x^I + T_{\exp x}$$

$$|i_x^I| \leqslant 9.54 \times 10^{-7}$$

$$|T_{\exp x}| \leqslant 4.69 \times 10^{-7}$$

$$|i_{\exp 5.7}| \leqslant |5.7 \times 9.54 \times 10^{-7} + 4.69 \times 10^{-7}| = 5.91 \times 10^{-6}$$

$$|\epsilon_{\exp 5.7}| \leqslant 5.91 \times 10^{-6} \times 5.7 = 3.37 \times 10^{-5}$$

1.4.5   Errors in Powers

If an instruction is written as

$$Y = X^{**}2 \tag{1.4.23}$$

the procedure for the machine is the same as

$$Y = X^*X \tag{1.4.24}$$

and the multiplication error theorem can be applied. However, for

$$Y = X^{**}2.7 \tag{1.4.25}$$

the machine converts the problem to

$$y = e^{2.7 \ln x} \tag{1.4.26}$$

which is considerably different. In (1.4.26), the computer finds the logarithm of x, multiplies the result by 2.7, then exponentiates, piling up quite a lot of error on the way. The process chart for such a problem is shown in Figure 1.4-3.

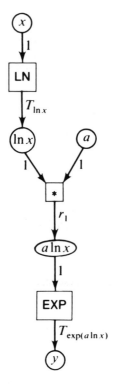

Fig. 1.4-3 Process chart for $y = x^a$, where a is not an integer.

The calculation of the input error is straightforward:

$$y = x^a \tag{1.4.27}$$

$$\Delta y = a x^{a-1} dx \tag{1.4.28}$$

$$\epsilon_y^I = a x^{a-1} \epsilon_x^I \tag{1.4.29}$$

We divide the left member by y and the right member by x^a:

$$i_y^I = a i_x^I \tag{1.4.30}$$

For the procedural error, we refer to Figure 1.4-3. As usual, we find the procedural error by assuming that the input errors (in this case i_x^I and i_a^I)

are 0:

$$i_y^O = (a \ln x)i_{a \ln x}^O + T_{\exp(a \ln x)} \tag{1.4.31}$$

$$i_{a \ln x}^O = i_{\ln x}^O + r_1 \tag{1.4.32}$$

$$i_{\ln x}^O = T_{\ln x} + i_x^O = T_{\ln x} \tag{1.4.33}$$

Therefore,

$$i_{a \ln x}^O = T_{\ln x} + r_1 \tag{1.4.34}$$

$$i_y^O = (a \ln x)(T_{\ln x} + r_1) + T_{\exp(a \ln x)} \tag{1.4.35}$$

TOTAL-ERROR THEOREM FOR x^a, a NOT AN INTEGER   For $y = x^a$, where a is not an integer,

$$i_y = ai_x^I + (a \ln x)(T_{\ln x} + r) + T_{\exp(a \ln x)} \tag{1.4.36}$$

where r is the roundoff error for the computer.

ILLUSTRATIVE PROBLEM 1.4-4   Compare the errors in calculating (a) $u = x^3$ and (b) $v = e^{3 \ln x}$, given that x is about 2.5. Assume that x enters with conversion error only, that the single-precision mode is used and that maximum r for the machine is 9.54×10^{-7}.

Solution   The process charts are shown in Figure 1.4-4. Although the statement of the problem gives information about input error, the solution does not require that

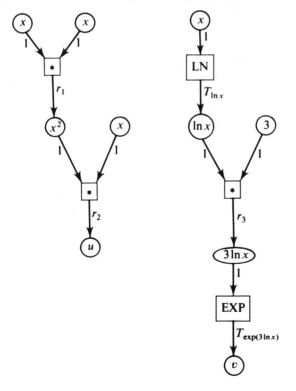

Fig. 1.4-4   Process charts for Illustrative Problem 1.4-4.

information because input errors are independent of procedure. We therefore concentrate on the procedural errors, including roundoff and truncation errors.

For the u procedure,

$$i_u^O = i_{x**2}^O + r_2$$

$$i_{x**2}^O = r_1$$

$$i_u^O = r_1 + r_2$$

$$|i_u^O| \leqslant 2R^O = 2(9.54 \times 10^{-7}) = 1.91 \times 10^{-6}$$

For the v procedure, we shall use the following approximations:

$$\ln 2.5 = 9.16 \times 10^{-1}$$
$$3 \ln 2.5 = 2.75$$
$$T_{3\ln x} = T_{\ln 15.265} = 1.05 \times 10^{-6}$$
$$T_{\exp(2.75)} = 4.69 \times 10^{-7}$$
$$i_v^O = i_{3\ln x}^O + T_{\exp(3\ln x)}$$
$$|i_v^O| \leqslant |i_{3\ln x}^O + 4.69 \times 10^{-7}|$$
$$i_{3\ln x}^O = i_{\ln x}^O + r_3$$
$$|i_{\ln x}^O| \leqslant T_{\ln x} = 1.05 \times 10^{-6}$$
$$|i_{3\ln x}^O| \leqslant 1.05 \times 10^{-6} + 9.54 \times 10^{-7} = 2.00 \times 10^{-6}$$
$$|i_v^0| \leqslant (2.00 \times 10^{-6}) + 4.69 \times 10^{-7} = 2.47 \times 10^{-6}$$

Note how much better the *u* procedure is. This leads to a very important guideline for programmers.

PRACTICAL RULE FOR EXPONENTIATION Avoid using exponents with decimal points. If they must be used, program in multiple precision.

Examples To find $x^{1.5}$, rewrite as $\sqrt{x^3}$. The square-root subroutine is excellent, as can be seen in Table 1.4-1. To find x^{25}, avoid X**25. Instead, use six steps:

$$Y = X*X \qquad \text{(this is } x^2\text{)}$$

$$Y = Y*Y \qquad \text{(this is } x^4\text{)}$$

$$Z = Y*Y \qquad \text{(this is } x^8\text{)}$$

$$Y = Z*Z \qquad \text{(this is } x^{16}\text{)}$$

$$Y = Y*Z \qquad \text{(this is } x^{24}\text{)}$$

$$Y = Y*X \qquad \text{(this is } x^{25}\text{)}$$

1.4.6 Square-Root Error

It is possible to find $\sqrt{x}$ by using the computer subroutine for $x^{0.5}$, but that would be quite inaccurate, for reasons explained in the previous section. The square-root process is so common that all computers have special subroutines for it. These subroutines are based upon a trial-and-error method to be explained in the next chapter. Although we have listed the error in Table 1.4-1 under *truncation*, it is a process that is truncated rather than a series. In any case, the error is quite small.

The input error is, as usual, easy to calculate:

$$y = x^{1/2} \tag{1.4.37}$$

$$\Delta y = \tfrac{1}{2} x^{-1/2} dx \tag{1.4.38}$$

$$\epsilon_y^I = \epsilon_x^I / (2\sqrt{x}\,) \tag{1.4.39}$$

We divide the left member by y and the right member by $\sqrt{x}$:

$$i_y^I = \epsilon_x^I / 2x \tag{1.4.40}$$

We now use

$$\epsilon_x^I = x i_x^I \tag{1.4.41}$$

to arrive at the following theorem:

TOTAL-ERROR THEOREM FOR SQUARE ROOT   If $y = \sqrt{x}$,

$$|i_y| \leqslant \frac{1}{2} |i_x^I| + T_{\sqrt{x}} \tag{1.4.42}$$

ILLUSTRATIVE PROBLEM 1.4-5   Compare $u = x^{0.5}$ and $v = \sqrt{x}$ when x is approximately 25. Assume the data are entered with conversion error only and that the problem is done in single precision with $|r| \leqslant 9.54 \times 10^{-7}$. Find the total errors.

Solution   The process charts are shown in Figure 1.4-5. For both procedures,

$$|i^I| = \tfrac{1}{2} |i_x^I| \leqslant \tfrac{1}{2} (9.54 \times 10^{-7}) = 4.77 \times 10^{-7}$$

$$|\epsilon^I| = |i^I| \times 5 = 2.39 \times 10^{-6}$$

For the u procedure, we use $0.5 \ln x = 1.61$, $\ln 25 = 3.22$:

$$i_u^O = i_{0.5\ln x}^O + T_{\exp(0.5\ln x)}$$

$$|i_u^O| \leqslant i_{0.5\ln x}^O + 4.69 \times 10^{-7}$$

$$i_{0.5\ln x}^O = i_{\ln x}^O + r_1$$

$$i_{\ln x}^O = T_{\ln x}$$

$$i_{0.5\ln x}^O = T_{\ln x} + r_1$$

$$|i_u^O| \leqslant |(T_{\ln x} + r_1) + 4.69 \times 10^{-7}|$$

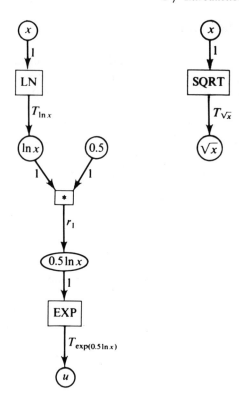

Fig. 1.4-5   Process charts for Illustrative Problem 1.4-5.

From the table, $T_{\ln x} \leqslant 1.05 \times 10^{-6}$. We use $|r_1| \leqslant 9.54 \times 10^{-7}$:

$$|i_u^O| \leqslant (1.05 \times 10^{-6} \times 3.22 + 9.54 \times 10^{-7}) + 4.69 \times 10^{-7}$$
$$|i_u^O| \leqslant 4.80 \times 10^{-6}$$
$$|\epsilon_u^O| = |i_u^O||u| \leqslant 4.80 \times 10^{-6} \times 5 = 2.40 \times 10^{-5}$$

Thus the total error is

$$|i_u| \leqslant 4.80 \times 10^{-6} + 4.77 \times 10^{-7} = 5.28 \times 10^{-6}$$
$$|\epsilon_u| \leqslant 2.64 \times 10^{-5}$$

For the v procedure,

$$i_v^O = T_{\sqrt{x}}$$
$$|i_v^O| \leqslant 8.70 \times 10^{-7}$$
$$|\epsilon_v^O| \leqslant 8.70 \times 10^{-7} + 2.39 \times 10^{-6} = 3.26 \times 10^{-6}$$

The total error is

$$|i_v| \leqslant 8.70 \times 10^{-7} + 4.77 \times 10^{-7} = 1.35 \times 10^{-6}$$
$$|\epsilon_v| \leqslant 5.65 \times 10^{-6}$$

As you can see, there is a sizable difference.

1.4.7   Errors in Trigonometric Functions

The errors for sines, cosines and tangents at input time can be calculated in the same fashion. Here is the calculation for the sine.

$$y = \sin x \tag{1.4.43}$$

$$\Delta y = \cos x \, dx \tag{1.4.44}$$

$$\epsilon_y^I = (\cos x)\epsilon_x^I \tag{1.4.45}$$

We divide the left member by y and the right member by $\sin x$:

$$i_y^I = (\cot x)\epsilon_x^I \tag{1.4.46}$$

We now use

$$\epsilon_x^I = x i_x^I \tag{1.4.47}$$

so that

$$i_y^I = (x \cot x)i_x^I \tag{1.4.48}$$

Similarly,

$$i_{\cos x}^I = -(x \tan x)i_x^I \tag{1.4.49}$$

$$i_{\tan x}^I = (x \sec x \csc x)i_x^I \tag{1.4.50}$$

These are left as exercises. To find the total error, one need merely add the truncation error from Table 1.4-1.

ILLUSTRATIVE PROBLEM 1.4-6   (a) Find the total error in computing $y = x \sin x$. (b) Find the maximum total error if x is about 0.4 and i_x^I is no greater than 5×10^{-9}. Assume IBM 360 single precision with $R \leqslant 9.54 \times 10^{-7}$.

Solution   The process chart is shown in Figure 1.4-6.

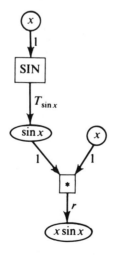

Fig. 1.4-6   Process chart for Problem 6.

To calculate the input error, since

$$y = x \sin x$$
$$\Delta y = (x \cos x + \sin x)\, dx$$

Dividing the left member by y, and the right member by $x \sin x$,

$$i_y^I = (1 + x \cot x)i_x^I$$

To find the procedural error, follow the chart:

$$i_{x\sin x}^O = i_{\sin x}^O + r$$
$$i_{\sin x}^O = T_{\sin x}$$
$$i_{x\sin x}^O = T_{\sin x} + r$$

(a) $\qquad i_{x\sin x} = (1 + x \cot x)i_x^I + T_{\sin x} + r$

(b) After the substitutions, we obtain

$$|i_{x\sin x}| \leqslant 2.83 \times 10^{-6}$$
$$|\epsilon_{x\sin x}| \leqslant 7.90 \times 10^{-9}$$

1.4.8 Errors in Inverse Trigonometric Functions

We offer one sample of the calculation for the error at input time of the arcsine function.

$$y = \arcsin x, \tag{1.4.51}$$
$$\sin y = x \tag{1.4.52}$$
$$\cos y \, \Delta y = \Delta x \tag{1.4.53}$$
$$\cos y \, \epsilon_y^I = \epsilon_x^I \tag{1.4.54}$$

$$|i_y^I| = \left| \frac{x i_x^I}{\sqrt{1-x^2}\,\arcsin x} \right| \tag{1.4.55}$$

Note that for values of $|x|$ near 1, this input error may become immense. Indeed, one of the values of this kind of calculation is that you can tell where the danger spots are.

The next two formulas are left as exercises.

$$|i_{\arccos x}^I| = \left| \frac{x i_x^I}{\sqrt{1-x^2}\,\arccos x} \right| \tag{1.4.56}$$

$$|i_{\arctan x}^I| = \left| \frac{x i_x^I}{(\sqrt{1+x^2}\,\arctan x)} \right| \tag{1.4.57}$$

The total errors are found, as usual, by adding the truncation relative errors from Table 1.4-1.

PRACTICAL RULE FOR SUBROUTINES   When working with logarithmic or inverse trigonometric functions, it is advisable to use double precision.

Our interests are not precisely the same as the manufacturer's interest, because computers are mainly sold to commercial enterprises whose main business is arithmetic with error-free integers. The manufacturer understandably wants to make claims for greater and greater speed of computation. In many cases, the greater speed of computation is meaningless because the program is held back by input-output speeds which may be thousands of times slower. At any rate, the difference in speed between single- and double-precision subroutines is miniscule. Table 1.4-2, certainly out of date by now but still instructive, is excerpted from IBM Manual C28-6596-4. We emphasize that it is displayed only to compare speeds on a rather low-speed IBM 360. The higher-speed machines in the IBM 360 and 370 series undoubtedly preserve the general relationship.

Table 1.4-2
Time Requirements for Selected Subroutines, in Microseconds

	SINGLE PRECISION	DOUBLE PRECISION	MAXIMUM DIFFERENCE
$\ln x$	9	14	5
$\cos^{-1} x$	16	24	8
$\sin^{-1} x$	16	12	4
$\tan^{-1} x$	11	13	2
$\cos x$	11	13	2
$\sin x$	11	12	1
$\tan x$	12	13	1
$\cosh x$	30	27	-3
$\sinh x$	32	28	-4
$\tanh x$	25	27	2
$\exp x$	16	15	-1
$\sqrt{x}$	7	7	0

For these reasons, all the programs in this book were done in double precision except when otherwise stated. Comparative tests with single precision showed no discernible time difference in the programs tested. The error ratio $(9.54 \times 10^{-7})/(2.22 \times 10^{-16})$ is 4,300,000,000—a very persuasive argument for double precision unless extreme speed is required.

For supplementary reading on the topics of this section, the following are highly recommended: Hamming, Chapters 2 and $(N+1)$, and Fox and Mayers, Chapters 1 and 2. The full references are given at the end of the book in the Selected Bibliography.)

PROBLEM SECTION 1.4

A. Verify the following formulas:
   (1) (1.4.49)   (3) (1.4.56)
   (2) (1.4.50)   (4) (1.4.57)

B. In the following, assume single-precision computing with chopping. (a) Compare the two procedures by calculating i^0 for both. (b) Derive the formula for i^1 for both. (c) Assuming that the approximation error at input $\leqslant 5 \times 10^{-9}$, calculate the maximum total error for the better procedure, if one is better.

(5) $u = \sqrt{x}$, $v = x/\sqrt{x}$, $x \sim 36$

(6) $u = x$, $v = \exp(\ln x)$, $x \sim 1.5$

(7) $u = x^{2.5}$, $v = (\sqrt{x})^5$, $x \sim 32$ [use $(\sqrt{x})^5 = x^2 \sqrt{x}$]

(8) $u = x^{1.5} - x^{0.5}$, $v = \sqrt{x}\,(x - 1)$, $x \sim 16$

(9) $u = \cos x \tan x$, $v = \sin x$, $x \sim 1.5$ radians

(10) $u = \sin 2x$, $v = 2 \sin x \cos x$, $x \sim 0.5$ radians

C. Find out which is better:

(11) $u = x(x + 2)$, $v = x^2 + 2x$

(12) $u = (x + 1)(x + 1)$, $v = 1 + x(2 + x)$

(13) $u = \cos 2x$, $v = 2 \cos^2 x - 1$

(14) $u = xe^x$, $v = e^{x + \ln x}$

1.5 CHAINS OF CALCULATIONS

1.5.1 A Statistical Example

By this time, you should be convinced that a chain of calculations inevitably causes a *propagation* of errors. The result of one calculation in a chain has an inherent error which becomes an input error for the next link in the chain. You have noticed that input errors become magnified. This can build up to a serious nuisance in a long sequence of computations.

Consider, for example, the calculation of the *standard deviation*, σ, defined by the equation

$$\sigma = \sqrt{\frac{\Sigma(X - M)^2}{n}} \tag{1.5.1}$$

where X is a *raw score* in a column of scores, M is the *mean* of the column, and n is the number of scores. Most beginners program the standard deviation exactly the way the formula is written. This is easy, but incorrect. The mean M is a calculated, rounded quantity and contains an error. Each difference, $X - M$, contains an error. When the quantity is squared, the error is increased. When the $(X - M)^2$ quantities are added, the error is increased again.

It is far better to avoid using derived quantities if a procedure can be developed which uses the original data. We shall derive a *computation formula* to replace (1.5.1):

$$\sigma^2 = \frac{\Sigma(X - M)^2}{n} \tag{1.5.2}$$

$$\sigma^2 = \frac{\Sigma(X^2 - 2MX + M^2)}{n} \tag{1.5.3}$$

$$= \frac{1}{n} \left[\Sigma(X^2) - 2M\Sigma X + nM^2 \right] \tag{1.5.4}$$

We now use the definition of the mean:

$$M = \Sigma X / n \tag{1.5.5}$$

$$\therefore \quad \sigma^2 = \frac{1}{n}\left[\Sigma(X^2) - \frac{2}{n}(\Sigma X)^2 + n\frac{(\Sigma X)^2}{n^2}\right] \tag{1.5.6}$$

$$= \frac{1}{n^2}\left[n\Sigma(X^2) - 2(\Sigma X)^2 + (\Sigma X)^2\right] \tag{1.5.7}$$

$$= \frac{1}{n^2}\left[n\Sigma(X^2) - (\Sigma X)^2\right] \tag{1.5.8}$$

$$\sigma = \frac{\sqrt{n\Sigma(X^2) - (\Sigma X)^2}}{n} \tag{1.5.9}$$

In (1.5.9), all the data are original data. As you will see in the exercises, we are not always that lucky.

1.5.2 Loop Control

In setting up a loop, it is a great temptation to write

```
X = X0
DO1I = 1,1000
X = X + DELTA
    .
    .
    .
1 ...
```

which means that the last value of X has 1000 errors of addition included. We suggest the following alternative:

```
X = X0
DO1I = 1,1000
A = FLOAT(I) or DFLOAT(I)
X = X + DELTA*A
    .
    .
    .
```

This avoids chaining the errors.

1.6 SOME PROGRAMMING HINTS

The following hints are based upon the theoretical discussion in Chapter 1. All the programs in the book use these as guidelines.

1. In programming a sum, rearrange the numbers so that smaller contributions are added first. For example, if $a > b > c$, then $(c + b) + a$ is better than $(a + b) + c$

2. Rearrange formulas so that differences close to zero are avoided. For example, if $\sqrt{b^2-4ac}$ and b have the same sign,

$$\frac{-2c}{b+\sqrt{b^2-4ac}} \quad \text{is better than} \quad \frac{-b+\sqrt{b^2-4ac}}{2a}$$

3. If possible, avoid expressions of the form a^b where b is a floating-point number. For example, $(\sqrt{a})^3$ is much better than $a^{1.5}$.

4. If you can possibly afford the time, use double precision for floating-point calculation. This is especially important for trigonometric and other functions represented by series.

5. If possible, rearrange your formulas to use original data rather than derived data. Examples from statistical analysis are given in the text.

6. If possible, rearrange your formulas to cut down on the number of arithmetic operations. Obviously, $x+y$ is better than $(x^2-y^2)/(x-y)$.

7. Avoid chaining operations which involve roundoff errors. For example,

```
        SUM=0
        DO1 I=1,100
        A=I
1       SUM=SUM+A
        SUM=SUM/3.
```

is better than

```
        SUM=0
        A=0
        B=1.0/3.0
        DO1 I=1,100
        A=A+B
1       SUM=SUM+A
```

8. Wherever possible and economically feasible, work with integers. This is relatively slow, but free of roundoff errors.

PROBLEM SECTION 1.5

In each of the following, derive the computation formula from the definition. Help, if needed, can be found in Dodes, I. A., *Introduction to Statistical Analysis*, Hayden, 1974.

(1) The third standard moment (skewness) of a distribution is defined as

$$\alpha_3 = \frac{\Sigma(z^3)}{n}$$

where $z=(X-M)/\sigma$. Show that

$$\alpha_3 = \frac{n^2\Sigma(X^3) - 3n\Sigma X\Sigma(X^2) + 2(\Sigma X)^3}{L^3}$$

where $L = \sqrt{n\Sigma(X^2) - (\Sigma X)^2}$

(2) The fourth standard moment (kurtosis) of a distribution is defined as

$$\alpha_4 = \frac{\Sigma(z^4)}{n}$$

Show that

$$\alpha_4 = \frac{n^3\Sigma(X^4) - 4n^2\Sigma X\Sigma(X^3) + 6n(\Sigma X)^2\Sigma(X^2) - 3(\Sigma X)^4}{L^4}$$

where L is defined as in problem (1).

(3) Pearson's r, the coefficient of linear correlation, is defined by

$$r_{xy} = \frac{\Sigma z_x z_y}{n}$$

Show that

$$r_{xy} = \frac{n\Sigma XY - \Sigma X\Sigma Y}{L_x L_y}$$

where L is defined as in problem (1).

(4) The regression coefficient of X_3 on X_1 and X_2 is defined by

$$b_{12.3} = \frac{r_{12.3}\sigma_{1.23}}{\sigma_{2.13}}$$

where

$$r_{12.3} = \frac{r_{12} - r_{13}r_{23}}{\sqrt{1 - (r_{13})^2}\sqrt{1 - (r_{23})^2}}$$

$$\sigma_{1.23} = \sigma_1\sqrt{1 - (r_{12})^2}\sqrt{1 - (r_{13.2})^2}$$

Show that

$$b_{12.3} = \frac{\sigma_1(r_{12} - r_{13}r_{23})}{\sigma_2(1 - r_{23}^2)}$$

Chapter 2

The Solution of an Equation

If you are starting with this chapter, be sure to read the summary of programming hints at the end of Chapter 1. If you are serious about expert programming, you should master Chapter 1 sooner or later. [In a short course, Section 2.1.4 may be omitted.]

Most people who are asked to state a difficult equation will think of something like

$$2.85437\, x^2 - 8.14576\, x + 11.91473 = 0$$

In fact, this is very easily solved by calculator, using the quadratic formula. There is nothing particularly difficult about it. In general, unless there are a great many quadratic equations to solve, it hardly pays to program a computer to solve it. Besides, as we explained in the previous chapter, computer implementation of the quadratic formula is unreliable under certain conditions.

Once we leave the quadratic and proceed to equations like

$$f(x) = 3\, x^7 - 5\, x^5 + 2\, x - 8 = 0$$

or

$$f(x) = e^x \ln x + \sin x - 2.8 = 0$$

no formula exists, and the peculiar capabilities of the computer can be employed profitably.

In this chapter, we shall describe two methods for finding the real roots of any equation. The simpler of the two methods requires no assumptions except that the root not be multiple. It is an excellent method for a computer. The 22 equations used in this chapter for illustrations and the problem section were all solved, correct to the nearest 10^{-9}, on a small IBM computer in a total of 2 minutes and 51 seconds, when this method was used.

If the function $f(x)$ is differentiable, a faster method (Newton-Raphson) can be employed. We include a complete discussion of this method because it is the basis of the computer square-root subroutine and of an important method for systems of non-linear equations.

Basic to all the many methods for solving an equation is the need for a *starter*. This is often the hardest part of the problem. We shall discuss this need in the first section.

2.1 STARTERS

2.1.1 Definitions

A *starter* is a value of x sufficiently close to a root of $f(x)$ and below it. For example, in

$$f(x) = x^2 - 5x + 6$$

for which the exact roots are $\{2, 3\}$, a starter for the smaller root might be 0 or 1 or 1.5 or 1.9, etc. A starter for the larger root might be 2.1 or 2.5 etc. In the equation

$$f(x) = x^2 - 2.3\,x + 1.32 = 0$$

for which the exact roots are $\{1.1, 1.2\}$, a starter for the smaller root might be 0 or 1 or 1.05 or 1.09 etc. A starter for the larger root might be 1.11, 1.15 etc.

We present two methods for finding starters graphically, the *double-graphical method* and the *single-graphical method*.

2.1.2 Double-Graphical Method for Starters

ILLUSTRATIVE PROBLEM 2.1-1   Find a starter for

$$x^2 - \sqrt{x} = 0.1$$

using the double-graphical method.

Solution   Here, $f(x) = x^2 - \sqrt{x} - 0.1 = 0$. We rewrite the equation as a pair of simultaneous equations:

$$\begin{cases} y = x^2 \\ y = \sqrt{x} + 0.1 \end{cases} \qquad (2.1.1)$$

each one easy to plot. We could equally well have chosen

$$\begin{cases} y = x^2 - 0.1 \\ y = \sqrt{x} \end{cases} \qquad (2.1.2)$$

but not

$$\begin{cases} y = x^2 - \sqrt{x} \\ y = 0.1 \end{cases} \qquad (2.1.3)$$

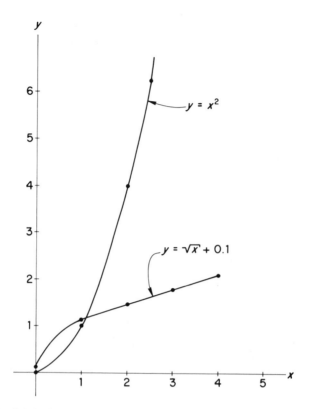

Fig. 2.1-1   Illustrative Problem 2.1-1: double-graphical method.

Returning to (2.1.1), note that at the point of intersection of the two curves (Figure 2.1-1),

$$x^2 = \sqrt{x} + 0.1 \tag{2.1.4}$$

which agrees with the original problem. From the figure, it should be clear that there is a root just above $x = 1$. Therefore, we can take $x = 1$ as a starter. A consideration of the slopes of the two graphs assures us that there is only one root.

ILLUSTRATIVE PROBLEM 2.1-2   Find a starter for

$$2x^4 - 9e^x = 20$$

Solution   We have $f(x) = 2x^4 - 9e^x - 20 = 0$. We rewrite this as

$$\begin{cases} y = 2x^4 - 20 \\ y = 9e^x \end{cases} \tag{2.15}$$

The double-graphical solution is shown in Figure 2.1-2. From the figure, it is evident that there is a root between -2 and -1 and another between 4 and 5. Our

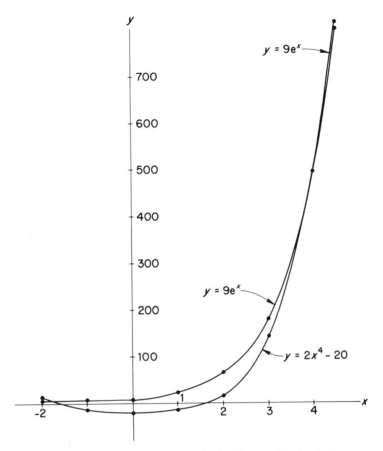

Fig. 2.1-2 Illustrative Problem 2: double-graphical solution.

starters are -2 and 4. The slopes of the curves seems to indicate that there are precisely two roots. This information is an important by-product of the double-graphical method.

2.1.3 The Single-Graphical Method

It is, of course, possible to work from a single graph, or even from a table of values. In the following illustrative problems, both the single-graphical and double-graphical methods are usable. The one striking advantage that the single-graphical method has is that it is possible to have a single graph drawn by the computer. However, there are also some disadvantages, as we shall see.

ILLUSTRATIVE PROBLEM 2.1-3   Find a starter for

$$\sqrt{x} - \ln x = 0.7$$

using the single-graphical method.

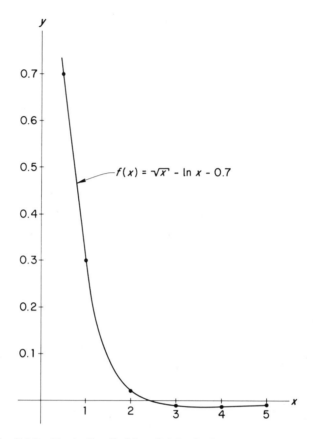

Fig. 2.1-3   Illustrative Problem 2.1-3: single-graphical method.

Solution   In this book, we are limiting ourselves to real numbers. Therefore we must take $x > 0$. A table of values for $f(x) = \sqrt{x} - \ln x - 0.7$ is as follows.

x	0.1	0.5	1	2	3	4	5	6
$f(x)$	1.92	0.70	0.30	0.02	-0.07	-0.09	-0.07	-0.04

We see from the table that there is a root between $x=2$ and $x=3$, so that $x=2$ is a suitable starter. The graph is shown in Figure 2.1-3.

If the graph continues to ascend as x increases, and if it crosses the x-axis, there is another root, perhaps more. But if we express the function as a pair of simultaneous equations:

$$\begin{cases} y = \sqrt{x} \\ y = \ln x + 0.7 \end{cases} \tag{2.1.6}$$

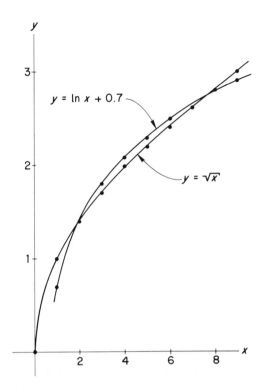

Fig. 2.1-4   Illustrative Problem 2.1-3: double-graphical method.

we can draw the system as shown in Figure 2.1-4. Then it is perfectly clear that there are precisely two roots. From Figure 2.1-3, there might have been no more, one more, or a dozen more.

ILLUSTRATIVE PROBLEM 2.1-4   Locate a starter between -5 and $+5$ for

$$7\sin x - e^{-x}\cos x = 0.7$$

Solution   Use $f(x) = 7\sin x - e^{-x}\cos x - 0.7 = 0$. Using the single-graphical method, there are roots in the intervals $(-5, -4)$, $(-3, -2)$, $(0, 1)$ and others. The graph is shown in Figure 2.1-5.

For comparison, we show the double-graphical approach (Figure 2.1-6) using

$$\begin{cases} y = 7\sin x - 0.7 \\ y = e^{-x}\cos x \end{cases} \tag{2.1.7}$$

The clear advantage of Figure 2.1-6 is that the periodicity of both curves is shown. It is obvious that there is an unlimited supply of roots above $x = 5$. A further advantage is that the curves are easily sketched. Also, everyone knows what the curves look like. There is less chance of an error.

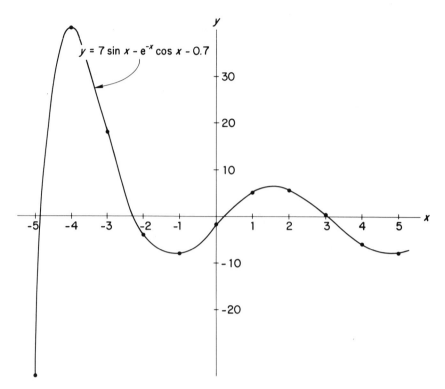

Fig. 2.1-5   Illustrative Problem 2.1-4: single-graphical method.

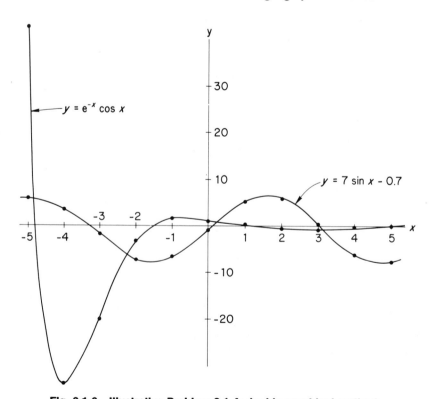

Fig. 2.1-6   Illustrative Problem 2.1-4: double-graphical method.

58

2.1.4 Sturm Functions

The graphical method is not very useful for polynomials of high degree. These arise in the calculation of *eigenvalues*, to be discussed later. We shall describe a method which works for any degree of polynomial. Consider

$$f(x) = a_n x^n + a_{n-1} x^{n-1} + \cdots + a_1 x + a_0 = 0 \qquad (2.1.7)$$

We now define a *Sturm sequence*.

STURM SEQUENCE OF POLYNOMIALS   A sequence $f_0(x)$, $f_1(x), \ldots, f_n(x)$, where

$$f_0(x) = f(x)$$

$$f_1(x) = f'(x)$$

and $f_2(x)$, $f_3(x), \ldots, f_n(x)$ are the negatives of the remainders in the process of finding the highest common factor of $f(x)$ and $f'(x)$ by Euclid's algorithm.

ILLUSTRATIVE PROBLEM 2.1-4.   Find the sequence of Sturm functions for

$$f(x) = x^4 - 4x^3 + x^2 + 6x + 2 = 0$$

Solution

$$f_0(x) = x^4 - 4x^3 + x^2 + 6x + 2$$
$$f_1(x) = 4x^3 - 12x^2 + 2x + 6$$

We shall be interested only in the *signs* of the functions when numbers are substituted, as you will see. Therefore, it is permissible to multiply or divide any function by a positive constant to make the algorithm easier. Instead of dividing f_0 by f_1, which would be the normal procedure in applying Euclid's algorithm, we shall first divide f_1 by 2, since that is a common factor, then multiply f_0 by 2. This has the effect of making the leading coefficients equal. Now we divide:

$$2x^3 - 6x^2 + x + 3 \overline{)\,2x^4 - 8x^3 + 2x^2 + 12x + 4}$$

The remainder is $-5x^2 + 10x + 7$. We change the sign:

$$f_2(x) = 5x^2 - 10x - 7$$

Making the same kind of alterations, instead of dividing f_1 by f_2, we divide as follows:

$$10x^2 - 20x - 14 \overline{)\,10x^3 - 30x^2 + 5x + 15}$$

The remainder is $-x + 1$. We change the sign:

$$f_3 = x - 1$$

Now we divide as follows:

$$5x - 5 \overline{)5x^2 - 10x - 7}$$

The remainder is -12, so that

$$f_4 = 12$$

The resulting expressions in the sequence are:

$$\begin{cases} f_0(x) = x^4 - 4x^3 + x^2 + 6x + 2 \\ f_1(x) = 4x^3 - 12x^2 + 2x + 6 \\ f_2(x) = 5x^2 - 10x - 7 \\ f_3(x) = x - 1 \\ f_4(x) = 12 \end{cases} \qquad (2.1.8)$$

We shall now explain some of the properties of the Sturm sequence. From the first division, we obtain

$$\frac{f(x)}{f_1(x)} = q_1 + \frac{R}{f_1(x)} \qquad (2.1.9)$$

where q_1 is the quotient, and R, the remainder, is actually $-f_2(x)$. Rewriting (2.1.9) and the other equations, we have

$$\begin{cases} f(x) = q_1 f_1(x) - f_2(x) \\ f_1(x) = q_2 f_2(x) - f_3(x) \\ f_2(x) = q_3 f_3(x) - f_4(x) \\ \qquad \vdots \\ f_{n-3}(x) = q_{n-2} f_{n-2}(x) - f_{n-1}(x) \\ f_{n-2}(x) = q_{n-1} f_{n-1}(x) - f_n(x) \end{cases} \qquad (2.1.10)$$

Lemma 2.1.1 *Iff $f(x)$ and $f'(x)$ have no equal roots, then $f(x)$ and $f'(x)$ have no common factor.*

Proof We shall investigate the special case

$$f(x) = x^3 + a_1 x^2 + a_2 x + a_3 = (x - r_1)(x - r_2)(x - r_3)$$
$$f'(x) = (x - r_1)(x - r_2) + (x - r_1)(x - r_3) + (x - r_2)(x - r_3)$$
$$\frac{f'(x)}{f(x)} = \frac{1}{x - r_1} + \frac{1}{x - r_2} + \frac{1}{x - r_3}$$

from which it is clear that there is no factor $(x - a)$ such that

$$f'(x) = (x - a)f(x) \text{ or } f(x) = (x - a)f'(x)$$

The extension to the general case is obvious. If $f(x)$ has root r with multiplicity m, it can be written as

$$f(x) = (x-r)^m g(x)$$

and a simple calculation shows that there is a common factor $(x-r)^{m-1}$ between $f(x)$ and $f'(x)$. This completes the proof. $\square$

Lemma 2.1.2 *If there are no equal roots, $f_n(x) \neq 0$.*

Proof We illustrate with a short sequence:

$$f(x) = q_1 f_1(x) - f_2(x)$$
$$f_1(x) = q_2 f_2(x) - f_3(x)$$
$$f_2(x) = q_3 f_3(x) - f_4(x)$$

Now, suppose that $f_4(x) = 0$. Then

$$f_2(x) = q_3 f_3(x)$$
$$f_1(x) = q_2 q_3 f_3(x) - f_3(x) = (q_2 q_3 - 1) f_3(x)$$
$$f(x) = q_1(q_2 q_3 - 1) f_3(x) - q_3 f_3(x)$$
$$= [q_1(q_2 q_3 - 1) - q_3] f_3(x)$$

Then

$$\frac{f'(x)}{f(x)} = \frac{(q_2 q_3 - 1) f_3(x)}{[q_1(q_2 q_3 - 1) - q_3] f_3(x)}$$

which would give $f'(x)$ and $f(x)$ a common factor, contrary to Lemma 2.1.1. The generalization is immediate. $\square$

Lemma 2.1.3 *If there are no equal roots, two consecutive functions after the first cannot vanish together.*

Proof Consider

$$f(x) = q_1 f_1(x) - f_2(x)$$
$$f_1(x) = q_2 f_2(x) - f_3(x)$$
$$f_2(x) = q_3 f_3(x) - f_4(x)$$
$$f_3(x) = q_4 f_4(x) - f_3(x)$$

and suppose that $f_1(x)$ and $f_2(x)$ vanish. Then, from the second and third equations,

$$f_3(x) = q_2 f_2(x) = 0$$
$$f_4(x) = q_3 f_3(x) = 0$$

Then

$$f_5(x) = q_4 f_4(x) - f_3(x) = 0$$

By Lemma 2.1.2, this is impossible. $\square$

Lemma 2.1.4 *If any $f_k(x)$ vanishes, $f_{k-1}(x)$ and $f_{k+1}(x)$ have opposite signs.*

Proof Consider the sequence (2.1.10), and let $f_k(x)$ vanish. Then

$$f_{k-1}(x) = q_k f_k(x) - f_{k+1} = -f_{k+1}$$

completing the proof. $\square$

Lemma 2.1.5 *If, as x increases, $f_k(x)$ passes through 0, the Sturm sequence loses one change of sign.*

Proof This is evident geometrically, because for a set of real, distinct roots, the graph of $f(x)$ crosses the x-axis once for every root.

    With all these lemmas proved, the following theorem can be proved. A complete proof can be found in Uspensky, J. V., *Theory of Equations*, McGraw-Hill, New York, 1948, pp. 143 ff.

THEOREM OF STURM   Let (a,b) be a permissible domain for $f(x)=0$, where neither a nor b is a root. Let $v(a)$ and $v(b)$ be the number of changes of sign at a and b. Then the number of roots on (a,b) is $v(a) - v(b)$.

ILLUSTRATIVE PROBLEM 2.1-5.   For the polynomial in Illustrative Problem 2.1-4, locate the roots.

Solution   We have the Sturm sequence from the previous problem. We form a table of *signs* as follows:

x	-2	-1	0	1	2	3	4
$f(x)$	$+$	$+$	$+$	$+$	$+$	$+$	$+$
$f_1(x)$	$-$	$-$	$+$	$+$	$-$	$+$	$+$
$f_2(x)$	$+$	$+$	$-$	$-$	$-$	$+$	$+$
$f_3(x)$	$-$	$-$	$-$	$+$	$+$	$+$	$+$
$f_4(x)$	$+$	$+$	$+$	$+$	$+$	$+$	$+$
v	4	4	2	2	2	0	0

There are two losses from -1 to 0; therefore there are two roots between -1 and 0. There are two losses from 2 to 3. There are two more roots between these values. (*Note:* Zeros in the table are disregarded in counting changes.)

PROBLEM SECTION 2.1

Each of the following has at least one root in $[-5, +5]$.
Find a starter for the smallest root in that region.

(1) $x^5 - 10x^3 = 4$

(2) $x^4 - 4x^3 = 6$

(3) $-2x^5 + 7x^4 = -100$

(4) $-3x^3 + 11x^4 = -40$

(5) $x^4 - 56\sqrt{x} = 2.5$

(6) $0.03x^5 - 15\sqrt{x} = 4$

(7) $x^2 + \sin x = 2$

(8) $x^3 - 4\cos x = 1$

(9) $2x^2 + \ln x = 3$

(10) $2x^2 - e^x = 0$

(11) $\sin x + 3\cos x = 0$

(12) $4\cos x - 6\sin x = 3$

(13) $25e^x - 150e^{-x} = 2$

(14) $\ln x + 30e^{-x} = 3$

(15) $\sqrt{x}\,\ln x + 7\sin x\cos x = 0.3$

(16) $(\sin x)(\ln x) + \sqrt{x}\,\cos x = 0.5$

(17) $x^3\sin x + x^2 e^{-x} = 15$

(18) $x^4\cos x + 6x^5 e^x = -1$

2.2 THE METHOD OF SIMPLE ITERATION

2.2.1 Checking a Root

Suppose you were asked to check roots as follows.

Case I. $f(x) = 9x^2 - 6x + 1 = 0,\ x = 0.305$

Case II. $f(x) = x - \sin x = 0,\ x = 3°10' = 0.05527$ radians

Case III. $f(x) = e^{-x}\cos x + 0.015 = 0,\quad x = 224°50' = 3.92408$
radians

Your first thought is to substitute the values of x to see whether these values "satisfy" the equation to the precision required. These values do check to the precision of the answer, yet they are all wrong. In fact, the first two have exact answers: $\frac{1}{3}$ and 0. We shall try to show that the method of substitution is often useless for checking. We do this by demonstrating that in each of these three cases a wide range of values "checks" to the precision of the answer.

Case I.   Any number between 0.31 and 0.35 will yield $y = 0.00$ correct to the nearest hundredth.

Case II. Any number between -0.142 and $+0.142$ radians $(-8°10'$ and $+8°10')$ yields $y = 0.000$ correct to the nearest thousandth.

Case III. Any number between 3.73 and 4.07 radians $(213°40'$ and $233°40')$ yields $y = 0.00$ correct to the nearest hundredth. Any number between 3.875 and 3.906 radians $(220°$ and $223°50')$ yields $y = 0.000$ correct to the nearest thousandth.

These examples show that a root cannot always be checked, in a practical sense, by mere substitution. If the function crosses the x-axis at a steep angle (like f_1 in Figure 2.2.1), a small error in x will cause a large

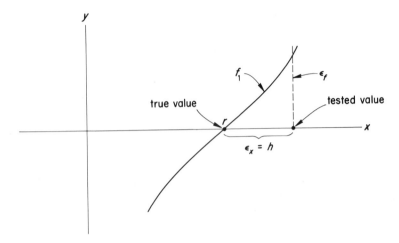

Fig. 2.2-1   A small error in x causes a large error in f_1.

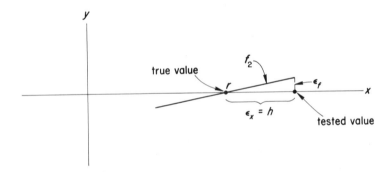

Fig. 2.2-2   A small error causes a small error in f_2.

error in f_1. For example, if $f(x) = e^{5x} - x - 1$, for which the exact root is 0, the substitution of $x = 0.1$ yields $f(x) = 0.5487$. No one could mistake $x = 0.1$ as a root. In this case, substitution is bearable.

On the other hand, if the function crosses the x-axis at a small angle, as in Figure 2.2-2 the very same error in x brings about a comparatively small error in $f(x)$.

2.2.2 Error and Slope

It is of some interest to investigate the relationship between an error in x, which we shall call h, and the corresponding error in $f(x)$, which we shall call Δf. Let

$$r = \text{the true root}$$

and

$$r + h = \text{the tested value of the computed root}$$

Then

$$f(r+h)-f(r)=(\text{the error in } f)=\Delta f$$

Using the mean-value theorem,[1]

$$f(r+h)-f(r)=hf'(\theta), \qquad \theta \in [r,r+h] \qquad (2.2.1)$$

If h is small, then

$$f'(\theta)\sim f'(r) \qquad (2.2.2)$$

and, by substituting,

$$\Delta f \sim hf'(r) \qquad (2.2.3)$$

$$\frac{\Delta f}{h}\sim f'(r) \qquad (2.2.4)$$

This shows that the error in f, divided by that in x, is approximately equal to the slope.

For example, consider any polynomial of the second degree:

$$f(x)=ax^2+bx+c$$

Then

$$f'(x)=2ax+b$$

At a root,

$$x=\frac{-b\pm\sqrt{b^2-4ac}}{2a}$$

which means that

$$2ax=-b\pm\sqrt{b^2-4ac}$$
$$f'(x)=\pm\sqrt{b^2-4ac} \qquad (2.2.5)$$

In absolute value, this is a minimum when $b^2-4ac=0$, i.e., when the polynomial is a perfect square. Therefore, substitution is least reliable for quadratics which are perfect squares.

In case II, $f'(x)=1-\cos x$, which is 0 at the exact root. From (2.2.4), substitution is not reliable at this root.

---

[1] $\in$ is a mathematical symbol of inclusion. The meaning of $x\in[-5,+5]$ is that x is in the closed interval from -5 to $+5$. *Closed* means that both endpoints are included. The symbol $x\in(-5,+5)$ would indicate that neither endpoint is included, i.e., x is neither -5 nor $+5$. This would be an *open* interval.

In case III, $f'(x) = -e^{-x}(\sin x + \cos x)$. At the root, f' has a value of about 0.03. From (2.2.4), Δf is about 0.03 times the error in the computed value. Substitution is not reliable.

These examples should be convincing. Substitution alone is not a satisfactory method for finding or checking a root.

2.2.3 The Simplest Computer Method

An examination of the curves in Figures 2.2-1 and 2.2-2 shows that in each case f is negative on one side of the root and positive on the other. This is, of course, necessary if f is to cross the x-axis. This thought leads to a method of solution which is extremely safe and easy on a computer, although laborious by hand. Since it involves no assumptions except that $f(x)$ crosses the x-axis at the root, we recommend it for computer solutions.

The method we are about to demonstrate is called *simple iteration*. We shall explain the name after the demonstrations.

ILLUSTRATIVE PROBLEM 2.2-1   Solve to the nearest thousandth by simple iteration for the smallest root between -5 and $+5$ in

$$f(x) = x^2 - \sqrt{x} - 0.1$$

Solution   In Illustrative Problem 2.1-1 we found that $x = 1$ is a starter. We start with $\Delta x = 1$:

x	$f(x)$
1	-0.1
2	$+7.5$

Actually, we don't care about the values of $f(x)$. It is only important for us to see that one is positive and one is negative (either order). Then there is a root between $x = 1$ and $x = 2$. We now test using $\Delta x = 0.1$:

x	$f(x)$
1.0	-0.1
1.1	$+5.1$

which shows that the root is between 1.0 and 1.1. We divide Δx by 10:

x	$f(x)$
1.0	-0.1
1.01	-0.08
1.02	-0.07
1.03	-0.05
$\vdots$	$\vdots$
1.06	-0.006
1.07	$+5.01$

Reduce the increment to 0.001:

x	$f(x)$
1.060	-0.006
1.061	-0.004
.	.
.	.
1.063	-0.001
1.064	$+5.00$

We know the root is between 1.063 and 1.064. Which is closer? We test for 1.0635:

x	$f(x)$
1.0630	$-$
1.0635	$-$
1.064	$+$

which shows that the root is between 1.0635 and 1.064. Therefore, to the nearest thousandth, $x = 1.064$.

ILLUSTRATIVE PROBLEM 2.2-2 Solve to the nearest thousandth by simple iteration for the smallest root $\in [-5, +5]$:

$$f(x) = 2x^4 - 9e^x - 20$$

Solution From Illustrative Problem 2.1-2, we know there is a starter at $x = -2$. Table 2.2-1 shows the convergence to the root. To the nearest thousandth, the root is -1.810.

Table 2.2-1

x	$f(x)$	ROOT BETWEEN
-2	$+$	
-1	$-$	-2 and -1
-2.0	$+$	
-1.9	$+$	
-1.8	$-$	-1.9 and -1.8
-1.90	$+$	
.	.	
.	.	
-1.82	$+$	
-1.81	$-$	-1.82 and -1.81
-1.820	$+$	
-1.819	$+$	
.	.	
.	.	
-1.811	$+$	
-1.810	$-$	-1.811 and -1.810
-1.8105	$-$	-1.8195 and -1.8100

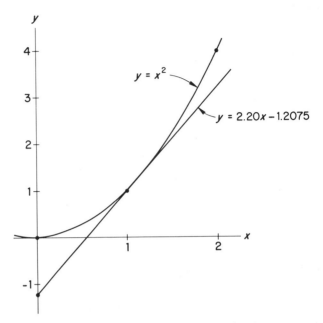

Fig. 2.2-3   Illustrative Problem 2.2-3: $f(x = x^2 - 2.20x + 1.2075)$.

ILLUSTRATIVE PROBLEM 2.2-3   Find, correct to the nearest hundredth, the smaller root of

$$f(x) = x^2 - 2.20x + 1.2075$$

Solution   A portion of the graph is shown in Figure 2.2-3. It can be seen that there is a root between 1 and 2, possibly two of then.

A careless test of x and $f(x)$ might be

x	$f(x)$
1	$+0.008$
2	$+0.81$

apparently showing no roots. The increment should be much smaller. Using $\Delta x = 0.1$, we have

x	$f(x)$
1.0	$+$
1.1	$-$

showing another root between 1.0 and 1.1. Continuing, the calculation yields 1.05 as the smaller root. The other root turns out to be 1.15.

2.2.4   Iteration

The word *iteration* is computer language for *repetition*. It refers to a series of trials where each trial is based upon the one before it. In the computer method just demonstrated, the iteration is "simple" because only the *signs* were considered in successive trials.

The greatest value of the "simple" iterative method is the fact that *errors are not propagated in the sequence of trials*. For example, in Illustrative

Problem 2.2-2 there were 27 trials. However, the answer was completely determined by the last two. Even if all the other 25 trials were incorrect, these two correct results locate the root.

2.2.5 Comment

Students looking at these problems invariably wish to try half-way points, and some authors recommend them. In other words, they try

x	$f(x)$
-1.90	$+$
-1.85	$+$
-1.80	$-$

thus eliminating many trials. Other students, noting that at $x = -1.90$, $f(x) = 9.71809$, reason that the root is closer to -1.80. Therefore they start testing at the other end of the interval with a negative value of Δx.

In fact, the straightforward process is easy to program and extremely fast, so that there is no compelling reason to look for small improvements —no harm either. Remember that the computer has no eyeballs.

2.2.6 Computer Program

A typical computer program, developed by a class, is shown in Figure 2.2-4.

```
C SET INITIAL CONDITIONS, INCLUDING STARTING POINT, STARTING INCREMENT,
C AND CRITERION.
      DOUBLE PRECISION X,Y,YP,DELTA,C
      X=1.0
      DELTA=1.0
      C=10.0**(-9)
      YP=0.0
C COMPUTATION BOX
1     Y=X*DLOG10(X)-1.2
C IS Y ZERO  IF IT IS,THIS IS AN EXACT ROOT AT X.
      IF(Y)2,7,2
C IS YP ZERO  IF IT IS, THIS IS THE FIRST TIME.
2     IF(YP)4,3,4
C IS Y*YP NEGATIVE  IF IT IS, WE HAVE PASSED A ROOT.
4     IF(Y*YP)5,3,3
3     YP=Y
C STEPPING X. RETURN TO COMPUTATION BOX.
6     X=X+DELTA
      GO TO 1
C IF THE CRITERION IS NOT SATISFIED, GO TO STEP BOX.
5     IF(DELTA-0.1*C)7,7,8
8     X=X-DELTA
      DELTA = 0.1 * DELTA
      GO TO 6
C PRINT
7     WRITE(3,100)X,Y
100   FORMAT(1H1,'THE SOLUTION IS',F12.9, 5X,'WHEN THIS VALUE IS SUBSTIT
     1UTED, THE VALUE OF Y IS',F12.9)
      STOP
      END
```

Fig. 2.2-4   A class program for the method of simple iteration.

PROBLEM SECTION 2.2

For each of the problems for Section 2.1, find the smallest root in $[-5, +5]$ to the nearest thousandth, using the method of simple iteration, if this is being done on a calculator. If it is being done on a computer, find the root correct to the nearest 10^{-9}.

2.3 THE NEWTON-RAPHSON METHOD (PART I)

2.3.1 The Problem

We have already illustrated a simple iteration which will find a single root of any equation, with no assumptions except that $f(x)$ must cross the x-axis. This suffices for almost all computer-science applications, whether or not the function is differentiable. However, there are differentiable functions which occur so often that a saving of microseconds is worth some extra trickery. One such problem is that of finding $\sqrt{N}$. To do this, the equation

$$x^2 - N = 0$$

is solved for the positive root. We shall show how this is done. Other such problems exist that are solved by the Newton-Raphson method.

Many really clever methods have been devised for speeding up the solution of an equation. Most of the useful methods use iteration. These iterations involve not only the *signs* of the previous trials but also the values. In the Newton-Raphson method, our starter is x_0. From this we find x_1 and $f(x_1)$, then x_2 and $f(x_2)$, and so on. These hopefully converge to the real root, $x = r$.

We shall discuss the two obvious subproblems: (a) How do we get from x_0 to x_1 to $x_2 \cdots$ to x_k to x_{k+1} to $\cdots$? (b) If the root is r, how do we know that as k increases x_k is coming closer to r? If x_k is approaching r as $k \rightarrow \infty$, we say that x_k *converges* to r, or that

$$\lim_{k \to \infty} x_k = r$$

2.3.2 Geometric Analysis

Consider the function in Figure 2.3-1, for which a root is r. Therefore $f(r) = 0$, this being the definition of a root. We start with a *guess* at x_0, the guess being a starter arrived at in the usual fashion. A vertical line locates point P which has coordinates $[x_0, f(x_0)]$. At P, we construct tangent T which has slope $f'(x_0)$. This intersects the x-axis at point Q and gives us x_1, the starting value for the next trial. In the diagram, one more trial has been shown.

Under certain conditions which we shall investigate, the sequence $\{x_k\}$ or $x_0, x_1, x_2, \ldots, x_k, x_{k+1}$ converges to r.

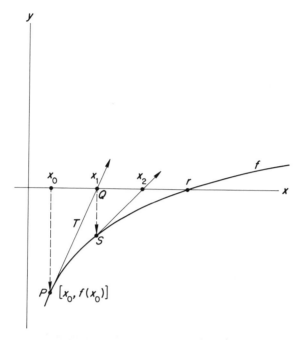

Fig. 2.3-1   Geometric analysis of the Newton-Raphson method.

Let us see what the geometric technique corresponds to, algebraically. Assume (Figure 2.3-2) that we started with x_0 and arrived, after k iterations, at x_k. We are about to find x_{k+1}. The coordinates of P are $[x_k, f(x_k)]$. The tangent line, T, has slope $f'(x_k)$. Let (x_T, y_T) be the coordinates of any point on T. Then, the slope of T is given by

$$\frac{y_T - f(x_k)}{x_T - x_k} = f'(x_k) \tag{2.3.1}$$

When $x_T = x_{k+1}$, then $y_T = 0$. Substituting,

$$\frac{0 - f(x_k)}{x_{k+1} - x_k} = f'(x_k) \tag{2.3.2}$$

Solving for x_{k+1}, we have the *Newton-Raphson formula*:

$$x_{k+1} = x_k - \frac{f(x_k)}{f'(x_k)} \tag{2.3.3}$$

This formula answers our first question, namely, how we get from one value of x to the next. For example,

$$x_1 = x_0 - f(x_0)/f'(x_0) \tag{2.3.4}$$

$$x_2 = x_1 - f(x_1)/f'(x_1) \tag{2.3.5}$$

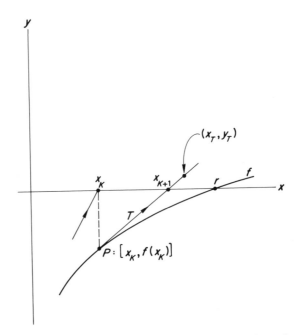

Fig. 2.3-2   **Derivation of the Newton-Raphson formula.**

2.3.3 Slopes

Before continuing to the second question (when will the method work?) we review some elementary concepts of calculus. In the following, f' stands for the first derivative and f'' for the second derivative. For example, if $f(x)=3x^4$, then $f'(x)=12x^3$ and $f''(x)=36x^2$.

Consider Figure 2.3-3. The signs of f' and f'' at certain points and on the intervals between them are as follows:

Between P and A	$f'<0$	$f''>0$
A (relative minimum)	$f'=0$	$f''>0$
Between A and Q	$f'>0$	$f''>0$
Q (point of inflection)	$f'>0$	$f''=0$
Between Q and B	$f'>0$	$f''<0$
B (relative maximum)	$f'=0$	$f''<0$
Between B and R	$f'<0$	$f''<0$
R (point of inflection)	$f'<0$	$f''=0$

To summarize briefly:

1. If the curve is rising from left to right, f' is positive.
2. If the curve is falling, f' is negative.
3. At a relative minimum or maximum, $f'=0$.
4. At a point of inflection, $f''=0$.

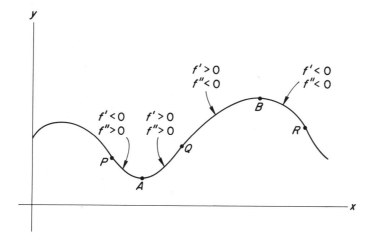

Fig. 2.3-3 A study of slopes.

5. If the curve is concave downwards, $f'' < 0$.
6. If the curve is concave upwards, $f'' > 0$.

The converse of statement (4) is not true.

2.3.4 Convergence

We shall prove certain (sufficient) conditions under which the Newton-Raphson method will work. First we point out some cases in which it does not work.

In Figure 2.3-4, the tangent line is parallel to the x-axis. Therefore, it will never meet the x-axis, and we will never get from x_k to x_{k+1}. The moral is that f' must never be 0 during a Newton-Raphson process.

In Figure 2.3-5, the the second tangent, T_2, produces a value of x_{k+2} which is worse than x_{k+1}, i.e., it is farther from r.

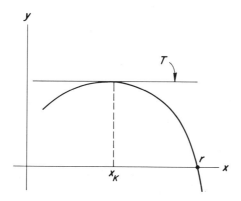

Fig. 2.3-4 The tangent is parallel to the x-axis.

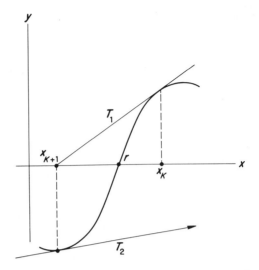

Fig. 2.3-5 The new estimate is worse than the previous one.

In Figure 2.3-6, although the curve is the same, the sequence *will* converge to *r*. The moral is that you cannot be sure, with an S-shaped curve, that the method will work. If x_0 is sufficiently close to *r* to avoid the reverse curvature of the S, the method will work.

In Figure 2.3-7, there is a double root. At this point, $f' = 0$. Also, f' approaches 0 as x_k approaches the root. Under the circumstances, the formula

$$x_{k+1} = x_k - \frac{f(x_k)}{f'(x_k)} \tag{2.3.6}$$

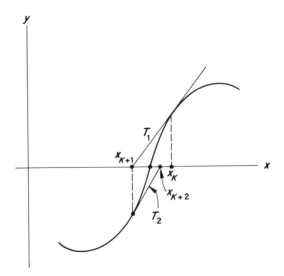

Fig. 2.3-6 Here the set of values will converge to the root.

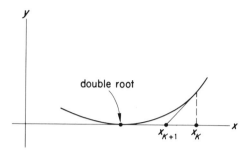

Fig. 2.3-7 Multiple roots present a problem in all methods.

has a denominator which is approaching zero. This may or may not cause practical difficulties, depending upon the rate at which the numerator, f, is approaching 0. We shall show a method for dealing with this problem, at least in part.

In a qualitative way, we may guess that the Newton-Raphson method will always work for the four cases shown in Figure 2.3-8. In each, the first two iterations are shown.

The method will also be successful, obviously, if the graphs are turned upside down. There are therefore at least eight cases which can be reduced to the four shown. Note, too, that Figure 2.3-8(b) and (c) represent the

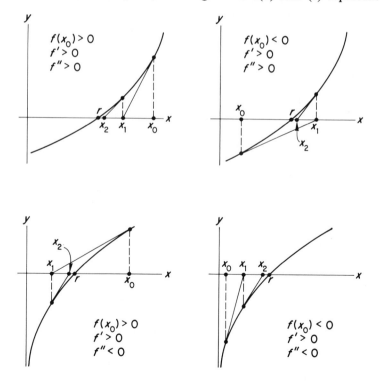

Fig. 2.3-8 Four cases where the Newton-Raphson method will work.

same situation reflected through the x-axis. Figure 2.3-8(a) and (d) represent the same situation for the same reason. This reduces the number of possible cases to two, namely (c) and (d). Finally, we shall show that (c) becomes (d) after one iteration. Therefore, we can embrace the eight cases by discussing only one, namely Figure 2.3-8(c).

2.3.5 Proof of a Theorem on the Newton-Raphson Method

We shall now prove that under the conditions of Fig. 2.3-8(c) the sequence $\{x_k\}$ must converge to r, the root. We are given that over the entire interval where iterations are taking place $f'(x)>0$ and $f''(x)<0$. Also, $f(x_0)>r>0$. In the following, MVT means mean-value theorem.

Part I. We shall prove that if x_0 is right of r, then x_1 is to the left of r, i.e., that $x_1-r<0$.

1. Consider *any* point $c_1 \in (r,x_0)$. By the MVT,

$$f'(x_0)-f'(c_1)=(x_0-c_1)f''(c_2), \qquad \text{where} \quad c_2 \in (c_1,x_0)$$

2. But $x_0-c_1>0$, and $f''<0$.
3. Therefore $f'(x_0)-f'(c_1)<0$, i.e., $f'(x_0)<f'(c_1)$ for *any* $c_1 \in (r,x_0)$.
4. Using the Newton-Raphson formula,

$$x_1=x_0-\frac{f(x_0)}{f'(x_0)}$$

5. Therefore $x_1-r=x_0-r-f(x_0)/f'(x_0)$.
6. By the MVT, $f(x_0)-f(r)=(x_0-r)f'(c_3)$, where $c_3 \in (r,x_0)$.
7. But $f(r)=0$.
8. Therefore $x_0-r=f(x_0)/f'(c_3)$.
9. Substituting in step 5,

$$x_1-r=\frac{f(x_0)}{f'(c_3)}-\frac{f(x_0)}{f'(x_0)}=f(x_0)\left[\frac{f'(x_0)-f'(c_3)}{f'(c_3)f'(x_0)}\right]$$

10. By step 3, $f'(x_0)-f'(c_3)<0$. Also, $f(x_0)>0$, $f'(c_3)>0$, $f'(x_0)>0$
11. Therefore, $x_1-r<0$, i.e., x_1 is left of r.

Part II. We shall now show that x_2 is to the right of x_1, i.e., $x_2>x_1$.

1. By the MVT,

$$f(x_1)-f(r)=(x_1-r)f'(c_4), \qquad \text{where} \quad c_4 \in (x_1,r)$$

2. But $f(r)=0$, $f'>0$, and $(x_1-r)<0$.

3. Therefore, $f(x_1) < 0$.
4. Using the Newton-Raphson formula,

$$x_2 = x_1 - \frac{f(x_1)}{f'(x_1)}$$

5. Since $f(x_1)$ is negative and $f'(x_1)$ is positive, $x_2 > x_1$, i.e., x_2 is to the right of x_1.

Part III. Now we show that $x_2 \leqslant r$, i.e., x_2 does not hop over to the right side of r.

1. Consider *any* point, c, between x_1 and r. By the MVT,

$$f'(x_1) - f'(c) = (x_1 - c)f''(c_4), \qquad \text{where} \quad c_4 \in (x_1, c)$$

2. But $x_1 \leqslant c$, and $f'' < 0$.
3. Therefore, $f'(x_1) - f'(c) \geqslant 0$, i.e., $f'(x_1) \geqslant f'(c)$ for all $c \in (x_1, r)$.
4. By the MVT,

$$f(r) - f(x_1) = (r - x_1)f'(c_5), \qquad \text{where} \quad c_5 \in (x_1, r)$$

5. By step 3, $f'(x_1) \geqslant f'(c_5)$. Also $f(r) = 0$.
6. Therefore,

$$-f(x_1) \leqslant (r - x_1)f'(x_1)$$

7. Therefore, $-f(x_1)/f'(x_1) \leqslant r - x_1$.
8. $x_2 = x_1 - f(x_1)/f'(x_1) \leqslant x_1 + (r - x_1) = r$, which proves that $x_2 \leqslant r$.

Summary. We have proved that even if x_0 is taken to the right of r, x_1 will be left of r. Once x_1 is left of r, x_2 (and therefore all subsequent x_k) will fall between x_1 and r. Each x_{k+1} is trapped between the previous x_k and the root r. This completes the proof of the convergence of $\{x_k\}$ to r, given the conditions at the beginning of the proof. As we mentioned previously, all eight cases are precisely the same after reflection through the proper axis. Thus we have the

> THEOREM If iterations are taken over a segment of a curve passing through the x-axis in such a way that $f' \neq 0$ and neither f' nor f'' changes sign on the interval, then the sequence of values produced by application of the Newton-Raphson formula will converge to the root.

The conditions given in the theorem are sufficient, not necessary. In other words, the sequence may sometimes converge to the root even if the conditions are not satisfied.

2.3.6   The Square-Root Subroutine

We shall compute $x = \sqrt{N}$ by finding a positive root for

$$f(x) = x^2 - N \qquad (2.3.7)$$

We have

$$f'(x) = 2x \qquad (2.3.8)$$

Using the Newton-Raphson formula,

$$x_{k+1} = x_k - \frac{f(x_k)}{f'(x_k)} \qquad (2.3.9)$$

$$x_{k+1} = x_k - \frac{x_k^2 - N}{2x_k} \qquad (2.3.10)$$

$$x_{k+1} = \frac{1}{2}\left(x_k + \frac{N}{x_k}\right) \qquad (2.3.11)$$

Note that for $x > 0$, $f'(x) > 0$. Also $f''(x) = 2 > 0$. Therefore, we always fulfill the sufficient conditions of the theorem and we are assured of a root. This is the formula used iteratively in the computer subroutine.

ILLUSTRATIVE PROBLEM 2.3-1   Using (2.3.11), find $\sqrt{41}$ correct to 8 decimal places.

Solution   The following calculations were done on a calculator.

$$x_0 = \text{guess} = 6$$

$$x_1 = \frac{1}{2}\left(x_0 + \frac{41}{x_0}\right) = 6.4166666667$$

$$x_2 = \frac{1}{2}\left(x_1 + \frac{41}{x_1}\right) = 6.4031385525$$

$$x_3 = \frac{1}{2}\left(x_2 + \frac{41}{x_2}\right) = 6.4031242435$$

$$x_4 = x_3$$

The error (roundoff) is only 2×10^{-9}, an excellent result for very little calculation.

The iterations are continued until two successive values differ by less than the required amount. Note that if there is an error prior to the last two values calculated, it will not matter. The errors do not propagate in this type of iteration.

ILLUSTRATIVE PROBLEM 2.3-2 (a) Use the Newton-Raphson formula to invent a formula for the cube root. (b) Find $\sqrt[3]{41}$, correct to 10^{-5}.

Solution

(a)
$$f(x) = x^3 - N$$
$$f'(x) = 3x^2$$
$$x_{k+1} = x_k - \frac{x_k^3 - N}{3x^2}$$

This formula is correct but has 4 multiplications, 1 division and 2 addition/subtractions. We can alter it slightly to remove one addition/subtraction:

$$x_{k+1} = \tfrac{1}{3}\left(2x_k + N/x_k^2\right) \tag{2.3.12}$$

(b) We start with a guess of 3, carrying at least 6 decimal places:

$$x_0 = 3$$
$$x_1 = \frac{1}{3}\left(2x_0 + \frac{41}{x_0^2}\right) = 3.518519$$
$$x_2 = 3.449613$$
$$x_3 = 3.448218$$
$$x_4 = 3.448217$$

The last two agree to five decimal places when rounded, completing the solution: $\sqrt[3]{41} \sim 3.44822$.

2.3.7 Application to Polynomials

The Newton-Raphson method is very fast and quite convenient for polynomials which, of course, are always differentiable. We illustrate.

ILLUSTRATIVE PROBLEM 2.3-3 Find a root of $x^5 - 3x^3 - 100$ between 2.5 and 3, correct to 5 decimal places.

Solution

$$f(x) = x^5 - 3x^3 - 100$$
$$f'(x) = 5x^4 - 9x^2$$
$$x_{k+1} = x_k - \frac{x_k^5 - 3x_k^3 - 100}{5x_k^4 - 9x_k^2} \tag{2.3.13}$$
$$x_{k+1} = \frac{4x_k^5 - 6x_k^3 + 100}{5x_k^4 - 9x_k^2} \tag{2.3.14}$$

The formula (2.3.13) has 13 multiplications, 1 division and 4 additions/subtractions. The formula (2.3.14) has 14 multiplications, 1 division and 3 additions. We rewrite the latter:

$$x_{k+1} = \frac{x_k^3(4x_k^2 - 6) + 100}{x_k^2(5x_k^2 - 9)} \tag{2.3.15}$$

which has only 9 multiplications, 1 division and 3 addition/subtractions. In programming this formula, we naturally avoid doing anything twice. In double precision, it becomes

```
Z=X*X
X=(Z*X*(4.D0*Z-6.D0)+1.D2)/(Z*(5.D0*Z-9.D0))
```

The initial starter is 2.5. The convergents are as follows by calculator:

$$x_0 = 2.5$$
$$x_1 = 2.853933$$
$$x_2 = 2.778102$$
$$x_3 = 2.773043$$
$$x_4 = 2.7730222$$
$$x_5 = 2.7730222$$

We therefore state that $x = 2.77302$ correct to 5 decimal places.

2.3.8 Multiple Roots

At a multiple root, $f' = 0$ and the convergence of the Newton-Raphson method depends upon the nature of the fraction f/f'. We demonstrate one case where the process works and one where it does not.

ILLUSTRATIVE PROBLEM 2.3-4   Investigate the solution of $f(x) = x^3 - 7x^2 + 15x - 9$ near $x = 3$. The exact roots are $\{1, 3, 3\}$.

Investigation:

$$f(x) = x^3 - 7x^2 + 15x - 9 = (x-3)^2(x-1)$$
$$f'(x) = 3x^2 - 14x + 15 = (x-3)(3x-5)$$

Applying the Newton-Raphson formula and simplifying,

$$x_{k+1} = \frac{2x_k^2 - x_k - 3}{3x_k - 5}$$

Start with $x = 2$:

$$x_0 = 2, \qquad x_1 = 3, \qquad x_2 = 3$$

leading to a satisfactory result immediately.

ILLUSTRATIVE PROBLEM 2.3-5 Investigate the solution of

$$f(x) = x^2 - 6x + 9$$

near $x = 3$. The exact roots are $\{3, 3\}$.

Investigation

$$f'(x) = 2x - 6$$

Applying the Newton-Raphson formula and simplifying,

$$x_{k+1} = \frac{x_k + 3}{2}$$

Start with $x = 2$:

$$x_0 = 2$$
$$x_1 = 2.5$$
$$\vdots$$
$$x_5 = 2.96875$$
$$\vdots$$
$$x_{10} = 2.999023$$

It is evident that the method is not efficient in this case. We can improve it somewhat. In the diagram of Figure 2.3-9,

$$x_{k+1} = x_k - QP = x_k - \frac{f(x_k)}{f'(x_k)} \qquad (2.3.16)$$

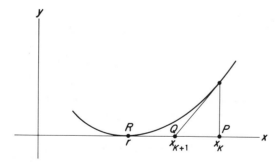

Fig. 2.3-9 The situation at a double root.

If QP is small, $\{x_k\}$ will converge very slowly. One solution is to multiply QP by the *multiplicity* of the root. In other words, if the root is a double root (if it has a multiplicity of 2), use

$$x_{k+1} = x_k - \frac{2f(x_k)}{f'(x_k)} \qquad (2.3.17)$$

To prove that this works, let m be the multiplicity of the root, and

$$f(x) = (x - r)^m \tag{2.3.18}$$

Then

$$f'(x) = m(x - r)^{m-1} \tag{2.3.19}$$

and

$$\frac{f}{f'} = \frac{x - r}{m} \tag{2.3.20}$$

Substituting in the Newton-Raphson formula

$$x_{k+1} = x_k - \frac{mf(x_k)}{f'(x_k)} \tag{2.3.21}$$

we have

$$x_{k+1} = x_k - (x_k - r) = r \tag{2.3.22}$$

In practical problems, it is unlikely that we will know the multiplicity of a root. The constant m must be large enough to speed convergence, yet not so large that x_{k+1} shoots to the wrong side of r. This is happening if $|x_{k+1} - x_k|$ is getting larger instead of smaller.

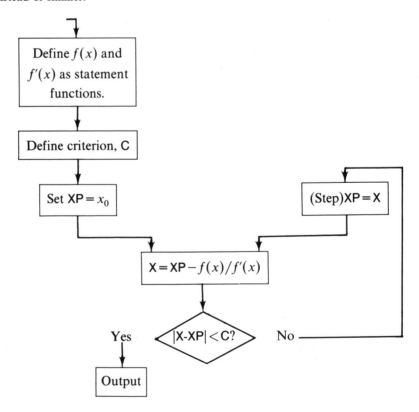

Fig. 2.3-10   Flow chart for Newton-Raphson method.

2.3.9 Computer Program

The flow chart (block diagram) is shown in Figure 2.3-10. The programming is quite simple, and the program extremely fast. In the problem section, set the criterion by writing

$$C = 5.D - 7$$

to be sure the previous values (XP) and the new values (X) agree to the necessary number of places. It is advisable to add a counter so that the program does not go through the loop endlessly. The count is started in the set box and stepped by 1 in the step box. A terminal value must be defined along with the definition of C.

PROBLEM SECTION 2.3

A. Using a suitable Newton-Raphson formula, find each of the following correct to the nearest millionth.

(1) $\sqrt{37}$ (4) $\sqrt[3]{545}$

(2) $\sqrt{148}$ (5) $\sqrt[4]{375}$

(3) $\sqrt[3]{286}$ (6) $\sqrt[4]{598}$

B. Find a root by the Newton-Raphson method, correct to the nearest millionth. The numbers in parentheses are the starters to be used.

(7) $x^5 - 10x^3 - 4 = 0$, (-4) (10) $3x^5 - 11x^4 - 40 = 0$, (3)

(8) $x^4 - 4x^3 - 6 = 0$, (-2) (11) $x^5 - 19x^2 + 19 = 0$, (-1)

(9) $2x^5 - 7x^4 - 100 = 0$, (3)   (12) $x^4 - 4x^3 + 2 = 0$, (1)

C. Find the smallest root between -5 and $+5$, using the Newton-Raphson method.

(13) $2x^5 + 3x^4 - 5x^3 - 11 = 0$

(14) $3x^5 + 7x^4 + 6x^3 + 1 = 0$

(15) $x^5 - 2x^4 - 13x^3 - 100 = 0$

(16) $x^5 + 5x^4 - 22x^3 + 17 = 0$

(17) $x^5 + x^4 + x^3 - 3x^2 + 17 = 0$

(18) $2x^5 - 5x^4 + 7x^3 - 35x^2 + 4 = 0$

2.4 NEWTON-RAPHSON METHOD (PART II)

2.4.1 Error Analysis

We shall not do a complete error analysis, but merely indicate the procedure. We start with an illustrative problem.

ILLUSTRATIVE PROBLEM 2.4-1. Find a root of

$$f(x) = x^2 - \sqrt{x} - 0.1$$

correct to 7 decimal places, using $x_0 = 1$.

Solution

$$f'(x) = 2x - \tfrac{1}{2}x^{-1/2}$$

$$x_{k+1} = x_k - \frac{x_k^2 - x_k^{1/2} - 0.1}{2x_k - \tfrac{1}{2}x_k^{-1/2}}$$

We improve this formula slightly to

$$x_{k+1} = \frac{2x_k^2 + x_k^{1/2} + 0.2}{4x_k - x_k^{-1/2}}$$

This can be programmed as follows:

```
Y=DSQRT(X)
X=(2.D0*X*X+Y+2.D−1)/(4.D0*X−1.D0/Y)
```

The convergents are as follows:

$$x_0 = 1$$
$$x_1 = 1.066666667$$
$$x_2 = 1.063645694$$
$$x_3 = 1.063639489$$
$$x_4 = 1.063639508$$

x_3 and x_4 agree to 7 decimal places after rounding.

From Figure 2.3-10, it becomes clear that the purpose of the program is really to find $|x_{k-1} - x_k|$. The process chart for this calculation is shown in Figure 2.4-1. It is quite complex. It is therefore to be expected that the maximum possible error will be large even though the process is very fast.

In contrast, the process chart for simple iteration (Figure 2.4-2) is quite short, involving much less error. However, it is slightly slower. The word used by computer-science professionals is *trade-off*. Here there is a trade-off of accuracy for speed.

PROBLEM SECTION 2.4

In each case, find a root correct to the nearest millionth, using the value in parentheses as x_0. Check several variations of the Newton-Raphson formula to make sure you are using an efficient one.

(1) $x^4 - 56\sqrt{x} - 2.5 = 0$, (3)
(2) $0.03x^5 - 15\sqrt{x} - 4 = 0$, (4)
(3) $x^2 + \sin x = 2$, (-2)
(4) $x^3 - 4\cos x = 1$, (1)
(5) $2x^2 + \ln x = 3$, (1)
(6) $2x^2 - e^x = 0$, (-1)
(7) $\sin x + 3\cos x = 0$, (-5)

(8) $4\cos x - 6\sin x = 3$, (-3)
(9) $25\,e^x - 150\,e^{-x} = 2$, (0)
(10) $\ln x + 30\,e^{-x} = 3$, (2)
(11) $\sqrt{x}\,\ln x + 7\sin x \cos x = 0.3$, (1)
(12) $(\sin x)\ln x + \sqrt{x}\,(\cos x) = 0.5$, (0)
(13) $x^3 \sin x + x^2 e^{-x} = 15$, (-2)
(14) $x^4 \cos x + 6x^5 e^x = -1$, (-5)

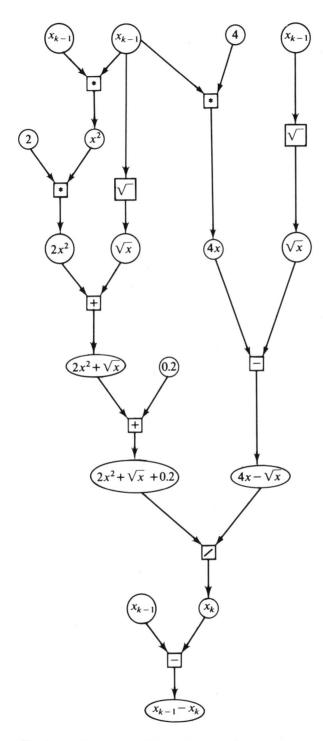

Fig. 2.4-1   Process chart for Illustrative Problem 2.4-1.

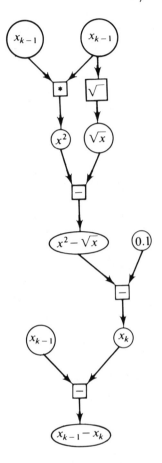

Fig. 2.4-2 The process chart for simple iteration.

Chapter 3

Systems of Linear Equations

Large systems of simultaneous equations are quite common in science and industry. Consequently, the problem of solving these equations is an important one. The methods taught in elementary courses—addition and subtraction, substitution, and the use of determinants—are usually inconvenient for more than two or three simultaneous equations unless the coefficients are small whole numbers.

In this chapter, we shall explore an excellent method, the *Gauss-Jordan method*; a method for evaluating the convergence, the *Seidel method*; matrix inversion; and eigenvalue problems.

We assume that the reader is familiar with the properties of matrices. In particular, we recall the following:

1. The sum of two $m \times n$ matrices is an $m \times n$ matrix. The product of an $m \times n$ matrix and a $n \times p$ matrix is an $m \times p$ matrix. Figure 3-1 displays a computer program developed by a computer-science class for the multiplication of two matrices, and Figure 3-2 displays the results.

2. *Property of associativity*: $A + (B + C) = (A + B) + C$, and

$$(AB)C = A(BC).$$

3. *Identity properties*: There is a unique zero matrix such that $A + 0 = 0 + A = A$. For *square* matrices, there is a unique unit matrix such that $AI = IA = A$.

4. *Inverse properties*: Every matrix has a unique negative such that $A + (-A) = (-A) + A = 0$. Some matrices, called *non-singular*, have unique (multiplicative) inverses such that $AA^{-1} = A^{-1}A = I$.

5. *Commutativity*: $A + B = B + A$. It is not true in general that $AB = BA$.

6. *Distributivity*: $A(B + C) = AB + AC$.

We shall also need to refer to the following theorems.

MULTIPLICATION THEOREM If A, B, L and R are matrices, and if $A = B$, then $LA = LB$ and $AR = BR$.

LEFT CANCELLATION THEOREM If A, B and C are matrices, if $CA = CB$, and if there exists a matrix C^{-1} such that $C^{-1}C = I$, then $A = B$.

87

> RIGHT CANCELLATION THEOREM If A, B and C are matrices, if $AC = BC$, and if there exists a matrix C^{-1} such that $CC^{-1} = I$, then $A = B$.

We note that if $AB = 0$, neither A nor B need be 0.

For use throughout the chapter, we shall recall some of the elementary techniques in the handling of simultaneous linear equations. The operations we are about to discuss are called *elementary row operations* (ERO) when they are applied to matrix equations.

Multiplication ERO: Consider the system of equations in the following problem. As a first step in eliminating x_1, we might multiply the third row by -5:

$$\begin{cases} 5x_1 + 6x_2 - 7x_3 = 0 \\ 3x_1 - 4x_2 + 8x_3 = 2 \\ x_1 \qquad\;\; -5x_3 = 7 \end{cases} \tag{3.1}$$

$$\begin{cases} 5x_1 + 6x_2 - 7x_3 = \quad 0 \\ 3x_1 - 4x_2 + 8x_3 = \quad 2 \\ -5x_1 \qquad\;\; +25x_3 = -35 \end{cases} \tag{3.2}$$

We know that this changes the system, but not the answers. In matrix form, (3.2) becomes

$$\begin{pmatrix} 5 & 6 & -7 \\ 3 & -4 & 8 \\ -5 & 0 & 25 \end{pmatrix} \cdot \begin{pmatrix} x_1 \\ x_2 \\ x_3 \end{pmatrix} = \begin{pmatrix} 0 \\ 2 \\ -35 \end{pmatrix} \tag{3.3}$$

$$A \qquad\qquad \cdot X \;=\; C$$

What we are pointing out is that it is permissible to multiply a row in the A matrix by a constant provided we also multiply the corresponding row in the C matrix by the same constant. Note that the X-vector is unchanged.

Addition ERO: Carrying this a step farther, let us add the third equation of (3.2) to the first equation:

$$\begin{cases} 0x_1 + 6x_2 + 18x_3 = -35 \\ 3x_1 - 4x_2 + 8x_3 = \quad 2 \\ -5x_1 \qquad\;\; +25x_3 = -35 \end{cases} \tag{3.4}$$

Again, the system has been changed, but the solutions are those of the original system. In matrix form, we have

$$\begin{pmatrix} 0 & 6 & 18 \\ 3 & -4 & 8 \\ -5 & 0 & 25 \end{pmatrix} \cdot \begin{pmatrix} x_1 \\ x_2 \\ x_3 \end{pmatrix} = \begin{pmatrix} -35 \\ 2 \\ -35 \end{pmatrix} \tag{3.5}$$

```
        DOUBLE PRECISION A(5,5),B(5,5),C(5,5)
C THIS PROGRAM MULTIPLIES TWO CONFORMABLE MATRICES UP TO 5 BY 5.
        WRITE(3,101)
101     FORMAT(1H1)
C THE LEAD CARD HAS THE NUMBER OF ROWS IN CC 1 AND THE NUMBER OF COLUMNS
C IN CC 2 FOR MATRIX A.
        READ(1,102)NRA,NCA
102     FORMAT(2I1)
C DATA CARDS. 2 DECIMAL PLACES, CC 1,11,21,31,41
        DO 9 J=1,NCA
9       READ(1,103)(A(I,J),I=1,NRA)
103     FORMAT(5F10.0)
C ECHO OF MATRIX A
        WRITE(3,104)
104     FORMAT(1H0,T18,'MATRIX A',//)
        DO 3 I=1,NRA
3       WRITE(3,105)(A(I,J),J=1,NCA)
105     FORMAT(1H ,5F11.4)
C SAME SET OF STEPS FOR MATRIX B
        READ(1,102)NRB,NCB
        DO 10 J=1,NCB
10      READ(1,103)(B(I,J),I=1,NRB)
        WRITE(3,106)
106     FORMAT(1H0,T18,'MATRIX B'//)
        DO 4 I=1,NRB
4       WRITE(3,105)(B(I,J),J=1,NCB)
C SET ALL C(I,J) TO ZERO
        DO 5 I=1,NRA
        DO 5 J=1,NCB
5       C(I,J)=0.D0
C COMPUTE
        DO 6 I=1,NRA
        DO 6 J=1,NCB
        DO 6 K=1,NCA
6       C(I,J)=C(I,J)+A(I,K)*B(K,J)
C DISPLAY PRODUCT MATRIX
        WRITE(3,107)
107     FORMAT(1H0,T15,'PRODUCT MATRIX'//)
        DO 7 I=1,NRA
7       WRITE(3,105)(C(I,J),J=1,NCB)
        END
```

Fig. 3-1 Matrix multiplication.

```
                        MATRIX A


            5.4100      -4.7800
            9.8400      -0.4100
           -3.3600       5.7300
           -1.5400       9.5400


                        MATRIX B


            7.0000      -6.2900       9.0800
           -3.9800      -5.6100       0.9700


                    PRODUCT MATRIX


           56.8944      -7.2131      44.4862
           70.5118     -59.5935      88.9495
          -46.3254     -11.0109     -24.9507
          -48.7492     -43.8328      -4.7294
```

Fig. 3-2 The results.

Thus it is permissible to add one row of matrix A to another row, provided that the same operation is performed with the corresponding elements of C. Again, the X-vector is unchanged.

Interchange ERO: The following system obviously has the same solutions as (3.1); we have merely interchanged the first and second equations:

$$\begin{cases} 3x_1 - 4x_2 + 8x_3 = 2 \\ 5x_1 + 6x_2 - 7x_3 = 0 \\ \quad x_1 \qquad - 5x_3 = 7 \end{cases} \qquad (3.6)$$

This corresponds to the matrix equation

$$\begin{bmatrix} 3 & -4 & 8 \\ 5 & 6 & -7 \\ 1 & 0 & -5 \end{bmatrix} \cdot \begin{bmatrix} x_1 \\ x_2 \\ x_3 \end{bmatrix} = \begin{bmatrix} 2 \\ 0 \\ 7 \end{bmatrix} \qquad (3.7)$$

It is permitted to interchange rows of the matrix A provided that the same rows of C are interchanged. The X column vector is unchanged.

We define an elementary matrix as follows.

ELEMENTARY MATRIX   Any matrix obtained from a square identity matrix by the use of one ERO.

We shall use the letters H, M and F for three types of elementary matrix corresponding to the three ERO discussed.

Permutation Matrix, H: For $I_{3\times3}$, if rows 1 and 2 are interchanged, we obtain H_{21}:

$$H_{21} = \begin{bmatrix} 0 & 1 & 0 \\ 1 & 0 & 0 \\ 0 & 0 & 1 \end{bmatrix}$$

Compression Matrix, M: For $I_{3\times3}$, if row 3 is multiplied by 5, we obtain $M_{3(5)}$:

$$M_{3(5)} = \begin{bmatrix} 1 & 0 & 0 \\ 0 & 1 & 0 \\ 0 & 0 & 5 \end{bmatrix}$$

Shear Matrix, F: If twice the third row of $I_{3\times3}$ is added to row 1, we obtain $F_{13(2)}$:

$$F_{13(2)} = \begin{bmatrix} 1 & 0 & 2 \\ 0 & 1 & 0 \\ 0 & 0 & 1 \end{bmatrix}$$

We shall now show the effect of pre-multiplication by these elementary matrices. Consider the system

$$\begin{cases} a_{11}x_1 + a_{12}x_2 + a_{13}x_3 = c_1 \\ a_{21}x_1 + a_{22}x_2 + a_{23}x_3 = c_2 \\ a_{31}x_1 + a_{32}x_2 + a_{33}x_3 = c_3 \end{cases} \tag{3.8}$$

This corresponds to the matrix equation

$$\begin{bmatrix} a_{11} & a_{12} & a_{13} \\ a_{21} & a_{22} & a_{23} \\ a_{31} & a_{32} & a_{33} \end{bmatrix} \cdot \begin{bmatrix} x_1 \\ x_2 \\ x_3 \end{bmatrix} = \begin{bmatrix} c_1 \\ c_2 \\ c_3 \end{bmatrix} \tag{3.9}$$

$$A \qquad\quad \cdot \quad X \ = \ C$$

Now, let us pre-multiply by H_{21}. In symbols,

$$H_{21}(AX) = H_{21}C \tag{3.10}$$

This is permissible by the multiplication theorem. We now use the property of associativity:

$$(H_{21}A)X = H_{21}C \tag{3.11}$$

$$H_{21}A = \begin{bmatrix} 0 & 1 & 0 \\ 1 & 0 & 0 \\ 0 & 0 & 1 \end{bmatrix} \cdot \begin{bmatrix} a_{11} & a_{12} & a_{13} \\ a_{21} & a_{22} & a_{23} \\ a_{31} & a_{32} & a_{33} \end{bmatrix} = \begin{bmatrix} a_{21} & a_{22} & a_{23} \\ a_{11} & a_{12} & a_{13} \\ a_{31} & a_{32} & a_{33} \end{bmatrix} \tag{3.12}$$

$$H_{21}C = \begin{bmatrix} 0 & 1 & 0 \\ 1 & 0 & 0 \\ 0 & 0 & 1 \end{bmatrix} \cdot \begin{bmatrix} c_1 \\ c_2 \\ c_3 \end{bmatrix} = \begin{bmatrix} c_2 \\ c_1 \\ c_3 \end{bmatrix} \tag{3.13}$$

Replacing (3.12) and (3.13) in (3.11), we have

$$\begin{bmatrix} a_{21} & a_{22} & a_{23} \\ a_{11} & a_{12} & a_{13} \\ a_{31} & a_{32} & a_{33} \end{bmatrix} \cdot \begin{bmatrix} x_1 \\ x_2 \\ x_3 \end{bmatrix} = \begin{bmatrix} c_2 \\ c_1 \\ c_3 \end{bmatrix} \tag{3.14}$$

or

$$\begin{cases} a_{21}x_1 + a_{22}x_2 + a_{23}x_3 = c_2 \\ a_{11}x_1 + a_{12}x_2 + a_{13}x_3 = c_1 \\ a_{31}x_1 + a_{32}x_2 + a_{33}x_3 = c_3 \end{cases} \tag{3.15}$$

This result can be generalized to the following statement.

CONCLUSION 1 When a matrix equation is pre-multiplied by a permutation elementary matrix, the equations are interchanged in exactly the same way as the elementary matrix.

By similar steps, the following can just as easily be shown:

CONCLUSION 2 When a matrix equation is pre-multiplied by a compression matrix, the equations are multiplied in exactly the same way as the elementary matrix.

CONCLUSION 3 When a matrix equation is pre-multiplied by a shear matrix, the equations are multiplied and added in exactly the same way as the elementary matrix.

ILLUSTRATIVE PROBLEM 3-1 What is an elementary matrix which, when used to pre-multiply, will interchange the second and fourth rows of a 4×4 matrix?

Solution A four-row H_{24} elementary matrix.

ILLUSTRATIVE PROBLEM 3-2 What is an elementary matrix which, when used to pre-multiply, will multiply the third row of a 3×3 matrix by -4.7?

Solution

$$M_{3(-4.7)} = \begin{bmatrix} 1 & 0 & 0 \\ 0 & 1 & 0 \\ 0 & 0 & -4.7 \end{bmatrix}$$

ILLUSTRATIVE PROBLEM 3-3 What is an elementary matrix which, when used to pre-multiply, will add three times the first row to the second row of a 3×3 matrix?

Solution

$$F_{21(3)} = \begin{bmatrix} 1 & 0 & 0 \\ 3 & 1 & 0 \\ 0 & 0 & 1 \end{bmatrix}$$

ILLUSTRATIVE PROBLEM 3-4 What does the following matrix do to matrix A when it is used to pre-multiply?

$$\begin{bmatrix} 1 & 0 & 0 \\ 0 & 1 & -3 \\ 0 & 0 & 1 \end{bmatrix}$$

Solution Subtracts three times the third row from the second row.

3.1 THE GAUSS-JORDAN METHOD

3.1.1 The General Idea

Suppose we are faced with the following problem:

$$\begin{cases} 2.630x_1+5.210x_2-1.694x_3+0.938x_4= \quad 4.230 \\ 3.160x_1-2.950x_2+0.813x_3-4.210x_4= -0.716 \\ 5.360x_1+1.880x_2-2.150x_3-4.950x_4= \quad 1.280 \quad (3.1.1) \\ 1.340x_1+2.980x_2-0.432x_3-1.768x_4= \quad 0.419 \end{cases}$$

This can be rewritten in matrix form as follows.

$$\begin{bmatrix} 2.630 & 5.210 & -1.694 & 0.938 \\ 3.160 & -2.950 & 0.813 & -4.210 \\ 5.360 & 1.880 & -2.150 & -4.950 \\ 1.340 & 2.980 & -0.432 & -1.768 \end{bmatrix} \cdot \begin{bmatrix} x_1 \\ x_2 \\ x_3 \\ x_4 \end{bmatrix} = \begin{bmatrix} 4.230 \\ -0.716 \\ 1.280 \\ 0.419 \end{bmatrix} \quad (3.1.2)$$

$$A \qquad \cdot \quad X \;=\; C$$

There are various methods for solving this problem. However we go about it, one possible end product we might want is

$$\begin{array}{lcl} 1x_1 & = & 1.038 \\ \quad 1x_2 & = & 0.209 \\ \qquad 1x_3 & = & 0.226 \\ \qquad\quad 1x_4 & = & 0.847 \end{array} \quad (3.1.3)$$

This can be rewritten in matrix form:

$$\begin{bmatrix} 1 & 0 & 0 & 0 \\ 0 & 1 & 0 & 0 \\ 0 & 0 & 1 & 0 \\ 0 & 0 & 0 & 1 \end{bmatrix} \cdot \begin{bmatrix} x_1 \\ x_2 \\ x_3 \\ x_4 \end{bmatrix} = \begin{bmatrix} 1.038 \\ 0.209 \\ 0.226 \\ 0.847 \end{bmatrix} \quad (3.1.4)$$

Of course, this is not the only satisfactory form of the answer. There are 4! or 24 rearrangements which are equally satisfactory. One of them is

$$\begin{array}{lcl} \qquad\quad 1x_4 & = & 0.847 \\ \quad 1x_2 & = & 0.209 \\ 1x_1 & = & 1.038 \\ \qquad 1x_3 & = & 0.226 \end{array} \quad (3.1.5)$$

or the matrix equivalent

$$\begin{bmatrix} 0 & 0 & 0 & 1 \\ 0 & 1 & 0 & 0 \\ 1 & 0 & 0 & 0 \\ 0 & 0 & 1 & 0 \end{bmatrix} \cdot \begin{bmatrix} x_1 \\ x_2 \\ x_3 \\ x_4 \end{bmatrix} = \begin{bmatrix} 0.847 \\ 0.209 \\ 1.038 \\ 0.226 \end{bmatrix} \tag{3.1.6}$$

To summarize: our problem is to get from a matrix equation like (3.1.3) to one of the 24 variations, like (3.1.4) or (3.1.6).

The Gauss-Jordan method is a systematic procedure for accomplishing the purpose.[1] The most important points to notice are (a) the final form, whether like (3.1.4) or like (3.1.6), has exactly one 1 in each row and column, the other entries in the matrix being 0, and (b) the column vector of unknowns is the same in all forms—it is untouched by the operations to be accomplished.

3.1.2 An Illustration with Exact Numbers

Before plunging into the solution of (3.1.1), we shall demonstrate the solution with exact numbers in order to clarify the thinking behind the process. The precise steps we are going to demonstrate are not the ones we use for a computer program, but the idea is the same. We shall display the computer program in section 3.1.7.

ILLUSTRATIVE PROBLEM 3.1-1   Solve the system

$$\begin{cases} 3x + y + z = 2 \\ -x + y - 2z = 7 \\ 2x \quad - z = 5 \end{cases} \tag{3.1.7}$$

Solution   First, rewrite in matrix form:

$$\begin{pmatrix} 3 & 1 & 1 \\ -1 & 1 & -2 \\ 2 & 0 & -1 \end{pmatrix} \cdot \begin{pmatrix} x \\ y \\ z \end{pmatrix} = \begin{pmatrix} 2 \\ 7 \\ 5 \end{pmatrix} \tag{3.1.8}$$

Remembering that the column vector of unknowns will not change, we omit the column and rewrite in *augmented matrix* form:

$$\begin{bmatrix} 3 & 1 & 1 & 2 \\ -1 & 1 & -2 & 7 \\ 2 & 0 & -1 & 5 \end{bmatrix} \tag{3.1.9}$$

The left three columns represent the *coefficient matrix*, and the rightmost column is the column vector of constants, often called the *stub*.

---

[1]There are several variations of the Gauss-Jordan method. We chose the one easiest to program and easiest to make an error analysis for.

We want the first column to have a 1 in one row and 0s in all the others. For the present, let us choose to make $a_{11}=1$ with $a_{21}=a_{31}=0$. We can make $a_{11}=1$ by dividing the first row, straight across, by 3. This is the same as multiplying the first equation by $\frac{1}{3}$, which, you may remember, is the same as pre-multiplying by $M_{1(1/3)}$, a permissible ERO.

$$\begin{bmatrix} \boxed{1} & \frac{1}{3} & \frac{1}{3} & \bigm| & \frac{2}{3} \\ -1 & 1 & -2 & \bigm| & 7 \\ 2 & 0 & -1 & \bigm| & 5 \end{bmatrix} \tag{3.1.10}$$

The boxed number is called a *pivot*, and row 1 is the *pivot row*. If we add row 1 to row 2, a permissible ERO, we obtain

$$\begin{bmatrix} \boxed{1} & \frac{1}{3} & \frac{1}{3} & \bigm| & \frac{2}{3} \\ 0 & \frac{4}{3} & -\frac{5}{3} & \bigm| & \frac{23}{3} \\ 2 & 0 & -1 & \bigm| & 5 \end{bmatrix} \tag{3.1.11}$$

We now wish to subtract twice row 1 from row 3. For convenience, we do this in two stages. First we multiply row 1 by -2 and write it as a *dummy row* underneath, as follows:

$$\begin{bmatrix} \boxed{1} & \frac{1}{3} & \frac{1}{3} & \bigm| & \frac{2}{3} \\ 0 & \frac{2}{3} & -\frac{5}{3} & \bigm| & \frac{23}{3} \\ 2 & 0 & -1 & \bigm| & 5 \end{bmatrix} \tag{3.1.12}$$

$$-2I \qquad -2 \quad -\frac{2}{3} \quad -\frac{2}{3} \quad -\frac{4}{3} \qquad \text{(dummy row)}$$

Second, we add the dummy to row 3, obtaining

$$\begin{bmatrix} \boxed{1} & \frac{1}{3} & \frac{1}{3} & \bigm| & \frac{2}{3} \\ 0 & \frac{4}{3} & -\frac{5}{3} & \bigm| & \frac{23}{3} \\ 0 & -\frac{2}{3} & -\frac{5}{3} & \bigm| & \frac{11}{3} \end{bmatrix} \tag{3.1.13}$$

The first column has now been completely processed. It is in final form.

We now choose a_{22} as the pivot, so that the second row is the pivot row. (This is *not* what we shall do later.) To make the pivot equal to 1, we multiply row 2 by $\frac{3}{4}$. The resulting augmented matrix is as follows:

$$\begin{bmatrix} 1 & \frac{1}{3} & \frac{1}{3} & \bigm| & \frac{2}{3} \\ 0 & \boxed{1} & -\frac{5}{4} & \bigm| & \frac{23}{4} \\ 0 & -\frac{2}{3} & -\frac{5}{3} & \bigm| & \frac{11}{3} \end{bmatrix} \tag{3.1.14}$$

Proceeding as in column 1, the result is

$$\begin{bmatrix} 1 & 0 & \frac{3}{4} & \bigm| & -\frac{5}{4} \\ 0 & \boxed{1} & -\frac{5}{4} & \bigm| & \frac{23}{4} \\ 0 & 0 & -\frac{5}{2} & \bigm| & \frac{15}{2} \end{bmatrix} \tag{3.1.15}$$

The third step begins by multiplying row 3 by $-\frac{2}{5}$:

$$\begin{bmatrix} 1 & 0 & \frac{3}{4} & \vline & -\frac{5}{4} \\ 0 & 1 & -\frac{5}{4} & \vline & \frac{23}{4} \\ 0 & 0 & \boxed{1} & \vline & -3 \end{bmatrix} \tag{3.1.16}$$

Proceeding as in column 1, we obtain

$$\begin{bmatrix} 1 & 0 & 0 & \vline & 1 \\ 0 & 1 & 0 & \vline & 2 \\ 0 & 0 & 1 & \vline & -3 \end{bmatrix} \tag{3.1.17}$$

The answers can be read directly from this augmented matrix.
We can, if we like, rewrite (3.1.17) as a matrix equation:

$$\begin{pmatrix} 1 & 0 & 0 \\ 0 & 1 & 0 \\ 0 & 0 & 1 \end{pmatrix} \cdot \begin{pmatrix} x \\ y \\ z \end{pmatrix} = \begin{pmatrix} 1 \\ 2 \\ -3 \end{pmatrix} \tag{3.1.18}$$

or in the original form:

$$\begin{cases} 1x + 0y + 0z = 1 \\ 0x + 1y + 0z = 2 \\ 0x + 0y + 1z = -3 \end{cases} \tag{3.1.19}$$

3.1.3 Justification of the Process

From a matrix point of view we started with the problem

$$AX = C \tag{3.1.20}$$

where we had to find column X. Then the series of elementary row operations actually changed the A-matrix into an I-matrix without affecting X. Diagrammatically,

$$\begin{array}{ccc} AX & = & C \\ A^{-1} \downarrow & & A^{-1} \downarrow \\ IX & = & X \end{array} \tag{3.1.21}$$

where the A^{-1} at the left of the arrow is intended to indicate that we pre-multiplied by the inverse of A. In other words, when we changed A to I (in effect, multiplying by A^{-1}), C had to change to X.

The diagram can be written as a set of matrix equations as follows:

$$AX = C \qquad \text{(given)} \qquad (3.1.22)$$

$$A^{-1}(AX) = A^{-1}C \qquad \text{(multiplication theorem)} \qquad (3.1.23)$$

$$(A^{-1}A)X = A^{-1}C \qquad \text{(associativity)} \qquad (3.1.24)$$

$$IX = A^{-1}C \qquad (A^{-1}A = I) \qquad (3.1.25)$$

$$X = A^{-1}C \qquad (IX = I). \qquad (3.1.26)$$

It was not obvious that the Gauss-Jordan method really accomplished a pre-multiplication by A^{-1}, but Equations (3.1.22)–(3.1.26) show that this is what happened, and that it was justified.

3.1.4 Zero Pivots

It is evident that a system like

$$\begin{cases} 0x_1 + 3x_2 + 7x_3 = 9 \\ 2x_1 + 0x_2 - x_3 = 4 \\ x_1 + 5x_2 + 0x_3 = 5 \end{cases} \qquad (3.1.27)$$

cannot be solved exactly like the previous problem. The corresponding matrix equation would be

$$\begin{pmatrix} 0 & 3 & 7 \\ 2 & 0 & -1 \\ 1 & 5 & 0 \end{pmatrix} \cdot \begin{pmatrix} x_1 \\ x_2 \\ x_3 \end{pmatrix} = \begin{pmatrix} 9 \\ 4 \\ 5 \end{pmatrix} \qquad (3.1.28)$$

and the augmented matrix would be

$$\begin{bmatrix} 0 & 3 & 7 & 9 \\ 2 & 0 & -1 & 4 \\ 1 & 5 & 0 & 5 \end{bmatrix} \qquad (3.1.29)$$

Since we cannot use 0 as a denominator, a_{11} cannot be a pivot. What can we do?

We've mentioned already that from a theoretical standpoint it doesn't matter where the 1s and 0s are as long as there is precisely *one* 1 in each row and column and 0s for all the other elements. So we could take a_{21} as the first pivot; or a_{31}; or, for that matter, a_{13}.

Does it make a practical difference? Yes, it may make a great deal of difference. Consideration of the best way to accomplish a program—i.e., maximum accuracy and minimum possible error—is one reason for the existence of computer science. In computer science, finding the best sequence of steps is called *optimizing*. Optimizing can be with reference to accuracy, speed, or space. We shall try to optimize the Gauss-Jordan method for accuracy.

3.1.5 Error Considerations

Let us retrace our steps with a very simple case in order to evaluate the propagated errors. We start with

$$\begin{cases} ax_1 + bx_2 = p \\ bx_1 + dx_2 = q \end{cases} \tag{3.1.30}$$

which is rewritten as

$$\begin{bmatrix} a & b & \vdots & p \\ c & d & \vdots & q \end{bmatrix} \tag{3.1.31}$$

We divide row 1 by a:

$$\begin{bmatrix} \boxed{1} & b/a & \vdots & p/a \\ c & d & \vdots & q \end{bmatrix} \tag{3.1.32}$$

We form a dummy by multiplying row 1 by $-c$:

$$\begin{bmatrix} -c & -c(b/a) & \vdots & -c(p/a) \end{bmatrix} \tag{3.1.33}$$

and add the dummy to row 2:

$$\begin{bmatrix} \boxed{1} & b/a & \vdots & p/a \\ 0 & d-c(b/a) & \vdots & q-c(p/a) \end{bmatrix} \tag{3.1.34}$$

Now we divide row 2 by a_{22}:

$$\begin{bmatrix} 1 & b/a & \vdots & p/a \\ 0 & 1 & \vdots & \dfrac{q-c(p/a)}{d-c(b/a)} \end{bmatrix} \tag{3.1.35}$$

We stop at this point and examine the value of x_2:

$$x_2 = \frac{q-c(p/a)}{d-c(b/a)} \tag{3.1.36}$$

Figure 3.1-1 shows the process chart for the determination of x_2. As a kind

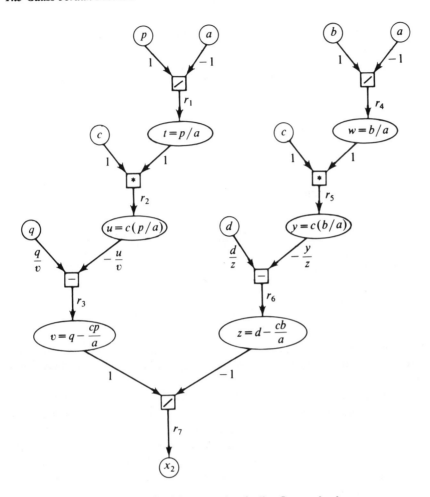

Fig. 3.1-1   Process chart for one step in the Gauss-Jordan process.

of review, we go through the calculations for $i_{x_2}^O$:

$$i_{x_2}^O = i_v^O - i_z^O + r_7 \tag{3.1.37}$$

$$i_v^O = (-u/v)i_u^O + r_3 \tag{3.1.38}$$

$$i_u^O = i_t^O + r_2 \tag{3.1.39}$$

$$i_t^O = r_1 \tag{3.1.40}$$

$$i_u^O = r_1 + r_2 \tag{3.1.41}$$

$$i_v^O = \frac{-cp}{aq - cp}(r_1 + r_2) + r_3 \tag{3.1.42}$$

Since the two sides of the process chart are identical except for the literal symbols, we merely make the changes in letters and find that

$$i_z^O = \frac{-cb}{ad-bc}(r_4+r_5)+r_6 \tag{3.1.43}$$

Therefore

$$i_{x_2}^O = -\frac{cp}{aq-cp}(r_1+r_2)+r_3+\frac{bc}{ad-bc}(r_4+r_5)+r_6+r_7 \tag{3.1.44}$$

Let f equal the fraction a/c. Then

$$i_{x_2}^O = \frac{-p}{fq-p}(r_1+r_2)+\frac{b}{fd-b}(r_4+r_5)+r_3+r_6+r_7 \tag{3.1.45}$$

Using R as the maximum of $|r|$,

$$|i_{x_2}^O| \leqslant \left(\frac{2|p|}{|fq-p|}+\frac{2|b|}{|fd-b|}+3\right)R \tag{3.1.46}$$

If f dominates the denominators—i.e., if $|a|\gg|c|$—the maximum relative error will be small. Therefore, our practical rule is to have $|a|$ larger than $|c|$.

What is true for a system of two equations is true for any system because, in effect, we are solving two at a time no matter how large the system is. The moral is that we should choose as pivot in each column the coefficient with maximum absolute value, subject to the condition that there is only one pivot in each row and column. A more general method of error determination based upon a more sophisticated analysis can be found in Fox and Mayers, p. 84.

3.1.6 The Optimized Procedure

We now return to the system (3.1.1) to solve it properly. We shall use the same general procedure as that in Illustrative Problem 3.1-1, except that if possible we shall choose as pivot in each column the element with the largest absolute value. The following numerical example was done on a calculator which rounds to 10 decimal digits. For the purpose of illustration, only 5 significant figures are printed in the book. On a computer, because of the rapid propagation of error, we strongly advise the use of double precision even if the original problem does not warrant it.

ILLUSTRATIVE PROBLEM 3.1-2 Solve the system for which the following is the augmented matrix:

$$\begin{bmatrix} 2.630 & 5.210 & -1.694 & 0.938 & \vdots & 4.230 \\ 3.160 & -2.950 & 0.813 & -4.210 & \vdots & -0.716 \\ 5.360 & 1.880 & -2.150 & -4.950 & \vdots & 1.280 \\ 1.340 & 2.980 & -0.432 & -1.768 & \vdots & 0.419 \end{bmatrix} \tag{3.1.47}$$

Solution We choose a_{31} as the pivot. Dividing the third row by 5.360, we obtain

$$\begin{bmatrix} 2.630 & 5.210 & -1.694 & 0.938 & \vdots & 4.230 \\ 3.160 & -2.950 & 0.813 & -4.210 & \vdots & -0.716 \\ \boxed{1} & 0.35075 & -0.40112 & -0.92351 & \vdots & 0.23881 \\ 1.340 & 2.980 & -0.432 & -1.768 & \vdots & 0.419 \end{bmatrix} \qquad (3.1.48)$$

To obtain a 0 in position a_{11}, we can form a dummy row by multiplying the pivot row by -2.630. This would yield the dummy row

$$[-2.630 \quad -0.92247 \quad 1.05495 \quad 2.42883 \; \vdots \; -0.62807]$$

which we should add to row 1 to obtain

$$[0 \quad 4.28753 \quad -0.63906 \quad 3.36682 \; \vdots \; 3.60194]$$

as a new first row.

This step-by-step procedure is tedious. Perhaps we can analyze it and develop a computer algorithm. Consider a square formed from a_{11}, a_{31}, a_{34} and a_{14}:

$$\begin{matrix} 2.630 & \cdots & 0.938 \\ \vdots & & \vdots \\ 5.360 & \cdots & -4.950 \end{matrix} \qquad (3.1.49)$$

First we divide by 5.360 to obtain

$$\begin{matrix} 2.630 & \cdots & 0.938 \\ \vdots & & \vdots \\ 1 & \cdots & -4.950/5.360 \end{matrix} \qquad (3.1.50)$$

Then we form a dummy by multiplying by -2.630:

$$\begin{matrix} 2.630 & \cdots & 0.938 \\ \vdots & & \vdots \\ -2.630 & \cdots & \dfrac{-2.630 \times -4.950}{5.360} \end{matrix} \qquad (3.1.51)$$

Finally we add to obtain, in the first row,

$$0 \quad \cdots \quad 0.938 + \frac{(-2.630)(-4.950)}{5.360} \qquad (3.1.52)$$

In general, if we start with a square

$$\begin{matrix} a_{iq} & \cdots & a_{ij} \\ \vdots & & \vdots \\ a_{pq} & \cdots & a_{pj} \end{matrix} \qquad (3.1.53)$$

where p is the pivot row and q the pivot column, then a_{iq} becomes 0 (requiring no computation), and the new a_{ij} is given by

$$a_{ij} = a_{ij} - \frac{a_{pj}}{a_{pq}} a_{iq} \qquad (3.1.54)$$

Following this procedure, we obtain, successively,

$$
\begin{bmatrix}
0 & 4.28753 & -0.63906 & 3.36682 & 3.60194 \\
0 & -4.05836 & 2.08054 & -1.29172 & -1.47063 \\
\boxed{1} & 0.35075 & -0.40112 & -0.92351 & 0.23881 \\
0 & 2.51000 & 0.10550 & -0.53050 & 0.09900
\end{bmatrix}
\quad (3.1.55)
$$

$$
\begin{bmatrix}
0 & \boxed{1} & -0.14905 & 0.78526 & 0.84010 \\
0 & 0 & 1.47564 & 1.89514 & 1.93879 \\
1 & 0 & -0.34884 & -1.19894 & -0.05585 \\
0 & 0 & 0.47962 & -2.50150 & -2.00964
\end{bmatrix}
\quad (3.1.56)
$$

$$
\begin{bmatrix}
0 & 1 & 0 & 0.97668 & 1.03593 \\
0 & 0 & \boxed{1} & 1.28428 & 1.31386 \\
1 & 0 & 0 & -0.75093 & 0.40248 \\
0 & 0 & 0 & -3.11747 & -2.63980
\end{bmatrix}
\quad (3.1.57)
$$

For the last pivot, we no longer have a choice. We would have to choose a_{44} even if it were not the largest in absolute value. The result is

$$
\begin{bmatrix}
0 & 1 & 0 & 0 & 0.20890 \\
0 & 0 & 1 & 0 & 0.22636 \\
1 & 0 & 0 & 0 & 1.03835 \\
0 & 0 & 0 & \boxed{1} & 0.84678
\end{bmatrix}
\quad (3.1.58)
$$

This means that

$$
\begin{pmatrix}
0 & 1 & 0 & 0 \\
0 & 0 & 1 & 0 \\
1 & 0 & 0 & 0 \\
0 & 0 & 0 & 1
\end{pmatrix}
\cdot
\begin{pmatrix} x_1 \\ x_2 \\ x_3 \\ x_4 \end{pmatrix}
=
\begin{pmatrix} 0.20890 \\ 0.22636 \\ 1.03835 \\ 0.84678 \end{pmatrix}
\quad (3.1.59)
$$

from which

$$
x_1 = 1.03835, \quad x_2 = 0.20890, \quad x_3 = 0.22636, \quad x_4 = 0.84678 \quad (3.1.60)
$$

3.1.6 Checking the Answers

You are, by this time, sophisticated enough to suspect that you cannot check the precision of the answers by substitution. We shall discuss this in the section on Seidel iteration.

3.1.7 A Computer Program for the Gauss-Jordan Method

Figure 3.1-2 shows the main program developed by a class for the Gauss-Jordan reduction of up to 10 simultaneous equations. Figure 3.1-3 shows the subroutine used to find the maximum absolute value in each column. Figure 3.1-4 shows the output.

```
      DOUBLE PRECISION A(10,11),X(10),D(10),C(10),R
      COMMON A,L(10),K(10)
C THE PURPOSE OF THIS PROGRAM IS TO SOLVE UP TO 10 SIMULTANEOUS EQUATIONS
C BY THE GAUSS-JORDAN METHOD.  N IS THE NUMBER OF DECIMAL PLACES DESIRED
C JT IS THE NUMBER OF COLUMNS, INCLUDING THE COLUMN OF CONSTANTS.
C
      READ(1,100)JT,N
100   FORMAT(I2,1X,I1)
      IF(N.NE.0)GO TO 3
      N=6
3     R=10.D0**(-N)
C IT(I TOTAL) IS THE NUMBER OF ROWS.
      IT=JT-1
C READ IN THE MATRIX BY COLUMNS, NO MORE THAN 5 ENTRIES PER CARD.
      DO 4 J=1,JT
4     READ(1,101)(A(I,J),I=1,IT)
101   FORMAT(5F10.0)
C SAVE ORIGINAL DATA ON TAPE FOR CHECKING.
      WRITE(7,102)((A(I,J), I=1,IT),J=1,JT)
102   FORMAT(D22.16)
      REWIND 7
C SET ROW INDEXES, L(I).
      DO 5 I=1,IT
5     L(I)=I
      DO 6 J=1,IT
C THE SUBROUTINE FINDS THE PIVOT ROW FOR EACH COLUMN.
      CALL MAX(J,IT,M)
      K(J)=M
C THE PIVOT IS ZERO OR CLOSE TO ZERO.  THE METHOD IS INAPPLICABLE.
      IF(DABS(A(M,J)).LE.R)GO TO 60
C REDUCE ROW M, THE PIVOT ROW, STARTING AT COLUMN J+1.
7     JI=J+1
      DO 8 JP=JI,JT
8     A(M,JP)=A(M,JP)/A(M,J)
C REDUCE OTHER ROWS, STARTING IN COLUMN JI.
      DO 6 IP=1,IT
C THE NEXT STEP SKIPS THE PIVOT ROW.
      IF(IP.EQ.M)GO TO 6
      DO 6 JP=JI,JT
      A(IP,JP)=A(IP,JP)-A(M,JP)*A(IP,J)
6     CONTINUE
C STORAGE OF THE ANSWERS.
      DO 10 I=1,IT
      M=K(I)
10    X(I)=A(M,JT)
C DISPLAY THE ANSWERS.
      WRITE(3,103)
103   FORMAT(1H1)
      DO 11 I=1,IT
      GO TO(51,52,53,54,55,56,57,58,59),N
51    WRITE(3,104)I,X(I)
104   FORMAT(1H0,'X(',I2,' ) = ',F 8.1)
      GO TO 11
52    WRITE(3,105)I,X(I)
```

Fig. 3.1-2   (Continued on next page)

```
105      FORMAT(1H0,'X(',I2,' ) = ',F 9.2)
         GO TO 11
53       WRITE(3,106)I,X(I)
106      FORMAT(1H0,'X(',I2,' ) = ',F10.3)
         GO TO 11
54       WRITE(3,107)I,X(I)
107      FORMAT(1H0,'X(',I2,' ) = ',F11.4)
         GO TO 11
55       WRITE(3,108)I,X(I)
108      FORMAT(1H0,'X(',I2,' ) = ',F12.5)
         GO TO 11
56       WRITE(3,109)I,X(I)
109      FORMAT(1H0,'X(',I2,' ) = ',F13.6)
         GO TO 11
57       WRITE(3,110)I,X(I)
110      FORMAT(1H0,'X(',I2,' ) = ',F14.7)
         GO TO 11
58       WRITE(3,111)I,X(I)
111      FORMAT(1H0,'X(',I2,' ) = ',F15.8)
         GO TO 11
59       WRITE(3,112)I,X(I)
112      FORMAT(1H0,'X(',I2,' ) = ',F16.9)
11       CONTINUE
         GO TO 12
C THE ERROR MESSAGE CONCERNING A ZERO PIVOT.
60       WRITE(3,114)
114      FORMAT(1H1,'BECAUSE OF A ZERO PIVOT, THIS SET OF EQUATIONS CANNOT
        1BE SOLVED BY THE GAUSS JORDAN METHOD')
         GO TO 17
C ECHO AND CHECK PROCEDURE.  INITIAL COEFFICIENTS ARE RESTORED IN CORE.
12       READ(7,102)((A(I,J),I=1,IT),J=1,JT)
         REWIND 7
         WRITE(3,115)
115      FORMAT(1H1,T15,'ECHO MATRIX'//)
C THE FOLLOWING MUST BE ADJUSTED IF INPUT IS WORSE THAN (3).(5) AND SIGN.
         DO 13 I=1,IT
13       WRITE(3,117)(A(I,J),J=1,JT)
117      FORMAT(1H ,11F10.5)
C CALCULATION OF PERTURBED CONSTANTS, USING  X(I).  DISPLAY OF ERRORS.
         DO 14 I=1,IT
         C(I)=0.D0
         DO 15 J=1,IT
15       C(I)=C(I)+A(I,J)*X(J)
14       D(I)=A(I,JT)-C(I)
         WRITE(3,119)
119      FORMAT(1H1,T2,'ROW',T18,'ACTUAL CONSTANTS',T44,'CALCULATED CONSTAN
        1TS',T72,'DIFFERENCES'//)
         DO 16 I=1,IT
16       WRITE(3,120)I,A(I,JT),C(I),D(I)
120      FORMAT(1H ,T2,I2,T12,1PD22.15,T42,D22.15,T72,D9.2)
17       STOP
         END
```

Fig. 3.1-2 Gauss-Jordan method (main program).

```
      SUBROUTINE MAX(J,IT,M)
      DOUBLE PRECISION A(10,11),Y(10)
      COMMON A,L(10),K(10)
C FOR ROWS WHICH HAVE ALREADY BEEN PIVOTS, SET COEFFICIENTS TO ZERO.
C FOR OTHER ROWS, FIND ABSOLUTE VALUES.
      DO 1 I=1,IT
      IF(L(I).NE.0)GO TO 3
      Y(I)=0.D0
      GO TO 1
3     Y(I)=DABS(A(I,J))
1     CONTINUE
C SET INITIAL VALUES FOR COMPARING COEFFICIENTS.
      IA=1
      IB=2
C NOW COMPARE IN PAIRS.
4     IF(Y(IB).GT.Y(IA))GO TO 6
      M=IA
      GO TO 7
6     M=IB
C TEST FOR TERMINATION.
7     IF(IB.GE.IT)GO TO 9
      IA=M
      IB=IB+1
      GO TO 4
9     L(M)=0
      RETURN
      END
```

Fig. 3.1-3   The subroutine to find the maximum absolute value in a column.

```
                    ECHO MATRIX


        2.63000     5.21000    -1.69400     0.93800     4.23000
        3.16000    -2.95000    -4.21000    -0.71600
        5.36000     1.88C00    -2.15000    -4.95000     1.28000
        1.34000     2.98000    -0.43200    -1.76800     0.41980



                    X( 1 ) =        1.038142

                    X( 2 ) =        0.209148

                    X( 3 ) =        0.226688

                    X( 4 ) =        0.846517
```

ROW	ACTUAL CONSTANTS	CALCULATED CONSTANTS	DIFFERENCES
1	4.230000000000000D 00	4.229999999999999D 00	1.55D-15
2	-7.160000000000000D-01	-7.160000000000006D-01	6.38D-16
3	1.280000000000000D 00	1.279999999999998D 00	1.78D-15
4	4.198000000000000D-01	4.197999999999993D-01	7.08D-16

Fig. 3.1-4   Output of the Gauss-Jordan program.

After finding the roots, the computer substituted them in the original equation in order to compare the calculated constants with the actual constants. Although the agreement does not tell anything about the precision of the answers, it is a useful calculation in a program to be debugged, because if the constants do *not* agree fairly well the answers are surely incorrect.

PROBLEM SECTION 3.1

The following are augmented matrices. The missing column vector contains x_1, x_2, x_3 and x_4. Solve to the nearest millionth by the Gauss-Jordan method.

(1)
$$\begin{bmatrix} 1.03690 & 3.01760 & 6.60230 & 6.19460 & | & 1.18320 \\ 4.20540 & 4.89790 & 3.82770 & 4.87900 & | & 0.43440 \\ 6.82570 & 9.21530 & 7.45230 & 1.16020 & | & 9.55410 \\ 1.43850 & 3.84160 & 7.11180 & 8.30430 & | & 2.03660 \end{bmatrix}$$

(2)
$$\begin{bmatrix} 3.01770 & 9.97760 & 9.73440 & 4.02380 & | & 5.34790 \\ 4.79670 & 7.57230 & 7.03280 & 4.05770 & | & 8.11150 \\ 9.37930 & 0.31720 & 5.81160 & 3.93510 & | & 9.80360 \\ 8.66930 & 4.31120 & 9.19640 & 4.32110 & | & 1.22170 \end{bmatrix}$$

(3)
$$\begin{bmatrix} 1.98390 & 7.32600 & 5.52860 & 0.48110 & | & 1.39560 \\ 9.06300 & 5.68770 & 3.35910 & 6.48920 & | & 9.88990 \\ 7.18630 & 4.07940 & 6.19650 & 9.63460 & | & 9.23150 \\ 9.50530 & 1.39480 & 5.17230 & 7.90650 & | & 6.57830 \end{bmatrix}$$

(4)
$$\begin{bmatrix} 1.31830 & 8.19820 & 7.86570 & 4.39670 & | & 8.18740 \\ 5.06520 & 1.45380 & 0.21840 & 6.34850 & | & 8.33390 \\ 9.48720 & 2.61620 & 2.97150 & 9.35720 & | & 1.49880 \\ 2.82570 & 2.48990 & 0.43340 & 8.07530 & | & 9.99370 \end{bmatrix}$$

(5)
$$\begin{bmatrix} 3.49140 & 2.87080 & 4.98430 & 4.25830 & | & 5.49090 \\ 9.45020 & 8.40880 & 1.14420 & 3.63350 & | & 0.99760 \\ 3.93740 & 6.55350 & 6.66820 & 6.00680 & | & 7.65800 \\ 3.41850 & 4.42580 & 3.60550 & 0.40440 & | & 0.26450 \end{bmatrix}$$

(6)
$$\begin{bmatrix} 0.57930 & 1.34860 & 5.85150 & 8.25300 & | & 2.58200 \\ 0.29000 & 4.69180 & 4.99200 & 9.83990 & | & 9.61980 \\ 6.34980 & 6.46830 & 0.39010 & 2.63870 & | & 6.65180 \\ 0.07820 & 0.74110 & 2.65970 & 0.28360 & | & 7.83140 \end{bmatrix}$$

(7)
$$\begin{bmatrix} 4.45370 & 6.49820 & 0.19080 & 3.04630 & | & 3.36990 \\ 6.44280 & 6.08340 & 4.77960 & 2.78560 & | & 2.36730 \\ 3.54410 & 8.53190 & 6.57960 & 6.77980 & | & 4.58840 \\ 2.83180 & 4.78140 & 4.42300 & 1.68370 & | & 4.15150 \end{bmatrix}$$

(8)
$$\begin{bmatrix} 2.25140 & 6.07960 & 7.86000 & 3.11510 & | & 6.09080 \\ 0.40600 & 7.66390 & 3.69240 & 5.82950 & | & 8.41080 \\ 9.45110 & 3.01570 & 5.99620 & 4.08230 & | & 5.53420 \\ 4.46120 & 4.02950 & 6.81910 & 4.13300 & | & 4.84790 \end{bmatrix}$$

(9)
$$\begin{bmatrix} 9.38820 & 2.83340 & 4.06660 & 2.72660 & | & 2.66360 \\ 4.91920 & 1.53680 & 4.33280 & 2.74030 & | & 8.39030 \\ 4.48760 & 4.24810 & 8.73790 & 9.75200 & | & 4.47220 \\ 4.71850 & 6.03120 & 8.64180 & 2.33340 & | & 6.92100 \end{bmatrix}$$

(10)
$$\begin{bmatrix} 7.14940 & 2.90460 & 4.36440 & 3.88560 & | & 5.95260 \\ 9.54010 & 0.13010 & 4.62480 & 8.00480 & | & 6.92550 \\ 3.40490 & 5.53430 & 5.32050 & 5.99730 & | & 2.62400 \\ 0.48510 & 6.57320 & 9.48680 & 7.33680 & | & 7.75780 \end{bmatrix}$$

3.2 SEIDEL ITERATION

3.2.1 The Problem

In the preceding section, we demonstrated the Gauss-Jordan method for solving a system of linear equations. Unfortunately, the set of answers

$$x_1 = 1.03835$$
$$x_2 = 0.20890$$
$$x_3 = 0.22636 \tag{3.2.1}$$
$$x_4 = 0.84678$$

does not tell us how many significant figures we actually have. They do not match the computer result.

If we actually substitute in the original, we find

$$\begin{bmatrix} 2.630 & 5.210 & -1.694 & 0.938 \\ 3.160 & -2.950 & 0.813 & -4.210 \\ 5.360 & 1.880 & -2.150 & -4.950 \\ 1.340 & 2.980 & -0.432 & -1.768 \end{bmatrix} \cdot \begin{bmatrix} 1.03835 \\ 0.20890 \\ 0.22636 \\ 0.84678 \end{bmatrix} = \begin{bmatrix} 4.230055 \\ -0.715982 \\ 1.280053 \\ 0.419016 \end{bmatrix} \tag{3.2.2}$$

From a practical point of view, the results in (3.2.2) may be perfectly satisfactory. From a computer-science point of view, we still do not know, from (3.2.2), how precise the results are, because experience tells us that wrong answers will also check satisfactorily from time to time.

3.2.2 An Iterative Scheme

You will find, throughout our work, that computer scientists are extremely fond of iterative schemes. Iteration is lengthy and tedious in hand or calculator operations, but the mindless tedium of the repetitions is precisely what makes it easy to program. (Computers don't get bored.) There are two other reasons for iteration: (1) even if we start with a wrong guess, we eventually get the right answer, provided the guess is not too far off, and (2) the difference in answers in successive iterations often leads to an estimate of the error.

To illustrate this, we are going to perform a *Seidel iteration* in the problem we just solved, but we are going to start, on purpose, with incorrect values. We start with

$$x_1^{(0)} = 1.035 \quad \text{instead of} \quad 1.03835$$
$$x_2^{(0)} = 0.209 \quad \text{instead of} \quad 0.20890$$
$$x_3^{(0)} = 0.227 \quad \text{instead of} \quad 0.22636 \tag{3.2.3}$$
$$x_4^{(0)} = 0.846 \quad \text{instead of} \quad 0.84678$$

The superscript is used to number the trials. (0) is the number of the original guess. Substituting these values in the X-vector of the original problem, we obtain

$$
\begin{bmatrix}
2.630 & 5.210 & -1.694 & 0.938 \\
3.160 & -2.950 & 0.813 & -4.210 \\
5.360 & 1.880 & -2.150 & -4.950 \\
1.340 & 2.980 & -0.432 & -1.768
\end{bmatrix}
\cdot
\begin{bmatrix}
1.035 \\
0.209 \\
0.227 \\
0.846
\end{bmatrix}
=
\begin{bmatrix}
4.219950 \\
-0.723059 \\
1.264770 \\
0.415928
\end{bmatrix}
$$

(3.2.4)

$$ A \qquad\qquad\cdot\quad X^{(0)} \ = \qquad C^{(0)} $$

The original problem was

$$
\begin{bmatrix}
2.630 & 5.210 & -1.694 & 0.938 \\
3.160 & -2.950 & 0.813 & -4.210 \\
5.360 & 1.880 & -2.150 & -4.950 \\
1.340 & 2.980 & -0.432 & -1.768
\end{bmatrix}
\cdot
\begin{bmatrix}
x_1 \\
x_2 \\
x_3 \\
x_4
\end{bmatrix}
=
\begin{bmatrix}
4.230 \\
-0.716 \\
1.280 \\
0.419
\end{bmatrix}
$$

(3.2.5)

$$ A \qquad\qquad\cdot\quad X \ = \qquad C $$

Using the addition theorem and the property of distributivity, we have

$$
\begin{aligned}
AX - AX^{(0)} &= C - C^{(0)} \\
A(X - X^{(0)}) &= C - C^{(0)}
\end{aligned}
$$

(3.2.6)

Let the first error in X be given by

$$ X - X^{(0)} = E^{(1)} $$

(3.2.7)

By subtraction,

$$
C - C^{(0)} =
\begin{bmatrix}
0.010050 \\
0.007059 \\
0.015230 \\
0.003072
\end{bmatrix}
$$

(3.2.8)

Substituting in (3.2.6), we have a new set of equations:

$$
\begin{bmatrix}
2.630 & 5.210 & -1.694 & 0.938 \\
3.160 & -2.950 & 0.813 & -4.210 \\
5.360 & 1.880 & -2.150 & -4.950 \\
1.340 & 2.980 & -0.432 & -1.768
\end{bmatrix}
\cdot
\begin{bmatrix}
\epsilon_1^{(1)} \\
\epsilon_2^{(1)} \\
\epsilon_3^{(1)} \\
\epsilon_4^{(1)}
\end{bmatrix}
=
\begin{bmatrix}
0.010050 \\
0.007059 \\
0.015230 \\
0.003072
\end{bmatrix}
$$

(3.2.9)

where, for example, $\epsilon_1^{(1)} = x - x_1^{(0)}$. We solve system (3.2.9) by the Gauss-

Jordan method, with the following result:

$$\epsilon_1^{(1)} = x_1 - x_1^{(0)} = +0.00333\ 5733$$
$$\epsilon_2^{(1)} = x_2 - x_2^{(0)} = -0.00010\ 2296 \tag{3.2.10}$$
$$\epsilon_3^{(1)} = x_3 - x_3^{(0)} \quad -0.00064\ 1211$$
$$\epsilon_4^{(1)} = x_4 - x_4^{(0)} = +0.00077\ 4913$$

Using these equations, we solve for x to obtain a new estimate, which we shall call $x^{(1)}$:

$$x_1^{(1)} = 1.035 + 0.00333\ 5733 = 1.03833\ 6$$
$$x_2^{(1)} = 0.208899\ 8 \tag{3.2.11}$$
$$x_3^{(1)} = 0.226359\ 9$$
$$x_4^{(1)} = 0.846777\ 5$$

Furthermore, the size of $\epsilon^{(1)}$ in (3.2.10) gives us an estimate of the error. We repeat the process:

$$\begin{bmatrix} 2.630 & 5.210 & -1.694 & 0.938 \\ 3.160 & -2.950 & 0.813 & -4.210 \\ 5.360 & 1.880 & -2.150 & -4.950 \\ 1.340 & 2.980 & -0.432 & -1.768 \end{bmatrix} \cdot \begin{bmatrix} 1.038336 \\ 0.208898 \\ 0.226359 \\ 0.846775 \end{bmatrix} = \begin{bmatrix} 4.23000\ 5064 \\ -0.71600\ 0223 \\ 1.28000\ 1100 \\ 0.41900\ 0992 \end{bmatrix} \tag{3.2.12}$$

Subtracting from (3.2.5), we obtain

$$\begin{bmatrix} 2.630 & 5.210 & -1.694 & 0.938 \\ 3.160 & -2.950 & 0.813 & -4.210 \\ 5.360 & 1.880 & -2.150 & -4.950 \\ 1.340 & 2.980 & -0.432 & -1.768 \end{bmatrix} \cdot \begin{bmatrix} \epsilon_1^{(2)} \\ \epsilon_2^{(2)} \\ \epsilon_3^{(2)} \\ \epsilon_4^{(2)} \end{bmatrix} = \begin{bmatrix} -0.00000\ 5064 \\ 0.00000\ 0223 \\ -0.00000\ 0100 \\ -0.00000\ 0992 \end{bmatrix} \tag{3.2.13}$$

Solving this by the Gauss-Jordan method, the results are

$$\epsilon_1^{(2)} = -0.00000\ 14384$$
$$\epsilon_2^{(2)} = -0.0000004473$$
$$\epsilon_3^{(2)} = -0.00000\ 11692 \tag{3.2.14}$$
$$\epsilon_4^{(2)} = -0.0000009973$$

These are the errors in $x^{(1)}$. We add them to find

$$x_1^{(2)} = 1.038335$$
$$x_2^{(2)} = 0.208898$$
$$x_3^{(2)} = 0.226358 \tag{3.2.15}$$
$$x_4^{(2)} = 0.846774$$

If we wished to find the errors in these, we would repeat the process. Actually, if (3.2.15) is substituted in the original, the column of constants turns out to be

$$\begin{bmatrix} 4.2299999999996 \\ -0.716000000000005 \\ 1.2799999999995 \\ 0.4189999999996 \end{bmatrix} \tag{3.2.16}$$

PROBLEM SECTION 3.2

For the systems in the problems for Section 3.1, (a) estimate the error in each x by performing a Seidel iteration, and (b) perform Seidel iterations until the error is less than 10^{-6}.

3.3   MATRIX INVERSION

3.3.1   The Inverse of a Matrix

We mentioned that if a square matrix, A, has an inverse A^{-1}, then this inverse is unique and

$$AA^{-1} = A^{-1}A = I \tag{3.3.1}$$

where I is the appropriate identity matrix. We can find the inverse of a small matrix by very elementary methods.

ILLUSTRATIVE PROBLEM 3.3-1   Find the inverse of

$$A = \begin{pmatrix} -6 & 7 \\ 7 & -8 \end{pmatrix}$$

Solution   Let

$$\begin{pmatrix} -6 & 7 \\ 7 & -8 \end{pmatrix} \cdot \begin{pmatrix} a & b \\ c & d \end{pmatrix} = \begin{pmatrix} 1 & 0 \\ 0 & 1 \end{pmatrix} \tag{3.3.2}$$

Then after multiplication in the left member,

$$\begin{pmatrix} -6a+7c & -6b+7d \\ 7a-8c & 7b-8d \end{pmatrix} = \begin{pmatrix} 1 & 0 \\ 0 & 0 \end{pmatrix} \tag{3.3.3}$$

By the definition of equal matrices,

$$\begin{cases} -6a+7c=1, \\ 7a-8c=0, \end{cases} \begin{cases} -6b+7d=0 \\ 7b+8d=1 \end{cases} \tag{3.3.4}$$

Solving by elementary methods,

$$\begin{aligned} a &= 8 \\ b &= 7 \\ c &= 7 \\ d &= 6 \end{aligned} \tag{3.3.5}$$

Therefore

$$A^{-1} = \begin{pmatrix} 8 & 7 \\ 7 & 6 \end{pmatrix} \tag{3.3.6}$$

Checking by either AA^{-1} or $A^{-1}A$, we find that the product is indeed I.

If the equations (3.3.4) have no solution, then A has no inverse, i.e., A is singular.

We are discussing matrix inverses in this chapter not because they are useful in solving systems of linear equations, but because they are important in their own right and because the computer method of finding an inverse is so much like that for solving a system of linear equations.

3.3.2 The Simplest Case

We know that

$$AA^{-1} = I \tag{3.3.7}$$

If we can transform A to I without disturbing A^{-1} (which is unknown at this point), the matrix I will apparently have to change to A^{-1}, as shown in the following diagram:

$$\begin{array}{ccc} A & A^{-1} & = & I \\ \downarrow & & & \downarrow \\ I & A^{-1} & = & A^{-1} \end{array} \tag{3.3.8}$$

We justify the procedure by the following set of steps:

$$AA^{-1}=I \qquad \textit{(definition of inverse)} \tag{3.3.9}$$

$$A^{-1}(AA^{-1})=A^{-1}I \qquad \textit{(multiplication theorem)} \tag{3.3.10}$$

$$(A^{-1}A)A^{-1}=A^{-1}I \qquad \textit{(associativity)} \tag{3.3.11}$$

$$IA^{-1}=A^{-1}I \qquad \textit{(definition of inverse)} \tag{3.3.12}$$

$$IA^{-1}=A^{-1} \qquad \textit{(definition of identity)} \tag{3.3.13}$$

Here it can be seen that when A transforms to I, I transforms to A^{-1}. Fortunately, we already know that in the Gauss-Jordan process A is transformed to I, so that we can use the process to find A^{-1}.

ILLUSTRATIVE PROBLEM 3.3-2   Find A^{-1} if

$$A = \begin{pmatrix} 4 & 3 \\ 2 & 1 \end{pmatrix}$$

Solution   We start with

$$\begin{pmatrix} 4 & 3 \\ 2 & 1 \end{pmatrix} A^{-1} = \begin{pmatrix} 1 & 0 \\ 0 & 1 \end{pmatrix} \tag{3.3.14}$$

which corresponds to [3.3.9]. Of course, we don't know A^{-1}. We write the tableau.

$$\begin{bmatrix} \boxed{4} & 3 & \vdots & 1 & 0 \\ 2 & 1 & \vdots & 0 & 1 \end{bmatrix} \tag{3.3.15}$$
$$ A I$$

Using a_{11} as a pivot, we divide the first row by 4 and obtain

$$\begin{bmatrix} \boxed{1} & \frac{3}{4} & \vdots & \frac{1}{4} & 0 \\ 2 & 1 & \vdots & 0 & 1 \end{bmatrix} \tag{3.3.16}$$

Multiplying the first row by -2 and adding to the second row, the result is

$$\begin{bmatrix} \boxed{1} & \frac{3}{4} & \vdots & \frac{1}{4} & 0 \\ 0 & -\frac{1}{2} & \vdots & -\frac{1}{2} & 1 \end{bmatrix} \tag{3.3.17}$$

Multiplying the second row by -2, we have

$$\begin{bmatrix} 1 & \frac{3}{4} & \vdots & \frac{1}{4} & 0 \\ 0 & \boxed{1} & \vdots & 1 & -2 \end{bmatrix} \tag{3.3.18}$$

Multiplying the second row by $-\frac{3}{4}$ and adding to the first row,

$$\begin{bmatrix} 1 & 0 & \vdots & -\frac{1}{2} & \frac{3}{2} \\ 0 & 1 & \vdots & 1 & -2 \end{bmatrix} \tag{3.3.19}$$
$$ I A^{-1}$$

We have transformed A to I. Therefore, we must have transformed A to A^{-1}. Then

$$A^{-1} = \begin{pmatrix} -\frac{1}{2} & \frac{3}{2} \\ 1 & -2 \end{pmatrix} \tag{3.3.20}$$

Checking by multiplication, we find that $A^{-1}A = I$.

Note that in each step, we really are multiplying by an elementary matrix, as explained in the section of the Gauss-Jordan method.

3.3.3 Optimized Procedure

As you already know, if a_{11} is 0 or if, during the procedure, any diagonal element, $a_{22}, a_{33}, \ldots$, becomes 0, the simple method used in Illustrative Problem 3.3-2 cannot be used. Also, in a real situation with inherent error problems, it is advantageous to start in each column with the element having the largest absolute value. Let us see what happens if we try this with the matrix of Illustrative Problem 3.3-1. We set up the tableau with the right side empty (because we don't know in advance where the pivots will be):

$$\left[\begin{array}{cc|cc} -6 & 7 & & \\ 7 & -8 & & \end{array}\right] \tag{3.3.21}$$

We note that 7 is the largest element in column 1. Therefore, the matrix at the right must be *permuted*:

$$\left[\begin{array}{cc|cc} -6 & 7 & 0 & 1 \\ \boxed{7} & -8 & 1 & 0 \end{array}\right] \tag{3.3.22}$$

Then

$$\left[\begin{array}{cc|cc} 0 & \boxed{\frac{1}{7}} & \frac{6}{7} & 1 \\ 1 & -\frac{8}{7} & \frac{1}{7} & 0 \end{array}\right] \tag{3.3.23}$$

$$\left[\begin{array}{cc|cc} 0 & 1 & 6 & 7 \\ 1 & 0 & 7 & 8 \end{array}\right] \tag{3.3.24}$$

Here it becomes clear that mathematical step-by-step procedure is better than guessing. Superficially, it might appear that if we interchanged rows (multiplied by H_{21}), the result would be

$$\left[\begin{array}{cc|cc} 1 & 0 & 7 & 8 \\ 0 & 1 & 6 & 7 \end{array}\right] \tag{3.3.25}$$

and that the right member of this tableau would be the inverse. However, it is not, as we know from Illustrative Problem 3.3-1. To get the correct inverse, we must also interchange columns, which leaves us with

$$\left[\begin{array}{cc|cc} 0 & 1 & 8 & 7 \\ 1 & 0 & 7 & 6 \end{array}\right] \tag{3.3.26}$$

The right member of this tableau *is* the correct inverse, but throws no light on the reason.

The problem is too small to illuminate the method. Let us find the inverse of

$$A = \begin{bmatrix} 1 & -2 & 3 \\ 3 & -1 & 4 \\ 2 & 1 & -2 \end{bmatrix} \tag{3.3.27}$$

by our optimized method. The correct solution is

$$A^{-1} = \begin{bmatrix} \frac{2}{15} & \frac{1}{15} & \frac{1}{3} \\ -\frac{14}{5} & \frac{8}{15} & -\frac{1}{3} \\ -\frac{1}{3} & \frac{1}{3} & -\frac{1}{3} \end{bmatrix} \tag{3.3.28}$$

We start with

$$\begin{bmatrix} 1 & -2 & 3 & \vdots & & & \\ 3 & -1 & 4 & \vdots & & & \\ 2 & 1 & -2 & \vdots & & & \end{bmatrix} \tag{3.3.29}$$

Noting that 3 will be the first pivot, we can begin to fill in the right half of the tableau (at this moment, we don't know where the next pivot will be):

$$\begin{bmatrix} 1 & -2 & 3 & \vdots & 0 & & \\ \boxed{3} & -1 & 4 & \vdots & 1 & 0 & 0 \\ 2 & 1 & -2 & \vdots & 0 & & \end{bmatrix} \tag{3.3.30}$$

The next calculation yields

$$\begin{bmatrix} 0 & -\frac{5}{3} & \frac{5}{3} & \vdots & -\frac{1}{3} & & \\ \boxed{1} & -\frac{1}{3} & \frac{4}{3} & \vdots & \frac{1}{3} & 0 & 0 \\ 0 & \frac{5}{3} & -\frac{14}{3} & \vdots & -\frac{2}{3} & & \end{bmatrix} \tag{3.3.31}$$

We arbitrarily choose a_{32} as pivot (a_{12} is just as good):

$$\begin{bmatrix} 0 & -\frac{5}{3} & \frac{5}{3} & \vdots & -\frac{1}{3} & 0 & \\ 1 & -\frac{1}{3} & \frac{4}{3} & \vdots & \frac{1}{3} & 0 & 0 \\ 0 & \boxed{\frac{5}{3}} & -\frac{14}{3} & \vdots & -\frac{2}{3} & 1 & \end{bmatrix} \tag{3.3.32}$$

Because the next pivot has to be a_{13}, we can actually complete the right half in this particular case (if the matrix were larger, we would have to wait):

$$\begin{bmatrix} 0 & -\frac{5}{3} & \frac{5}{3} & \bigm| & -\frac{1}{3} & 0 & 1 \\ 1 & -\frac{1}{3} & \frac{4}{3} & \bigm| & \frac{1}{3} & 0 & 0 \\ 0 & \boxed{\frac{5}{3}} & -\frac{14}{3} & \bigm| & -\frac{2}{3} & 1 & 0 \end{bmatrix} \qquad (3.3.33)$$

This shows that we could have started (if we had known) with

$$\begin{bmatrix} 1 & -2 & 3 & \bigm| & 0 & 0 & 1 \\ 3 & 1 & 4 & \bigm| & 1 & 0 & 0 \\ 2 & 1 & -2 & \bigm| & 0 & 1 & 0 \end{bmatrix} \qquad (3.3.34)$$

The matrix

$$J^{-1} = \begin{pmatrix} 0 & 0 & 1 \\ 1 & 0 & 0 \\ 0 & 1 & 0 \end{pmatrix} \qquad (3.3.35)$$

is an elementary matrix which we shall discuss presently.
Continuing from (3.3.33), we have

$$\begin{bmatrix} 0 & 0 & -3 & \bigm| & -1 & 1 & 1 \\ 1 & 0 & \frac{2}{5} & \bigm| & \frac{1}{5} & \frac{1}{5} & 0 \\ 0 & \boxed{1} & -\frac{14}{5} & \bigm| & -\frac{2}{5} & \frac{3}{5} & 0 \end{bmatrix} \qquad (3.3.36)$$

$$\underbrace{\begin{bmatrix} 0 & 0 & \boxed{1} & \bigm| & \frac{1}{3} & -\frac{1}{3} & -\frac{1}{3} \\ 1 & 0 & 0 & \bigm| & \frac{1}{15} & \frac{1}{3} & \frac{2}{15} \\ 0 & 1 & 0 & \bigm| & \frac{8}{15} & -\frac{1}{3} & -\frac{14}{5} \end{bmatrix}}_{\displaystyle J^{-1} \qquad\qquad B} \qquad (3.3.37)$$

We shall call B a *scrambled inverse*, and we shall show how to unscramble it. We shall justify the procedure afterwards. Our order of operations was

Step:	1	2	3
Pivot Row:	2	3	1

(3.3.38)

The rule for unscrambling is as follows:

UNSCRAMBLING PROCEDURE   Unscramble rows *up* and unscramble columns *down* the order table.

Using (3.3.38), we rewrite the rows of B in (3.3.37) so that

row 2 becomes row 1,
row 3 becomes row 2,
row 1 becomes row 3.

The result is as follows:

$$
\begin{bmatrix}
1 & 0 & 0 & \vdots & \frac{1}{15} & \frac{1}{3} & \frac{2}{15} \\
0 & 1 & 0 & \vdots & \frac{8}{15} & -\frac{1}{3} & -\frac{14}{15} \\
0 & 0 & 1 & \vdots & \frac{1}{3} & -\frac{1}{3} & -\frac{1}{3}
\end{bmatrix}
\tag{3.3.39}
$$

This is unscrambling *up* the order table. Now, using (3.3.38), we rewrite the columns of (3.3.39) so that

column 1 becomes column 2,
column 2 becomes column 3,
column 3 becomes column 1:

$$
\begin{pmatrix}
\frac{2}{15} & \frac{1}{15} & \frac{1}{3} \\
-\frac{14}{15} & \frac{8}{15} & -\frac{1}{3} \\
-\frac{1}{3} & \frac{1}{3} & -\frac{1}{3}
\end{pmatrix}
\tag{3.3.40}
$$

$$A^{-1}$$

Comparing this with (3.3.28), we see that it is the correct inverse.

3.3.4 Properties of Permutation Matrices

We have already pointed out that if we take an identity matrix such as

$$
\begin{bmatrix}
1 & 0 & 0 \\
0 & 1 & 0 \\
0 & 0 & 1
\end{bmatrix}
\tag{3.3.41}
$$

and interchange two rows, such as the second and third, then the resulting matrix

$$
H_{23} =
\begin{bmatrix}
1 & 0 & 0 \\
0 & 0 & 1 \\
0 & 1 & 0
\end{bmatrix}
\tag{3.3.42}
$$

is an elementary matrix called a *permutation matrix*. When H_{23} pre-multiplies another matrix A,

$$\begin{bmatrix} 1 & 0 & 0 \\ 0 & 0 & 1 \\ 0 & 1 & 0 \end{bmatrix} \cdot \begin{bmatrix} a_{11} & a_{12} & a_{13} \\ a_{21} & a_{22} & a_{23} \\ a_{31} & a_{32} & a_{33} \end{bmatrix} = \begin{bmatrix} a_{11} & a_{12} & a_{13} \\ a_{31} & a_{32} & a_{33} \\ a_{21} & a_{22} & a_{23} \end{bmatrix} \qquad (3.3.43)$$

$$H_{23} \qquad \cdot \qquad A \qquad = \qquad A'$$

it makes the same interchange in A as was previously made in I, i.e., it permutes the same two rows of A.

Now, suppose we wish to change A' back to A. This means that we have to interchange the same rows again. In other words,

$$HA' = A \qquad (3.3.44)$$

We use this fact to prove a theorem.

THEOREM   Every elementary permutation matrix is its own inverse.

Proof

$$A' = HA \qquad (3.3.45)$$

$$\therefore \ HA' = HHA \qquad (3.3.46)$$

$$\therefore \quad A = HHA \qquad (3.3.47)$$

$$\therefore \ HH = I \qquad (3.3.48)$$

$$\therefore \quad H = H^{-1} \qquad (3.3.49)$$

This means that to undo the effect of pre-multiplying by H, you pre-multiply by H again. Since every permutation matrix is its own inverse, the following corollary is immediate.

THEOREM   Every elementary permutation matrix has an inverse. (Every elementary premutation matrix is non-singular.)

ILLUSTRATIVE PROBLEM 3   A matrix A has had row interchanges in the order $H_{23}, H_{35}, H_{13}, H_{24}$. How do we get the original matrix back?

Solution

$$A' = H_{24}H_{35}H_{13}H_{23}A \qquad (3.3.50)$$

$$H_{24}A' = H_{35}H_{13}H_{23}A \qquad (3.3.51)$$

$$H_{35}H_{24}A' = H_{13}H_{23}A \qquad (3.3.52)$$

$$H_{13}H_{35}H_{24}A' = H_{23}A \qquad (3.3.53)$$

$$H_{23}H_{13}H_{35}H_{24}A' = A \qquad (3.3.54)$$

To clarify the concept, let

$$J = H_{24}H_{35}H_{13}H_{23} \tag{3.3.55}$$

Then Equation (3.3.50) can be rewritten as

$$A' = JA \tag{3.3.56}$$

Then, from (3.3.54),

$$A = J^{-1}A \tag{3.3.57}$$

where

$$J^{-1} = H_{23}H_{13}H_{35}H_{24} \tag{3.3.58}$$

This is easily generalized:

THEOREM   If $J = H_1H_2H_3\ldots H_n$, then $J^{-1} = H_nH_{n-1}\ldots H_3H_2H_1$.

We shall need one other kind of permutation. We demonstrate:

$$\begin{bmatrix} a_{11} & a_{12} & a_{13} \\ a_{21} & a_{22} & a_{23} \\ a_{31} & a_{32} & a_{33} \end{bmatrix} \cdot \begin{bmatrix} 1 & 0 & 0 \\ 0 & 0 & 1 \\ 0 & 1 & 0 \end{bmatrix} = \begin{bmatrix} a_{11} & a_{13} & a_{12} \\ a_{21} & a_{23} & a_{22} \\ a_{31} & a_{33} & a_{32} \end{bmatrix} \tag{3.3.59}$$

Note that if H_{23} pre-multiplies, it interchanges *rows* 2 and 3; if H_{23} post-multiplies, it interchanges *columns* 2 and 3.

ILLUSTRATIVE PROBLEM 3.3-4   If

$$J = H_{23}H_{12} \quad \text{and} \quad A = \begin{pmatrix} 1 & 2 & 3 \\ 4 & 5 & 6 \\ 7 & 8 & 9 \end{pmatrix}$$

find (a) JA, (b) AJ, (c) JAJ.

Solution

(a) $\quad JA = H_{23}(H_{12}A) = H_{23}\begin{pmatrix} 4 & 5 & 6 \\ 1 & 2 & 3 \\ 7 & 8 & 9 \end{pmatrix} = \begin{pmatrix} 4 & 5 & 6 \\ 7 & 8 & 9 \\ 1 & 2 & 3 \end{pmatrix}$

(b) $\quad AJ = (AH_{23})H_{12} = \begin{pmatrix} 1 & 3 & 2 \\ 4 & 6 & 5 \\ 7 & 9 & 8 \end{pmatrix}H_{12} = \begin{pmatrix} 3 & 1 & 2 \\ 6 & 4 & 5 \\ 9 & 7 & 8 \end{pmatrix}$

(c) $\quad JAJ = (JA)J = \begin{pmatrix} 4 & 5 & 6 \\ 7 & 8 & 9 \\ 1 & 2 & 3 \end{pmatrix}H_{23}H_{12} = \begin{pmatrix} 4 & 6 & 5 \\ 7 & 9 & 8 \\ 1 & 3 & 2 \end{pmatrix}H_{12} = \begin{pmatrix} 6 & 4 & 5 \\ 9 & 7 & 8 \\ 3 & 1 & 2 \end{pmatrix}$

3.3.5 Justification of Unscrambling

We now return to a proof that our unscrambling procedure is correct. Remember that, in the tableau, A^{-1} (which is unknown, as explained in Illustrative Problem 3.3-2) is omitted, and the manipulations of the Gauss-Jordan process are equivalent to multiplications (pre- and post-) by elementary matrices. We start with

$$AA^{-1} = I \qquad (3.3.60)$$

and post-multiply by J^{-1}:

$$AA^{-1}J^{-1} = IJ^{-1} \qquad (3.3.61)$$

In the tableau, at this point, we have $[A|J^{-1}]$, which corresponds to (3.3.34). Then

$$(AA^{-1})J^{-1} = J^{-1} \qquad (3.3.62)$$

$$(J^{-1}A^{-1})(AA^{-1})J^{-1} = (J^{-1}A^{-1})J^{-1} \qquad (3.3.63)$$

$$J^{-1}(AA^{-1})A^{-1}J^{-1} = J^{-1}A^{-1}J^{-1} \qquad (3.3.64)$$

$$J^{-1}A^{-1}J^{-1} = J^{-1}A^{-1}J^{-1} \qquad (3.3.65)$$

In the tableau, at this point, we have just $[J^{-1}|J^{-1}A^{-1}J^{-1}]$. This corresponds to (3.3.37) with $B = J^{-1}A^{-1}J^{-1}$.

To unscramble, we first pre-multiply B by J:

$$J(J^{-1}A^{-1}J^{-1}) = A^{-1}J^{-1} \qquad (3.3.66)$$

The tableau corresponding to this is $[JJ'|A^{-1}J^{-1}]$, which corresponds to (3.3-39).

Finally, we post-multiply by J, which leaves A^{-1} alone in the right half of the tableau. A summary of the unscrambling steps is as follows.

	Left Half	Not Shown	Right Half
	A	$A^{-1} =$	I
Post-multiply by J^{-1}	A	$A^{-1}J^{-1} =$	J^{-1}
Pre-multiply by $J^{-1}A^{-1}$	J^{-1}	$A^{-1}J^{-1} =$	$J^{-1}A^{-1}J^{-1}$
Pre-multiply by J	I	$A^{-1}J^{-1} =$	$A^{-1}J^{-1}$
Post-multiply by J	I	$A^{-1} =$	A^{-1}

Note that post-multiplication of the left half of a tableau is hidden because it takes place in the "not shown" part of the tableau.

3.3.6 A Practical Example

We shall find the inverse of the matrix used in the previous subsection.

ILLUSTRATIVE PROBLEM 3.3-5   Find A^{-1} if A is

$$\begin{bmatrix} 2.630 & 5.210 & -1.694 & 0.938 \\ 3.160 & -2.950 & 0.813 & -4.210 \\ 5.360 & 1.880 & -2.150 & -4.950 \\ 1.340 & 2.980 & -0.432 & -1.768 \end{bmatrix}$$

Solution We note that the best pivot for column 1 is a_{31}. Therefore, the first (partial) tableau is

$$\begin{bmatrix} 2.630 & 5.210 & -1.694 & 0.938 & | & 0 \\ 3.160 & -2.950 & 0.813 & -4.210 & | & 0 \\ \boxed{5.360} & 1.880 & -2.150 & -4.950 & | & 1 & 0 & 0 & 0 \\ 1.340 & 2.980 & -0.432 & -1.768 & | & 0 \end{bmatrix}$$

After Gauss-Jordan reduction of the first column, we have

$$\begin{bmatrix} 0 & \boxed{4.28753} & -0.63906 & 3.36682 & | & -0.49067 & 1 & 0 & 0 \\ 0 & -4.05836 & 2.08054 & -1.29172 & | & -0.58955 & 0 \\ 1 & 0.35075 & -0.40112 & -0.92351 & | & 0.18657 & 0 & 0 & 0 \\ 0. & 2.51000 & 0.10550 & -0.53050 & | & -0.25000 & 0 \end{bmatrix}$$

The next reduction yields

$$\begin{bmatrix} 0 & 1 & -0.14905 & 0.78526 & | & -0.11444 & 0.23323 & 0 & 0 \\ 0 & 0 & \boxed{1.47564} & 1.89514 & | & -1.05399 & 0.94655 & 1 & 0 \\ 1 & 0 & -0.34884 & -1.19894 & | & 0.22671 & -0.08181 & 0 \\ 0 & 0 & 0.47962 & -2.50150 & | & 0.03725 & -0.58542 & 0 \end{bmatrix}$$

The reduction of column 3 leads to

$$\begin{bmatrix} 0 & 1 & 0 & 0.97668 & | & -0.22090 & 0.32884 & 0.10101 & 0 \\ 0 & 0 & 1 & 1.28428 & | & -0.71426 & 0.64145 & 0.67767 & 0 \\ 1 & 0 & 0 & -0.75093 & | & -0.02245 & 0.14195 & 0.23640 & 0 \\ 0 & 0 & 0 & \boxed{-3.11747} & | & 0.37982 & -0.89307 & -0.32503 & 1 \end{bmatrix}$$

The final reduction leads to the scrambled matrix in the right half of the tableau. The steps were as follows:

Step:	1	2	3	4
Pivot row:	3	1	2	4

First we unscramble *rows up*, so that 3→1, 1→2, 2→3 and 4→4:

$$\begin{bmatrix} -0.11394 & 0.35707 & 0.31469 & -0.24088 \\ -0.10191 & 0.04905 & -0.00082 & 0.31329 \\ -0.55779 & 0.27354 & 0.54377 & 0.41196 \\ -0.12184 & 0.28647 & 0.10426 & -0.32077 \end{bmatrix}$$

then we scramble *columns down*, so that 1→3, 2→1, 3→2 and 4→4:

$$\begin{bmatrix} 0.35707 & 0.31469 & -0.11394 & -0.24088 \\ 0.04905 & -0.00082 & -0.10191 & 0.31329 \\ 0.27354 & 0.54377 & -0.55779 & 0.41196 \\ 0.28647 & 0.10426 & -0.12184 & -0.32077 \end{bmatrix}$$

3.3.7 Computer Program

Figure 3.3-1 shows a computer program which finds the inverse by unscrambling, and Figure 3.3-2 shows the output of the computer program. Subroutine MAX, used in this program, is in Figure 3.1-3.

This program, like most programs which find an inverse, also calculates the determinant of the matrix. This will be discussed in the section which follows.

3.3.8 Iterative Improvement of the Process

We shall merely sketch an iterative method for improving the Gauss-Jordan result. Let this result be $A_{(0)}^{-1}$. Then an estimate of its accuracy can be found by computing $A_{(0)}^{-1}A$. This should be I, but probably is not. We define an error matrix as

$$R = A^{-1}A - A_{(0)}^{-1}A \tag{3.3.67}$$

$$R = I - A_{(0)}^{-1}A \tag{3.3.68}$$

R can easily be computed. We must now make use of the identity

$$(I-R)(I+R+R^2+R^3+\cdots)=I \tag{3.3.69}$$

which is true when the summation in the left member converges. Then

$$(I+R+R^2+\cdots)=(I-R)^{-1} \tag{3.3.70}$$

Returning to (3.3.68), we have

$$I-R = A_{(0)}^{-1}A \tag{3.3.71}$$

$$A^{-1}=(I-R)^{-1}A_{(0)}^{-1} \tag{3.3.72}$$

This gives us a new value for A^{-1}, which we call $A_{(1)}^{-1}$. Using (3.3.70) for $(I-R)^{-1}$, we have

$$A_{(1)}^{-1}=(I+R+R^2+\cdots)A_{(0)}^{-1} \tag{3.3.73}$$

Under ordinary circumstances, R is quite small and the iteration can proceed using only $I+R+R^2$.

```
         DOUBLE PRECISION A(5,5),Z(5,5),C(5),P(5),DET,R
         COMMON A,L(5),K(5)
         R=1.D-6
C IT IN CC 1 IS THE NUMBER OF ROWS OR COLUMNS IN MATRIX A.
C IF INDEX = 1 IN CC 2, THE PROGRAM REPEATS.
C R IS USED TO TEST THE PIVOT.
         KOUNT=-1
1        READ(1,100)IT,INDEX
100      FORMAT(2I1)
C READ IN MATRIX A. ENTRIES IN CC 1,11,21,31,41 AT MOST.
         DO 2 J=1,IT
2        READ(1,101)(A(I,J),I=1,IT)
101      FORMAT(5F10.0)
C DISPLAY MATRIX.
         KOUNT=KOUNT+1
         WRITE(3,102)KOUNT
102      FORMAT(1H1,T10,'MATRIX FOR PROBLEM ',I2/)
         DO 3 I=1,IT
3        WRITE(3,103)(A(I,J),J=1,IT)
103      FORMAT(1H0,5F10.5)
C SAVE ORIGINAL DATA FOR CHECKING.
         WRITE(7,104)((A(I,J),I=1,IT),J=1,IT)
104      FORMAT(D23.16)
         REWIND 7
C SET IDENTITY MATRIX, Z.
         DO 4 I=1,IT
         DO 4 J=1,IT
4        Z(I,J)=0.D0
C SET ROW INDEXES,L(I).
         DO 7 I=1,IT
7        L(I)=1
C THE FOLLOWING LOOP IS THE G-J REDUCTION.
         DO 8 J=1,IT
C A SUBROUTINE FINDS THE PIVOT ROW FOR EACH COLUMN.
         CALL MAX(J,IT,M)
C K(J) KEEPS A RECORD OF THE PIVOT ORDER.  P(J) IS THE PIVOT.
         K(J)=M
         P(J) = A(M,J)
C IF THE PIVOT IS TOO SMALL, THE G-J METHOD IS INAPPLICABLE.
         IF(DABS(A(M,J)-R))9,9,10
C MESSAGE
9        WRITE(3,105)M,J
105      FORMAT(1H0,'ZERO PIVOT. METHOD INAPPLICABLE.',T60,'M=',I1,T70,'J='
        1I1)
         GO TO 25
C REDUCE ROW M, THE PIVOT ROW, STARTING IN COLUMN J+1.
10       JI=J+1
         DO 11 JP=JI,IT
11       A(M,JP)=A(M,JP)/A(M,J)
         Z(M,J)=1.D0
C REDUCE IDENTITY MATRIX, SKIPPING ZEROS. THIS IS THE PIVOT ROW.
         DO 12 JP=1,IT
         IF(Z(M,JP))13,12,13
13       Z(M,JP)=Z(M,JP)/A(M,J)
12       CONTINUE
C REDUCE OTHER ROWS, STARTING IN COLUMN JI.
         DO 8  IP=1,IT
C THE NEXT STEP CAUSES THE PIVOT ROW TO BE SKIPPED.
         IF(IP-M)14,8,14
14       DO 15 JP=JI,IT
15       A(IP,JP)=A(IP,JP)-A(M,JP)*A(IP,J)
         DO 16 JP=1,IT
16       Z(IP,JP)=Z(IP,JP)-Z(M,JP)*A(IP,J)
8        CONTINUE
C REARRANGEMENT OF THE SCRAMBLED Z MATRIX.
C READ THE SCRAMBLED Z MATRIX ON TOP OF THE A MATRIX.
         DO 17 I=1,IT
         DO 17 J=1,IT
```

```
17      A(I,J)=Z(I,J)
        WRITE(3,699)KOUNT
699     FORMAT(1H1,T10,'SCRAMBLED INVERSE, PROBLEM ',I2)
        DO 700 I=1,IT
700     WRITE(3,107)(A(I,J),J=1,IT)
C UNSCRAMBLE ROWS. K(I) GIVES THE PIVOT ROW ORDER. THE RESULT REPLACES Z.
        DO 18 I=1,IT
        M=K(I)
        DO 18 J=1,IT
18      Z(I,J)=A(M,J)
        WRITE(3,800)
800     FORMAT(1H0,//)
        WRITE(3,801)
801     FORMAT(1H0,T10,'ROWS UNSCRAMBLED',/)
        DO 701 I=1,IT
701     WRITE(3,107)(Z(I,J),J=1,IT)
C UNSCRAMBLE COLUMNS. THE RESULT REPLACES A
        DO 20 J=1,IT
        M=K(J)
        DO 20 I=1,IT
20      A(I,M)=Z(I,J)
C       PRINT INVERSE
        WRITE(3,106)KOUNT
        WRITE(3,800)
106     FORMAT(1H1,T20,'INVERSE, PROBLEM ',I2/)
        DO 21 I=1,IT
21      WRITE(3,107)(A(I,J),J=1,IT)
107     FORMAT(1H0,5F12.6)
C ORDER OF THE PIVOTS.
        WRITE(3,181)
181     FORMAT(1H0,///)
        WRITE(3,111)
111     FORMAT(1H0,T20,'ORDER OF PIVOTS'/)
        WRITE(3,500)(K(II),II=1,IT)
500     FORMAT(1H0,T20,'K',5I6,///)
C CHECKING BY MULTIPLICATION.
C READ ORIGINAL MATRIX BACK INTO CORE.  THIS REPLACES Z, NO LONGER NEEDED.
        READ(7,104)((Z(I,J),I=1,IT),J=1,IT)
        REWIND 7
C CALCULATION OF THE DETERMINANT
C NUMBER OF INVERSIONS = KT
        KT=0
        N=1
28      NP=N+1
30      IF(K(N)-K(NP))32,32,31
31      KT=KT+1
32      IF(NP-IT)38,27,27
38      NP=NP+1
        GO TO 30
27      IF(N-IT+1)33,29,29
33      N=N+1
        GO TO 28
C MULTIPLICATION OF PIVOTS
29      DET=1.0
        DO 34 I=1,IT
34      DET=DET*P(I)
C CORRECTION FOR SCRAMBLING OF ROWS
        DET=DET*(-1.D0)**KT
C   PRINT DETERMINANT
        WRITE(3,1000)
1000    FORMAT(1H0,/)
37      WRITE(3,110)DET
110     FORMAT(1H0,T25,'THE DETERMINANT IS',F20.6)
C THE NEXT STEP CHECKS FOR RETURN INSTRUCTIONS.
25      IF(INDEX-1)26,1,26
26      STOP
        END
```

Fig. 3.3-1   A computer program for finding an inverse by unscrambling.

```
                        MATRIX FOR PROBLEM   0


            2.63000      5.21000     -1.69400      0.93800

            3.16000     -2.95000      0.81300     -4.21000

            5.36000      1.88C00     -2.15000     -4.95000

            1.34000      2.98000     -0.43200     -1.76800




                        SCRAMBLED INVERSE, PROBLEM   0


           -0.101907     0.049050    -0.000820     0.313294

           -0.557791     0.273537     0.543774     0.411965

           -0.113947     0.357076     0.314690    -0.240876

           -0.121836     0.286472     0.104258    -0.320774




                        ROWS UNSCRAMBLED


           -0.113947     0.357076     0.314690    -0.240876

           -0.101907     0.049050    -0.000820     0.313294

           -0.557791     0.273537     0.543774     0.411965

           -0.121836     0.286472     0.104258    -0.320774




                        INVERSE, PROBLEM   0


            0.357076     0.314690    -0.113947    -0.240876

            0.049050    -0.000820    -0.101907     0.313294

            0.273537     0.543774    -0.557791     0.411965

            0.286472     0.104258    -0.121836    -0.320774




        ORDER OF PIVOTS

        K       3       1       2       4




            THE DETERMINANT IS           -105.719285
```

Fig. 3.3-2 Output of the matrix-inverse program.

3.3.9 Solving Matrix Equations by Inverses

We do not recommend the following procedure for solving linear systems. It is mentioned as a matter of interest. If

$$AX = C \qquad (3.3.74)$$

then

$$A^{-1}AX = A^{-1}C \qquad (3.3.75)$$

$$X = A^{-1}C \qquad (3.3.76)$$

This shows that a system can be solved by finding A^{-1}, then pre-multiplying C, the column of constants. The Gauss-Jordan procedure previously displayed actually does the same thing without the intermediate step of writing the inverse.

PROBLEM SECTION 3.3

A. Find the resulting matrix without multiplying:

(1)

$$\begin{pmatrix} 0 & 1 & 0 \\ 1 & 0 & 0 \\ 0 & 0 & 1 \end{pmatrix}\begin{pmatrix} 1 & 2 & 3 \\ 4 & 5 & 6 \\ 7 & 8 & 9 \end{pmatrix}\begin{pmatrix} 0 & 1 & 0 \\ 1 & 0 & 0 \\ 0 & 0 & 1 \end{pmatrix}$$

(2)

$$\begin{pmatrix} 1 & 0 & 0 \\ 0 & 0 & 1 \\ 0 & 1 & 0 \end{pmatrix}\begin{pmatrix} 1 & 2 & 3 \\ 4 & 5 & 6 \\ 7 & 8 & 9 \end{pmatrix}\begin{pmatrix} 1 & 0 & 0 \\ 0 & 0 & 1 \\ 0 & 1 & 0 \end{pmatrix}$$

(3)

$$\begin{pmatrix} 0 & 0 & 1 \\ 0 & 1 & 0 \\ 1 & 0 & 0 \end{pmatrix}\begin{pmatrix} 1 & 2 & 3 \\ 4 & 5 & 6 \\ 7 & 8 & 9 \end{pmatrix}\begin{pmatrix} 0 & 0 & 1 \\ 0 & 1 & 0 \\ 1 & 0 & 0 \end{pmatrix}$$

(4)

$$\begin{pmatrix} 0 & 0 & 1 \\ 1 & 0 & 0 \\ 0 & 1 & 0 \end{pmatrix}\begin{pmatrix} 1 & 2 & 3 \\ 4 & 5 & 6 \\ 7 & 8 & 9 \end{pmatrix}\begin{pmatrix} 0 & 0 & 1 \\ 1 & 0 & 0 \\ 0 & 1 & 0 \end{pmatrix}$$

B. Find a permutation matrix which will transform

$$A = \begin{bmatrix} 1 & 2 & 3 & 4 \\ 5 & 6 & 7 & 8 \\ 9 & 10 & 11 & 12 \\ 13 & 14 & 15 & 16 \end{bmatrix}$$

into the matrix shown, by using HAH:

(5)   (7)

$$\begin{bmatrix} 1 & 3 & 2 & 4 \\ 9 & 7 & 6 & 8 \\ 5 & 11 & 10 & 12 \\ 13 & 15 & 14 & 16 \end{bmatrix} \qquad \begin{bmatrix} 11 & 10 & 9 & 12 \\ 7 & 6 & 5 & 8 \\ 3 & 2 & 1 & 4 \\ 15 & 14 & 13 & 16 \end{bmatrix}$$

(6)   (8)

$$\begin{bmatrix} 1 & 4 & 3 & 2 \\ 13 & 16 & 15 & 14 \\ 9 & 12 & 11 & 10 \\ 5 & 8 & 7 & 6 \end{bmatrix} \qquad \begin{bmatrix} 1 & 2 & 4 & 3 \\ 5 & 6 & 8 & 7 \\ 13 & 14 & 16 & 15 \\ 9 & 10 & 12 & 11 \end{bmatrix}$$

C. Find J^{-1}:

(9) $J = H_{21}H_{32}H_{41}$ (11) $J = H_{27}H_{36}H_{45}H_{23}$
(10) $J = H_{13}H_{14}H_{25}H_{35}$ (12) $J = H_{12}H_{21}H_{13}H_{25}$

D. Find inverses of the following matrices:

(13)

$$\begin{pmatrix} 6.92600 & 2.05500 & -6.37800 \\ 7.75800 & 6.91200 & -9.70100 \\ 5.96400 & 9.88500 & 8.14300 \end{pmatrix}$$

(14)

$$\begin{pmatrix} 0.80750 & -1.56800 & 7.52300 \\ 5.59400 & 4.46900 & -0.47560 \\ -9.65800 & 9.90000 & -6.25800 \end{pmatrix}$$

(15)

$$\begin{pmatrix} 3.30680 & -1.42110 & 1.04850 \\ -3.50970 & 4.27700 & -7.11820 \\ 3.68380 & -7.72300 & -6.59140 \end{pmatrix}$$

(16)

$$\begin{pmatrix} -2.14590 & 8.73770 & 0.86200 \\ 4.87110 & 7.36100 & -5.27280 \\ -7.94360 & -3.20970 & 9.13480 \end{pmatrix}$$

E. Find inverses of the following matrices:

(17)

$$\begin{bmatrix} 6.07960 & 2.83340 & 1.34860 & 0.19080 \\ 7.66390 & 1.53680 & 4.69180 & 4.77960 \\ 3.01570 & 4.24810 & 6.46830 & 6.57960 \\ 4.02950 & 6.03120 & 0.74110 & 4.42300 \end{bmatrix}$$

(18)

$$\begin{bmatrix} 3.49140 & 2.25140 & 2.81050 & 0.34830 \\ 9.45020 & 0.40600 & 0.48140 & 5.73150 \\ 3.93740 & 9.45110 & 8.51700 & 6.31740 \\ 3.41850 & 4.46120 & 8.64900 & 7.19020 \end{bmatrix}$$

(19)

$$\begin{bmatrix} 5.92310 & 7.14940 & 8.74370 & 2.59860 \\ 4.50280 & 9.54010 & 8.27580 & 4.60050 \\ 0.11730 & 3.40490 & 7.10930 & 4.28400 \\ 0.88480 & 0.48510 & 3.68330 & 8.16830 \end{bmatrix}$$

(20)

$$\begin{bmatrix} 2.90460 & 4.36440 & 6.20350 & 1.03690 \\ 0.13010 & 4.62480 & 7.18860 & 4.20540 \\ 5.53430 & 5.32050 & 9.45060 & 6.82570 \\ 6.57320 & 9.48680 & 1.52630 & 1.43850 \end{bmatrix}$$

3.4 DETERMINANTS

3.4.1 An Operational Definition

At one time, before computers, it was considered appropriate to solve systems of linear equations by the use of determinants (which we shall presently define). As a result, a large body of literature arose, including different, but compatible, definitions and many interesting theorems. In the computer age, the use of determinants for solving systems—except for systems of two, or perhaps three, linear equations—is not practical. However, the study of determinants, including their evaluation, is still important for other purposes. Consequently, we shall start with one of the many definitions and develop just enough theory to explain how to evaluate a determinant by the Gauss-Jordan method. Then we shall apply this briefly to the solution of a small system, although, as we said, that is no longer an important practical use of determinants.

Consider the general $n \times n$ square matrix

$$A = \begin{bmatrix} a_{11} & a_{12} & a_{13} & \cdots & a_{1n} \\ a_{21} & a_{22} & a_{23} & \cdots & a_{2n} \\ \vdots & \vdots & \vdots & & \vdots \\ a_{n1} & a_{n2} & a_{n3} & \cdots & a_{nn} \end{bmatrix} \tag{3.4.1}$$

A is merely a tabulation of numbers and has no numerical value as such. Let us confer a numerical value upon it by the following device:

1. Choose any element in row 1, any element in row 2 *not in the same column*, any element in row 3 *in a column different from the other two,*

and continue until you have a set of numbers chosen in such a way that each element is alone in its row and column. For example, in the following matrix,

$$\begin{bmatrix} a_{11} & a_{12} & a_{13} \\ a_{21} & a_{22} & a_{23} \\ a_{31} & a_{32} & a_{33} \end{bmatrix} \tag{3.4.2}$$

here are all the possibilities. We shall explain the symbols at the right in a moment.

POSSIBILITIES	PERMUTATION	
$\{a_{11}, a_{22}, a_{33}\}$	123	
$\{a_{11}, a_{23}, a_{32}\}$	132	
$\{a_{12}, a_{21}, a_{33}\}$	213	(3.4.3)
$\{a_{12}, a_{23}, a_{31}\}$	231	
$\{a_{13}, a_{21}, a_{32}\}$	312	
$\{a_{13}, a_{22}, a_{31}\}$	321	

You can see that *permutation* refers to the order of the second subscripts when the first subscripts are in natural order, $1, 2, 3, \cdots$.

2. Calculate the number, t, of *inversions* for the permutation. The number of inversions is the number of interchanges needed to bring the permutation back to natural order. For example, consider the permutation 321. We interchange 3 and 2, yielding 231. That's one inversion. Then we interchange 3 and 1, yielding 213. That's two inversions. Finally, we interchange 2 and 1, yielding 3 inversions. We say that $t = 3$ in this case. Actually, if the interchanges are done inefficiently, waste motions may cause t to be 5, 7, 9, etc., each waste motion having to be undone. Actually, we are only concerned with the *parity* of t, i.e., whether it is odd or even. Here is a table for the permutations in (3.4.3).

PERMUTATION	t	PARITY	INTERCHANGES	
123	0	even		
132	1	odd	2 and 3	
213	1	odd	1 and 2	
231	2	even	3 and 1, then 1 and 2	(3.4.4)
312	2	even	3 and 1, then 3 and 2	
321	3	odd	3 and 2, then 3 and 1, then 2 and 1	

Before proceeding, we mention an easier way to count inversions. Proceed from left to right along the permutation, counting, each time, how many digits *to the right* are less than the one being considered. Here is a diagrammatic explanation for two cases:

$$
\begin{array}{ccccc}
2 & 3 & 1 & \qquad & 3 & 2 & 1 \\
\downarrow & \downarrow & & , & \downarrow & \downarrow & \\
1 & + \ 1 & = \ 2 & & 2 & + \ 1 & = \ 3
\end{array}
\qquad (3.4.5)
$$

In the case of *231*, there is one digit to the right of the *2* that is less than 2, and one digit to the right of *3* that is less than 3. In the case of *321*, there are two digits to the right of the *3* that are less than 3, and one digit to the right of the *2* that is less than 2. For the permutation 53124, the number of inversions, by the same method, is $4+2+0+0=6$.

3. We form the product of the elements chosen in step 1 and append to each a $+$ if the number of inversions is even and a $-$ if it is odd.

For the matrix (3.4.2) with products from (3.4.3) and parity from (3.4.4), we obtain the sum

$$
(a_{11}a_{22}a_{33}+a_{12}a_{23}a_{31}a_{13}a_{21}a_{32})-(a_{11}a_{23}a_{32}+a_{12}a_{21}a_{33}+a_{13}a_{22}a_{31}) \quad (3.4.6)
$$

which is defined as the determinant of matrix (3.4.2). Symbols for the determinant of matrix A are $|A|$, $D(A)$ or det (A); also we write

$$
|A| = \begin{vmatrix} a_{11} & a_{12} & a_{13} \\ a_{21} & a_{22} & a_{23} \\ a_{31} & a_{32} & a_{33} \end{vmatrix}
\qquad (3.4.7)
$$

Using the symbol Σ to mean *the sum of*, the general definition of a determinant can be written as follows.

DETERMINANT The determinant of a square matrix A is

$$
|A| = \Sigma(-1)^t a_{1i_1} a_{2i_2} a_{3i_3} \cdots a_{ni_n}
\qquad (3.4.8)
$$

where the summation is taken over all i from 1 to n, and t is the number of inversions.

ILLUSTRATIVE PROBLEM 3.4-1 One contribution in the calculation of a matrix turns out to be $a_{23}a_{42}a_{31}a_{15}a_{54}$. (a) Find $i_1, i_2, \ldots, i_5$. (b) Find t. (c) Is a $+$ or a $-$ adjoined?

Solution (a) We rearrange the terms so that the first subscripts are in natural order: $a_{15}a_{23}a_{31}a_{42}a_{54}$. Then $i_1 = 5$, $i_2 = 3$, etc. (b) As explained previously, $t = 6$. (c) Because t is even, a $+$ should be appended. In the formula (3.4.8), $(-1)^6 = +1$, which has the same effect.

ILLUSTRATIVE PROBLEM 3.4-2 Find $\begin{vmatrix} a_{11} & a_{12} \\ a_{21} & a_{22} \end{vmatrix}$.

Solution The pairs are $\{a_{11}a_{22}\}$ with permutation $(1,2)$ and parity 0, and $\{a_{12}, a_{21}\}$ with permutation $(2,1)$ and parity 1. Therefore, the determinant is $a_{11}a_{22}-a_{12}a_{21}$.

ILLUSTRATIVE PROBLEM 3.4-3   Evaluate $\begin{vmatrix} 1 & 2 \\ -3 & 4 \end{vmatrix}$.

Solution   Using the previous problem, this is $(1)(4)-(2)(-3)=4+6=10$.

Except for very small matrices (e.g., 2×2 and perhaps 3×3), the definition is not very useful. It is chiefly useful for theoretical work.

3.4.2   Elementary Row Operations on Determinants

Before generalizing, we shall illustrate several theorems by use of 2×2 determinants.

1. What is the effect of interchanging two rows? Compare

$$|A|=\begin{vmatrix} a & b \\ c & d \end{vmatrix} \quad \text{and} \quad |B|=\begin{vmatrix} c & d \\ a & b \end{vmatrix}$$

   Then $|A|=ad-bc$, and $|B|=bc-ad$. We shall prove that if two rows of any determinant are interchanged, the sign of the determinant is changed.
2. What is the effect of multiplying a row by a constant? Compare

$$|A|=\begin{vmatrix} a & b \\ c & d \end{vmatrix} \quad \text{and} \quad |B|=\begin{vmatrix} ka & kb \\ c & d \end{vmatrix}$$

   Then $|A|=ad-bc$, and $|B|=kad-kbc=k|A|$. We shall prove that if a row of a determinant is multiplied by a constant, the determinant is multiplied by that constant.
3. What is the value of $|A|$ if two rows are equal? Consider

$$|A|=\begin{vmatrix} a & b \\ a & b \end{vmatrix}=ab-ab=0$$

   We shall prove that any determinant with two equal rows has a value of 0.
4. What is the effect of adding a multiple of one row to another row? Consider

$$|A|=\begin{vmatrix} a & b \\ c & d \end{vmatrix} \quad \text{and} \quad |B|=\begin{vmatrix} a+kc & b+kd \\ c & d \end{vmatrix}$$

   Then $|A|=ad-bc$, and $|B|=(a+kc)d-(b+kd)c=ad+kcd-bc-kcd=|A|$. We shall prove that in any determinant, addition of a multiple of one row to another row leaves the value of the determinant unchanged.

3.4.3   Four Theorems on Determinants

ROW-INTERCHANGE THEOREM   If any two rows of a determinant are interchanged, the sign of the determinant is changed.

Proof By definition,

$$|A| = \Sigma(-1)^t a_{1i_1} a_{2i_2} \cdots a_{pi_p} \cdots a_{qi_q} \cdots a_{ni_n}$$

Now consider

$$|B| = \Sigma(-1)^{t'} a_{12_1} a_{2i_2} \cdots a_{qi_q} \cdots a_{pi_p} \cdots a_{ni_n}$$

where rows p and q have been interchanged. Except for t and t', the contributions are precisely the same because the product is made up of numbers, which are associative and commutative. The difference between t and t' is exactly 1, because there is one more interchange in B than in A. Therefore, whenever t is odd, t' will be even and vice versa. This completes the proof. □

Row-Multiplication Theorem   If any row of a determinant is multiplied by k, the determinant is multiplied by k.

Proof By definition,

$$|A| = \Sigma(-1)^t a_{1i_1} a_{2i_2} \cdots a_{pi_p} \cdots a_{ni_n}$$

Now consider

$$|B| = \Sigma(-1)^t a_{1i_1} a_{2i_2} \cdots k a_{pi_p} \cdots a_{ni_n}$$

where each element in the pth row has been multiplied by k. Since this is a sum with k in each term, the k can be factored. Then

$$|B| = k \sum (-1)^t a_{1i_1} a_{2i_2} \cdots a_{pi_p} \cdots a_{ni_n} = k|A| \quad □$$

Equal-Rows Theorem   If two rows of a determinant are equal, the determinant is 0.

Proof Let $|A|$ be a determinant with two equal rows. Now interchange the equal rows. By the row-interchange theorem, this results in $-|A|$. However, the value of a determinant certainly cannot be changed by interchanging two equal rows. Therefore $|A| = -|A|$, from which $|A|$ must equal 0, since zero is the only number which equals its own negative. □

Row Addition Theorem   If a multiple of one row is added to another row of a determinant, the value is unchanged.

Proof By definition,

$$|A| = \Sigma(-1)^t a_{1i_1} a_{2i_2} \cdots a_{pi_p} \cdots a_{qi_q} \cdots a_{ni_n}$$

Now consider

$$|B| = \Sigma(-1)^t a_{1i_1} a_{2i_2} \cdots \left(a_{pi_p} + k a_{qi_q}\right) \cdots a_{qi_q} \cdots a_{ni_n}$$

where k times an element of the qth row has been added to the corresponding element of the pth row.

Before continuing, let us take a specific example for $|B|$ and expand it according to the definition. Let

$$|B| = \begin{vmatrix} a & b & c \\ d+3g & e+3h & f+3i \\ g & h & i \end{vmatrix}$$

$$= \left[a(e+3h)i + b(f+3i)g + c(d+2q)h \right]$$

$$\quad - \left[a(f+3i)h + b(d+3q)i + c(e+3h)g \right]$$

$$= \left[aei + 3ahi + bfg + 3big + cdh + 3cgh \right]$$

$$\quad - \left[afh + 3aih + bdi + 3bgi + ceg + 3chg \right]$$

$$= \left[(aei + bfg + cdh) - (afh + bdi + ceg) \right]$$

$$\quad + \left[(3ahi + 3big + cgh) - (3aih + 3bgi + 3chg) \right]$$

Note that the second set of sums vanishes.

What we are pointing out is that in the expansion we have obtained six contributions, each of which contains a sum. These can be separated into two sets, as we have done. Returning to the original proof, we do the same, expanding $|B|$ into the sums

$$\Sigma(-1)^t a_{1i_1} a_{2i_2} \cdots a_{pi_p} \cdots a_{qi_q} \cdots a_{ni_n} + k\Sigma(-1)^t a_{1i_1} a_{2i_2} \cdots a_{qi_q} \cdots a_{qi_q} \cdots a_{ni_n}$$

The second summation has two equal rows and is therefore 0. The first summation is the same as $|A|$, completing the proof. $\square$

3.4.4 The Principal-Diagonal Theorem

Consider

$$|A| = \begin{vmatrix} a_{11} & 0 & 0 & \cdots & 0 \\ 0 & a_{22} & 0 & \cdots & 0 \\ 0 & 0 & a_{33} & \cdots & 0 \\ \vdots & \vdots & \vdots & & \vdots \\ 0 & 0 & 0 & \cdots & a_{nn} \end{vmatrix}$$

where all the elements are 0 except those in the principal diagonal. Then it is obvious that

$$|A| = a_{11}a_{22}a_{33} \cdots a_{nn}$$

since every other contribution will have at least one zero as a factor. □
We call this a *diagonal determinant*.

PRINCIPAL-DIAGONAL THEOREM.   The value of a diagonal determinant is the product of the elements of the principal diagonal.

Corollary   *The value of the determinant of an identity matrix is* 1.

3.4.5 The Gauss-Jordan Method for a Determinant

We shall introduce the method by two examples.

Case I.

$$|A| = \begin{vmatrix} 4 & 3 \\ 2 & 1 \end{vmatrix}$$

By elementary methods, $|A| = (4)(1) - (3)(2) = -2$. Now let us go through a Gauss-Jordan reduction and see what happens to the determinant.

First, we divide the first row by 4, which also divides the determinant by 4. Then

$$\tfrac{1}{4}|A| = \begin{vmatrix} 1 & \tfrac{3}{4} \\ 2 & 1 \end{vmatrix}$$

Then we add -2 times the first row to the second row. This should not change the value of the determinant:

$$\tfrac{1}{4}|A| = \begin{vmatrix} 1 & \tfrac{3}{4} \\ 0 & -\tfrac{1}{2} \end{vmatrix}$$

Now we multiply the second row by -2. This multiplies the determinant by -2. Therefore,

$$(-2)(\tfrac{1}{4})|A| = \begin{vmatrix} 1 & \tfrac{3}{4} \\ 0 & 1 \end{vmatrix}$$

Now we add $-\tfrac{3}{4}$ times the second row to the first row. This does not change the determinant:

$$(-2)(\tfrac{1}{4})|A| = \begin{vmatrix} 1 & 0 \\ 0 & 1 \end{vmatrix} = 1$$

$$|A| = (1)(4)(-\tfrac{1}{2}) = -2.$$

Note that $|A|$ is the product of the pivots.

Case II.

$$|A| = \begin{vmatrix} -6 & 7 \\ 7 & -8 \end{vmatrix}$$

From the definition, $|A| = -1$. If we go through the Gauss-Jordan process properly, beginning with a_{21} as pivot, we first divide the second row by 7 and continue as in case I. Unfortunately, the result is 1 instead of -1. (The steps are left as an exercise.) Why is this incorrect? The reason is that we actually interchanged rows 1 and 2 when we started with the second row. Therefore, by the row-interchange theorem, the result was multiplied by -1, explaining the discrepancy.

In these two cases, the Gauss-Jordan method was hardly an improvement over direct computation. We proceed to another case, which shows the power of the method.

ILLUSTRATIVE PROBLEM 3.4-4   Find the determinant of the matrix in Sec. 3.3.6.

Solution   The pivots were 5.360, 4.28753, 1.47564 and -3.11747. The order of rows was 3124, which means that there were 2 inversions. Therefore

$$|A| = (-1)^2(5.360)(4.28753)(1.47564)(-3.11747)$$
$$= -105.7193936$$

You have probably noticed that since only the pivots and the number of inversions are needed, the entire Gauss-Jordan process is not needed if one wishes the value of the determinant. However, the evaluation of a determinant on a computer is usually made the tag-end of the inversion of a matrix.

3.4.6 Transposes

We promised to say something about the solution of linear systems by determinants. Before doing so, we shall prove informally an important theorem in the theory of determinants.

> TRANSPOSE THEOREM   The value of a determinant is not changed if the rows and columns are interchanged.

Here is an example:

$$\begin{vmatrix} 1 & 0 & 2 \\ 0 & 3 & 4 \\ 5 & 6 & 0 \end{vmatrix} = -54, \qquad \begin{vmatrix} 1 & 0 & 5 \\ 0 & 3 & 6 \\ 2 & 4 & 0 \end{vmatrix} = -54$$

The two determinants are said to be *transposes* of each other. If the symbol for the first one is $|A|$, then the symbol for the second is $|A^T|$.

Informal Proof Because the value of a determinant is found as a sum of contributions formed by taking precisely one factor from each row and column, the contributions are the same when they are taken precisely once from each column and row. □

An immediate consequence of this theorem is that all the theorems on rows apply equally well to columns. For example, we have:

COLUMN-INTERCHANGE THEOREM   If any two columns of a determinant are interchanged, the sign of the determinant is changed.

COLUMN MULTIPLICATION THEOREM   If any column of a determinant is multiplied by k, the determinant is multiplied by k.

EQUAL-COLUMNS THEOREM   If two columns of a determinant are equal, the determinant equals 0.

COLUMN ADDITION THEOREM   If a multiple of one column is added to another column of a determinant, the value is unchanged.

3.4.7 Cramer's Rule

Consider the system

$$a_{11}x_1 + a_{12}x_2 + \ldots + a_{1k}x_k + \ldots + a_{1n}x_n = c_1$$
$$a_{21}x_1 + a_{22}x_2 + \ldots + a_{2k}x_k + \ldots + a_{in}x_n = c_2$$
$$\vdots$$
$$a_{n1}x_1 + a_{n2}x_2 + \ldots + a_{nk}x_k + \ldots + a_{nn}x_n = c_n$$

We can write this as a matrix equation in the usual way:

$$AX = C \qquad (3.4.9)$$

where A is the matrix of coefficients, X is a column vector for the xs and C is a column vector for the constants. We shall find it convenient to express matrix A as a set of column vectors:

$$A = (A^1, A^2, \ldots, A^k, \ldots, A^n) \qquad (3.4.10)$$

where the superscripts denote the column number. For example, the first column is

$$A^1 = \begin{bmatrix} a_{11} \\ a_{21} \\ \vdots \\ a_{k1} \\ \vdots \\ a_{n1} \end{bmatrix}$$

Then, in place of (3.4.9), we could write

$$\begin{pmatrix} A^1 & A^2 & \dots & A^k & \dots & A^n \end{pmatrix} \begin{bmatrix} x_1 \\ x_2 \\ \vdots \\ x_k \\ \vdots \\ x_n \end{bmatrix} = C \qquad (3.4.11)$$

from which

$$A^1 x_1 + A^2 x_2 + \dots + A^k x_k + \dots A^n x_n = C \qquad (3.4.12)$$

Now, consider the determinant

$$\begin{vmatrix} A^1 & A^2 & \dots & C & \dots & A^n \end{vmatrix} \qquad (3.4.13)$$

where C has replaced the kth column. Substituting from (3.4.12), this is equal to

$$\begin{vmatrix} A^1 & A^2 & \dots & (x_1 A^1 + x_2 A^2 + \dots + x_k A^k + \dots + x_n A^n) & \dots A^n \end{vmatrix}$$
$$(3.4.14)$$

By the column addition and multiplication theorems, this is equal to

$$x_1 |A^1 A^2 \dots A^1 \dots A^n| + x_2 |A^1 \quad A^2 \dots A^2 \quad \dots A^n|$$
$$+ \dots + x_k |A^1 \quad A^2 \dots A^k \quad \dots A^n| \qquad (3.4.15)$$
$$+ \dots + x_n |A^1 \quad A^2 \dots A^n \quad \dots A^n|$$

All of these but one have two equal columns and are therefore equal to 0. Thus the total value of (3.4.15) is

$$x_k \begin{vmatrix} A^1 & A^2 & \dots & A^k & \dots & A^n \end{vmatrix} \qquad (3.4.16)$$

which is the same as

$$x_k |A| \qquad (3.4.17)$$

Setting (3.4.13) and (3.4.17) equal, assuming $|A| \neq 0$, we have

$$x_k = \frac{\begin{vmatrix} A^1 & A^2 & \ldots & C & \ldots & A^n \end{vmatrix}}{|A|} \qquad (3.4.18)$$

which completes the proof of Cramer's rule. $\square$

THEOREM: CRAMER'S RULE   If $AX = C$ and $|A| \neq 0$, then the value of x_k is the fraction (3.4.18), where the denominator is the determinant of A and the numerator is the determinant formed from A by replacing the kth column with the column vector of constants.

ILLUSTRATIVE PROBLEM 3.4-5   Solve

$$1.2356x - 4.2382y = 17.2846$$
$$3.2639x + 2.4188y = 2.3562$$

Solution

$$x = \frac{\begin{vmatrix} 17.2846 & -4.2382 \\ 2.3562 & 2.4188 \end{vmatrix}}{\begin{vmatrix} 1.2356 & -4.2382 \\ 3.2639 & 2.4188 \end{vmatrix}} = \frac{51.79403732}{16.82173016} = 3.078995830$$

$$y = \frac{\begin{vmatrix} 1.2356 & 17.2846 \\ 3.2639 & 2.4188 \end{vmatrix}}{\begin{vmatrix} 1.2356 & -4.2382 \\ 3.2639 & 2.4188 \end{vmatrix}} = \frac{-53.50388522}{16.82173016} = -3.180641016$$

Note that the denominator is the same for x and y and need be calculated only once, no matter how many unknowns there are.

PROBLEM SECTION 3.4

Find the determinants of the matrices which were inverted in the problems for Section 3.3.

3.5 EIGENVALUE PROBLEMS

3.5.1 Background

If a column vector is pre-multiplied by a matrix, not the identity matrix, the result is a column vector which differs in size, in direction, or in both:

$$\begin{pmatrix} a_{11} & a_{12} \\ a_{21} & a_{22} \end{pmatrix} \begin{pmatrix} x_1 \\ x_2 \end{pmatrix} = \begin{pmatrix} a_{11}x_1 + a_{12}x_2 \\ a_{21}x_1 + a_{22}x_2 \end{pmatrix} \qquad (3.5.1)$$

We shall be interested in the case where there is no change in direction, i.e.,

$$AX = \lambda X \qquad (3.5.2)$$

The factor of stretching or shrinking, λ, is called an *eigenvalue* (or characteristic root, proper root, latent root), and the corresponding value of the vector X is called the eigenvector (or characteristic vector, proper vector, latent vector). The alternative names come from practical problems in electrical theory, mechanical theory, quantum mechanics and statistical analysis. A useful computer program must be able to find eigenvectors and eigenvalues of a matrix with hundreds of rows and columns.

Equation (3.5.2) can be written as

$$AX = \lambda IX$$
$$(A - \lambda I)X = 0 \qquad (3.5.3)$$

The matrix $A - \lambda I$ appears as follows:

$$\begin{bmatrix} a_{11}-\lambda & a_{12} & a_{13} & a_{14} & \cdots \\ a_{21} & a_{22}-\lambda & a_{23} & a_{24} & \cdots \\ a_{31} & a_{32} & a_{33}-\lambda & a_{34} & \cdots \\ \vdots & \vdots & \vdots & \vdots & \end{bmatrix} \qquad (3.5.4)$$

and the result of the multiplication (3.5.3) is a set of *homogeneous linear equations*:

$$\begin{aligned} (a_{11}-\lambda)x_1 & + a_{12}x_2 & + a_{13}x_3 + \cdots &= 0 \\ a_{21}x_1 + (a_{22}-\lambda)x_2 & & + a_{23}x_3 + \cdots &= 0 \qquad (3.5.5) \\ a_{31}x_1 & + a_{32}x_2 + (a_{33}-\lambda)x_3 + \cdots &= 0 \\ & & \vdots & \end{aligned}$$

Using Cramer's rule (see section 3.4.7), we see that the determinant of the denominator, namely, the determinant of matrix (3.5.5), must vanish if there is to be a non-trivial solution, i.e., a solution other than $x_1 = x_2 = \cdots = 0$.

3.5.2 Calculation from the Definition

We shall start with a hand calculation, mainly to reenforce the definitions.

ILLUSTRATIVE PROBLEM 3.5.1   Find the eigenvalues and eigenvectors for

$$A = \begin{pmatrix} 2 & 1 & 1 \\ 1 & 2 & 1 \\ 0 & 0 & 1 \end{pmatrix} \qquad (3.5.6)$$

Solution   From (3.5.3), we have

$$\begin{pmatrix} 2-\lambda & 1 & 1 \\ 1 & 2-\lambda & 1 \\ 0 & 0 & 1-\lambda \end{pmatrix} \begin{pmatrix} x_1 \\ x_2 \\ x_3 \end{pmatrix} = \begin{pmatrix} 0 \\ 0 \\ 0 \end{pmatrix} \qquad (3.5.7)$$

For non-trivial solutions,

$$\begin{vmatrix} 2-\lambda & 1 & 1 \\ 1 & 2-\lambda & 1 \\ 0 & 0 & 1-\lambda \end{vmatrix} = 0 \qquad (3.5.8)$$

from which

$$\lambda^3 - 5\lambda^2 + 7\lambda - 3 = 0. \qquad (3.5.9)$$

This equation is called the *characteristic equation*. This equation must have three roots, since it is a cubic; but these may be real or complex, distinct or multiple. In this case, we can factor:

$$(\lambda - 1)(\lambda - 1)(\lambda - 3) = 0 \qquad (3.5.10)$$

$$\lambda = 1, 1, 3 \qquad (3.5.11)$$

We have two distinct eigenvalues for the matrix A, of which one is a double root. We note in passing that the sum of the three eigenvalues in (3.5.11) is 5, and this agrees with the *trace* of the matrix A in (3.5.6). (The trace is the sum of the elements in the principal diagonal.) It can be proved that this is always so.

To find the eigenvectors, we substitute each value of λ in (3.5.7). For $\lambda = 1$,

$$\begin{pmatrix} 1 & 1 & 1 \\ 1 & 1 & 1 \\ 0 & 0 & 0 \end{pmatrix} \begin{pmatrix} x_1 \\ x_2 \\ x_3 \end{pmatrix} = \begin{pmatrix} 0 \\ 0 \\ 0 \end{pmatrix} \qquad (3.5.12)$$

from which

$$x_1 + x_2 + x_3 = 0. \qquad (3.5.13)$$

This has an infinite set of solutions. Arbitrarily, we choose $x_1 = 1$. Then x_2 can equal -1, whence $x_3 = 0$. One possible eigenvector is then $\begin{pmatrix} 1 \\ -1 \\ 0 \end{pmatrix}$. Note that $\begin{pmatrix} c \\ -c \\ 0 \end{pmatrix}$ satisfies (3.5.12) equally well. If we start with $x_2 = 1$, we can arrive at $\begin{pmatrix} 0 \\ 1 \\ -1 \end{pmatrix}$. If we start with $x_3 = 1$, we can arrive at $\begin{pmatrix} -1 \\ 0 \\ 1 \end{pmatrix}$. However, the three possibilities are not independent, since

$$\begin{pmatrix} 1 \\ -1 \\ 0 \end{pmatrix} + \begin{pmatrix} -1 \\ 0 \\ 1 \end{pmatrix} = \begin{pmatrix} 0 \\ -1 \\ 1 \end{pmatrix} = -\begin{pmatrix} 0 \\ 1 \\ -1 \end{pmatrix} \qquad (3.5.14)$$

We arbitrarily choose any two of the possible *independent* solutions and write the eigenvector as follows:

$$c_1 \begin{pmatrix} 1 \\ -1 \\ 0 \end{pmatrix} + c_2 \begin{pmatrix} 0 \\ 1 \\ -1 \end{pmatrix} \tag{3.5.15}$$

For $\lambda = 3$, using the same technique, we arrive at

$$-x_1 + x_2 + x_3 = 0 \tag{3.5.16}$$

from which some possibilities are

$$\begin{pmatrix} 1 \\ 1 \\ 0 \end{pmatrix}, \begin{pmatrix} 0 \\ 1 \\ -1 \end{pmatrix}, \begin{pmatrix} 1 \\ 0 \\ 1 \end{pmatrix}$$

Each eigenvalue contributes *one* eigenvector. We must choose one of these that is independent of (3.5.15). Our result is

$$V = c_1 \begin{pmatrix} 1 \\ -1 \\ 0 \end{pmatrix} + c_2 \begin{pmatrix} 0 \\ 1 \\ -1 \end{pmatrix} + c_3 \begin{pmatrix} 1 \\ 0 \\ 1 \end{pmatrix} \tag{3.5.17}$$

where c_1, c_2 and c_3 are any constants.

There are various conventions about the expression of an eigenvector. For example, some people adjust the components to make the largest one equal to 1. Others adjust them so that the square root of the sum of their squares is 1. Still others leave them as they come out of the calculation. From a purely mathematical point of view it makes no difference unless answers are being compared, and then there must be some sort of agreement.

3.5.3 The Method of Jacobi

The hand method illustrated above is obviously not suited for large matrices, and certainly not to a computer algorithm. At this time, there are two excellent methods in use: the method of Givens, and the method of Householder. Both these methods yield all the eigenvalues of a real symmetric matrix. In addition, there are some special methods which give only the *largest* eigenvalue. We shall concentrate on the methods which give all the eigenvalues. It will be easier to explain the two current methods if we pause to comment on an older method, that of Jacobi.

Most practical problems deal with *real symmetric* matrices, i.e., matrices in which the elements are real and in which

$$A^T = A \tag{3.5.18}$$

Here A^T is the *transpose* of A, i.e., the matrix with rows and columns interchanged. An example of a real symmetric matrix is

$$\begin{bmatrix} 1 & 4 & -5 \\ 4 & 2 & 6 \\ -5 & 6 & 3 \end{bmatrix}$$

For this type of matrix, there always exists another matrix, B, such that

$$BB^T = I, \quad \text{or} \quad B^T = B^{-1} \tag{3.5.19}$$

and also that

$$B^{-1}AB = D \tag{3.5.20}$$

where D is a *diagonal matrix*, a matrix in which all the elements off the principal diagonal are 0. (We shall demonstrate.) Any matrix which satisfies (3.5.19) is said to be *orthogonal*. The pre- and post-multiplication in (3.5.20) is called a *similarity transformation*. We shall now prove that A and D in (3.5.20) have the same eigenvalues.

THEOREM   Eigenvalues are preserved under a similarity transformation.

Proof Let

$$C = B^{-1}AB$$

Then

$$BC = (BB^{-1})AB = AB \quad \text{and} \quad BCB^{-1} = A(BB^{-1}) = A$$

Then

$$
\begin{aligned}
A - \lambda I &= BCB^{-1} - \lambda I \\
&= BCB^{-1} - \lambda BB^{-1} \\
&= B(CB^{-1} - \lambda B^{-1}) \\
&= B(C - \lambda I)B^{-1} \\
\therefore \quad |A - \lambda I| &= |B(C - \lambda I)B^{-1}| \\
&= |BB^{-1}| \cdot |C - \lambda I| \\
&= |I| \cdot |C - \lambda I| = |C - \lambda I|
\end{aligned}
$$

Since the determinants are equal, they lead to the same characteristic equation, and the eigenvalues must be equal. □

How do we find the eigenvalues of the diagonal matrix? Consider

$$D = \begin{pmatrix} d_{11} & 0 & 0 \\ 0 & d_{22} & 0 \\ 0 & 0 & d_{33} \end{pmatrix} \tag{3.5.21}$$

To find the eigenvalues, we write

$$|D - \lambda I| = \begin{vmatrix} d_{11} - \lambda & 0 & 0 \\ 0 & d_{22} - \lambda & 0 \\ 0 & 0 & d_{33} - \lambda \end{vmatrix} = 0 \tag{3.5.22}$$

from which

$$(\lambda - d_{11})(\lambda - d_{22})(\lambda - d_{33}) = 0 \tag{3.5.23}$$

$$\lambda = d_{11}, d_{22}, d_{33} \tag{3.5.24}$$

The argument is easily generalized to prove

THEOREM   The principal diagonal elements of a diagonal matrix are its eigenvalues.

We shall need one more definition before describing the method of Jacobi:

ROTATION MATRIX   An identity matrix in which any four elements at the vertices of a rectangle have been replaced by $\cos\theta, -\sin\theta, \sin\theta, \cos\theta$ in positions pp, pq, qq, qp.

An example is

$$B = \begin{pmatrix} 1 & 0 & 0 & 0 & 0 \\ 0 & \cos\theta & 0 & -\sin\theta & 0 \\ 0 & 0 & 1 & 0 & 0 \\ 0 & \sin\theta & 0 & \cos\theta & 0 \\ 0 & 0 & 0 & 0 & 1 \end{pmatrix}$$

Here $p = 2$ and $q = 4$. We leave as an exercise the demonstration that $B^T B = I$, so that $B^T = B^{-1}$.

We return to the method of Jacobi. Our purpose is to pre- and post-multiply a real symmetric matrix A in such a way that the result is a diagonal matrix. Then the elements in the principal diagonal will be the eigenvalues. The problem is to find a B which behaves like (3.5.20).

Jacobi's solution was to form B as a product of rotation matrices. We illustrate with a 3×3 matrix, in which $c = \cos\theta$ and $s = \sin\theta$:

$$\begin{bmatrix} c & s & 0 \\ -s & c & 0 \\ 0 & 0 & 1 \end{bmatrix} \begin{bmatrix} a_{11} & a_{12} & a_{13} \\ a_{21} & a_{22} & a_{23} \\ a_{31} & a_{32} & a_{33} \end{bmatrix} \begin{bmatrix} c & -s & 0 \\ s & c & 0 \\ 0 & 0 & 1 \end{bmatrix} \tag{3.5.25}$$

Performing the multiplication, and using the fact that the A matrix is symmetric (i.e., $a_{ij} = a_{ji}$), we obtain

$$
\begin{bmatrix}
a_{11}c^2 + 2a_{12}sc + a_{22}s^2 & (a_{22} - a_{11})sc + a_{12}(c^2 - s^2) & a_{13}c + a_{23}s \\
(a_{22} - a_{11})sc + a_{12}(c^2 - s^2) & a_{22}c^2 - 2a_{12}sc + a_{11}s^2 & a_{23}c - a_{13}s \\
a_{13}c + a_{32}s & a_{23}c - a_{13}s & a_{33}
\end{bmatrix}
\tag{3.5.26}
$$

As a start towards diagonalization, we let the element in row 2, column 1 become 0. Then

$$
(a_{22} - a_{11})sc + a_{12}(c^2 - s^2) = 0 \tag{3.5.27}
$$

that is,

$$
(a_{22} - a_{11})\sin\theta\cos\theta = -a_{12}(\cos^2\theta - \sin^2\theta) \tag{3.5.28}
$$

$$
(a_{22} - a_{11})\tfrac{1}{2}\sin 2\theta = -a_{12}\cos 2\theta \tag{3.5.29}
$$

$$
\tan 2\theta = \frac{2a_{12}}{a_{11} - a_{22}} \tag{3.5.30}
$$

More generally, if the substitutions are made so that

$$
b_{pp} = \cos\theta, \quad b_{pq} = -\sin\theta, \quad b_{qq} = \cos\theta, \quad b_{qp} = \sin\theta, \tag{3.5.31}
$$

then

$$
\tan 2\theta = \frac{2a_{pq}}{a_{pp} - a_{qq}} \tag{3.5.32}
$$

If we use the usual trigonometric identities, we obtain the following, where $R_1, R_2, \ldots, R_5$ have been defined for calculating convenience:

$$
R_1 = a_{pp} - a_{qq} \tag{3.5.33}
$$

$$
R_2 = \sqrt{R_1^2 + 4a_{pq}^2} \tag{3.5.34}
$$

$$
R_3 = R_1/R_2 \; (= \cos 2\theta) \tag{3.5.35}
$$

$$
R_4 = \sqrt{2(1 + R_3)} \tag{3.5.36}
$$

$$
R_5 = 2a_{pq}/R_2 \; (= \sin 2\theta) \tag{3.5.37}
$$

$$
s = |R_5|/R_4 \tag{3.5.38}
$$

$$
c = \pm(1 + R_3)/R_4 \tag{3.5.39}
$$

The sign of c is chosen to agree with the sign of a_{pq}.

When matrix A is pre- and post-multiplied as in (3.5.25), using the values of s and c in (3.5.38) and (3.5.39), the off-diagonal elements a_{pq} and a_{qp}

are *annihilated*, except for possible roundoff. We illustrate:

ILLUSTRATIVE PROBLEM 3.5-2   Find the eigenvalues of

$$A = \begin{pmatrix} 1 & 2 & 0 \\ 2 & 2 & 2 \\ 0 & 2 & 3 \end{pmatrix}$$

Solution (partial).   We start with a_{12} and a_{21}, so that the rotation matrix is as shown in (3.5.25). Using (3.5.33)–(3.5.39), the result of the rotation is

$$\begin{pmatrix} 3.56157 & -0.00002 & 1.57642 \\ -0.00002 & -0.56156 & 1.23082 \\ 1.57642 & 1.23082 & 3 \end{pmatrix}$$

Note that the zeros in positions $(1,3)$ and $(3,1)$ have been spoiled.

We now shift to another pair of positions. To reduce error, it is desirable to annihilate the element largest in absolute value. We therefore choose the entry 1.57642. (That is where the s and $-s$ elements will operate.) The result of the rotation is

$$\begin{pmatrix} 4.88198 & 0.79033 & 0 \\ 0.79033 & -0.56156 & 0.94355 \\ 0 & 0.94355 & 1.67954 \end{pmatrix}$$

We have the zeros back in positions $(3,1)$ and $(1,3)$, but now the zeros in $(1,2)$ and in $(2,1)$ have been spoiled.

This is the fly in the ointment. Actually, by going through the matrix again and again until the maximum off-diagonal entry is small enough to be neglected, the eigenvalues can be found to a satisfactory degree of accuracy. The next method is a considerable improvement.

3.5.4   The Givens Reduction

The trouble with the method of Jacobi is that annihilation takes place at the corners of the rectangle, but every other element of the matrix is also altered. Therefore, as soon as the rectangle is moved, a previously gained zero is spoiled. The Givens reduction does not disturb zeros already formed. But there is a price—a trade-off. The Givens method does not lead to a pleasant diagonal matrix. It leads to a *tridiagonal matrix*. This means that all elements except those on the principal diagonal and its two neighbors are annihilated. (We shall illustrate.)

To introduce the method, consider the following, where A is given as real and symmetric:

$$\begin{bmatrix} 1 & 0 & 0 & 0 & 0 \\ 0 & c & s & 0 & 0 \\ 0 & -s & c & 0 & 0 \\ 0 & 0 & 0 & 1 & 0 \\ 0 & 0 & 0 & 0 & 1 \end{bmatrix} \begin{bmatrix} a_{11} & a_{12} & a_{13} & a_{14} & a_{15} \\ a_{21} & a_{22} & a_{23} & a_{24} & a_{25} \\ a_{31} & a_{32} & a_{33} & a_{34} & a_{35} \\ a_{41} & a_{42} & a_{43} & a_{44} & a_{45} \\ a_{51} & a_{52} & a_{53} & a_{54} & a_{55} \end{bmatrix} \begin{bmatrix} 1 & 0 & 0 & 0 & 0 \\ 0 & c & -s & 0 & 0 \\ 0 & s & c & 0 & 0 \\ 0 & 0 & 0 & 1 & 0 \\ 0 & 0 & 0 & 0 & 1 \end{bmatrix}$$

$$(3.5.40)$$

The result is

$$
\begin{bmatrix}
a_{11} & ca_{12}+sa_{13} & \boxed{ca_{13}-sa_{12}} & a_{14} & a_{15} \\
ca_{21}+sa_{31} & c^2a_{22}+2sca_{23}+s^2a_{33} & (c^2-s^2)a_{23}+sc(a_{33}-a_{22}) & ca_{24}+sa_{34} & ca_{25}+sa_{35} \\
\boxed{ca_{31}-sa_{21}} & (c^2-s^2)a_{23}+sc(a_{33}-a_{22}) & c^3a_{33}-2sca_{23}+s^2a_{22} & ca_{34}-sa_{24} & ca_{35}-sa_{25} \\
a_{41} & ca_{42}+sa_{43} & ca_{43}-sa_{42} & a_{44} & a_{45} \\
a_{51} & ca_{52}+sa_{53} & ca_{53}-sa_{22} & a_{54} & a_{55}
\end{bmatrix}
$$

$$(3.5.41)$$

We now annihilate the element in position $(1,3)$ by letting

$$
\begin{aligned}
ca_{13}-sa_{12} &= 0 \\
a_{12}\sin\theta &= a_{13}\cos\theta \\
\tan\theta &= a_{13}/a_{12}
\end{aligned}
$$

$$(3.5.42)$$

From this, defining

$$
S=\sqrt{(a_{12})^2+(a_{13})^2}
$$

we can write

$$
\sin\theta = |a_{13}|/S, \qquad \cos\theta = \pm\sqrt{1-\sin^2\theta} \tag{3.5.43}
$$

where the sign of the cosine is chosen to satisfy (3.5.42). We then repeat the process using a larger rectangle, as shown in (3.5.44). Note that one corner remains at $(2,2)$. The primes show the changed elements, in accordance with (3.5.41):

$$
\begin{bmatrix}
1 & 0 & 0 & 0 & 0 \\
0 & c' & 0 & s' & 0 \\
0 & 0 & 1 & 0 & 0 \\
0 & -s' & 0 & c' & 0 \\
0 & 0 & 0 & 0 & 0
\end{bmatrix}
\begin{bmatrix}
a_{11} & a'_{12} & 0 & a_{14} & a_{15} \\
a'_{21} & a'_{22} & a'_{23} & a'_{24} & a'_{25} \\
0 & a'_{32} & a'_{33} & a'_{34} & a'_{35} \\
a_{41} & a'_{42} & a'_{43} & a_{44} & a_{45} \\
a_{51} & a'_{52} & a'_{53} & a_{54} & a_{55}
\end{bmatrix}
\begin{bmatrix}
1 & 0 & 0 & 0 & 0 \\
0 & c' & 0 & -s' & 0 \\
0 & 0 & 1 & 0 & 0 \\
0 & s' & 0 & c' & 0 \\
0 & 0 & 0 & 0 & 1
\end{bmatrix}
$$

$$(3.5.44)$$

The result of the second rotation is shown in the following. The double prime shows the changed elements:

$$
\begin{bmatrix}
a_{11} & a''_{12} & 0 & a''_{14} & a_{15} \\
a''_{21} & a''_{22} & a''_{23} & a''_{24} & a''_{25} \\
0 & a''_{32} & a''_{33} & a''_{34} & a'_{35} \\
a''_{41} & a''_{42} & a''_{43} & a''_{44} & a''_{45} \\
a_{51} & a''_{52} & a'_{53} & a''_{45} & a_{55}
\end{bmatrix}
$$

We are interested in the elements in positions $(1,4)$ and $4,1)$. Their value is given by

$$a''_{14} = a''_{41} = a_{14} \cos\theta' - a_{12}\sin\theta'$$

These elements can be eliminated by choosing

$$\tan\theta' = a_{14}/a_{12}$$

This yields two zeros in the first row and two zeros in the first column. Note that the previous zeros are unchanged.

Now, by taking a larger rotation rectangle, a_{15} can be annihilated. In general, to place zeros in row and column 1, we use

$$\tan\theta = a_{1k}/a_{12} \tag{3.5.45}$$

To place zeros in row and column 2, we start the rotation rectangle at $(3,3)$, and so on.

ILLUSTRATIVE PROBLEM 3.5-3   Using the Givens reduction procedure, reduce the following matrix to tridiagonal form:

$$\begin{bmatrix} 10 & 7 & 8 & 7 \\ 7 & 5 & 6 & 5 \\ 8 & 6 & 10 & 9 \\ 7 & 5 & 9 & 10 \end{bmatrix}$$

Solution   We form

$$\begin{bmatrix} 1 & 0 & 0 & 0 \\ 0 & c & s & 0 \\ 0 & -s & c & 0 \\ 0 & 0 & 0 & 1 \end{bmatrix} \begin{bmatrix} 10 & 7 & 8 & 7 \\ 7 & 5 & 6 & 5 \\ 8 & 6 & 10 & 9 \\ 7 & 5 & 9 & 10 \end{bmatrix} \begin{bmatrix} 1 & 0 & 0 & 0 \\ 0 & c & -s & 0 \\ 0 & s & c & 0 \\ 0 & 0 & 0 & 1 \end{bmatrix}$$

where $\tan\theta = a_{13}/a_{12} = \frac{8}{7}$. Let $S = \sqrt{64+49} \sim 10.63015$, $s = \sin\theta = 8/S \sim 0.75258$, $c = \cos\theta = 7/S \sim 0.65850$. The result of the first rotation is

$$\begin{bmatrix} 10 & 10.63017 & 0 & 7 \\ 10.63017 & 13.77884 & 1.68140 & 10.06574 \\ 0 & 1.68140 & 1.22124 & 2.16364 \\ 7 & 10.06574 & 2.16364 & 10 \end{bmatrix}$$

We now form

$$\begin{bmatrix} 1 & 0 & 0 & 0 \\ 0 & c & 0 & s \\ 0 & 0 & 1 & 0 \\ 0 & -s & 0 & c \end{bmatrix} \begin{bmatrix} 10 & 10.63017 & 0 & 7 \\ 10.63017 & 13.77884 & 1.68140 & 10.06574 \\ 0 & 1.68140 & 1.22124 & 2.16364 \\ 7 & 10.06574 & 2.16364 & 10 \end{bmatrix} \begin{bmatrix} 1 & 0 & 0 & 0 \\ 0 & c & 0 & -s \\ 0 & 0 & 1 & 0 \\ 0 & s & 0 & c \end{bmatrix}$$

where $\tan\theta = a_{14}/a_{12} = 7/10.63017$. Let $S = \sqrt{7^2 + (10.63017)^2} \sim 12.72794$,

$$s = \sin\theta = 7/S \sim 0.54997$$
$$c = \cos\theta \sim 10.63017/S \sim 0.83518$$

The result of the second rotation is

$$\begin{bmatrix} 10 & 12.20159 & 0 & 0 \\ 12.72790 & 21.88264 & 2.59421 & 2.24085 \\ 0 & 2.59421 & 1.22124 & 0.88231 \\ 0 & 2.24085 & 0.88231 & 1.89602 \end{bmatrix}$$

We move the rotation rectangle so that its corner is at $(3,3)$ instead of $(2,2)$ and form

$$\begin{bmatrix} 1 & 0 & 0 & 0 \\ 0 & 1 & 0 & 0 \\ 0 & 0 & c & s \\ 0 & 0 & -s & c \end{bmatrix}\begin{bmatrix} 10 & 12.72790 & 0 & 0 \\ 12.72790 & 21.88264 & 2.59421 & 2.24085 \\ 0 & 2.59421 & 1.22124 & 0.88231 \\ 0 & 2.24085 & 0.88231 & 1.89602 \end{bmatrix}\begin{bmatrix} 1 & 0 & 0 & 0 \\ 0 & 1 & 0 & 0 \\ 0 & 0 & c & -s \\ 0 & 0 & s & c \end{bmatrix}$$

where $\tan\theta = a_{24}/a_{23} = 2.24085/2.59421$. Let $S = \sqrt{(2.24085)^2 + (2.59421)^2} \sim 3.42802$, $s = \sin\theta = 2.24085/S \sim 0.65369$, $c = \cos\theta = 2.59421\sqrt{S} \sim 0.75677$. The result of the third rotation is

$$\begin{bmatrix} 10 & 12.20159 & 0 & 0 \\ 12.20159 & 21.88264 & 3.42804 & 0 \\ 0 & 3.42804 & 2.38254 & 0.46209 \\ 0 & 0 & 0.46209 & 0.73476 \end{bmatrix} \tag{3.5.46}$$

which is in the required tridiagonal form.

At this point, we leave the Givens method to discuss another method which arrives at the tridiagonal form more quickly. After that, we shall explain how the eigenvalues are extracted from the tridiagonal form.

3.5.5 The Householder Reduction Method

The outstanding advantages of the Householder method, to be demonstrated, are: (1) it reduces an entire row and column in one calculation, (2) it is faster, and (3) it requires less storage because it can be programmed to operate a vector at a time. However, the calculation is a little more complicated.

We present the method for a 4×4 matrix, but the extension to any size of matrix is immediate. Consider the real symmetric matrix

$$A = \begin{bmatrix} a_{11} & a_{12} & a_{13} & a_{14} \\ a_{12} & a_{22} & a_{23} & a_{24} \\ a_{13} & a_{23} & a_{33} & a_{34} \\ a_{14} & a_{24} & a_{34} & a_{44} \end{bmatrix}$$

which we wish to reduce to a tridiagonal form via a similarity transformation. In particular, in the first step, we want the first row and column to look like this:

$$\begin{bmatrix} a'_{11} & a'_{12} & 0 & 0 \\ a'_{12} & & & \\ 0 & & & \\ 0 & & & \end{bmatrix}$$

We would like to accomplish this, if possible, by pre- and post-multiplying by a real, symmetric, orthogonal matrix, i.e., a matrix P_1 such that $P_1 = P_1^T = P_1^{-1}$. A solution, according to Householder, can be attained by starting with a matrix of the form

$$P_1 = I - 2ww^I \tag{3.5.47}$$

where

$$w = \begin{bmatrix} w_1 \\ w_2 \\ w_3 \\ w_4 \end{bmatrix} \quad \text{and} \quad w^T = (w_1, w_2, w_3, w_4)$$

We now impose the condition that $w^T w = 1$, which makes P_1 orthogonal, as shown in the following:

$$\begin{aligned} PP^T &= (I - 2ww^T)(I - 2ww^T) \\ &= I - 4ww^T + 4w(w^T w)w^T \\ &= I - 4ww^T + 4ww^T = I \end{aligned}$$

Now we shall explore further to see what else must be true about w. Note that

$$ww^T = \begin{bmatrix} w_1^2 & w_1 w_2 & w_1 w_3 & w_1 w_4 \\ w_1 w_2 & w_2^2 & w_2 w_3 & w_2 w_4 \\ w_1 w_3 & w_2 w_3 & w_3^2 & w_3 w_4 \\ w_1 w_4 & w_2 w_4 & w_3 w_4 & w_4^2 \end{bmatrix}$$

$$\therefore \quad P_1 = \begin{bmatrix} 1 - 2w_1^2 & -2w_1 w_2 & -2w_1 w_3 & -2w_1 w_4 \\ 1 - 2w_1 w_2 & 1 - 2w_2^2 & -2w_2 w_3 & -2w_2 w_4 \\ -2w_1 w_3 & -2w_2 w_3 & 1 - 2w_3^2 & -2w_3 w_4 \\ -2w_1 w_4 & -2w_2 w_4 & -2w_3 w_4 & 1 - 2w_4^2 \end{bmatrix}$$

Looking back at the Givens reduction, we see that a_{11} remained unchanged by the similarity transformation. Therefore we would like the element in position $(1,1)$ of P_1 to be 1. Therefore, $w_1 = 0$. The resulting matrix P_1 is

$$P_1 = \begin{bmatrix} 1 & 0 & 0 & 0 \\ 0 & 1-2w_2^2 & -2w_2w_3 & -2w_2w_4 \\ 0 & -2w_2w_3 & 1-2w_3^2 & -2w_3w_4 \\ 0 & -2w_3w_4 & -2w_3w_4 & 1-2w_4^2 \end{bmatrix} \qquad (3.5.48)$$

We now calculate the product P_1AP_1. We shall not show the entire matrix (we don't need it). We are interested in the following new elements, indicated by primes:

$$a_{11}' = a_{11}$$

$$a_{12}' = a_{12} - 2vw_2 \qquad (3.5.49)$$

$$a_{13}' = a_{13} - 2vw_3 \qquad (3.5.50)$$

$$a_{14}' = a_{14} - 2vw_4 \qquad (3.5.51)$$

where

$$v = w_2a_{12} + w_3a_{13} + w_4a_{14}$$

Squaring and adding (3.5.49), (3.5.50) and (3.5.51), and using the fact that $w^Tw = w_2^2 + w_3^2 + w_4^2 = 1$, we find that

$$(a_{12}')^2 + (a_{13}')^2 + (a_{14}')^2 = (a_{12})^2 + (a_{13})^2 + (a_{14})^2 \qquad (3.5.52)$$

We want a_{13}' and a_{14}' to vanish, so the three equations to determine w_2, w_3 and w_4 are derived from (3.5.49), (3.5.50) and (3.5.51):

$$a_{13} - 2vw_3 = 0$$

$$a_{14} - 2vw_4 = 0$$

$$a_{12}' = a_{12} - 2vw_2 = \pm\sqrt{(a_{12})^2 + (a_{13})^2 + (a_{14})^2} \qquad (3.5.53)$$

The solution of the system can be stated as follows:

$$S = \pm\sqrt{(a_{12})^2 + (a_{13})^2 + (a_{14})^2} \qquad (3.5.54)$$

$$w_2^2 = \tfrac{1}{2}(1 + a_{12}/S) \qquad (3.5.55)$$

$$w_3 = a_{13}/(2w_2S) \qquad (3.5.56)$$

$$w_4 = a_{14}/(2w_2S) \qquad (3.5.57)$$

The sign of S is taken to be the same as that of a_{12}, so that w_2 in (3.5.55) is as large as possible. This reduces the error in (3.5.56) and (3.5.57).

The result of using (3.5.54)–(3.5.57) is a matrix with a'_{13}, a'_{14}, a'_{31} and a'_{41} annihilated. This new matrix is now operated on by

$$P_2 = \begin{bmatrix} 1 & 0 & 0 & 0 \\ 0 & 1 & 0 & 0 \\ 0 & 0 & 1-2w_3^2 & -2w_3 w_4 \\ 0 & 0 & -2w_3 w_4 & 1-2w_4^2 \end{bmatrix} \qquad (3.5.58)$$

using

$$S = \pm \sqrt{(a'_{23})^2 + (a'_{24})^2} \qquad (3.5.59)$$

$$w_3^2 = \tfrac{1}{2}(1 + a'_{23}/S) \qquad (3.5.60)$$

$$w_4 = a'_{24}/(2w_3 S) \qquad (3.5.61)$$

where S and a'_{23} agree in sign.

ILLUSTRATIVE PROBLEM 3.5-4   Reduce the matrix of Illustrative Problem 3.5-3 to tridiagonal form by the method of Householder.

Solution   We calculate the first set of ws using (3.5.54)–(3.5.57):

$$S = \sqrt{7^2 + 8^2 + 7^2} \sim 12.72792$$

$$w_2^2 = \tfrac{1}{2}[1 + 7/S] \sim 0.77499$$

$$w_2 \sim 0.88033$$

$$w_3 = 8/(2w_2 S) \sim 0.35699$$

$$w_4 = 7/(2w_2 S) \sim 0.31237$$

The matrix P_1 is

$$\begin{bmatrix} 1 & 0 & 0 & 0 \\ 0 & -0.54998 & -0.62854 & -0.54998 \\ 0 & -0.62854 & 0.74512 & -0.22303 \\ 0 & -0.54998 & -0.22303 & 0.80485 \end{bmatrix}$$

The result of the first reduction is

$$\begin{bmatrix} 10 & -12.72804 & 0 & 0 \\ -12.72804 & 21.88311 & -1.90159 & -2.85204 \\ 0 & -1.90159 & 0.81524 & 0.58280 \\ 0 & -2.85204 & 0.58280 & 2.30198 \end{bmatrix}$$

We now calculate the second set of ws, using (3.5.59)–(3.5.61):

$$S \sim -3.42785, \qquad w_3^2 \sim 0.77738, \qquad w_3 \sim 0.88169, \qquad w_4 \sim 0.47183.$$

The matrix P_2 is

$$\begin{bmatrix} 1 & 0 & 0 & 0 \\ 0 & 1 & 0 & 0 \\ 0 & 0 & -0.55475 & -0.83202 \\ 0 & 0 & -0.83202 & 0.55475 \end{bmatrix}$$

and the result of the reduction is

$$\begin{bmatrix} 10 & -12.72804 & 0 & 0 \\ -12.72804 & 21.88311 & 3.42786 & 0 \\ 0 & 3.42786 & 2.38245 & -0.46213 \\ 0 & 0 & -0.46213 & 0.73479 \end{bmatrix} \qquad (3.5.62)$$

This is *not* the same as the result of the Givens technique, but should have the same eigenvalues, as will now be discussed.

3.5.6 Finding the Eigenvalues: Algebraic Method

We shall start the discussion with the tridiagonal matrix

$$\begin{bmatrix} a_1 & b_2 & 0 & 0 \\ b_2 & a_2 & b_3 & 0 \\ 0 & b_3 & a_3 & b_4 \\ 0 & 0 & b_4 & a_4 \end{bmatrix} \qquad (3.5.63)$$

This matrix is the result of similarity transformations and therefore has the same eigenvalues as the original. We form the determinant:

$$f_4(\lambda) = \begin{vmatrix} a_1-\lambda & b_2 & 0 & 0 \\ b_2 & a_2-\lambda & b_3 & 0 \\ 0 & b_3 & a_3-\lambda & b_4 \\ 0 & 0 & b_4 & a_4-\lambda \end{vmatrix} = 0 \qquad (3.5.64)$$

There are several ways of proceeding. We choose to expand by minors in the last row. Then

$$f_4(\lambda) = (a_4-\lambda) \begin{vmatrix} a_1-\lambda & b_2 & 0 \\ b_2 & a_2-\lambda & b_3 \\ 0 & b_3 & a_3-\lambda \end{vmatrix} - b_4 \begin{vmatrix} a_1-\lambda & b_2 & 0 \\ b_2 & a_2-\lambda & 0 \\ 0 & b_3 & b_4 \end{vmatrix}$$

$$(3.5.65)$$

The first determinant in this equation will be defined as $f_3(\lambda)$, i.e., the determinant formed from f_4 by starting at the top left and taking three rows and columns. The second determinant can be expanded by the minors of its last row:

$$\begin{vmatrix} a_1-\lambda & b_2 & 0 \\ b_2 & a_2-\lambda & 0 \\ 0 & b_3 & b_4 \end{vmatrix} = b_4 \begin{vmatrix} a_1-\lambda & b_2 \\ b_2 & a_2-\lambda \end{vmatrix} - b_3 \begin{vmatrix} a_1-\lambda & 0 \\ b_2 & 0 \end{vmatrix} \quad (3.5.66)$$

The last determinant in this equation vanishes. The other two-row determinant is $f_2(\lambda)$. Substituting in (3.5.65),

$$f_4(\lambda) = (a_4-\lambda)f_3(\lambda) - (b_4)^2 f_2(\lambda) \quad (3.5.67)$$

Although we have demonstrated this relationship for a 4×4 matrix, it is perfectly general. We then have the recurrence relation

$$\begin{cases} f_k(\lambda) = (a_k-\lambda)f_{k-1}(\lambda) - (b_k)^2 f_{k-2}(\lambda), \quad k=2,3,\dots,n \quad (3.5.68) \\ f_1(\lambda) = a_1-\lambda, \quad f_0(\lambda) = 1 \quad (3.5.69) \end{cases}$$

where (3.5.69) has been appended to complete the recursion. Starting with $f_0(\lambda) = 1$, we can eventually find $f_n(\lambda) = 0$, which is the characteristic equation.

ILLUSTRATIVE PROBLEM 3.5-5 Find the characteristic equation for (3.5.62), the result of the Householder reduction.

Solution

$$f_0(\lambda) = 1$$
$$f_1(\lambda) = 10-\lambda$$
$$f_2(\lambda) = (a_2-\lambda) f_1(\lambda) - (b_2)^2 f_0(\lambda)$$
$$= (21.88311-\lambda)(10-\lambda) - (-12.72804)^2 \cdot 1$$
$$= 56.8281 - 31.88311\lambda + \lambda^2$$
$$f_3(\lambda) = (a_3-\lambda)f_2(\lambda) - (b_3)^2 f_1(\lambda)$$
$$= (2.38245-\lambda)f_2(\lambda) - (3.42786)^3(10-\lambda)$$
$$= 101.1115 - 129.36014\lambda + 34.26556\lambda^2 - \lambda^3$$
$$f_4(\lambda) = (0.73479-\lambda)f_3(\lambda) - 0.21356 f_2(\lambda)$$
$$= 62.15952 - 185.3550\lambda + 154.32457\lambda^2 - 35.00035\lambda^3 + \lambda^4$$

The characteristic equation is then

$$\lambda^4 - 35.000\lambda^3 + 154.32\lambda^2 - 189.36\lambda + 62.160 = 0$$

It is not very accurate, because the roundoff error is quite large, as it must be in a calculation which involves so much subtractive cancellation. If the same technique is applied to the result of the Givens reduction (3.5.46), the result is

$$\lambda^4 - 35.000\lambda^3 + 159.12\lambda^2 - 224.14\lambda + 82.326 = 0$$

The discrepancy in the two characteristic equations (which should be precisely the same) is due to roundoff. If this algebraic technique is used, it is imperative that multiple precision be employed throughout.

The characteristic equation can be solved by one of the methods explained in Chapter 2.

3.5.7 Finding the Eigenvalues: Computer Technique

We have already mentioned that the eigenvalues of the tridiagonal matrix are the same as those of the original. A suitable computer technique is to start at some trial value $\lambda = a$, calculate $f_0, f_1, f_2, \ldots, f_n$ at a, and then step λ using one of the methods of Chapter 2. While this is not very fast, it is very accurate.

To make sure that no values are missed, it is important to have a preliminary run in which a table of values, like that in Section 2.1.4, is printed. It can be proved (see Ralston, pp. 494ff.) that $f_n, f_{n-1}, f_{n-2}, \ldots, f_0$ (note the reverse order of the subscripts) form a Sturm sequence, so that observation of the signs will guide the mathematician in the choice of starters and intervals.

3.5.8 Zeros in the Tridiagonal Matrix

If one of the b_k in (3.5.68) is zero, the formula fails. In such a case, the determinant can be partitioned and each subdeterminant handled separately. The final characteristic polynomial is the product of the two characteristic polynomials found. In some computer centers, provision is made in the program to replace such zeros by arbitrary small numbers which, it is hoped, will not perturb the eigenvalues significantly, yet will allow the recursion to take place. Actually, because of roundoff, true zeros are seldom found in results based upon floating-point calculation. This is one case where errors are accepted without much grief.

PROBLEMS SECTION 3.5

A. Using the definition, find the eigenvalues and a set of eigenvectors for each of the following:

(1) $\begin{pmatrix} 1 & 2 & -1 \\ 1 & 0 & 1 \\ 4 & -4 & 5 \end{pmatrix}$
(3) $\begin{pmatrix} 1 & 2 & 1 \\ 0 & -2 & -4 \\ 0 & 3 & 5 \end{pmatrix}$

(2) $\begin{pmatrix} 2 & -2 & 3 \\ 1 & 1 & 1 \\ 1 & 3 & -1 \end{pmatrix}$
(4) $\begin{pmatrix} 2 & 4 & 1 \\ 1 & -2 & -1 \\ 0 & 0 & 0 \end{pmatrix}$

B. Using (a) the method of Givens, (b) the method of Householder, find the eigenvalues of the following, correct to the nearest thousandth:

(5) $\begin{bmatrix} 4 & 2 & 0 \\ 2 & 5 & 3 \\ 0 & 3 & 6 \end{bmatrix}$ (7) $\begin{bmatrix} 1 & 1 & 0.5 \\ 1 & 1 & 0.25 \\ 0.5 & 0.25 & 2 \end{bmatrix}$

(6) $\begin{bmatrix} 2 & 3 & 5 \\ 3 & 4 & 6 \\ 5 & 6 & 7 \end{bmatrix}$ (8) $\begin{bmatrix} 4 & -1 & -1 & -1 \\ -1 & 4 & -1 & -1 \\ -1 & -1 & 4 & -1 \\ -1 & -1 & -1 & 4 \end{bmatrix}$

Chapter 4

Non-linear Systems

In many practical problems, the approximate location of a desired root is known, and the laborious procedure described in Section 4.1 is unnecessary. In a short course, start with Section 4.2.

The solution of linear systems is so straightforward and well developed that it comes as a shock to find that there are no comparable methods for solving systems which are not linear. We mention at the outset that the methods are woefully inadequate. Research, particularly in the approximate location of roots, is needed.

Linear systems deal with the intersections of lines, planes and (in higher dimensions) hyperplanes. Non-linear systems deal with the intersections of surfaces and hypersurfaces which are not plane. Even if these surfaces, defined by $F(x,y,z,\ldots), G(x,y,z,\ldots), H(x,y,z,\ldots),\ldots$, are perfectly smooth, the problem of finding starters is difficult. We shall attack this problem in Section 1. The second section deals with theory needed to justify the Newton-Raphson method for higher dimensions, and the last section explains the algorithm.

4.1. STARTERS

4.1.1 Case I. Reduction to a Single Equation

The simplest non-linear system is typified by

$$\begin{cases} x^2 + y^2 = 7.43 \\ x^2 - y = 8.45 \end{cases} \tag{4.1.1}$$

Substituting from the second equation into the first,

$$x^2 + (x^2 - 8.45)^2 = 7.43 \tag{4.1.2}$$

$$x^4 + 15.90x^2 + 63.9725 = 0 \tag{4.1.3}$$

which can be solved by methods already explained.

4.1.2   Case II. Reduction to a Quasi-linear System

The following system is not linear:

$$\begin{cases} x^2+y+z^3=30 \\ 3x^2-y+2z^3=55 \\ -9x^2+3y-z^3=-30 \end{cases} \tag{4.1.4}$$

but, if we use $s=x^2$, $t=z^3$, it becomes

$$\begin{cases} s+y+t=30 \\ 3s-y+2t=55 \\ -9s+3y-t=-30 \end{cases} \tag{4.1.5}$$

which can be solved for s, y and t by the Gauss-Jordan method.

4.1.3   Case III. The Double-Graphical Method

Consider the two functions

$$\begin{cases} x=\cos y +0.85 \\ y=\sin x +1.32 \end{cases} \tag{4.1.6}$$

The reduction to a single equation leads to

$$f(x)=x-\cos(\sin x+1.32)-0.85=0 \tag{4.1.7}$$

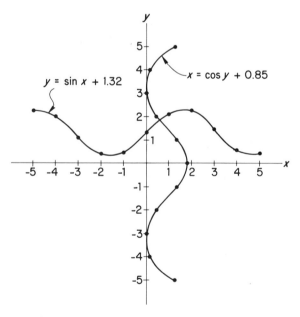

Fig. 4.1-1   Case III: double-graphical method for locating a starter.

which can be solved by simple iteration. However, some thought about the process chart and the propagated errors leads one to seek another method. In any case, we need a starter. Figure 4.1-1 shows the graphs of the two functions. From the figure, (0.6, 1.8) seems to be a suitable starter. Also, from the graphs it appears that there is one and only one real root of the system.

4.1.4   Case IV. The Computer Sieve Method

Unfortunately, none of these easy methods will work for a system like

$$\begin{cases} 4.75 \sin 2x - 3.14 e^y = 0.4945 \\ 3.63 \cos 3x + \quad \sin y = 0.3995 \end{cases} \tag{4.1.8}$$

or

$$\begin{cases} \cos x + \sin y + \ln z = 1.3667 \\ \quad e^x + \sin 2y - z^2 = -11.2813 \\ \quad e^{-x} + \sin y + e^{-z} = -0.1813 \end{cases} \tag{4.1.9}$$

for both of which we will find starters and solutions in this chapter.

Systems like (4.1.9) often have an infinite number of real roots; the three-dimensional graphs, although possible, would not be practical. Figure 4.1-2 is an impression of portions of three surfaces, F, G and H, meeting in space. [They are *not* the system (4.1.9).] If we go to four or more

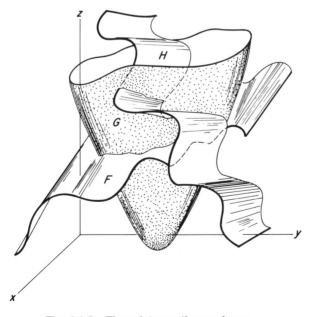

Fig. 4.1-2   Three intersecting surfaces.

dimensions, the graphical method is clearly impossible. (However, it is theoretically possible to *project* the surfaces down to two dimensions to find the intersections.)

In actual problems of science and technology, we are interested in a solution within a specific neighborhood, (say, $x \in [1,2]$, $y \in [-3, -2]$, $z \in [7,8]$), and as computer scientists we can limit our search to the region requested. In this chapter, we shall restrict ourselves arbitrarily to the interval $[-5, +5]$ for all variables.

We now illustrate a *computer sieve* method for the system (4.1.9). First, they were rewritten

$$\begin{cases} F(x,y,z) = \cos x + \sin y + \ln z - 1.3667 \\ G(x,y,z) = e^x + \sin 2y - z^2 + 11.2813 \\ H(x,y,z) = e^{-x} + \sin y + e^{-z} + 0.1813 \end{cases} \quad (4.1.10)$$

Then we examined the situation in the neighborhood of a root (Figure 4.1-3). At the root, it must be true that

$$\begin{cases} F = 0 \\ G = 0 \\ H = 0 \end{cases} \quad (4.1.11)$$

and also

$$\begin{cases} F - G = 0 \\ F - H = 0 \\ G - H = 0 \end{cases} \quad (4.1.12)$$

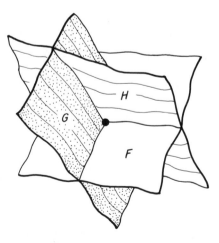

Fig. 4.1-3 At the root.

Near the root, the six left members of (4.1.11) and (4.1.12) are, except for unusual cases, somewhere near 0.

Here's how we tried to locate the roots.

1. The initial values of x, y and were set at $(-5, -5, 0.1)$. The value of z could not be started at -5 because of the presence of $\ln z$ in the system.

2. The computer was programmed to calculate $F, G, H, F-G, F-H$ and $G-H$.

3. A decision had to be made about the *mesh size* of the sieve. The program was started with a mesh size of 0.5. If the absolute values of all the six quantities in (4.1.11) and (4.1.12) were $\leqslant 0.5$, the computer printed the actual values of the six quantities.

4. Otherwise, the computer returned and incremented x, y and z. An *increment size* of 0.5 was chosen arbitrarily. The three variables were incremented one at a time until they all reached 5.

5. At this point, the computer counted the number of times it had printed out values for the specific mesh size. If fewer than 10 values had been printed, the mesh size was doubled and the program run over again, starting at the original values of x, y and z. In one case, ten values were not reached until the mesh size had been multiplied by 64.

6. As a result of these runs, a rough idea was gained about the location of the roots. The program was then rerun starting closer to the actual roots and using a smaller mesh size.

Figure 4.1-4 shows the computer printout for the system (4.1.8), and Figure 4.1-5 shows that for (4.1.9). The computer was programmed to print results on adjacent lines if the values of x, y and z were adjacent, and otherwise to skip a line.

In Figure 4.1-4, it seems very likely that there is a root in the neighborhood of $(4.5, -1.5)$ because of the accumulation of values there. There are also possibilities at $(-1.5, -3.5)$, $(+1.5, -4.0)$ and other places. One of the things to look for is obviously a point of accumulation. Another is a pair of points where F or G or their difference changes sign, showing (if they are continuous) that F or G or both have passed through 0.

In Figure 4.1-5, there are obvious clusters about $(0.5, -1.5, 3.5)$ and $(0.5, 5.0, 3.5)$.

Unfortunately, not all the apparent roots are actual roots, and unless the mesh size is very small, roots are easy to miss.

4.1.5 The Computer Sieve Program

Figure 4.1-6 shows the computer program used to find starters for all the problems of this chapter. Only part of the rather lengthy program has been shown, enough to show the logic. The first part took care of the five three-dimensional programs and, beginning with $N = 1$ on page 163, the two-dimensional programs. The entire job took 33 minutes on the IBM 360/30.

X	Y	F	G	(F-G)
-1.5	-5.0	-1.19	-0.21	-0.98
1.5	-5.0	0.15	-0.21	0.36
-1.5	-4.5	-1.20	-0.19	-1.01
1.5	-4.5	0.14	-0.19	0.33
-1.5	-4.0	-1.22	-0.41	-0.81
1.5	-4.0	0.12	-0.41	0.53
-1.5	-3.5	-1.26	-0.81	-0.45
1.5	-3.5	0.08	-0.81	0.89
-1.5	-3.0	-1.32	-1.31	-0.02
1.5	-3.0	0.02	-1.31	1.33
4.5	-2.5	1.21	1.16	0.04
4.5	-2.0	1.04	0.85	0.19
4.5	-1.5	0.76	0.76	-0.00
4.5	-1.0	0.31	0.92	-0.61
-2.5	0.0	0.92	0.86	0.06
0.5	0.0	0.36	-0.14	0.51

Fig. 4.1-4   The printout for the system (4.1.8).

X	Y	Z	F	G	H	(F-G)	(F-H)	(G-H)
0.5	-3.0	3.5	0.62	0.96	0.68	-0.34	-0.05	0.28
0.0	-2.5	3.5	0.29	0.99	0.61	-0.70	-0.33	0.38
0.0	-2.0	3.5	-0.02	0.79	0.30	-0.81	-0.33	0.49
0.0	-1.5	3.5	-0.11	-0.11	0.21	-0.00	-0.33	-0.32
0.5	-1.5	3.5	-0.23	0.54	-0.18	-0.77	-0.05	0.72
0.5	-1.0	3.5	-0.08	-0.23	-0.02	0.15	-0.05	-0.21
0.5	-0.5	3.5	0.28	-0.16	0.34	0.45	-0.05	-0.50
1.0	-0.5	3.5	-0.05	0.91	0.10	-0.96	-0.15	0.81
0.5	0.0	3.5	0.76	0.68	0.82	0.08	-0.05	-0.14
0.5	3.0	3.5	0.90	0.40	0.96	0.50	-0.05	-0.56
0.0	3.5	3.5	0.54	0.69	0.86	-0.15	-0.33	-0.17

X	Y	Z	F	G	H	(F-G)	(F-H)	(G-H)
0.0	4.5	3.5	-0.09	0.44	0.23	-0.53	-0.33	0.21
0.0	5.0	3.5	-0.07	-0.51	0.25	0.44	-0.33	-0.77
0.5	5.0	3.5	-0.20	0.14	-0.14	-0.33	-0.05	0.28
0.5	5.5	3.5	0.06	-0.32	0.11	0.38	-0.05	-0.43
1.5	-3.0	4.0	-0.05	0.04	0.28	-0.09	-0.33	-0.24
1.5	-1.5	4.0	-0.91	-0.38	-0.57	-0.53	-0.33	0.20
1.5	0.0	4.0	0.09	-0.24	0.42	0.33	-0.33	-0.66
1.5	0.5	4.0	0.57	0.60	0.90	-0.03	-0.33	-0.30
1.5	3.5	4.0	-0.26	0.42	0.07	-0.68	-0.33	0.35
1.5	5.0	4.0	-0.87	-0.78	-0.54	-0.09	-0.33	-0.24
2.0	-2.5	4.5	-0.88	-0.62	-0.27	-0.26	-0.61	-0.35
2.0	3.5	4.5	-0.63	-0.92	-0.02	0.29	-0.61	-0.90
2.5	3.5	5.0	-0.91	-0.88	-0.08	-0.03	-0.83	-0.80
3.0	-4.0	5.5	0.10	0.13	0.99	-0.02	-0.89	-0.86
3.0	-3.5	5.5	-0.30	0.46	0.59	-0.76	-0.89	-0.13
3.0	2.5	5.5	-0.05	0.16	0.83	-0.21	-0.89	-0.68

Fig. 4.1-5 The printout for the system (4.1.9).

```
C THE PURPOSE IS TO SEARCH FOR ROOTS IN NON-LINEAR SYSTEMS
C R IS THE CRITERION, N THE PROBLEM NUMBER AND K THE COUNT OF RESULTS.
      N=1
      K=0
C XP,YP AND ZP ARE 'PREVIOUS ROOTS'
      XP=500.
      YP=500.
      ZP=500.
1     WRITE(3,100)
100   FORMAT(1H1,T4,'X',T12,'Y',T19,'Z',T29,'F',T39,'G',T49,'H',T58,'(F-
     1G)',T68,'(F-H)',T78,'(G-H)')
      IF(K)110,110,109
109   K=0
      GO TO 111
110   GO TO(601,602,603,604,605),N
601   R=1.5
111   GO TO(201,202,203,204,205),N
201   XA=-5.0
      YA=-5.0
      Z=-5.0
      X=XA
      Y=YA
301   F= SIN(X)+ COS(Y)+ EXP(Z)-1.3461
      G= COS(X)- SIN(Y)+ EXP(-Z)-5.3288
      H= EXP(X)+ EXP(Y)- SIN(Z)-8.9769
      GO TO 2
```

```
602     R=0.5
202     XA=-5.0
        YA=-5.0
        Z=1.0
        X=XA
        Y=YA
302     F= COS(X)+ SIN(Y)+ALOG(Z)-1.3667
        G= EXP(X)+ SIN(2.0*Y)-Z*Z+11.2813
        H= EXP(-X)+ SIN(Y)+ EXP(-Z)+0.1813
        GO TO 2
2       A=F-G
        B=F-H
        C=G-H
        FP=ABS(F)
        GP=ABS(G)
        HP=ABS(H)
        S=ABS(A)
        T=ABS(B)
        U=ABS(C)
        IF(FP-R)3,3,9
3       IF(GP-R)4,4,9
4       IF(HP-R)5,5,9
5       IF(S-R)6,6,9
6       IF(T-R)7,7,9
7       IF(U-R)8,8,9
8       K=K+1
        XPA=ABS(X-XP)
        YPA=ABS(Y-YP)
        ZPA=ABS(Z-ZP)
        IF(XPA-0.5)61,61,63
61      IF(YPA-0.5)62,62,63
62      IF(ZPA-0.5)64,64,63
63      WRITE(3,163)
163     FORMAT(1H0)
64      WRITE(3,101)X,Y,Z,F,G,H,A,B,C,N
101     FORMAT(1H ,F4.1,2F8.1,6F10.2,I10)
        XP=X
        YP=Y
        ZP=Z
9       IF(X-5.0)12,12,10
10      IF(Y-5.0)13,13,11
11      IF(Z-5.0)15,15,16
16      IF(R-64)551,558,558
551     IF(K-10)554,56,56
554     IF(K)552,552,553
552     R=R+0.5
        GO TO 111
553     R=R+0.5
        GO TO 1
558     WRITE(3,158)
158     FORMAT(1H0,'NO VALUES FOR R=64')
        GO TO 17
56      IF(N-5)17,14,14
17      N=N+1
        K=0
        GO TO 1
12      X=X+0.5
        GO TO 19
13      X=XA
        Y=Y+0.5
        GO TO 19
15      X=XA
        Y=YA
        Z=Z+0.5
```

```
19        GO TO(301,302,303,304,305),N
14        N=1
          K=0
21        WRITE(3,102)
102       FORMAT(1H1,T4,'X',T12,'Y',T20,'F',T30,'G',T40,'(F-G)')
          IF(K)210,210,209
209       K=0
          GO TO 211
210       GO TO(701,702,703,704,705,706,707,708,709),N
701       R=39.0
211       GO TO(401,402,403,404,405,406,407,408,409),N
401       XA=-5.0
          Y=-5.0
          X=XA
501       F=3.0*X+Y*Y-9.0493
          G=Y-7.0*X**3+151.0341
          GO TO 22
702       R=2.5
402       XA=-5.0
          Y=-5.0
          X=XA
502       F=X-Y*Y+2.2623
          G=Y+2.0*X**3+6.4504
          GO TO 22
22        A=F-G
          FP=ABS(F)
          GP=ABS(G)
          S=ABS(A)
          IF(FP-R)23,23,29
23        IF(GP-R)24,24,29
24        IF(S-R)28,28,29
28        K=K+1
          XPA=ABS(X-XP)
          YPA=ABS(Y-YP)
          IF(XPA-0.5)71,71,72
71        IF(YPA-0.5)73,73,72
72        WRITE(3,163)
73        WRITE(3,103)X,Y,F,G,A,N
103       FORMAT(1H ,F4.1,F8.1,3F10.2,I10)
          XP=X
          YP=Y
29        IF(X-5.0)33,33,30
30        IF(Y-5.0)34,34,31
31        IF(R-64)571,578,578
571       IF(K-10)574,51,51
574       IF(K)572,572,573
572       R=R+0.5
          GO TO 211
573       R=R+0.5
          GO TO 21
578       WRITE(3,158)
          GO TO 32
51        IF(N-9)32,35,35
32        N=N+1
           K=0
          GO TO 21
33        X=X+0.5
          GO TO 39
34        X=XA
          Y=Y+0.5
          GO TO 39
39        GO TO(501,502,503,504,505,506,507,508,509),N
35        STOP
          END
```

Fig. 4.1-6 The computer sieve program.

PROBLEM SECTION 4.1

For each of the printouts shown, try to guess at the locations of the roots. (No answers are given for this problem section. The way to check is shown later.)

A. Two-dimensional:

(1)

X	Y	F	G	(F−G)
2.5	−5.0	23.45	36.66	−13.21
2.5	−4.5	18.70	37.16	−18.46
2.5	−4.0	14.45	37.66	−23.21
2.5	−3.5	10.70	38.16	−27.46
2.5	−3.0	7.45	38.66	−31.21
3.0	−0.5	0.20	−38.47	38.67
3.0	0.0	−0.05	−37.97	37.92
3.0	0.5	0.20	−37.47	37.67
3.0	1.0	0.95	−36.97	37.92
3.0	1.5	2.20	−36.47	38.67

(2)

X	Y	F	G	(F−G)
−5.0	2.6	0.07	−1.93	2.00
−1.5	2.6	−1.89	−2.15	0.26
−1.0	2.6	−1.73	−1.68	−0.05
−0.5	2.6	−1.37	−1.34	−0.03
0.0	2.6	−0.89	−1.22	0.33
0.5	2.6	−0.41	−1.34	0.93
1.0	2.6	−0.05	−1.68	1.63
1.5	2.6	0.11	−2.15	2.25
4.5	2.6	−1.87	−2.43	0.56
5.0	2.6	−1.85	−1.93	0.08
5.5	2.6	−1.60	−1.51	−0.09

(3)

X	Y	F	G	(F−G)
0.1	1.0	−0.38	1.37	−1.75
0.6	1.0	−0.72	2.66	−3.38
4.6	1.0	−2.59	0.70	−3.29
5.1	1.0	−1.61	0.30	−1.91
1.6	1.5	2.86	1.84	1.02
2.1	1.5	1.91	1.36	0.55
2.6	1.5	1.21	0.83	0.38
3.1	1.5	0.92	0.25	0.67
3.6	1.5	1.13	−0.35	1.48
4.1	1.5	1.77	−0.97	2.74

(4)

X	Y	F	G	(F-G)
-3.5	-4.5	-1.75	1.70	-3.45
-3.5	-4.0	-1.74	0.37	-2.10
-3.5	-3.5	-1.71	-0.48	-1.23
-3.5	-3.0	-1.67	-0.64	-1.03
-3.5	-2.5	-1.61	-0.07	-1.53
-3.5	-2.0	-1.50	1.08	-2.58
-3.5	-1.5	-1.33	2.54	-3.87
-3.5	1.0	3.66	3.95	-0.29
-3.0	3.0	-0.69	-1.13	0.44
-2.5	3.5	1.66	-1.43	3.09

(5)

X	Y	F	G	(F-G)
-2.5	-5.0	1.08	-3.92	5.00
-2.5	-3.5	4.68	-0.74	5.42
-2.5	-3.0	1.87	-0.33	2.20
-2.5	-2.5	-0.17	-1.68	1.51
-2.5	-2.0	0.44	-3.55	3.99
-2.5	0.0	2.71	-0.25	2.96
-2.5	0.5	0.19	-1.17	1.35
-2.5	1.0	-0.02	-3.08	3.06
-2.5	3.0	3.55	-0.33	3.88
-2.5	3.5	0.74	-0.74	1.48
-2.5	4.0	-0.26	-2.54	2.28

(6)

X	Y	F	G	(F-G)
2.5	-3.5	-0.52	-1.00	0.48
2.0	-3.0	-0.97	-0.44	-0.52
2.5	-3.0	-0.47	-0.50	0.03
3.0	-3.0	0.03	-0.53	0.56
2.5	-2.5	-0.66	0.00	-0.66
3.0	-2.5	-0.16	-0.03	-0.13
3.5	-2.5	0.34	-0.05	0.39
4.0	-2.5	0.84	-0.06	0.90
3.5	-2.0	-0.04	0.45	-0.49
4.0	-2.0	0.46	0.44	0.02
4.5	-2.0	0.96	0.43	0.53
4.0	-1.5	-0.03	0.94	-0.97
4.5	-1.5	0.47	0.93	-0.46
5.0	-1.5	0.97	0.93	0.04

B. Three-dimensional:

(7)

X	Y	Z	F	G	H	(F-G)	(F-H)	(G-H)
2.0	-5.0	-2.0	-0.02	0.69	-0.67	-0.70	0.65	1.36
2.0	-4.5	-2.0	-0.51	0.67	-0.67	-1.18	0.16	1.33
2.0	0.5	-2.0	0.58	1.16	0.97	-0.59	-0.39	0.19
1.5	1.5	-2.0	-0.14	1.13	0.90	-1.28	-1.04	0.24
-4.0	2.0	-2.0	-0.87	0.50	-0.66	-1.37	-0.21	1.16
-3.5	2.0	-2.0	-1.28	0.21	-0.65	-1.49	-0.63	0.86
2.0	-3.0	-1.5	-1.20	-1.12	-0.54	-0.08	-0.66	-0.58
2.0	-2.5	-1.5	-1.01	-0.66	-0.51	-0.35	-0.51	-0.16
2.0	-2.0	-1.5	-0.63	-0.35	-0.46	-0.28	-0.17	0.10
2.0	-1.5	-1.5	-0.14	-0.27	-0.37	0.12	0.22	0.10
2.0	-1.0	-1.5	0.33	-0.42	-0.22	0.75	0.55	-0.20
2.0	-0.5	-1.5	0.66	-0.78	0.02	1.45	0.65	-0.80
1.0	1.5	-1.5	-0.21	-1.30	-0.78	1.09	0.57	-0.52
-5.0	2.0	-1.5	-0.58	-1.47	-0.58	0.89	0.00	-0.89

(8)

X	Y	Z	F	G	H	(F-G)	(F-H)	(G-H)
0.1	0.1	0.6	-1.19	-1.18	-0.27	-0.01	-0.92	-0.91
0.6	0.1	0.6	0.60	-0.47	0.40	1.07	0.19	-0.87
0.1	0.6	0.6	-0.88	0.61	0.08	-1.49	-0.97	0.53
0.6	0.6	0.6	0.91	1.32	0.76	-0.42	0.15	0.56
0.1	0.1	1.1	-1.07	-0.68	0.33	-0.39	-1.40	-1.02
0.6	0.1	1.1	0.72	0.03	1.01	0.69	-0.29	-0.98
1.1	0.1	1.1	1.33	1.21	-0.12	0.11	1.45	1.34
0.1	0.1	1.6	-0.20	-0.18	0.71	-0.02	-0.91	-0.89
0.1	0.1	2.1	0.61	0.32	0.98	0.29	-0.37	-0.66
0.1	0.1	2.6	0.62	0.82	1.19	-0.19	-0.57	-0.38

(9)

X	Y	Z	F	G	H	(F-G)	(F-H)	(G-H)
2.0	-4.0	0.1	1.25	-0.24	0.70	1.49	0.55	-0.93
2.0	-5.0	0.6	1.41	0.32	0.05	1.08	1.36	0.27
2.0	-4.5	0.6	0.59	-0.79	0.55	1.38	0.03	-1.35
2.5	-4.5	0.6	1.31	0.86	1.05	0.45	0.26	-0.19
1.5	-4.0	0.6	0.99	0.50	0.55	0.49	0.44	-0.05

(9) (Continued)

X	Y	Z	F	G	H	(F-G)	(F-H)	(G-H)
0.0	-3.0	0.6	0.26	1.43	0.05	-1.17	0.21	1.38
0.5	-3.0	0.6	0.51	1.40	0.55	-0.89	-0.04	0.85
1.0	-3.0	0.6	1.10	0.97	1.05	0.13	0.05	-0.08
-2.0	-1.5	0.6	-0.21	-0.59	-0.45	0.38	0.24	-0.14
-1.0	-1.5	0.6	1.39	0.61	0.55	0.78	0.83	0.05
-0.5	-1.5	0.6	0.79	1.04	1.05	-0.24	-0.26	-0.02
-2.0	-0.5	0.6	1.39	-0.06	0.55	1.44	0.83	-0.61
-2.0	0.0	0.6	1.42	0.19	1.05	1.23	0.36	-0.86
-3.5	1.0	0.6	1.40	0.95	0.55	0.45	0.85	0.40
-4.0	1.5	0.6	0.26	-0.89	0.55	1.15	-0.29	-1.44
-3.5	1.5	0.6	0.23	1.02	1.05	-0.78	-0.82	-0.04
-4.0	2.0	0.6	0.23	0.64	1.05	-0.41	-0.82	-0.41
-5.0	3.0	0.6	0.13	1.42	1.05	-1.29	-0.92	0.37
2.0	-4.5	1.1	-0.02	0.39	0.77	-0.41	-0.79	-0.38
-2.0	-1.5	1.1	-0.82	0.60	-0.23	-1.41	-0.59	0.83
-1.5	-1.5	1.1	0.72	0.62	0.27	0.10	0.45	0.35
-2.0	-1.0	1.1	0.35	0.53	0.27	-0.18	0.08	0.26
-2.0	-0.5	1.1	0.78	1.13	0.77	-0.35	0.01	0.36
-2.0	0.0	1.1	0.81	1.37	1.27	-0.56	-0.46	0.10
-4.0	0.5	1.1	1.25	0.82	-0.23	0.43	1.48	1.05
-4.0	1.0	1.1	0.82	0.23	0.27	0.59	0.55	-0.04
-4.5	1.5	1.1	0.93	1.42	0.27	-0.49	0.66	1.15
-4.0	1.5	1.1	-0.35	0.29	0.77	-0.64	-1.12	-0.48

(10)

X	Y	Z	F	G	H	(F-G)	(F-H)	(G-H)
0.1	2.1	-3.5	-3.36	-4.96	2.98	1.60	-6.34	-7.94
0.1	2.6	-3.5	-3.15	-5.31	-2.32	2.16	-0.83	-2.99
0.6	2.6	-3.5	-0.89	-5.48	-0.53	4.59	-0.37	-4.96
1.1	2.6	-3.5	0.04	-5.85	0.08	5.89	-0.04	-5.93
1.6	2.6	-3.5	0.52	-6.34	0.45	6.86	0.07	-6.79
2.1	2.6	-3.5	0.66	-6.81	0.73	7.47	-0.07	-7.54
4.6	2.6	-3.5	-0.41	-6.42	1.51	6.01	-1.92	-7.93
5.1	2.6	-3.5	-0.24	-5.93	1.61	5.69	-1.86	-7.54
3.6	3.1	-3.5	0.07	-7.68	-7.47	7.75	7.54	-0.21
4.1	3.1	-3.5	-0.18	-7.36	-7.34	7.18	7.16	-0.02
4.6	3.1	-3.5	-0.24	-6.89	-7.22	6.66	6.99	0.33
5.1	3.1	-3.5	-0.07	-6.40	-7.12	6.34	7.05	0.72

4.2. THEORY FOR THE MANY-DIMENSIONAL NEWTON-RAPHSON FORMULA

4.2.1 Review

In the two-dimensional case (Figure 4.2-1), we assumed that after k iterations we arrived at point P where $x = x_k$ and $y = f(x_k)$. At this point, we can easily calculate $f'(x_k)$, the slope of tangent T to the curve f.

By definition, if (x_T, y_T) is any point on the tangent,

$$f'(x_k) = \frac{y_T - f(x_k)}{x_T - x_k} \tag{4.2.1}$$

At point Q, $x_T = x_{k+1}$ and $y_T = 0$. Therefore,

$$f'(x_k) = \frac{-f(x_k)}{x_{k+1} - x_k} \tag{4.2.2}$$

and consequently,

$$x_{k+1} = x_k - \frac{f(x_k)}{f'(x_k)} \tag{4.2.3}$$

which is the two-dimensional Newton-Raphson formula.

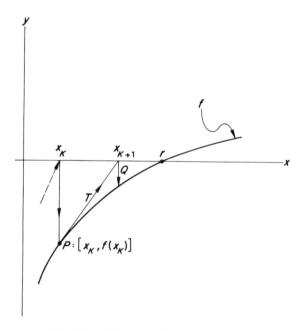

Fig. 4.2-1   The two-dimensional case.

4.2.2 Algebraic Derivation for Two Dimensions

Because we are going to be working in higher-dimensional spaces, the geometric approach is impractical. An algebraic derivation has the advantage that it can be extended to any number of dimensions.

We start with the mean-value theorem:

$$f(x_{k+1}) - f(x_k) = (x_{k+1} - x_k) f'(c) \qquad (4.2.4)$$

where $c \in (x_k, x_{k+1})$. We don't know c. In fact, the only value we know in the interval is at x_k. We use this known value as an approximation of the unknown value at c. Then, approximately,

$$f(x_{k+1}) - f(x_k) = (x_{k+1} - x_k) f'(x_k) \qquad (4.2.5)$$

Now, for $f(x_{k+1}) = 0$, we obtain (4.2.3), as desired.

4.2.3 Two Simultaneous Non-linear Equations

We shall apply the same kind of reasoning to lead up to the general formula for the Newton-Raphson method. In Figure 4.2-2, we have two surfaces, F and G, intersecting at r. At this point, $F = 0$, $G = 0$ and $F = G$, so that all the conditions for a root of two equations are satisfied.

Now, let us consider surface F alone (Figure 4.2-3). Suppose we have reached (x_k, y_k) at P by iteration. Then, by analogy with the two-dimensional case, a tangent should take us to point Q.

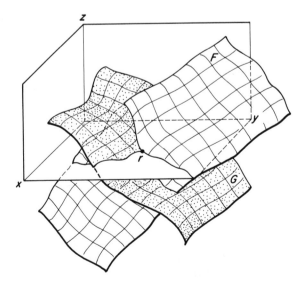

Fig. 4.2-2 Two intersecting surfaces.

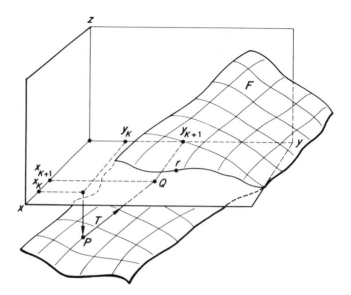

Fig. 4.2-3   A portion of the surface *F*.

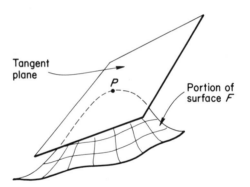

Fig. 4.2-4   The tangents to *F* at *P*.

Unfortunately, there is an infinity of tangent lines at point *P* (Figure 4.2-4). In fact, there is a *plane* tangent to *F* at *P*. Every line on this plane is tangent to *F*. We need one that is in the proper direction.

4.2.4   The Mean-Value Theorem for Many Variables

Consider Figure 4.2-5, where we have drawn a vector from $P(1,2,3)$ to $Q(3,6,9)$. The vector $\overrightarrow{PQ}$ is defined as the row vector $(2,4,6)$, the coordinates being obtained by subtracting: the coordinates of Q minus the coordinates of P. Q is called the *terminal point*, and P is called the *initial point*.

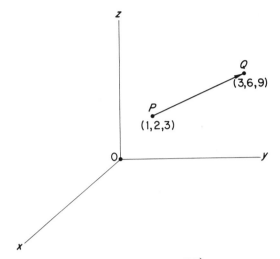

Fig. 4.2-5   Vector $\overrightarrow{PQ}$.

VECTOR BETWEEN GIVEN POINTS   Given $P(a,b,c,\ldots)$ and $Q(r,s,t,\ldots)$, then $\overrightarrow{PQ}$ is the vector $(r-a,s-b,t-c,\ldots)$.

ILLUSTRATIVE PROBLEM 4.2-1   Express the four-dimensional vector (a) from $(0, -2,3,8)$ to $(-2,4,5,1)$; (b) from $\begin{pmatrix} 2 \\ 7 \end{pmatrix}$ to $\begin{pmatrix} -5 \\ 1 \end{pmatrix}$.

Solution   (a) $(-2,6,2,-7)$, (b)$\begin{pmatrix} -7 \\ -6 \end{pmatrix}$.

For two dimensions, the mean-value theorem can be written as

$$f(x_{k+1})-f(x_k)=\Delta x\, f'(\theta), \qquad \theta \in (x_k,x_{k+1}) \tag{4.2.6}$$

where $\Delta x = x_{k+1}-x_x$, and both $f(x_{k+1})$ and θ are unknown. In our use of (4.2.6) for the Newton-Raphson algorithm, we hoped· that $f(x_{k+1})$ was approximately 0 and that $f'(\theta)$ was approximately $f'(x_k)$. We now present the MVT for many dimensions.

MEAN-VALUE THEOREM FOR MANY DIMENSIONS   If f is differentiable in neighborhood N, then for any two points $\mathbf{P}(x_k,y_k,z_k,\ldots)$ and $\mathbf{Q}(x_k,y_k,z_k,\ldots)$, there is a point $\mathbf{P}_\theta(x_\theta,y_\theta,z_\theta,\ldots)$ on the segment PQ such that

$$f(\mathbf{Q})-f(\mathbf{P})=\Delta\mathbf{X}\cdot\nabla\mathbf{P}_\theta \tag{4.2.7}$$

In the statement of the theorem, $\mathbf{P}$ refers to the vector from the origin O to the point P, and $\mathbf{Q}$ to the vector from O to Q. It is customary to consider the terminal points of vectors from the origin as synonymous with the vectors. Also

$$\Delta \mathbf{X} = (x_{k+1} - x_k, y_{k+1} - y_k, z_{k+1} - z_k, \dots)$$

$$\nabla f = \begin{pmatrix} f_x \\ f_y \\ f_z \\ \vdots \end{pmatrix}$$

where the partials are evaluated at $(x_\theta, y_\theta, z_\theta, \dots)$.

The symbol ∇f is read as *del f* or as the *gradient of f*. ∇ is called the *del operator* and is defined as a row or column vector of partial derivatives:

$$\nabla = \left(\frac{\partial}{\partial x}, \frac{\partial}{\partial y}, \frac{\partial}{\partial z}, \dots \right) \tag{4.2.8}$$

The dot between $\Delta \mathbf{X}$ and ∇f in (4.2.7) means that an inner product is to be formed.

ILLUSTRATIVE PROBLEM 4.2-2   Write the MVT for two dimensions.

Solution

$$f(x_{k+1}, y_{k+1}) - f(x_k, y_k) = (\Delta x, \Delta y) \cdot \begin{pmatrix} \partial/\partial x \\ \partial/\partial y \end{pmatrix} f \tag{4.2.9}$$

$$= (\Delta x, \Delta y) \cdot \begin{pmatrix} f_x \\ f_y \end{pmatrix} \tag{4.2.10}$$

$$= f_x \Delta x + f_y \Delta y \tag{4.2.11}$$

ILLUSTRATIVE PROBLEM 4.2-3   Write the MVT for five dimensions.

Solution   Looking at the previous problem, we see that we can omit the first step. Then

$$f(x_{k+1}, y_{k+1}, z_{k+1}, w_{k+1}, v_{k+1}) - f(x_k, y_k, z_k, w_k, v_k)$$

$$= (\Delta x, \Delta y, \Delta z, \Delta w, \Delta v) \cdot \begin{pmatrix} f_x \\ f_y \\ f_z \\ f_w \\ f_v \end{pmatrix} \tag{4.2.12}$$

$$= f_x \Delta x + f_y \Delta y + f_z \Delta z + f_w \Delta w + f_v \Delta v \tag{4.2.13}$$

The column vector in (4.2.12) is, of course, ∇f. It is supposed to be measured at an intermediate point, θ, but in practice this is almost always unknown, and ∇f is measured at the nearest known point, usually $(x_k, y_k, \dots)$.

ILLUSTRATIVE PROBLEM 4.2-4   Write the MVT for one dimension.

Solution

$$f(x_{k+1}) - f(x_k) = \Delta x \; \frac{\partial f}{\partial x} \tag{4.2.14}$$

However, because f is a function of only one variable, $\partial f / \partial x = df/dx = f'(x)$, so that this becomes

$$f(x_{k+1}) - f(x_k) = \Delta x \; f'(x_\theta) \tag{4.2.15}$$

as expected.

We now proceed with the proof of the theorem.

Proof   We define

$$\mathbf{R} = (x + th, y + tk, z + tl, \dots) = \mathbf{P} + t\mathbf{H} \tag{4.2.16}$$

where $t\epsilon[0,1]$ and

$$\mathbf{H} = \mathbf{Q} - \mathbf{P} = (\Delta x, \Delta y, \dots) \tag{4.2.17}$$

Note that $\mathbf{R}$ describes every point on the segment from P to Q, because when $t = 0$, $\mathbf{R} = \mathbf{P}$, but when $t = 1$, $\mathbf{R} = \mathbf{P} + \mathbf{H}$, which [from (4.2.17)] is $\mathbf{Q}$. We now invent a function of the variable t:

$$g(t) = f(\mathbf{R}) \tag{4.2.18}$$

where $\mathbf{R}$ is a function of t. Using the elementary MVT for one variable on the interval $t\epsilon[0,1]$, we have

$$g(1) - g(0) = (1 - 0) g'(\theta), \qquad \theta \in (0, 1) \tag{4.2.19}$$

Hence, using (4.2.18), $g(1) = f(\mathbf{Q})$ and $g(0) = f(\mathbf{P})$. It remains to identify $g'(\theta)$. We start with

$$g'(t) = f'\left[\mathbf{R}(t)\right] \tag{4.2.20}$$

The *complete differential* of f, from elementary calculus, is given by

$$df = \frac{\partial f}{\partial x} dx + \frac{\partial f}{\partial y} dy + \cdots \tag{4.2.21}$$

$$\frac{df}{dt} = f_x \frac{dx}{dt} + f_y \frac{dy}{dt} + \cdots \tag{4.2.22}$$

which can be rewritten as

$$\frac{df}{dt} = \left(\frac{dx}{dt}, \frac{dy}{dt}, \dots \right) \cdot \begin{pmatrix} f_x \\ f_y \\ \vdots \end{pmatrix} \tag{4.2.23}$$

The column vector is clearly ∇f. From (4.2.16), the row vector is

$$\frac{d\mathbf{R}}{dt} = \left(\frac{dx}{dt}, \frac{dy}{dt}, \dots\right) = \mathbf{H} \tag{4.2.24}$$

The last equation is derived from (4.2.16) by differentiating with respect to t, since $\mathbf{P}$ and $\mathbf{H}$ are not functions of t. In (4.2.7) we wrote $\Delta \mathbf{X}$ for $\mathbf{H}$. This completes the proof. $\square$

4.2.5 The Newton-Raphson Method

To round off the discussion, we shall sketch the method for two equations, deferring the detailed treatment to Section 4.3.

We are given two equations

$$\begin{cases} F(x,y,z)=0 \\ G(x,y,z)=0 \end{cases} \tag{4.2.25}$$

For example, the original problem may be

$$\begin{cases} 4.75 \ \sin 2x - 3.14 e^y = 0.4945 \\ 3.63 \ \cos 3x + \sin y \ = 0.3995 \end{cases} \tag{4.2.26}$$

which we rewrite as

$$\begin{cases} F(x,y) = 4.75 \sin 2x - 3.14 e^y - 0.4945 = 0 \\ G(x,y) = 3.63 \cos 3x + \sin y \ - 0.3995 = 0 \end{cases} \tag{4.2.27}$$

Using the MVT once for F and once for G, and abbreviating $\mathfrak{h} = x_{k+1} - x_k$, $\mathfrak{k} = y_{k+1} - y_k$, we have

$$\begin{cases} F(x_{k+1}, y_{k+1}) = F(x_k, y_k) + (\mathfrak{h}, \mathfrak{k}) \cdot \begin{pmatrix} F_x \\ F_y \end{pmatrix} = 0 \\ \\ G(x_{k+1}, y_{k+1}) = G(x_k, y_k) + (\mathfrak{h}, \mathfrak{k}) \cdot \begin{pmatrix} G_x \\ G_y \end{pmatrix} = 0 \end{cases} \tag{4.2.28}$$

We know the values of F, G, F_x, F_y, G_x and G_y at (x_k, y_k). Thus (4.2.28) can be written as an easy set of simultaneous equations with $\mathfrak{h}$ and $\mathfrak{k}$ as unknowns:

$$\begin{cases} F_x \mathfrak{h} + F_y \mathfrak{k} = -F \\ G_x \mathfrak{h} + G_y \mathfrak{k} = -G \end{cases} \tag{4.2.29}$$

After solving for $\mathfrak{h}$ and $\mathfrak{k}$, we iterate by using

$$\begin{cases} x_{k+1} = x_k + \mathfrak{h} \\ y_{k+1} = y_k + \mathfrak{k} \end{cases} \tag{4.2.30}$$

and thus get a new estimate for x and y. This will become clear in the next section.

PROBLEM SECTION 4.2

A. Form the vector from A to B:

    (1) $A = (2,3)$, $B = (4,5)$ (5) $A = (3, -5), B = (-4, 8)$
    (2) $A = (1,3)$, $B = (2,4)$ (6) $A = (5, -7)$, $B(-3, 9)$
    (3) $A = (-2,6), B = (3, -5)$ (7) $A = (-1, -3)$, $B = (-2, -5)$
    (4) $A = (-3,7)$, $B = (-5, -8)$ (8) $A = (-2, -5)$, $B = (-5, -4)$

B. For each of the following, find F_x and F_y:

    (9) $(2x - y)^4$ (14) $(x^2 + y^2)^{-1/3}$
    (10) $(2x + 3y^2)^{3/2}$ (15) $(x^2 + y^2)^{1/2}$
    (11) $(x^2 + y^2)/(xy)$ (16) $\ln (x^2 - y^2)$
    (12) $e^y \sin x$ (17) $x + 3\ln x - y^2$
    (13) $e^y \cos x$ (18) $2x^2 - xy - 5x + 1$

C. Write the mean-value theorem for
    (19) three variables: x, y, z
    (20) four variables: x, y, z, t

4.3 THE NEWTON-RAPHSON ALGORITHM FOR MANY EQUATIONS

4.3.1 Two Equations (Theory)

We write the two equations in the form

$$\begin{cases} F(x,y) = 0 \\ G(x,y) = 0 \end{cases} \tag{4.3.1}$$

Then we use the MVT once for each equation:

$$\begin{cases} F(x_{k+1}, y_{k+1}) - F(x_k, y_k) = (\mathfrak{h}, \mathfrak{k}) \cdot \begin{pmatrix} F_x \\ F_y \end{pmatrix} \\ \\ G(x_{k+1}, y_{k+1}) - G(x_k, y_k) = (\mathfrak{h}, \mathfrak{k}) \cdot \begin{pmatrix} G_x \\ G_y \end{pmatrix} \end{cases} \tag{4.3.2}$$

Using $F(x_{k+1}, y_{k+1})$ and $G(x_{k+1}, y_{k+1}) = 0$, we obtain

$$\begin{cases} \mathfrak{h}F_x + \mathfrak{k}F_y = -F \\ \mathfrak{h}G_x + \mathfrak{k}G_y = -G \end{cases} \qquad (4.3.3)$$

all computations being done at (x_k, y_k). We solve for $\mathfrak{h}$ and $\mathfrak{k}$, either by determinants (easy for two equations) or by the Gauss-Jordan method. Then we use

$$\begin{cases} x_{k+1} = x_k + \mathfrak{h} \\ y_{k+1} = y_k + \mathfrak{k} \end{cases} \qquad (4.3.4)$$

to obtain a new estimate. The procedure is repeated until $\mathfrak{h}$ and $\mathfrak{k}$ (the errors) are as small as required.

4.3.2 Two Equations (Practice)

Consider

$$\begin{cases} 4.75 \sin 2x - 3.14 e^y = 0.4945 \\ 3.63 \cos 3x + \sin y = 0.3995 \end{cases} \qquad (4.3.5)$$

We rewrite this to look like (4.3.1):

$$\begin{cases} F(x,y) = 4.75 \sin 2x - 3.14 e^y - 0.4945 = 0 \\ G(x,y) = 3.63 \cos 3x + \sin y - 0.3995 = 0 \end{cases} \qquad (4.3.6)$$

Then

$$F_x = 9.50 \cos 2x, \qquad\qquad G_x = -10.89 \sin 3x$$
$$F_y = -3.14 e^y, \qquad\qquad G_y = \cos y \qquad (4.3.7)$$

To find a starter, we look at Figure 4.1-4. Three possible starters are $(4.5, -1.5)$, $(-1.5, -3.5)$ and $(1.5, -4.5)$. In actual practice, we would try all three. For our purposes, we arbitrarily choose

$$x_0 = 1.5 \qquad y_0 = -4.5 \qquad (4.3.8)$$

Substituting, we have

$$F(1.5, -4.5) = 4.75 \sin 3.0 - 3.14 e^{-4.5} - 0.4945 = 0.14094\,07434$$
$$G(1.5, -4.5) = 3.63 \cos 4.5 + \sin(-4.5) - 0.3995 = -0.18716\,21736$$
$$F_x(1.5, -4.5) = 9.50 \cos 3.0 = -9.40492\,7895$$
$$F_y(1.5, -4.5) = -3.14 e^{-4.5} = -0.03488\,2491$$
$$G_x(1.5, -4.5) = -10.89 \sin 4.5 = 10.64530\,298$$
$$G_y(1.5, -4.5) = \cos(-4.5) = -0.21079\,57994$$

Substituting these values into (4.3.3), we have

$$\begin{cases} -9.40492\,7895\mathfrak{h} - 0.03488\,2491\mathfrak{f} = -0.14094\,07434 \\ 10.64530\,298\mathfrak{h} - 0.21079\,5799\,4\mathfrak{f} = 0.18716\,21736 \end{cases} \qquad (4.3.9)$$

from which

$$\mathfrak{h} = 0.01539\,5026$$
$$\mathfrak{f} = -0.11040\,8773 \qquad (4.3.10)$$

Then

$$x_1 = x_0 + \mathfrak{h} = 1.5 + 0.01539\,5026 = 1.51539\,5026$$

$$y_1 = y_0 + \mathfrak{f} = -4.5 - 0.11040\,8773 = -4.61040\,8723 \qquad (4.3.11)$$

This gives us a new estimate for (x,y) and an estimate $(\mathfrak{h},\mathfrak{f})$ of the error of this estimate. On a computer, using double precision (16 sf), three iterations lead to errors $(\mathfrak{h},\mathfrak{f})$ less than a thousandth, and six iterations bring $(\mathfrak{h},\mathfrak{f})$ down to less than 10^{-7}. The results of the iterations are as follows:

x	y
1.50000\,0000	$-4.50000\,0000$
1.51539\,5026	$-4.61040\,8773$
1.51548\,2050	$-4.65271\,5718$
1.51551\,0361	$-4.66256\,4264$
1.51551\,2090	$-4.66316\,3627$
1.51551\,2097	$-4.66316\,5863$

4.3.3 Three Equations (Theory)

We write the three equations in the form

$$\begin{cases} F(x,y,z) = 0 \\ G(x,y,z) = 0 \\ H(x,y,z) = 0 \end{cases} \qquad (4.3.12)$$

Then we apply the mean-value theorem once for each equation. We are using

$$\begin{cases} \mathfrak{h} = x_{k+1} - x_k \\ \mathfrak{f} = y_{k+1} - y_k \\ \mathfrak{l} = z_{k+1} - z_k \end{cases} \qquad (4.3.13)$$

Then

$$\left\{ \begin{array}{l} F(x_{k+1}, y_{k+1}, z_{k+1}) - F(x_k, y_k, z_k) = (\mathfrak{h}, \mathfrak{k}, \mathfrak{l}) \begin{bmatrix} F_x \\ F_y \\ F_z \end{bmatrix} \\[3em] G(x_{k+1}, y_{k+1}, z_{k+1}) - G(x_k, y_k, z_k) = (\mathfrak{h}, \mathfrak{k}, \mathfrak{l}) \begin{bmatrix} G_x \\ G_y \\ G_z \end{bmatrix} \\[3em] H(x_{k+1}, y_{k+1}, z_{k+1}) - H(x_k, y_k, z_k) = (\mathfrak{h}, \mathfrak{k}, \mathfrak{l}) \begin{bmatrix} H_x \\ H_y \\ H_z \end{bmatrix} \end{array} \right. \qquad (4.3.14)$$

This leads to the set of simultaneous equations

$$\left\{ \begin{array}{l} \mathfrak{h} F_x + \mathfrak{k} F_y + \mathfrak{l} F_z = -F \\ \mathfrak{h} G_x + \mathfrak{k} G_y + \mathfrak{l} G_z = -G \\ \mathfrak{h} H_x + \mathfrak{k} H_y + \mathfrak{l} H_z = -H \end{array} \right. \qquad (4.3.15)$$

We solve for $(\mathfrak{h}, \mathfrak{k}, \mathfrak{l})$, and then substitute in

$$\left\{ \begin{array}{l} x_{k+1} = x_k + \mathfrak{h} \\ y_{k+1} = y_k + \mathfrak{k} \\ z_{k+1} = z_k + \mathfrak{l} \end{array} \right. \qquad (4.3.16)$$

giving us a new estimate of the root. The procedure is repeated until $(\mathfrak{h}, \mathfrak{k}, \mathfrak{l})$ is as small as required.

4.3.4 Three Equations (Practice)

Consider

$$\left\{ \begin{array}{l} \cos x + \sin y + \ln z = \quad 1.3667 \\ e^x + \sin 2y - z^2 \quad = -11.2813 \\ e^{-x} + \sin y + e^{-z} = \quad -0.1813 \end{array} \right. \qquad (4.3.17)$$

First, we rewrite this in standard form:

$$\left\{ \begin{array}{l} F(x,y,z) = \cos x + \sin y + \ln z - 1.3667 = 0 \\ G(x,y,z) = e^x \quad + \sin 2y - z^2 + 11.2813 = 0 \\ H(x,y,z) = e^{-x} + \sin y + e^{-z} + 0.1813 = 0 \end{array} \right. \qquad (4.3.18)$$

Then we calculate the partials:

$$F_x = -\sin x, \qquad G_x = e^x, \qquad H_x = -e^{-x}$$

$$F_y = \cos y, \qquad G_y = 2\cos 2y, \qquad H_y = \cos y$$

$$F_z = 1/z, \qquad G_z = -2z, \qquad H_z = -e^{-z} \qquad (4.3.19)$$

To find a starter, we look at Figure 4.1-5. Possible starters are $(0.5, -1.5, 3.5)$ and $(0.5, 5.0, 3.5)$. We choose

$$\begin{cases} x_0 = 0.5 \\ y_0 = -1.5 \\ z_0 = 3.5 \end{cases} \qquad (4.3.20)$$

Substituting in (4.3.15), we solve for $(\mathfrak{h}, \mathfrak{k}, \mathfrak{l})$. The new estimates for (x, y, z) with double-precision computing are as follows:

X	Y	Z
0.500000000	-1.500000000	3.500000000
0.020621423	-2.917475943	3.865017185
0.248705556	-2.160728680	3.786617844
0.393259854	-2.084794993	3.691182545
0.440429798	-2.129328714	3.706143017
0.446505542	-2.133764772	3.707670580
0.446602837	-2.133844842	3.707699162

After only seven iterations, the results are correct to the nearest millionth.

Strangely enough, if we start at $(1.0, -0.5, 3.5)$, the result, after eight iterations, is far from the previous one. This is shown in the following table. This should underline the importance of starting as near as possible to the desired root.

1.000000000	-0.500000000	3.500000000
0.643948673	-0.764849803	3.450580666
3.928986353	1.605912000	4.425646635
4.628038954	35.331082930	5.639787661
4.094863634	34.621385134	8.140108875
4.099511555	34.755906501	8.491721820
4.100957533	34.756899900	8.489264612
4.100956789	34.756900030	8.489265294

4.3.5 Pitfalls

Investigation I: Consider the following problem

$$\begin{cases} 7e^x - 3\sin 2y = -2.1371 \\ 3e^{-x} + 2\cos 2y = 38.7932 \end{cases} \qquad (4.3.21)$$

According to the computer printout, some possible starters are

$$(-2.5, -2.5), \quad (-2.5, -2.0), \quad (-2.5, 0.5), \quad (-2.5, 3.5) \qquad (4.3.22)$$

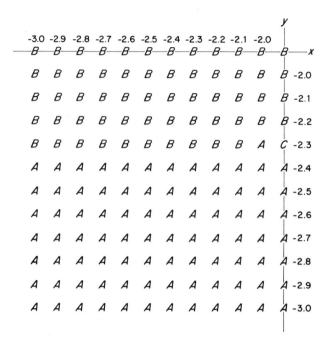

Fig. 4.2-6   A tiny difference in the starter can cause a different result.

We program the computer to solve (4.3.21) by the Newton-Raphson method with starters ranging from $(-3, -3)$ to $(-1.9, -1.9)$ incrementing by tenths.

Three different sets of answers were obtained. Calling these A, B and C, Figure 4.2-6 shows the answers obtained when different starters were used.

Observe that the starter $(-2.1, -2.3)$ led to set B answers, whereas $(-2.0, -2.3)$ led to set A, and $(-1.9, -2.3)$ led to set C.

In an actual problem, it may be important to try different starters in the same region to unearth the different roots.

Investigation II: The problem solved in section 4.3.2 reached the nearest thousandth in three iterations and the nearest hundred-millionth in six. This is very encouraging, but not a universal rule for Newton-Raphson iterations. Using the starter $(3, 0)$ for a set which looks much simpler,

$$\begin{cases} 3x + y^2 = \quad 9.0493 \\ x^3 - y = 151.0341 \end{cases} \tag{4.3.23}$$

it takes *eleven* iterations to reach

$$\begin{cases} x = 2.789000 \\ y = 0.826014 \end{cases} \tag{4.3.24}$$

which are correct to the nearest millionth.

Furthermore, if the system (4.3.21) is started with $x_0 = 2.5$, $y_0 = -2.0$, which looks like a point of accumulation on the computer printout, then the computer, after *35* iterations, reaches

$$\begin{cases} x = -2.583001 \\ y = 41.864313 \end{cases} \qquad (4.3.25)$$

correct to the nearest millionth, but nowhere near the starter.

PROBLEM SECTION 4.3

Solve by the Newton-Raphson method correct to the nearest millionth. The starter is given in parentheses.

(1) $\begin{cases} x + e^y = 68.0742, \\ \sin x - y = -3.5942 \end{cases}$ $(2.5, 4.0)$

(2) $\begin{cases} x - \cos y = 3.9578, \\ e^{-x} + y = -2.4204 \end{cases}$ $(3.0, -3.0)$

(3) $\begin{cases} 7e^x - 3\sin 2y = -2.1371, \\ 3e^{-x} + 2\cos 2y = 38.7932 \end{cases}$ $(-2.5, 0.5)$

(4) Same equations as (3); $(-2.5, -2.5)$

(5) Same equations as (3); $(-2.5, 3.5)$

(6) $\begin{cases} 2\cos x + 3e^y = 10.5234, \\ \ln x - xy = -3.7699 \end{cases}$ $(3.1, 1.5)$

(7) $\begin{cases} 3e^{-x} + 2e^y = 101.1182, \\ \sin x + 3\cos y = -1.9793 \end{cases}$ $(-3.5, -1.5)$

(8) $\begin{cases} x - y^2 = -2.2623, \\ 2x^3 + y = -6.4504 \end{cases}$ $(-1.0, -1.0)$

(9) $\begin{cases} \sin x + \cos y + e^z = 1.3461, \\ \cos x - \sin y + e^{-z} = 5.3288, \\ e^x + e^y - \sin z = 8.9769 \end{cases}$ $(1.5, 1.5, -2.0)$

(10) $\begin{cases} \ln x - \cos^2 y - \sin 2z = -3.0309, \\ e^x + \ln y + z = 0.5875, \\ \sin 3x - e^{-y} + \ln z = -0.8458 \end{cases}$ $(0.3, 0.3, 0.6)$

(11) $\begin{cases} \sin x + \ln x + \ln y - e^z = 1.87073, \\ \cos x + \sin y + e^{-z} = 39.93815, \\ \ln x - e^y + \sin z = -13.09764 \end{cases}$ $(0.6, 2.6, -3.5)$

(12) Same equations as (11); $(3.6, 3.1, -3.5)$

Chapter 5

Classical Interpolation Theory

This chapter deals mainly with classical methods which computer scientists are expected to know, even though they are only of theoretical value.

No doubt you are familiar with the elementary method of interpolation. For example, suppose two given values are

$$f(1.140) = \log 1.140 = 0.0569048513$$
$$f(1.150) = \log 1.150 = 0.0606978404$$

Then, assuming that $f(x)$ is linear between those values, we can proceed by ratio and proportion to obtain

$$f(1.145) = \log 1.145 \sim 0.0588013459$$

The true value (correct to the number of places cited) is

$$\log 1.145 = 0.0588054867$$

representing an error of 4.14×10^{-6} and a relative error of 7.04×10^{-5}. This may be satisfactory for some purposes, but one is hardly happy to interpolate in a 10-place table and emerge with a result correct to only 5 places.

Assuming that a mathematician wants to interpolate accurately in a table of values, the following are some of the considerations in choosing a procedure:

1. The data may be accurate to the number of digits displayed (as in a table of logarithms) or may have random errors (as the result of a scientific study).
2. The arguments may be equally spaced, as in most published tables, or may not be, as in some experimental studies.
3. It may be desirable to have *least-squares error* characteristics or *minimax* characteristics. These will be discussed in the chapters on Fourier and Chebyshev series.

4. The investigation may involve finding a few isolated values, or it may involve printing a new table with arguments spaced closer. The latter problem is called *subtabulation.*

In this chapter, we concern ourselves with a classical problem: interpolation in a table, using a polynomial, with the assumption that the data are precise. In particular, we attack two kinds of tables:

1. Given a table of values where the arguments may be unequally spaced, required to interpolate by means of a polynomial. In Table 5-1, we have simulated this situation by excerpting values from a full table of values for the error function, erf (x).

Table 5-1
The Error Function

x	erf(x)
1.00	0.84270 07929
1.10	0.88020 50696
1.32	0.93806 51551
1.48	0.96365 40654
1.59	0.97546 20158

2. Given a table of values where the arguments are equally spaced, required to interpolate in the table or to subtabulate the table by use of a polynomial. In Table 5-2, a few values are given for the sine integral, Si(x). A typical problem might be to recalculate the table with arguments spaced at 0.01 instead of 0.05.

Table 5-2   The Sine Integral

x	$S_i(x)$
0.50	0.49310 74180
0.55	0.54084 03951
0.60	0.58812 88096
0.65	0.63493 50541
0.70	0.68122 22391
0.75	0.72695 42472
0.80	0.77209 57855
0.85	0.81661 24372
0.90	0.86047 07107
0.95	0.90363 80880
1.00	0.94608 30704
1.05	0.98777 52233
1.10	1.02868 52187
1.15	1.06878 48757
1.20	1.10804 71990
1.25	1.14644 64157
1.30	1.18395 80091
1.35	1.22055 87513
1.40	1.25622 67328
1.45	1.29094 13902

In Section 5.1. we explore the classical Lagrange method, a method suitable for either table. In Sections 5.2 and 5.3, we discuss the *difference operator* Δ together with some related material. In the last section, we discuss repeated differences and the difference table.

5.1 THE INTERPOLATING POLYNOMIAL

5.1.1 The Problem

We consider a table of x and $f(x)$ in which the values of x may or may not be equally spaced, i.e., the table may resemble either Table 5-1 or Table 5-2. It is probably obvious that a polynomial of some kind can always be drawn through n distinct points. For example, if we have points $(x_0, f(x_0))$, $(x_1, f(x_1))$ and $(x_2, f(x_2))$, and if we assume that in the interval containing these points the function $f(x)$ can be approximated by a polynomial

$$P(x) = Ax^2 + Bx + C \tag{5.1.1}$$

then by direct substitution we obtain

$$\begin{cases} y_0 = f(x_0) = (x_0)^2 A + (x_0) B + C \\ y_1 = f(x_1) = (x_1)^2 A + (x_1) B + C \\ y_2 = f(x_2) = (x_2)^2 A + (x_2) B + C \end{cases} \tag{5.1.2}$$

The system is easily solved for the coefficients A, B and C by the Gauss-Jordan method. (We shall illustrate this.) It seems clear that if the system has a solution, it is a unique solution.

More generally, if we have $n+1$ given points, it appears that a polynomial of degree *no higher than n* can be passed precisely through them. In the system above, if $A \neq 0$, then (5.1.1) is of the second degree. If $A = 0$ and $B \neq 0$, the (5.1.1) is of first degree. If both A and B are 0, then $P(x)$ is of zero degree, i.e., a constant. We shall prove the following theorem.

INTERPOLATING POLYNOMIAL THEOREM   Given $n+1$ distinct points $(x_0, y_0), (x_1, y_1), \ldots, (x_n, y_n)$, it is always possible to construct a polynomial $P(x)$ of degree no higher than n such that $P(x_k) = y_k$ for $k = 0, 1, \ldots, n$. The polynomial is unique.

We shall prove the existence and the uniqueness separately.

5.1.2 Existence

As a guide to the demonstration of existence of a polynomial through a set of points, let us construct a polynomial through three points: (x_0, y_0),

(x_1, y_1) and (x_2, y_2). We want to be sure that $P(x_k) = y_k$ for $k = 0, 1, 2$. Consider the following clever formulation for three points, an example of *Lagrange's method*.

$$P(x) = \frac{(x - x_1)(x - x_2)}{(x_0 - x_1)(x_0 - x_2)} y_0 + \frac{(x - x_0)(x - x_2)}{(x_1 - x_0)(x_1 - x_2)} y_1 + \frac{(x - x_0)(x - x_1)}{(x_2 - x_0)(x_2 - x_1)} y_2$$

$$(5.1.3)$$

Note that when $x = x_0$, the coefficient of y_0 is 1 and the coefficients of y_1 and y_2 are 0. Thus

$$P(x_0) = y_0 \qquad (5.1.4)$$

Similarly, when $x = x_1$, the first and third coefficients are 0 and the second coefficient is 1. Thus

$$P(x_1) = y_1 \qquad (5.1.5)$$

Similarly

$$P(x_2) = y_2 \qquad (5.1.6)$$

This proves that $P(x)$ actually passes through the three given points.

Let us examine (5.1.3) to see how these coefficients were formed. It is easiest to start with the denominators. In the denominator of the coefficient of y_0, note that there are two binomials, each beginning with x_0 with the other given values of x_k subtracted. In the denominator of the coefficient of y_1, each binomial begins with x_1. In the denominator of the coefficient of y_2, each binomial begins with x_2. The numerators are like the denominators except that the first term of each binomial is x instead of x_0, x_1 or x_2.

What is the degree of the polynomial in (5.1.3)? Note that the subscripted variables x_0, x_1, y_0, y_1 and y_2 are constants. Therefore, all the denominators and all the y_k are constants. The first numerator, expanded, is

$$(x - x_1)(x - x_2) = x^2 - (x_1 + x_2)x + x_1 x_2 \qquad (5.1.7)$$

which is second degree. The other numerators are also second degree. When the three fractions are added, it is possible that the x^2 term, or an x term, will vanish, but it is not possible that a term with x^3 or higher degree will appear.

We omit the obvious generalization.

5.1.3 Uniqueness

The proof of uniqueness depends upon two rather obvious theorems from the theory of equations:

1. The difference of two polynomials, each of no more than nth degree, is a polynomial of no more than nth degree.

2. If a polynomial of no more than nth degree vanishes at $n+1$ points, it is identically zero. This means it vanishes for *all* values of the argument.

To illustrate statement 1, let $P_1 = ax^3 + bx^2 + cx + 7$, and let $P_2 = dx^3 + ex^2 + fx + 9$. Then $P_1 - P_2 = (a-d)x^3 + (b-e)x^2 + (c-f)x - 2$, a polynomial whose degree depends upon the relative values of a and d, b and e, and c and f.

To illustrate statement 2, let $P(x) = ax^2 + bx + c$ for $x = 1, 2$ and 3. Then

$$\begin{cases} P(1) = a + b + c = 0 \\ P(2) = 4a + 2b + c = 0 \\ P(3) = 9a + 3b + c = 0 \end{cases} \tag{5.1.8}$$

Subtracting the first equation from the second, and the second from the third, we have

$$\begin{cases} 3a + b = 0 \\ 5a + b = 0 \end{cases} \tag{5.1.9}$$

from which $a = 0$ and $b = 0$. Substituting in any of the three equations of (5.1.8), we have $c = 0$. Therefore $P(x)$ is identically zero.

Proof of Uniqueness   Suppose $P_1(x)$ and $P_2(x)$ are two *different* polynomials, both of which satisfy $f(x)$ for $n+1$ points. Then, by statement 1 above, there is a polynomial $P_3(x)$ such that

$$P_3(x) = P_1(x) - P_2(x) \tag{5.1.10}$$

Since $P_1(x)$ and $P_2(x)$ are of degree at most n, $P_3(x)$ is also of degree at most n.

We were given that $P_1(x)$ and $P_2(x)$ pass through the same $n+1$ points. Therefore, at each of these points,

$$P_1(x_k) = P_2(x_k), \qquad k = 0, 1, \ldots, n \tag{5.1.11}$$

Therefore, at each of these points,

$$P_3(x_k) = 0, \qquad k = 0, 1, \ldots, n \tag{5.1.12}$$

However, a polynomial of at most nth degree which vanishes at $n+1$ points has to be zero for all values of x. Therefore, substituting in (5.1.10) we have

$$P_1(x) - P_2(x) = 0 \tag{5.1.13}$$

which contradicts the hypothesis that there are two *different* interpolating polynomials through the points. This completes the proof of uniqueness. $\square$

ILLUSTRATIVE PROBLEM 5.1-1 What is the Lagrange polynomial through $(1,3)$, $(2,5)$ and $(3,7)$?

Solution

$$P(x) = \frac{(x-2)(x-3)}{(1-2)(1-3)}3 + \frac{(x-1)(x-3)}{(2-1)(2-3)}5 + \frac{(x-1)(x-2)}{(3-1)(3-2)}7$$

$$= 2x - 1$$

Observe that the x^2 term vanishes in this case because the three given points are collinear.

5.1.4 Error of the Interpolation Polynomial

Assume that we have approximated $f(x)$ by a polynomial of nth degree, $P_n(x)$. It doesn't matter what method was used since there is one and only one such polynomial through the same points, as proved above. Then the error is given by

$$E_n(x) = f(x) - P_n(x) \tag{5.1.14}$$

The error will be zero at $n+1$ points: $x_0, x_1, \ldots, x_n$. Then E_n can be represented by

$$E_n(x) = g(x) \cdot (x - x_0)(x - x_1) \cdots (x - x_n) \tag{5.1.15}$$

which obviously vanishes at the required points.

We now employ a trick to find out what $g(x)$ is. We invent the function

$$F(t) = E(t) - g(x)(t - x_0)(t - x_1) \cdots (t - x_n) \tag{5.1.16}$$

It is obvious that $F(t)$ is 0 when $t = x_0, x_1, \ldots, x_n$. This happens $n+1$ times. However, because of (5.1.15), it is also zero when $t = x$. Therefore, $F(t)$ vanishes at least $n+2$ times on any interval I which contains $x_0, x_1, \ldots, x_n$.

We now remind the reader of

ROLLE'S THEOREM Let g be a function continuous on $[a,b]$ and differentiable on (a,b), and let $g(a) = g(b) = 0$. Then, for some $\theta \in (a,b)$, $g'(\theta) = 0$.

In Figure 5.1-1, we see that the theorem guarantees that there is at least one value of x such that $g'(x) = 0$. In the figure, there are actually five such points.

Applying the theorem, we differentiate $F(t)$ once. Then there are at least $n+1$ horizontal tangents. Differentiating again, $F''(t)$ has at least n horizontal tangents. Skipping to the end, $F^{(n+1)}(t)$ has at least one value of t inside the interval $I = [a,b]$, such that $F^{(n+1)}(\theta) = 0$.

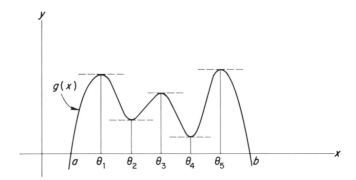

Fig. 5.1-1 Rolle's theorem guarantees that there is at least one θ such that $g'(\theta)=0$.

Now consider the right member of (5.1.16). The unknown function $g(x)$ is not a function of t and remains unchanged. The expression $(t-x_0)(t-x_1)\cdots(t-x_n)$ is a polynomial in t of degree $n+1$. Therefore, its $(n+1)$st derivative is a constant, namely $(n+1)!$. Now think about $E(t)$. From (5.1.14), the $(n+1)$st derivative of $E(t)$ is $f^{(n+1)}(t)-P^{(n+1)}(t)$. But P is a polynomial of degree n, and its $(n+1)$st derivative vanishes.

Putting these all together at $t=\theta$, we have

$$F^{(n+1)}(\theta)=0=f^{(n+1)}(\theta)-g(x)\cdot(n+1)! \tag{5.1.17}$$

$$g(x)=\frac{f^{(n+1)}(\theta)}{(n+1)!} \tag{5.1.18}$$

This proves the following very important theorem.

INTERPOLATING-POLYNOMIAL ERROR THEOREM   If $f(x)$ has $n+1$ derivatives on interval I and if it is approximated by a polynomial passing through $n+1$ points on I, then the error, $f(x)-P(x)$, is given by

$$E_n(x)=\frac{f^{(n+1)}(\theta)}{(n+1)!}\pi_n(x),\qquad \theta\in I \tag{5.1.19}$$

where

$$\pi_n(x)=(x-x_0)(x-x_1)\cdots(x-x_n) \tag{5.1.20}$$

ILLUSTRATIVE PROBLEM 5.1-2   Estimate the theoretical error of interpolation, given that five points were used, for

$$f(x)=\operatorname{erf}(x)=\frac{2}{\sqrt{\pi}}\int_0^x e^{-t^2}dt \tag{5.1.21}$$

Solution   Since there were five given points for the fourth-degree polynomial, we must find the fifth derivative of (5.1.21). By the Fundamental Theorem of the calculus,

$$f'(x) = \frac{2}{\sqrt{\pi}} e^{-x^2}$$

Then

$$f''(x) = \frac{2}{\sqrt{\pi}} e^{-x^2}(-2x) = -2xf'(x)$$

$$f'''(x) = -2f'(x) - 2xf''(x)$$

$$f^{iv}(x) = -4f''(x) - 2xf'''(x)$$

$$f^{v}(x) = -6f'''(x) - 2xf^{iv}(x)$$

For the interval containing the given points $(1.00, 1.59)$, the maximum value of the fifth derivative is about -8.3. Substituting in (5.1.19),

$$|E_4(x)| \leqslant \left| \frac{-8.3}{5!} (1.20 - 1.00)(1.20 - 1.10)(1.20 - 1.32)(1.20 - 1.48)(1.20 - 1.59) \right|$$

$$= 1.8 \times 10^{-5}$$

The actual error was 1.1×10^{-5}.

ILLUSTRATIVE PROBLEM 5.1-3 The Bessel function of order zero can be defined by

$$J_0(x) = \frac{1}{\pi} \int_0^{\pi} \cos(x \sin t) \, dt$$

Suppose it is approximated by a fifth-degree polynomial which agrees at $x = 0, 1, 2, 3, 4, 5$. Investigate the maximum errors of interpolation in this range.

Solution Here $n = 5$, so we need the sixth derivative. According to Leibniz's rule, if the integrand is continuous and has a continuous derivative, we can differentiate with respect to the parameter x under the integral sign. Then

$$J_0'(x) = \frac{-1}{\pi} \int_0^{\pi} \sin(x \sin t) \sin t \, dt$$

$$J_0''(x) = \frac{-1}{\pi} \int_0^{\pi} \cos(x \sin t) \sin^2 t \, dt$$

and so on. Then

$$|J_0^{vi}(x)| \leqslant \frac{1}{\pi} \int_0^{\pi} dt = 1$$

Also, $\pi(x) = (x - 0)(x - 1)(x - 2)(x - 3)(x - 4)(x - 5)$, and $(n + 1)! = 6! = 720$. The following table exhibits some of the estimates and, for comparison, the actual values of J_0.

x	$J_0(x)$	$|E(x)|$
0.5	0.93847	2.1×10^{-3}
1.5	0.51183	6.8×10^{-3}
2.5	-0.04838	4.9×10^{-3}
3.5	-0.38013	6.8×10^{-3}
4.5	-0.32054	2.1×10^{-3}

Evidently, the fifth-degree polynomial was not very successful. Errors as high as 6.8×10^{-3} are certainly nothing to be proud of. If a polynomial of degree 12 is chosen (using 13 values of x), the largest error is about 8×10^{-7}.

5.1.5 Computer Method for Simple Polynomial Approximation

There are two important results in the preceding discussion:

1. There is only one polynomial of degree n or less which passes through a given set of $n+1$ points, so the method for finding the polynomial is theoretically immaterial. Of course, the roundoff errors may be different.
2. If the function supplying the points is sufficiently smooth, the maximum error due to the use of polynomial as an approximation can be estimated by the formula (5.1.19).

Under the circumstances, we are free to use the most convenient method for computing. We choose to solve a system of simultaneous equations in the illustrative problems which follow.

We shall have different approaches to approximation in the chapters on Fourier series and Chebyshev series.

ILLUSTRATIVE PROBLEM 5.1-4   Given the following table of values for $\sin x$, find a polynomial which passes precisely through the given points. Find the errors at the midvalues (midway between the given points), using the computer value of $\sin x$ as the true value.

x	$f(x)$
0.0	0.0
0.200	1.98669 33079 50612D $-$01
0.400	3.89418 34230 86505D $-$01
0.500	4.79425 53860 42030D $-$01
0.750	6.81638 76002 33342D $-$01
0.830	7.37931 37110 99627D $-$01
1.000	8.41470 98480 78965D $-$01

Solution   We set up seven equations. The second equation (with numerical values abbreviated to ease reading) is $a_0 + a_1(0.2) + a_2(0.2)^2 + a_3(0.2)^3 + a_4(0.2)^4 + a_5(0.2)^5 + a_6(0.2)^6 = 0.198...$

Solving on the computer, using the Gauss-Jordan method, the coefficients are as follows.

$$A(0) = 0.0$$

$$A(1) = 1.0000043960389040\ 00$$

$$A(2) = -5.7560113872375240-05$$

$$A(3) = -1.6637667802057020-01$$

$$A(4) = -7.3286482798061550-04$$

$$A(5) = 9.3259509910718270-03$$

$$A(6) = -6.9225925965670290-04$$

To check at the midvalues, we calculate $\frac{1}{2}(0.0+0.2)$, $\frac{1}{2}(0.2+0.4)$, etc., and find the sine at these values. The actual error is true value minus machine value. The following table shows these selected values:

MIDVALUE	ERROR
.100	$-9.00D-08$
.300	$1.74D-08$
.450	$-3.03D-09$
.625	$1.22D-08$
.790	$-2.96D-09$
.915	$2.76D-08$

These are not necessarily the maximum errors. Figure 5.1-2 displays the program used to obtain these results. In the program, FGJDOD was a house program which called a subroutine to accomplish the Gauss-Jordan reduction. The parameter in parentheses is the total number of columns, including the column of constants.

```
        DOUBLE PRECISION X(10,11),A(10),CK(7),XM(6),FXM(6),P,A0,
       A1,A2,A3,A14,A5,A6,Y,Z(7),C(7)
        COMMON X,A
C
        P(Y)=A0+Y*(A1+Y*(A2+Y*(A3+Y*(A4+Y*(A5+Y*A6)))))
C
        DO 10 I=1,7
10      X(I,1)=1.D0
        X(1,2)=0.D0
        X(7,2)=1.D0
C
1       X(2,2)=2.D-1
        X(3,2)=4.D-1
        X(4,2)=5.D-1
        X(5,2)=7.5D-1
        X(6,2)=8.3D-1
C
        DO 11 I=1,7
        Y=X(1,2)
11      X(I,8)=DSIN(Y)
C
        DO 12 I=1,6
        L=I+1
12      XM(I)=(X(L,2)+X(I,2))/2.D0
        DO 13 I=1,6
13      FXM(I)=DSIN(XM(I))
        WRITE(3,101)
101     FORMAT(1H1,T12,'X',T21,'F(X)'//)
        DO 122 I=1,7
122     WRITE(3,102)X(I,2),X(I,8)
102     FORMAT(1H0,T10,F5.3,T19,1PD9.2)
C CALCULATION OF MATRIX ELEMENTS.
        DO 235 I=1,7
C THE NEXT TWO CARDS ARE TO PRESERVE GIVEN VALUES.
        Z(I)=X(I,2)
        C(I)=X(I,8)
C
        Y=Z(I)
        X(I,3)=Y**2
        X(I,4)=Y*X(I,3)
        X(I,5)=X(I,3)**2
        X(I,6)=X(I,5)*Y
235     X(I,7)=X(I,4)**2
```

Fig. 5.1-2 (Continued on next page)

```
C CALCULATION OF COEFFICIENTS.
      CALL FGJDOD(8)
C CHECKING AT THE GIVEN POINTS.
      A0=A(1)
      A1=A(2)
      A2=A(3)
      A3=A(4)
      A4=A(5)
      A5=A(6)
      A6=A(7)
C
      DO 24 I=1,7
      Y=Z(I)
24    CK(I)=C(I)-P(Y)
C PRINT-OUT OF COEFFICIENTS AND ERRORS AT GIVEN POINTS.
      WRITE(3,103)
103   FORMAT(1H1,T10,'COEFFICIENTS',T75,'ERRORS AT GIVEN POINTS'//)
      DO 26 I=1,7
      L=I-1
26    WRITE(3,104)L,A(I),CK(I)
104   FORMAT(1H0,T10,'A(',I1,') = ',1PD22.15,T75,D22.15)
25    CONTINUE
C ERRORS AT MIDVALUES.
      WRITE(3,105)
105   FORMAT(1H1,T10,'ERRORS AT MIDVALUES'//)
      WRITE(3,106)
106   FORMAT(1H0,T10,'MIDVALUE',T24,'ERROR'//)
      DO 28 I=1,6
      Y=XM(I)
      CK(I)=FXM(I)-P(Y)
28    WRITE(3,107)XM(I),CK(I)
107   FORMAT(1H0,T10,F4.3,T22,1PD9.2)
      END
```

Fig. 5.1-2   A computer program to send a polynomial through six given points.

In the program, the fourth line is the *nested* polynomial equivalent to $a_0 + a_1 x + \cdots + a_6 x^6$. We shall discuss it more thoroughly later. For the present, note that it minimizes the number of operations to do the computation. For example, the computation of

$$ax^3 + bx^2 + cx + d$$

requires six multiplications and three additions. The computation of

$$d + x(c + x(b + ax))$$

requires three multiplications and three additions.

Would the same coefficients be obtained for (1) a different set of points? (2) a different selection of points? The answer is "no" to both questions. We shall probe more deeply into these questions and into the question of estimating the error later in the book.

PROBLEM SECTION 5.1

A. Using Lagrange's formula, find the interpolating polynomial for each of the following sets of points:
   (1) $(-2, -126), (-1.5, -33), (0, -3)(4, 77)$
   (2) $(-2, -38), (-1, -7), (0, 4), (3, 97)$
   (3) $(-2.7, -21.08), (-0.8, 1.72), (0.9, 9.88), (4.6, 7.0)$
   (4) $(-2.6, -14.64), (-1.3, -10.22), (2.8, -3.72), (3.9, 7.46)$

B. Using Lagrange's method, (a) find an interpolating polynomial of fifth degree using only the points marked with an asterisk, (b) exhibit a table of the theoretical errors over the range, (c) find the actual errors at the five intermediate points.

(5) Erf(x) is defined as

$$\frac{2}{\sqrt{\pi}} \int_0^x e^{-t^2} dt$$

Using max $f'(x)$ as approximately 40 and the six values marked with an

Table 5.1-1

x	erf(x)
*0.0	0.0
0.2	0.22270
*0.4	0.42839
0.6	0.60386
*0.8	0.74210
1.0	0.84270
*1.2	0.91031
1.4	0.95229
*1.6	0.97635
1.8	0.98909
*2.0	0.99532

asterisk in Table 5.1-1, show that erf(x) is approximated by

$$p\left[\frac{1.74312}{x-0.4} - \frac{6.03923}{x-0.8} + \frac{7.40812}{x-1.2} - \frac{3.97278}{x-1.6} + \frac{0.809993}{x-2.0} \right]$$

where $p = x(x-0.4)(x-0.8)(x-1.2)(x-1.6)(x-2.0)$ and that the theoretical errors, according to (5.1.19), are given by

x	ERROR
0.2	3.4×10^{-3}
0.6	1.1×10^{-3}
1.0	0.8×10^{-3}
1.4	1.1×10^{-3}
1.8	3.4×10^{-3}

whereas the actual errors are given by

x	ERROR
0.2	70×10^{-5}
0.6	11×10^{-5}
1.0	1×10^{-5}
1.4	9×10^{-5}
1.8	43×10^{-5}

(6) The Jacobian elliptic function is defined by

$$K(m) = \int_0^{\pi/2} \frac{d\theta}{(1 - m\sin^2\theta)^{1/2}}$$

Proceed as in (5), using Table 5.1-2.

Table 5.1-2

m	$K(m)$
*0.00	1.57080
0.09	1.60805
*0.18	1.64968
0.27	1.69675
*0.36	1.75075
0.45	1.81388
*0.54	1.88953
0.63	1.98337
*0.72	2.10595
0.81	2.28055
*0.90	2.57809

C. (7) Sin x is approximated by a fifth-degree polynomial which agrees at $0°$, $5°$, $10°$, $15°$, $20°$, and $25°$. Estimate the theoretical maximum error of interpolation at the midvalues. (*Hint*: $|f^{vi}| \leqslant 1$.)

(8) Ln x is approximated by a sixth-degree polynomial which agrees at (a) $1, 2, 3, 4, 5, 6, 7$, (b) $10, 11, 12, 13, 14, 15, 16$. Estimate the theoretical maximum error of interpolation at the midvalues. (*Hint*) $|f^{vii}| \leqslant 6!/x^7$.)

D. For each of the following, use the Gauss-Jordan method to find the interpolating polynomial. Check at the midvalues, using the function named and a sixth-degree polynomial.

(9) $\tan x$, x from 0 to 1

(10) $\arcsin x$, x from 0 to 1

(11) $\arctan x$, x from 0 to 1

(12) $\sinh x$, x from 0 to 1

(13) $\tanh x$, x from 0 to 1

(14) e^x, x from 0 to 1

(15) e^{x^2}, x from 0 to 1

(16) $\ln(1-x)$, x from 0 almost to 1

5.2 FACTORIAL FUNCTIONS

5.2.1 The Binomial Theorem

Secondary-school algebra courses always include the following theorem, proved by induction for integral values of n.

BINOMIAL THEOREM   For integral n,

$$(x+y)^n = x^n + nx^{n-1}y + \frac{n(n-1)}{1\cdot 2}x^{n-2}y^2 + \cdots + y^n \qquad (5.2.1)$$

The coefficients

$$1, \quad n, \quad \frac{n(n-1)}{1\cdot 2}, \quad \frac{n(n-1)(n-2)}{1\cdot 2\cdot 3}, \quad \cdots$$

are called *binomial coefficients*. It is sometimes convenient to define them in terms of *factorials*:

FACTORIAL

$$0! = 1, \qquad n! = n[(n-1)!] \text{for } n \geqslant 1 \qquad (5.2.2)$$

This definition is said to be *recursive* because after the first one, each quantity is calculated from the preceding one. For example, $0! = 1$ is a starter. Then $1! = 1(0!) = 1$. Then $2! = 2(1!) = 2$. Then $3! = 3(2!) = 6$, and so on.

Binomial coefficients can be defined in terms of factorials as follows.

BINOMIAL COEFFICIENT

$$\binom{n}{k} = \frac{n!}{k!(n-k)!} \qquad (5.2.3)$$

The symbol $\binom{n}{k}$ can also be written ^nC_k, C_k^n, $C(n,k)$, and other ways. C stands for *combination*, a word from the theory of probability. The symbol, however it is written, may be read as n, C, k.

ILLUSTRATIVE PROBLEM 5.2-1   Find the numerical values of (a) $\binom{3}{0}$, (b) $\binom{5}{3}$, (c) $\binom{7}{2}$, (d) $\binom{8}{8}$.

Solution   (a) $\dfrac{3!}{0!\,3!} = 1$;

(b) $\dfrac{5!}{3!\,2!} = \dfrac{5 \times 4 \times 3 \times 2 \times 1}{3 \times 2 \times 1 \times 2 \times 1} = 10$;

(c) $\dfrac{7 \times 6 \times 5 \times 4 \times 3 \times 2 \times 1}{2 \times 1 \times 5 \times 4 \times 3 \times 2 \times 1} = 21$;

(d) $\dfrac{8!}{0!\,8!} = 1$.

5.2.2 Pascal's Triangle

Pascal's triangle, shown in Figure 5.2-1, is often used to find the numerical values of small binomial coefficients. In this triangle, a binomial coefficient is easily obtained by adding two adjacent coefficients in one line to form the one between them and below. For example, the next line of the triangle is $1, 5, 10, 10, 5, 1$. In the next problem, we prove a general relationship which is needed later.

$$
\begin{array}{c}
1 \\
1 \quad 1 \\
1 \quad 2 \quad 1 \\
1 \quad 3 \quad 3 \quad 1 \\
1 \quad 4 \quad 6 \quad 4 \quad 1
\end{array}
$$

Fig. 5.2-1 A portion of Pascal's triangle.

ILLUSTRATIVE PROBLEM 5.2-2   Prove that $\binom{x+1}{k} = \binom{x}{k} + \binom{x}{k-1}$.

Solution

$$\binom{x+1}{k} = \frac{(x+1)!}{k!(x-k+1)!}, \quad \binom{x}{k} = \frac{x!}{k!(x-k)!}, \quad \binom{x}{k-1} = \frac{x!}{(k-1)!(x-k+1)!}.$$

Using $(x+1)! = (x+1)x!$, $(x-k+1)! = (x-k+1)(x-k)!$, and $k! = k(k-1)!$, we have

$$\binom{x}{k} + \binom{x}{k-1} = \frac{(x+1)!(x-k+1)}{(x+1)k!(x-k+1)!} + \frac{(x+1)!k}{(x+1)k!(x-k+1)!}$$

$$= \frac{(x+1)!}{k!(x-k+1)!}\left[\frac{x-k+1}{x+1} + \frac{k}{x+1}\right]$$

$$= \frac{(x+1)!}{k!(x-k+1)!}$$

which completes the proof. $\square$

5.2.3   Factorial Functions

A somewhat more flexible computation of binomial coefficients is based upon *factorial function notation.* We pause to illustrate this notation before defining it.

$$5^{(3)} = 5 \times 4 \times 3 = 60$$
$$3^{(2)} = 3 \times 2 = 6$$
$$0^{(3)} = 0(-1)(-2) = 0$$
$$\left(\frac{1}{2}\right)^{(3)} = \frac{1}{2} \times \frac{-1}{2} \times \frac{-3}{2} = \frac{+3}{8}$$
$$n^{(0)} = 1$$
$$0^{(0)} = 1$$

We suggest that $5^{(3)}$ be read as "5 to three factors".

FACTORIAL FUNCTION $n^{(0)} = 1$, and for any real n and any non-negative integral m, $n^{(m)} = n(n-1)(n-2)\cdots$ to m factors. The last one is $\cdot (n-m+1)$

Note that

$$\binom{5}{2} = \frac{5 \times 4 \times 3 \times 2 \times 1}{2 \times 1 \times 3 \times 2 \times 1} = \frac{5 \times 4}{2 \times 1} = \frac{5^{(2)}}{2!}$$

Similarly

$$\binom{7}{3} = \frac{7^{(3)}}{3!}$$

In general, we can prove the

BINOMIAL-COEFFICIENT THEOREM   For integral n and k,

$$\binom{n}{k} = \frac{n^{(k)}}{k!} \tag{5.2.4}$$

Proof   Note that

$$n! = \left[n(n-1)(n-2) \cdots (n-k+1) \right]\left[(n-k)(n-k-1) \cdots 1 \right] \tag{5.2.5}$$

where the first bracket has k factors and the second has $n-k$ factors. Therefore

$$n! = n^{(k)}(n-k)! \tag{5.2.6}$$

But

$$\binom{n}{k} = \frac{n!}{k!(n-k)!} = \frac{n^{(k)}(n-k)!}{k!(n-k)!} = \frac{n^{(k)}}{k!} \tag{5.2.7}$$

which completes the proof. □

The binomial theorem can be stated compactly in either combinatorial or factorial-function form:

BINOMIAL THEOREM   For integral n and k,

$$(x+y)^n = \sum_{k=0}^{n} \binom{n}{k} x^{n-k} y^k = \sum_{k=0}^{n} \frac{n^{(k)}}{k!} x^{n-k} y^k \tag{5.2.8}$$

ILLUSTRATIVE PROBLEM 5.2-3   Find $(a+b)^5$.

Solution

$$(a+b)^5 = \sum_{k=0}^{5} \binom{5}{k} a^{5-k} b^k$$

$$= \binom{5}{0} a^5 b^0 + \binom{5}{1} a^4 b^1 + \binom{5}{2} a^3 b^2 + \binom{5}{3} a^2 b^3 + \binom{5}{4} a^1 b^4 + \binom{5}{5} a^0 b^5$$

$$= a^5 + 5a^4 b + 10a^3 b^2 + 10a^2 b^3 + 5ab^4 + b^5$$

ILLUSTRATIVE PROBLEM 5.2-4   Find the first three terms of $(p+q)^{12}$.

Solution

$$\binom{12}{0} p^{12} q^0 + \binom{12}{1} p^{11} q^1 + \binom{12}{2} p^{10} q^2 = p^{12} + 12p^{11} q + 66p^{10} q^2.$$

5.2.4 The Binomial Series

In (5.2.8), if $x = 1$, $y = a$, and n is any real number, the result is called a *binomial series*. The following can be proved; the proof is difficult.[1]

BINOMIAL-SERIES THEOREM   For $|a| < 1$, and n real but neither 0 nor a positive integer,

$$(1 + a)^n = \sum_{k=0}^{\infty} \frac{n^{(k)}}{k!} a^k \qquad (5.2.9)$$

Mathematicians have agreed to extend the definition of binomial coefficient to allow n to be any real number, so that the binomial-series theorem can also be expressed as

$$(1 + a)^n = \sum_{k=0}^{\infty} \binom{n}{k} a^k \qquad (5.2.10)$$

ILLUSTRATIVE PROBLEM 5.2-5   Find the first three terms of the binomial series expansion of $\sqrt{1.0373}$.

Solution

$$\sqrt{1.0370} = (1 + 0.0370)^{1/2}$$

$$= 1 + \frac{\left(\frac{1}{2}\right)^{(1)}}{1!}(0.0370)^1 + \frac{\left(\frac{1}{2}\right)^{(2)}}{2!}(0.0370)^2 + \cdots$$

$$= 1 + \tfrac{1}{2}(0.0370) + \left(-\tfrac{1}{8}\right)(0.00136\,900) + \cdots$$

$$= 1.0183 \qquad \text{rounded to 5 sf}$$

This happens to be correct for all the digits displayed, but the method is not a good computer procedure. It is merely practice in the symbolism so far as we are concerned.

PROBLEM SECTION 5.2

The following exercises are designed to develop skill in the use of symbols.

A. Evaluate the following:

(1) $\displaystyle\sum_{i=1}^{10} (2i - 1)$ (4) $\displaystyle\sum_{k=1}^{10} (k^2 + k - 5)$

(2) $\displaystyle\sum_{j=1}^{10} (j^2 + i)$ (5) $3! \times 4^{(2)}$

(3) $\displaystyle\sum_{k=1}^{10} (2k^2 - k + 3)$ (6) $5! \times 4^{(3)}$

---

[1]See A. E. Taylor, *Advanced Calculus*, Ginn, 1955, pp. 572–574.

(7) $5^{(2)}/4!$ (9) $(\tfrac{1}{2})^{(3)}/5!$

(8) $6^{(3)}/3!$ (10) $(\tfrac{2}{3})^{(2)}/(\tfrac{2}{3})^2$

B.

(11) Expand $(x+y)^{12}$ to 4 terms.

(12) Expand $(m+n)^{13}$ to 4 terms.

(13) Estimate $\sqrt[3]{1.152}$ by expansion to 3 terms. The true value is near 1.0483.

(14) Estimate $\sqrt[5]{1.289}$ by expansion to 3 terms. The true value is near 1.0521.

C. Prove the following:

(15) $\binom{n}{r+1} = \dfrac{n-r}{r+1}\binom{n}{r}$

(16) $r\binom{n}{r} = n\binom{n-1}{r-1}$

(17) $\binom{-n}{k} = (-1)^k \binom{n+k-1}{k}$

(18) $\displaystyle\sum_{j=0}^{n} \binom{n}{j}^2 = \binom{2n}{n}$

5.3 THE FORWARD DIFFERENCE OPERATOR

5.3.1 The Operation of Differencing

From calculus, we are familiar with the relation

$$\Delta f(x) = f(x+\Delta x) - f(x) \tag{5.3.1}$$

If we let $h = \Delta x$, this becomes

$$\Delta f(x) = f(x+h) - f(x) \tag{5.3.2}$$

Δf is defined as the *first forward difference* of f with respect to h. The operation is called *differencing* and Δ is defined as the *forward difference operator*. It is evident that if $f(x) = c$, where c is a constant, then $f(x) = f(x+h) = c$, so that $\Delta c = 0$.

We shall prove that Δ is a *linear operator*, i.e., if c is a constant, then

(a) $$\Delta\left[cf(x)\right] = c\,\Delta f(x) \tag{5.3.3}$$

(b) $$\Delta\left[f(x)+g(x)\right] = \Delta f(x) + \Delta g(x) \tag{5.3.4}$$

Proof

(a) $\Delta[cf(x)] = cf(x+h) - cf(x)$ (5.3.5)

$= c[f(x+h) - f(x)]$ (5.3.6)

$= c\Delta f(x)$ (5.3.7)

(b) $\Delta[f(x) + g(x)] = [f(x+h) + g(x+h)] - [f(x) + g(x)]$ (5.3.8)

$= [f(x+h) - f(x)] + [g(x+h) - g(x)]$ (5.3.9)

$= \Delta f(x) + \Delta g(x)$ (5.3.10)

ILLUSTRATIVE PROBLEM 5.3-1   Find $\Delta f(x)$ if $f(x) = x^2 - 3x + 7$.

Solution   Since Δ is a linear operator,

$$\Delta f(x) = \Delta x^2 - 3\Delta x + \Delta 7$$
$$= [(x+h)^2 - x^2] - 3[(x+h) - x] + 0$$
$$= 2hx + h^2 - 3h$$

ILLUSTRATIVE PROBLEM 5.3-2   The following table shows the numerical values of Δf when a table of values for $f(x)$ is given:

x	$f(x)$	$\Delta f(x)$
0	6	-5
1	1	-5
2	-4	13
3	9	73
4	82	199
5	281	415
6	696	

5.3.2   Differencing Sums, Products and Quotients

From the fact that Δ is a linear operator, we know immediately that if u and v are functions of x, then

$$\Delta(u+v) = \Delta u + \Delta v \qquad (5.3.11)$$

$$\Delta(u-v) = \Delta u - \Delta v \qquad (5.3.12)$$

For a *product*, we have

$$\Delta(uv) = u(x+h)v(x+h) - u(x)v(x) \qquad (5.3.13)$$

We shall use the following abbreviations from time to time when there is no danger of confusion:

$$u = u(x), \qquad\qquad v = v(x)$$
$$u_+ = u(x+h), \qquad\qquad v_+ = v(x+h) \tag{5.3.14}$$

Then we rewrite (5.3.13) as follows:

$$\Delta(uv) = u_+ v_+ - uv \tag{5.3.15}$$
$$= u_+ v_+ - u_+ v + u_+ v - uv \tag{5.3.16}$$
$$= u_+ (v_+ - v) + v(u_+ - u) \tag{5.3.17}$$
$$= u_+ \Delta v + v \Delta u \tag{5.3.18}$$

Alternatively, we could have written, in place of (5.3.16),

$$\Delta(uv) = u_+ v_+ - uv_+ + uv_+ - uv \tag{5.3.19}$$
$$= v_+ (u_+ - u) + u(v_+ - v) \tag{5.3.20}$$
$$= v_+ \Delta u + u \Delta v \tag{5.3.21}$$

Either (5.3.18) or (5.3.21) can be used, whichever is more convenient. Note the similarity to $d(uv) = u\,dv + v\,du$.

For a *quotient*, we have, using the same abbreviations,

$$\Delta\left(\frac{u}{v}\right) = \frac{u_+}{v_+} - \frac{u}{v} \tag{5.3.22}$$

$$= \frac{vu_+ - uv_+}{vv_+} \tag{5.3.23}$$

$$= \frac{vu_+ - vu + vu - uv_+}{vv_+} \tag{5.3.24}$$

$$= \frac{v \Delta u - u \Delta v}{vv_+} \tag{5.3.25}$$

Note the similarity to $d(u/v) = (v\,du - u\,dv)/v^2$.

5.3.3 $\Delta(x^n)$

$$\Delta(x^n) = (x+h)^n - x^n \tag{5.3.26}$$

$$= \left(x^n + nx^{n-1}h + \binom{n}{2}x^{n-2}h^2 + \cdots + h^n\right) - x^n \tag{5.3.27}$$

$$= nx^{n-1}h + \binom{n}{2}x^{n-2}h^2 + \binom{n}{3}x^{n-3}h^3 + \cdots + h^n \tag{5.3.28}$$

If n is small, it is often easier to derive the specific formula rather than resort to this general formula.

ILLUSTRATIVE PROBLEM 5.3-3   Find $\Delta(2x^3)$.

Solution

$$\Delta(2x^3) = 2\Delta(x^3)$$
$$= 2\left[(x+h)^3 - x^3\right]$$
$$= 2h[3x^2 + 3xh + h^2]$$

If $h = 1$, a very common case, there is a much easier method, which we shall attack in Section 5.3.8.

5.3.4 Trigonometric Differences

In the following, we shall need to use six identities from elementary trigonometry:

$$\sin A - \sin B = 2\cos\tfrac{1}{2}(A+B)\sin\tfrac{1}{2}(A-B) \qquad (5.3.29)$$

$$\cos A - \cos B = -2\sin\tfrac{1}{2}(A+B)\sin\tfrac{1}{2}(A-B) \qquad (5.3.30)$$

$$\tan A - \tan B = \frac{\sin(A-B)}{\cos A \cos B} \qquad (5.3.31)$$

$$\sec A - \sec B = \frac{2\sin\tfrac{1}{2}(A+B)\sin\tfrac{1}{2}(A-B)}{\cos A \cos B} \qquad (5.3.32)$$

$$\csc A - \csc B = \frac{-2\cos\tfrac{1}{2}(A+B)\sin\tfrac{1}{2}(A-B)}{\sin A \sin B} \qquad (5.3.33)$$

$$\cot A - \cot B = \frac{-\sin(A-B)}{\sin A \sin B} \qquad (5.3.34)$$

We shall demonstrate the derivation of the formula for the first forward difference of $\sin(ax+b)$, leaving the others as exercises for the problem section:

$$\Delta\sin(ax+b) = \sin(ax+ah+b) - \sin(ax+b) \qquad (5.3.35)$$

Using (5.3.29),

$$\Delta\sin(ax+b) = 2\cos\tfrac{1}{2}(ax+ah+b+ax+b)\sin\tfrac{1}{2}(ax+ah+b-ax-b)$$
$$\qquad (5.3.36)$$

$$= 2\cos\left[a\left(x+\frac{h}{2}\right)+b\right]\sin\tfrac{1}{2}ah \qquad (5.3.37)$$

We shall use the Greek letter ψ (psi) to represent the argument of the cosine in (5.3.37). Then the formula becomes

$$\Delta \sin(ax + b) = 2 \cos \psi \sin \tfrac{1}{2} ah \qquad (5.3.38)$$

Table A.4 in the Appendix lists the trigonometric difference formulas.

ILLUSTRATIVE PROBLEM 5.3-4   For $h = 1$, find the first forward difference of $f(x) = 2^x \sin x$.

Solution   Let $u = 2^x$, $v = \sin x$. Then

$$\Delta f = u_+ \Delta v + v \Delta u$$
$$u_+ = 2^{x+1}$$
$$\Delta v = 2 \sin \tfrac{1}{2} \cos\left(x + \tfrac{1}{2}\right)$$
$$v = \sin x$$
$$\Delta u = 2^{x+1} - 2^x = 2(2^x) - 2^x = 2^x$$
$$\Delta f = 2x\left[4 \sin \tfrac{1}{2} \cos\left(x + \tfrac{1}{2}\right) + \sin x\right]$$

5.3.5   Differencing Exponential Functions

$$\Delta(a^{cx+d}) = a^{cx+ch+d} - a^{cx+d} \qquad (5.3.39)$$
$$= a^{cx+d}(a^{ch} - 1) \qquad (5.3.40)$$

In particular,

$$\Delta a^x = a^{x+h} - a^x = a^x(a^h - 1) \qquad (5.3.41)$$

ILLUSTRATIVE PROBLEM 5.3-5   For $h = 1$, find the first forward difference of 3^{-2x}.

Solution $3^{-2x}(3^{-2} - 1) = 3^{-2x}\left(-\tfrac{8}{9}\right) = -8/(9^{x+1})$

5.3.6   Differencing Logarithmic Functions

$$\Delta \ln(ax + b) = \ln(ax + ah + b) - \ln(ax + b) \qquad (5.3.42)$$
$$= \ln \frac{ax + ah + b}{ax + b} \qquad (5.3.43)$$
$$= \ln\left(1 + \frac{ah}{ax + b}\right) \qquad (5.3.44)$$

If $b = 0$, we have the special case

$$\Delta \ln ax = \ln\left(1 + \frac{h}{x}\right) \qquad (5.3.45)$$

5.3.7   Differencing Factorial Functions

We shall find that the special case of differencing with $h = 1$ plays a very important part, particularly for polynomials. For $h = 1$,

$$\Delta x^{(1)} = \Delta x = (x+1) - x = 1 \tag{5.3.46}$$

$$\Delta x^{(2)} = \Delta \left[x(x-1) \right] \tag{5.3.47}$$

$$= \Delta x^2 - \Delta x \tag{5.3.48}$$

$$= \left[(x+1)^2 - x^2 \right] - \left[(x+1) - x \right] \tag{5.3.49}$$

$$= 2x \tag{5.3.50}$$

More generally, for $h = 1$,

$$\Delta x^{(n)} = \Delta \left[x(x-1)(x-2) \cdots (x-n+1) \right] \tag{5.3.51}$$

$$= (x+1)x(x-1) \cdots (x-n+2) - x(x-1)(x-2) \cdots (x-n+1) \tag{5.3.52}$$

$$= \left[(x+1) - (x-n+1) \right] x^{(n-1)} \tag{5.3.53}$$

$$= n x^{(n-1)} \tag{5.3.54}$$

ILLUSTRATIVE PROBLEM 5.3-6   Find $\Delta[x(x-1)(x-2)(x-3)]$

Solution   This is $\Delta x^{(4)} = 4x^{(3)} = 4x(x-1)(x-2)$

ILLUSTRATIVE PROBLEM 5.3-7   Find $\Delta[(x-5)(x-6)(x-7)]$

Solution $\Delta(x-5)^{(3)} = 3(x-5)^{(2)} = 3(x-5)(x-6)$.

5.3.8   Differencing Polynomials for $h = 1$

In Section 5.3.3, we showed how to difference x^n for any h. In a surprising number of practical cases, problems arise or can be modified so that $h = 1$. Then, if the polynomial can be written in factorial-function form, the easy method of Section 5.3.7 can be used to difference the polynomial.

We digress now to explain one way to change a polynomial to factorial-function form. The following is a review of *synthetic division*, an easy way to do certain kinds of algebraic long division. Consider the problem

$$(x^3 + 3x^2 - 13x - 16)/(x-3)$$

which is done the long way as follows:

$$x - 3 \overline{)x^3 + 3x^2 - 13x - 16} \qquad x^2 + 6x + 5 + \frac{-1}{(x-3)}$$

$$
\begin{array}{r}
x^2 + 6x + 5 + \dfrac{-1}{(x-3)} \\[4pt]
x - 3 \overline{)\, x^3 + 3x^2 - 13x - 16 \,} \\[2pt]
{}^{\ominus}x^3 \, {}^{\oplus}\!- 3x^2 \\[2pt]
\hline
6x^2 - 13x \\[2pt]
{}^{\ominus}6x^2 \, {}^{\oplus}\!- 18x \\[2pt]
\hline
5x - 16 \\[2pt]
{}^{\ominus}5x \, {}^{\oplus}\!- 15 \\[2pt]
\hline
-1
\end{array}
$$

In this long division, there is much repetition of coefficients and changing of signs. Let us agree that (1) we are dividing by a binomial of the form $x - r$; (2) the dividend and quotient are in descending order *with no missing terms*; and (3) the signs are changed all at once. Then the same long division can be rewritten as follows:

$$
\begin{array}{r}
1 \quad +6 \; + \; 5 \\[2pt]
3 \overline{)1 \quad +3 \; - 13 \; - 16} \\[2pt]
3 \\[2pt]
\hline
6 \\[2pt]
+18 \\[2pt]
\hline
5 \\[2pt]
+15 \\[2pt]
\hline
-1
\end{array}
$$

With a slight rearrangment to save space, we have the usual form of synthetic division:

$$
\begin{array}{r|rrrr}
3\rfloor & 1 & +3 & -13 & -16 \\
& & +3 & +18 & +15 \\
\hline
& 1 & +6 & +5 \,|\, & +1
\end{array}
$$

which should be compared with the original problem.

ILLUSTRATIVE PROBLEM 5.3-8   Divide $2x^2 + x - 21$ by $x + 3$ using synthetic division.

Solution

$$
\begin{array}{r|rrr}
-3\rfloor & 2 & 1 & -2 \\
& & -6 & +15 \\
\hline
& 2 & -5 \,|\, & +13
\end{array}
$$

The answer is

$$2x - 5 + \frac{13}{(x+3)}.$$

Explanation   Carry down the leading coefficient, 2. Multiply 2 by -3, and write the product, -6, under the next coefficient, 1. Add, obtaining -5. Multiply -5 by -3, obtaining $+15$. Write this under the next coefficient and add. In the answer (which, of course, is of degree one lower than the original dividend), the last sum is the numerator of the remainder fraction. Then -5 is the constant term, and working from right to left, the other results are multiplied by $x, x^2, x^3, \ldots$ as far as they go.

ILLUSTRATIVE PROBLEM 5.3-9   Divide $x^3 - 2$ by $x - 1$ using synthetic division.

Solution   The only difference in procedure is that zeros must be written for the missing terms:

$$
\underline{1\rfloor} \quad
\begin{array}{rrrr}
1 & 0 & 0 & -2 \\
  & 1 & 1 & 1 \\
\hline
1 & 1 & 1 & \!\!\!\big| -1
\end{array}
$$

The result is

$$x^2 + x + 1 - \frac{1}{(x-1).}$$

From this, we can write

$$x^3 - 2 = (x-1)(x^2 + x + 1) - 1 \tag{5.3.55}$$

If we now divide $x^2 + x + 1$ by $x - 2$, we have

$$\frac{x^2 + x + 1}{x - 2} = x + 3 + \frac{7}{x-2} \tag{5.3.56}$$

which can be rewritten as

$$x^2 + x + 1 = (x+3)(x-2) + 7 \tag{5.3.57}$$

Returning to (5.3.55) and entering (5.3.57),

$$x^3 - 2 = (x-1)[(x-2)(x+3) + 7] - 1 \tag{5.3.58}$$

Once again, dividing $x + 3$ by $x - 3$,

$$\frac{x+3}{x-3} = 1 + \frac{6}{x-3} \tag{5.3.59}$$

from which

$$(x+3) = 1(x-3) + 6 \tag{5.3.60}$$

Replacing this in (5.3.58), we obtain

$$x^3 - 2 = (x-1)\{(x-2)[(x-3)+6]+7\} - 1 \tag{5.3.61}$$

which can be rewritten as

$$x^3 - 2 = (x-1)(x-2)(x-3) + 6(x-1)(x-2) + 7(x-1) - 1 \qquad (5.3.62)$$

$$x^3 - 2 = 1(x-1)^{(3)} + 6(x-1)^{(2)} + 7(x-1)^{(1)} - 1 \qquad (5.3.63)$$

Note that the coefficients $1, 6, 7, -1$ are the last quotient and the remainders (in reverse order) for the successive synthetic divisions. The process can be written more efficiently, combining all the successive synthetic divisions.

ILLUSTRATIVE PROBLEM 5.3-10   Express $x^3 - 2$ as a factorial function of $x - 1$.

Solution

$$\therefore x^3 - 2 = 1(x-1)^{(3)} + 6(x-1)^{(2)} + 7(x-1)^{(1)} - 1.$$

ILLUSTRATIVE PROBLEM 5.3-11   Express $x^3 - 2$ as a factorial function of x.

Solution

$$\therefore x^3 - 2 = x^{(3)} + 3x^{(2)} + x^{(1)} - 2$$

In general, we prefer to transform polynomials into factorial functions of x. It will be understood that this has been done unless otherwise stated.

ILLUSTRATIVE PROBLEM 5.3-12   Transform $P(x) = x^4 + 6x^2 - 7x + 8$ to factorial-function form, and then difference it using $h = 1$.

Solution

$$
\begin{array}{r|rrrrr}
0| & 1 & 0 & 6 & -7 & 8 \\
  &   & 0 & 0 & 0 & 0 \\
\hline
1| & 1 & 0 & 6 & -7 & \boxed{8} \\
  &   & 1 & 1 & 7 & \\
\hline
2| & 1 & 1 & 7 & \boxed{0} & \\
  &   & 2 & 6 &   & \\
\hline
3| & 1 & 3 & \boxed{13} & & \\
  &   & 3 &   & & \\
\hline
1| & 6 & & & & \\
\end{array}
$$

$\therefore P(x) = x^{(4)} + 6x^{(3)} + 13x^{(2)} + 0x^{(1)} + 8$, and $\Delta P(x) = 4x^{(3)} + 18x^{(2)} + 26x^{(1)}$

5.3.9   Stirling Numbers

Consider the polynomial

$$P(x) = ax^3 + bx^2 + cx + d \qquad (5.3.64)$$

which we wish to rewrite in factorial-function form:

$$
\begin{array}{r|cccc}
0| & a & b & c & d \\
  &   & 0 & 0 & 0 \\
\hline
1| & a & b & c & \boxed{d} \\
  &   & a & a+b & \\
\hline
2| & a & a+b & \boxed{a+b+c} & \\
  &   & 2a & & \\
\hline
  & a & \boxed{3a+b} & & \\
\end{array}
$$

Then

$$P(x) = ax^{(3)} + (3a+b)x^{(2)} + (a+b+c)x^{(1)} + d \qquad (5.3.65)$$

It is evident that coefficients can be calculated to convert from $x^{(n)}$ to x^n, or from x^n to $x^{(n)}$. They are called Stirling numbers of the first kind and the second kind. Tables which accomplish this conversion are available; it is usually just as easy to do the calculation as to find and use the tables.

PROBLEM SECTION 5.3

A. Derive the following formulas

(1) $\Delta \cos(ax+b)$,
(2) $\Delta \tan(ax+b)$,
(3) $\Delta \sec(ax+b)$,
(4) $\Delta \csc(ax+b)$,
(5) $\Delta \cot(ax+b)$.

B. In the following, transform each polynomial into factorial-function form beginning with $x^{(n)}$; then difference it, assuming $h = 1$:
- (6) $5x^3 - 3x^2 + 4$
- (7) $2x^4 + x^2 - 7$
- (8) $x^5 - 2x^2 + 4$
- (9) x^6
- (10) x^7

C. Find the first forward difference, assuming $h = 1$:
- (11) $\sin u$
- (12) $a/2^x$
- (13) $2^x \sin(a/2^x)$
- (14) $\tan u$
- (15) $\tan(a/2^x)$

D. Derive the following formulas:
- (16) $\Delta \sinh (ax + b) = 2 \cosh \psi \sinh(ah/2)$
- (17) $\Delta \tanh (ax + b)\ 2 \sinh \psi \sinh(ah/2)$
- (18) $\Delta \tanh(ax + b) = \dfrac{\sinh ah}{\cosh(ax + ah + b)\cosh(ax + b)}$

5.4 REPEATED DIFFERENCES

5.4.1 Lagrange's Theorem

We define $\Delta^2 f$ as the *second forward difference* of f. It is found by differencing the first forward difference:

$$\Delta^2 f = \Delta(\Delta f) = \Delta f(x + h) - \Delta f \tag{5.4.1}$$

$$= \left[f(x + 2h) - f(x + h) \right] - \left[f(x + h) - f(x) \right] \tag{5.4.2}$$

$$= f(x + h) - 2f(x + h) + f(x) \tag{5.4.3}$$

Differencing once again, we obtain the *third forward difference*:

$$\Delta^3 f = \Delta(\Delta^2 f) = \Delta f(x + 2h) - 2\Delta f(x + h) + \Delta f(x) \tag{5.4.4}$$

$$= f(x + 3h) - 3f(x + 2h) + 3f(x + h) - f(x) \tag{5.4.5}$$

Similarly, the *fourth forward difference* is

$$\Delta^4 f = f(x + 4h) - 4f(x + 3h) + 6f(x + 2h) - 4f(x + h) + f(x) \tag{5.4.6}$$

A comparison of (5.4.3), (5.4.5) and (5.4.6) seems to indicate that the nth forward difference is given by

$$\Delta^n f = f(x + nh) - \binom{n}{1} f(x + [n - 1]h) + \binom{n}{2} f(x + [n - 2]h) + \ldots \tag{5.4.7}$$

It is customary to write this in reverse order:

LAGRANGE'S THEOREM

$$\Delta^n f = \sum_{k=0}^{n} (-1)^{n-k} \binom{n}{k} f(x + kh) \tag{5.4.8}$$

The following proof is worth studying because it exhibits the importance of dummy handling, a technique which we shall employ many times. Later in this section, we have an alternate proof using operators.

Proof For $n = 0$,

$$\Delta^0 f = (-1)^0 \binom{0}{0} f(x + 0h) = f(x) \tag{5.4.9}$$

so that the proposition is true for $n = 0$. Assume the proposition true for $n = r$:

$$\Delta^r f = \sum_{k=0}^{r} (-1)^{r-k} \binom{r}{k} f(x + kh) \tag{5.4.10}$$

We wish to show that this implies

$$\Delta^{r+1} f = \sum_{k=0}^{r+1} (-1)^{r-k+1} \binom{r+1}{k} f(x + kh) \tag{5.4.11}$$

Difference both members of (5.4.10) with reference to x:

$$\Delta(\Delta^r f) = \Delta \sum_{k=0}^{r} (-1)^{r-k} \binom{r}{k} f(x + kh) \tag{5.4.12}$$

The left member is simply $\Delta^{r+1} f$. In the right member, the operator Δ operates only on $f(x + kh)$ since only x is involved in the change:

$$\Delta^{r+1} f = \sum_{k=0}^{r} (-1)^{r-k} \binom{r}{k} [f(x + [k+1]h) - f(x + kh)] \tag{5.4.13}$$

We separate the summation into two summations:

$$\Delta^{r+1} f = \sum_{k=0}^{r} (-1)^{r-k} \binom{r}{k} f(x + [k+1]h) - \sum_{k=0}^{r} (-1)^{r-k} \binom{r}{k} f(x + kh) \tag{5.4.14}$$

In the first summation, we alter the dummy by letting $m = k + 1$, so that $k = m - 1$. The first summation is then

$$\sum_{m=1}^{r+1} (-1)^{r-m+1} \binom{r}{m-1} f(x + mh) \tag{5.4.15}$$

Now, since the letter used for the dummy makes no difference, we replace

m by k and rewrite (5.4.14) as

$$\Delta^{r+1}f = \sum_{k=1}^{r+1} (-1)^{r-k+1} \binom{r}{k-1} f(x+kh) - \sum_{k=0}^{r} (-1)^{r-k} \binom{r}{k} f(x+kh)$$

(5.4.16)

The first summation can be expressed as a sum of terms from $k=1$ to $k=r$, plus the $(r+1)$st contribution. Note that $(-1)^{r-k+1} = -(-1)^{r-k}$. Also, when $k=r+1$, then $r-k+1=0$. Thus the first summation becomes

$$\sum_{k=1}^{r} -(-1)^{r-k} \binom{r}{k-1} f(x+kh) + (-1)^0 \binom{r}{r} f(x+[r+1]h) \quad (5.4.17)$$

The second summation can also be expressed as a sum of terms from $k=1$ to $k=r$, plus the zeroth term:

$$(-1)^{r-0} \binom{r}{0} f(x+0h) + \sum_{k=1}^{r} (-1)^{r-k} \binom{r}{k} f(x+kh) \quad (5.4.18)$$

Substituting in (5.4.16) and collecting the summations,

$$\Delta^{r+1}f = f(x+[r+1]h)$$
$$- \sum_{k=1}^{r} \left[\binom{r}{k} + \binom{r}{k-1} \right] (-1)^{r-k} f(x+kh)$$
$$+ (-1)^{r+1} f(x) \quad (5.4.19)$$

We know from Illustrative Problem 5.2-2 that

$$\binom{r}{k} + \binom{r}{k-1} = \binom{r+1}{k} \quad (5.4.20)$$

Substituting in (5.4.19), we have

$$\Delta^{r+1}f = f(x+[r+1]h)$$
$$- \sum_{k=1}^{r} \binom{r+1}{k} (-1)^{r-k} f(x+kh)$$
$$+ (-1)^{r+1} f(x) \quad (5.4.21)$$

which is equivalent to (5.4.11), completing the induction. $\square$

5.4.2 The Shift Operator E

We define the *shift operator* so that

$$Ef(x) = f(x+h) \tag{5.4.22}$$

Then

$$E^2 f(x) = E[Ef(x)] = f(x+2h) \tag{5.4.23}$$

and in general we define

SHIFT OPERATOR   For all n,

$$E^n f(x) = f(x+nh) \tag{5.4.24}$$

In particular,

$$E^0 f(x) = f(x) \tag{5.4.25}$$

$$E^{-1} f(x) = f(x-h) \tag{5.4.26}$$

$$E^{1/2} f(x) = f\left(x + \tfrac{1}{2}h\right) \tag{5.4.27}$$

It is easy to show that E is a linear operator.
  Since

$$\Delta f(x) = f(x+h) - f(x) \tag{5.4.28}$$

we have

$$\Delta f(x) = Ef(x) - 1f(x) \tag{5.4.29}$$

where 1 is defined as the *identity operator*.

IDENTITY OPERATOR

$$1f(x) = f(x) \tag{5.4.30}$$

Then we can write *formally* the operational equation

$$\Delta = E - 1 \tag{5.4.31}$$

where all three symbols are operators.
  We pause to discuss the algebra of operators.

5.4.3 The Algebra of Operators

By (5.4.24), if n is an integer,

$$E^n = E \cdot E \cdot E \cdots \qquad \text{to } n \text{ "factors"} \tag{5.4.32}$$

This resembles

$$x^n = x \cdot x \cdot x \cdots \qquad \text{to } n \text{ factors} \qquad (5.4.33)$$

The temptation to see how far the analogy will go is overwhelming. If $A, B, C, \ldots$ are operators, we now define the "multiplication"

$$ABC \cdots f(x) \qquad (5.4.34)$$

as meaning that the operations are performed from right to left. For example,

$$EE^{-1}f(x) = E\big[E^{-1}f(x)\big] = E\big[f(x-h)\big] = \mathbf{1}f(x) \qquad (5.4.35)$$

so that, operationally,

$$EE^{-1} = \mathbf{1} \qquad (5.4.36)$$

We can also show, similarly, that

$$E^{-1}E = \mathbf{1} \qquad (5.4.37)$$

This is very much like

$$xx^{-1} = x^{-1}x = 1 \qquad (5.4.38)$$

so that we can write *formally*

$$E^{-1} = \frac{\mathbf{1}}{E} \qquad (5.4.39)$$

If A and B are operators, we further define *addition of operators* so that

$$(A + B)f(x) = Af(x) + Bf(x) \qquad (5.4.40)$$

We shall say that we have defined multiplication of operators and addition of operators *in transformation fashion*. Now it is easy to prove, for any operators which obey these rules for addition and multiplication, that many of the laws of normal algebra apply:

Operators	Numbers	
$\mathbf{1}A = A\mathbf{1} = A$	$1x = x1 = x$	(5.4.41)
$A(B + C) = AB + AC$	$x(y + z) = xy + xz$	(5.4.42)
$(A + B)C = AC + BC$	$(x + y)z = xz + yz,$	(5.4.43)
$A(BC) = (AB)C$	$x(yz) = (xy)z$	(5.4.44)

Proof of (5.4.42)

$$A(B+C)f(x) = A\big[Bf(x)+Cf(x)\big] \tag{5.4.45}$$

$$= [AB]f(x)+[AC]f(x) \tag{5.4.46}$$

$$= [AB+AC]f(x) \tag{5.4.47}$$

$$\therefore \quad A(B+C)=AB+AC \quad \square \tag{5.4.48}$$

In general, operators do *not* commute, i.e.,

$$ABf(x) \neq BAf(x) \tag{5.4.49}$$

An exception is shown in (5.4.41). Also,

$$E\Delta = \Delta E \tag{5.4.50}$$

Proof of (5.4.50)

$$E\Delta f(x) = E\big[f(x+h)-f(x)\big] \tag{5.4.51}$$

$$= f(x+2h)-f(x+h) \tag{5.4.52}$$

$$\Delta Ef(x) = \Delta\big[f(x+h)\big] \tag{5.4.53}$$

$$= f(x+2h)-f(x+h) \quad \square \tag{5.4.54}$$

One outcome of these considerations is that linear operators which obey these transformation definitions of multiplication and addition fulfill the formal requirements of the binomial theorem, so that $(A+B)^n$ can be expanded like $(x+y)^n$. This leads to a very simple formal derivation of the Lagrange Theorem:

$$\Delta^n = (E-1)^n \tag{5.4.55}$$

$$= E^n - \binom{n}{1}E^{n-1} + \binom{n}{2}E^{n-2} + \cdots \tag{5.4.56}$$

$$\Delta^n f(x) = f(x+nh) + \binom{n}{1}f(x+[n-1]h) + \binom{n}{2}f(x+[n-2]h) + \cdots \tag{5.4.57}$$

Note that Lagrange's theorem can be expressed as follows, in operational form:

Lagrange's Theorem

$$\Delta^n f(x) = (E-1)^n f(x) \tag{5.4.58}$$

5.4.4 Polynomials

Let

$$f(x) = a_n x^n + a_{n-1} x^{n-1} + \cdots + a_0 \qquad (5.4.59)$$

We recall from (5.4.28) that

$$\Delta x^n = n x^{n-1} h + \binom{n}{2} x^{n-2} h^2 + \binom{n}{3} x^{n-3} h^3 + \cdots \qquad (5.4.60)$$

We shall use the symbol $O(x^{n-2})$ to mean *terms in x of degree no higher than $n-2$*. The symbol is read "large O of x^{n-2}"; the letter O stands for *order*. Then

$$\Delta x^n = n x^{n-1} h + O(x^{n-2}) \qquad (5.4.61)$$

and, returning to (5.4.59),

$$\Delta f(x) = a_n n x^{n-1} h + O(x^{n-2}) \qquad (5.4.62)$$

$$\Delta^2 f(x) = a_n n(n-1) x^{n-2} h^2 + O(x^{n-3}) \qquad (5.4.63)$$

$$\vdots$$

$$\Delta^n f(x) = a_n n! x^{n-n} h^n = a_n n! h^n \qquad (5.4.64)$$

$$\Delta^{n+1} f(x) = 0$$

which completes the proof of the following theorem.

> POLYNOMIAL DIFFERENCE THEOREM   The nth difference of a polynomial of degree n is constant, and the $(n+1)$st difference is 0.

5.4.5 The Forward-Difference Table

The following table displays the repeated differences for

$$f(x) = x^4 - 3x^3 + 2x^2 - 5x + 6$$

x	$f(x)$	Δf	$\Delta^2 f$	$\Delta^3 f$	$\Delta^4 f$	$\Delta^5 f$
0	6	-5	0	18	24	0
1	1	-5	18	42	24	0
2	-4	13	60	66	24	
3	9	73	126	90		
4	82	199	216			
5	281	415				
6	696					

If we did not know $f(x)$, the fact that $\Delta^5 f$ vanishes would suggest that $f(x)$ might be a fourth-degree polynomial. [By Lagrange's theorem, $\Delta^{n+1} f = 0$ is *necessary* but not *sufficient* for $f(x)$ to be a polynomial of nth degree.]

5.4.6 Computer Program for a Forward-Difference Table

Figure 5.4-1 displays a computer program which produces a forward-difference table for $\sin x$, shown in Figure 5.4-2. In addition, a detailed table of forward differences is displayed in Figure 5.4-3.

```
      DOUBLE PRECISION DELTA(20,20),X(20),F(20),D,Y,Z,G
      COMMON DELTA,X,F
C THE FOLLOWING THREE CARDS MUST BE REPLACED FOR EACH FUNCTION.
      G(Y)=DSIN(Y)
100   FORMAT(1H1,T57,'SIN X'//)
      N=3
C
      KT=20
      IX=-5
C
      DO 8 I=1,KT
      IX=IX+5
      Y=IX
      Y=Y*1.D-2
      X(I)=Y
      Z=G(Y)
8     F(I)=Z
C
      WRITE(3,100)
      CALL TDDOD(KT,N)
C
      WRITE(3,100)
      WRITE(3,102)
102   FORMAT(1H0,T25,'I',T35,'X',T52,'F(X)'//)
      DO 9 I=1,KT
9     WRITE(3,103)I,X(I),F(I)
103   FORMAT(1H ,T24,I2,T33,F5.2,T50,F13.10)
      END
      SUBROUTINE TDDOD(KT,NN)
C THE MAIN PROGRAM MUST HAVE DOUBLE PRECISION DELTA(20,20),X(20)F(20)
C AND COMMON DELTA,X,F.
      DOUBLE PRECISION DELTA(20,20),X(20),F(20)
      COMMON DELTA,X,F
      DIMENSION N(20)
C DELTA(IP,I) HAS ORDER IP AND REFERS TO F(I).
      DO 7 I=1,20
7     N(I)=I
C
C FIRST DIFFERENCES
      JT=KT-1
      DO 1 I=1,JT
      K=I+1
1     DELTA(1,I)=F(K)-F(I)
C
C OTHER DIFFERENCES.
      DO 2 IP=2,JT
```

```
        J=KT-IP
        IPM=IP-1
        DO 2 I=1,J
        IPL=I+1
2       DELTA(IP,I)=DELTA(IPM,IPL)-DELTA(IPM,I)
C
        DO 6 KK=1,NN
        IF(KK.GT.1)WRITE(3,109)
109     FORMAT(1H1)
        WRITE(3,102)
102     FORMAT(1H0,T46,'TABLE OF FORWARD DIFFERENCES'//)
        IF(KT-10)21,21,22
21      L=KT-1
        GO TO 23
22      L=9
23      WRITE(3,103)(N(I),I=1,L)
103     FORMAT(1H0,T6,'I',T6,10I10//)
C
        L=KT-1
        DO 3 I=1,L
        M=KT-I
        IF(M.GT.9)M=9
3       WRITE(3,104)I,(DELTA(IP,I),IP=1,M)
104     FORMAT(1H ,T5,I2,T10,10(1PD10.2))
C
        IF(KT-10)5,5,35
35      WRITE(3,107)
        WRITE(3,107)
        WRITE(3,107)
C
        L=KT-1
        WRITE(3,103)(N(I),I=10,L)
C
        L=KT-10
        M=KT-1
        DO 4 I=1,L
        WRITE(3,104)I,(DELTA(IP,I),IP=10,M)
4       M=M-1
5       WRITE(3,107)
107     FORMAT(1H0)
        L=0
        WRITE(3,108)
        K=KT-1
        DO 6 I=1,K
        JT=KT-I
        DO 6 J=1,JT
        L=L+1
        IF(L-50)6,6,51
51      WRITE(3,108)
108     FORMAT(1H1,T2, ' LIST OF FORWARD DIFFERENCES',//)
        L=0
6       WRITE(3,106)I,J,DELTA(I,J)
106     FORMAT(1H ,T10,'DELTA(',I2,',',I2,')=',1PD22.15)
        RETURN
        END
```

Fig. 5.4-1 A computer program to supply a forward-difference table for $\sin x$. **This one provides three copies. To use for other functions, change three cards in the main program.**

SIN X

TABLE OF FORWARD DIFFERENCES

	1	2	3	4	5	6	7	8	9
1	5.00D-02	-1.25D-04	-1.25D-04	6.24D-07	3.10D-07	-2.33D-09	-7.69D-10	7.75D-12	1.90D-12
2	4.99D-02	-2.50D-04	-1.24D-04	9.34D-07	3.08D-07	-3.10D-09	-7.61D-10	9.66D-12	1.88D-12
3	4.96D-02	-3.74D-04	-1.23D-04	1.24D-06	3.04D-07	-3.86D-09	-7.51D-10	1.15D-11	1.85D-12
4	4.92D-02	-4.97D-04	-1.22D-04	1.55D-06	3.01D-07	-4.61D-09	-7.40D-10	1.34D-11	1.82D-12
5	4.87D-02	-6.18D-04	-1.20D-04	1.85D-06	2.96D-07	-5.35D-09	-7.26D-10	1.52D-11	1.78D-12
6	4.81D-02	-7.39D-04	-1.18D-04	2.14D-06	2.91D-07	-6.08D-09	-7.11D-10	1.70D-11	1.73D-12
7	4.74D-02	-8.57D-04	-1.16D-04	2.43D-06	2.85D-07	-6.79D-09	-6.94D-10	1.87D-11	1.69D-12
8	4.65D-02	-9.73D-04	-1.14D-04	2.72D-06	2.78D-07	-7.49D-09	-6.76D-10	2.04D-11	1.64D-12
9	4.55D-02	-1.09D-03	-1.11D-04	3.00D-06	2.70D-07	-8.16D-09	-6.55D-10	2.20D-11	1.58D-12
10	4.45D-02	-1.20D-03	-1.08D-04	3.27D-06	2.62D-07	-8.82D-09	-6.33D-10	2.36D-11	1.52D-12
11	4.33D-02	-1.31D-03	-1.05D-04	3.53D-06	2.53D-07	-9.45D-09	-6.09D-10	2.51D-11	1.46D-12
12	4.20D-02	-1.41D-03	-1.01D-04	3.78D-06	2.44D-07	-1.01D-08	-5.84D-10	2.66D-11	
13	4.05D-02	-1.51D-03	-9.76D-05	4.02D-06	2.34D-07	-1.06D-08	-5.58D-10		
14	3.90D-02	-1.61D-03	-9.35D-05	4.26D-06	2.23D-07	-1.12D-08			
15	3.74D-02	-1.70D-03	-8.93D-05	4.48D-06	2.12D-07				
16	3.57D-02	-1.79D-03	-8.48D-05	4.69D-06					
17	3.39D-02	-1.88D-03	-8.01D-05						
18	3.20D-02	-1.96D-03							
19	3.01D-02								

	10	11	12	13	14	15	16	17	18	19
1	-2.22D-14	-8.36D-15	7.37D-15	-1.45D-14	2.74D-14	-4.65D-14	6.67D-14	-6.56D-14	-2.14D-14	3.58D-13
2	-3.05D-14	-9.86D-16	-7.17D-15	1.29D-14	-1.91D-14	2.02D-14	1.10D-15	-8.70D-14	3.36D-13	
3	-3.15D-14	-8.16D-15	5.69D-15	-6.26D-15	1.04D-15	2.13D-14	-8.59D-14	2.49D-13		
4	-3.97D-14	-2.47D-15	-5.69D-16	-5.22D-15	2.23D-14	-6.47D-14	1.63D-13			
5	-4.21D-14	-3.04D-15	-5.79D-15	1.71D-14	-4.24D-14	9.86D-14				
6	-4.52D-14	-8.83D-15	1.13D-14	-2.53D-14	5.63D-14					
7	-5.40D-14	-2.47D-15	-1.40D-14	3.10D-14						
8	-5.15D-14	-1.15D-14	1.70D-14							
9	-6.30D-14	5.51D-15								
10	-5.75D-14									

Fig. 5.4-2 **Forward-difference table up to Δ^{19} for** sin x.

LIST OF FORWARD DIFFERENCES

```
DELTA( 1, 1)= 4.997916927067831D-02
DELTA( 1, 2)= 4.985424737614982D-02
DELTA( 1, 3)= 4.960471582677104D-02
DELTA( 1, 4)= 4.923119832146200D-02
DELTA( 1, 5)= 4.873462845946170D-02
DELTA( 1, 6)= 4.811624740681663D-02
DELTA( 1, 7)= 4.737760079411176D-02
DELTA( 1, 8)= 4.652053485319914D-02
DELTA( 1, 9)= 4.554719180257970D-02
DELTA( 1,10)= 4.446000449297278D-02
DELTA( 1,11)= 4.326169032645617D-02
DELTA( 1,12)= 4.195524446437617D-02
DELTA( 1,13)= 4.054393234100421D-02
DELTA( 1,14)= 3.903128150165149D-02
DELTA( 1,15)= 3.742107278564312D-02
DELTA( 1,16)= 3.571733087618855D-02
DELTA( 1,17)= 3.392431424076995D-02
DELTA( 1,18)= 3.204650448719067D-02
DELTA( 1,19)= 3.008859516189037D-02
DELTA( 2, 1)=-1.249218945284947D-04
DELTA( 2, 2)=-2.495315493787750D-04
DELTA( 2, 3)=-3.735175053090483D-04
DELTA( 2, 4)=-4.965986620002928D-04
DELTA( 2, 5)=-6.183810526450700D-04
DELTA( 2, 6)=-7.386466127048685D-04
DELTA( 2, 7)=-8.570659409126157D-04
DELTA( 2, 8)=-9.733430506194406D-04
DELTA( 2, 9)=-1.087187309606926D-03
DELTA( 2,10)=-1.198314166516612D-03
DELTA( 2,11)=-1.306445862079994D-03
DELTA( 2,12)=-1.411312123371963D-03
DELTA( 2,13)=-1.512650839352716D-03
DELTA( 2,14)=-1.610208716008379D-03
DELTA( 2,15)=-1.703741909454562D-03
DELTA( 2,16)=-1.793016635418601D-03
DELTA( 2,17)=-1.877809753579276D-03
DELTA( 2,18)=-1.957909325300308D-03
DELTA( 3, 1)=-1.246096545502802D-04
DELTA( 3, 2)=-1.239859559302733D-04
DELTA( 3, 3)=-1.230523566912445D-04
DELTA( 3, 4)=-1.218111906447772D-04
DELTA( 3, 5)=-1.202655600597985D-04
DELTA( 3, 6)=-1.184193282077472D-04
DELTA( 3, 7)=-1.162771097068249D-04
DELTA( 3, 8)=-1.138442589874860D-04
DELTA( 3, 9)=-1.111268569096851D-04
DELTA( 3,10)=-1.081316955633820D-04
DELTA( 3,11)=-1.048662612919693D-04
DELTA( 3,12)=-1.013387159807533D-04
DELTA( 3,13)=-9.755787665566284D-05
```

Fig. 5.4-3   (Continued on next page)

SIN X

I	X	F(X)
1	0.0	0.0
2	0.05	0.0499791693
3	0.10	0.0998334166
4	0.15	0.1494381325
5	0.20	0.1986693308
6	0.25	0.2474039593
7	0.30	0.2955202067
8	0.35	0.3428978075
9	0.40	0.3894183423
10	0.45	0.4349655341
11	0.50	0.4794255386
12	0.55	0.5226872289
13	0.60	0.5646424734
14	0.65	0.6051864057
15	0.70	0.6442176872
16	0.75	0.6816387600
17	0.80	0.7173560909
18	0.85	0.7512804051
19	0.90	0.7833269096
20	0.95	0.8134155048

Fig. 5.4-3   Some of the detailed output from the forward-difference program.

Chapter 6

Summation by Formula

The theory in this chapter is needed for explaining the derivation of the error formulas in the next chapter and other chapters. In an abbreviated course, just study Section 6.1.

One of the things that a computer does extremely quickly is finding the sum of a series. However, as you know from our discussion of error analysis, very large errors may be propagated in the process owing to the great number of additions, subtractions, multiplications and divisions involved.

It is so easy to program a summation—and so much harder to think—that all of us tend to use this computer capability even when the sum of the series can be found in closed form. Let us consider an example. Suppose we are asked to program

$$\sum_{k=0}^{n} (-1)^k \left(\frac{1}{2}\right)^k = 1 - \frac{1}{2} + \frac{1}{4} - \cdots \qquad (6.1)$$

where $n = 10^8$. This would take a few easy steps in FORTRAN or any other compiler language. However, it would require a sizeable chunk of computer time, and the answer would be full of propagated errors owing to about 100 million additions and 200 million multiplications.

How much better it would be to use a simple formula! In this case, you undoubtedly recognize (6.1) as a geometric series with the leading term a equal to 1 and the common ratio r equal to $-\frac{1}{2}$, so that the computation is trivial. The formula is

$$S = \frac{a(1 - r^n)}{1 - r} \qquad (6.2)$$

where n is the number of terms.

In this chapter, we shall devote ourselves to various methods for reducing the number of operations required to sum a series. We shall do this by some very clever tricks based upon the theory of finite differences.

6.1 SUMS OF DIFFERENCES

6.1.1 The Sum-Difference Theorem

A basic, simple, powerful theorem enables us to find in closed form the sum of a series like this which may look quite threatening:

$$\frac{1}{1\times2}+\frac{1}{2\times3}+\frac{1}{3\times4}+\cdots+\frac{1}{n(n+1)} \tag{6.1.1}$$

We start by seeking the sum of a difference:

$$\sum_{k=a}^{b-1}\Delta f(k) \tag{6.1.2}$$

Using the fact that, for $h=1$,

$$\Delta f(k)=f(k+1)-f(k) \tag{6.1.3}$$

and the additional fact that Σ is a linear operator, we have

$$\sum_{k=a}^{b-1}\Delta f(k)=\sum_{k=a}^{b-1}f(k+1)-\sum_{k=a}^{b-1}f(k) \tag{6.1.4}$$

We would like to have the same subscript in both summations in the right member. Therefore, we change the dummy of the first summation in the right member by writing

$$j=k+1 \tag{6.1.5}$$

which alters the summation as follows:

$$\sum_{k=a}^{b-1}f(k+1)=\sum_{j=a+1}^{b}f(j) \tag{6.1.6}$$

Since the name of the dummy doesn't matter, we can now change j back to k and substitute in (6.1.4):

$$\sum_{k=a}^{b-1}\Delta f(k)=\sum_{k=a+1}^{b}f(k)-\sum_{k=a}^{b-1}f(k) \tag{6.1.7}$$

Now we would like to make the limits of the summations equal. We therefore write

$$\sum_{k=a+1}^{b}f(k)=\sum_{k=a+1}^{b-1}f(k)+f(b) \tag{6.1.8}$$

$$\sum_{k=a}^{b-1}f(k)=f(a)+\sum_{k=a+1}^{b-1}f(k) \tag{6.1.9}$$

and substitute in (6.1.7):

$$\sum_{k=a}^{b-1} \Delta f(k) = \sum_{k=a+1}^{b-1} f(k) + f(b) - f(a) - \sum_{k=a+1}^{b-1} f(k) \qquad (6.1.10)$$

Since the two summations in the right member cancel, we have proved the following theorem.

SUM-DIFFERENCE THEOREM

$$\sum_{k=a}^{b-1} \Delta f(k) = f(b) - f(a) \qquad (6.1.11)$$

The similarity to

$$\int_a^b \frac{df(x)}{dx} dx = f(b) - f(a) \qquad (6.1.12)$$

is obvious. Note the difference in the upper limits.

6.1.2 Application to Polynomials

We have already proved that

$$\Delta x^{(N+1)} = (N+1)x^{(N)} \qquad (6.1.13)$$

From this,

$$x^{(N)} = \frac{1}{N+1} \Delta x^{(N+1)} \qquad (6.1.14)$$

Now, suppose we wish to find

$$\sum_{x=a}^{b-1} x^{(N)} = \frac{1}{N+1} \sum_{x=a}^{b-1} \Delta x^{(N+1)} \qquad (6.1.15)$$

Here $f(x) = x^{(N+1)}$. We apply the sum-difference theorem, which leads to the following theorem.

FACTORIAL SUM THEOREM

$$\sum_{x=a}^{b-1} x^{(N)} = \frac{1}{N+1} [b^{(N+1)} - a^{(N+1)}] \qquad (6.1.16)$$

ILLUSTRATIVE PROBLEM 6.1-1   Find $\sum_{k=1}^{n} k^2$.

Solution   By methods already explained, we have

$$k^2 = k^{(2)} + k^{(1)} \qquad (6.1.17)$$

$$\sum_{k=1}^{n} k^2 = \sum_{k=1}^{n} k^{(2)} + \sum_{k=1}^{n} k^{(1)} \qquad (6.1.18)$$

Using the factorial sum theorem twice, with $a = 1$, $b = n+1$,

$$\sum_{k=1}^{n} k^2 = \frac{1}{3}\left[(n+1)^{(3)} - 1^{(3)}\right] + \frac{1}{2}\left[(n+1)^{(2)} - 1^{(2)}\right] \tag{6.1.19}$$

$$= \frac{1}{3}\left[(n+1)n(n-1) - 1 \cdot 0 \cdot (-1)\right] + \frac{1}{2}\left[(n+1)n - 1 \cdot 0\right] \tag{6.1.20}$$

$$= \frac{1}{6}(n+1) \ n \ [2(n-1)+3] \tag{6.1.21}$$

$$= \frac{1}{6}(n+1)n(2n+1) \tag{6.1.22}$$

The result (6.1.22) requires 3 multiplications, 1 division and 2 additions. The brute-force program would have required n multiplications and $n-1$ additions. The advantage of the closed form is evident.

6.1.3   Reciprocals of Polynomials

We shall now attack the problem of finding the sum of a series like

$$\sum_{k=1}^{n} \frac{1}{k(k+1)} \tag{6.1.23}$$

The denominator is evidently $(k+1)^{(2)}$, and there is a temptation to assume that the fraction can be expressed as $(k+1)^{(-2)}$, but this is unfortunately not so.

In fact, we shall have to make a decision as to what we want a negative factorial superscript to mean. We approach the problem naively like this. Observe that

$$k^{(p)} = \left[k(k-1)\cdots(k-m+1)\right]\left[(k-m)(k-m-1)\cdots(k-p+1)\right] \tag{6.1.24}$$

where the first bracket has m factors and the second has $p-m$ factors. Then

$$k^{(p)} = k^{(m)}(k-m)^{(p-m)} \tag{6.1.25}$$

Now let $p = 0$. Then, since $k^{(0)} = 1$

$$1 = k^{(m)}(k-m)^{(-m)} \tag{6.1.26}$$

$$(k-m)^{(-m)} = \frac{1}{k^{(m)}} \tag{6.1.27}$$

Now, let $(k-m) = n$, so that $k = m+n$. Substituting in (6.1.27), we obtain the following *definition*:

FACTORIAL FUNCTION WITH NEGATIVE SUPERSCRIPT   For any k and positive integral m

$$n^{(-m)} = \frac{1}{(m+n)^{(m)}} \quad \text{and} \quad \frac{1}{k^{(m)}} = (k-m)^{(-m)} \tag{6.1.28}$$

At this point, most people wonder why this is a definition rather than a theorem. Didn't we prove it? No, we didn't. The reason is that our assumption ($p=0$) makes nonsense of (6.1.24). In fact, the purpose of the spurious derivation is to find a definition that makes negative factorial functions behave like the positive ones.

ILLUSTRATIVE PROBLEM 6.1-2   Find $5^{(-2)}$, $2^{(-1)}$, $0^{(-3)}$, $(-3)^{(-2)}$.

Solution $1/7^{(2)}=\frac{1}{42}$, $1/3^{(1)}=\frac{1}{3}$, $0^{(-3)}=1/3^{(3)}=\frac{1}{6}$, $1/(-1)^{(2)}=\frac{1}{2}$.

ILLUSTRATIVE PROBLEM 6.1-3   Convert to negative-factorial form

$$\frac{1}{k^{(3)}}, \quad \frac{1}{3^{(2)}}, \quad \frac{1}{(n+3)^{(4)}}$$

Solution $(k-3)^{(-3)}$, $(1)^{(-2)}$, $(n-1)^{(-4)}$.

We would naturally like to use the factorial sum theorem (6.1.16) for negative factorials, but we must make sure that it applies. Observe that the proof depends completely upon (6.1.13), which in turn depends upon the theorem

$$\Delta x^{(n)} = n x^{(n-1)} \tag{6.1.29}$$

Is this true for negative n? Let $n=-m$, where $m>0$. Then we would be content if we could prove that

$$\Delta x^{(-m)} = (-m)x^{(-m-1)}, \qquad m>0 \tag{6.1.30}$$

Using (6.1.28), this is the same as proving that

$$\Delta x^{(-m)} = (-m)\frac{1}{(x+m+1)^{(m+1)}} \tag{6.1.31}$$

We investigate the left member of this equation to see whether it leads to the right member, as we hope:

$$\Delta x^{(-m)} = \Delta \frac{1}{(x+m)^{(m)}} \tag{6.1.32}$$

$$= \frac{1}{(x+1+m)^{(m)}} - \frac{1}{(x+m)^{(m)}} \tag{6.1.33}$$

But

$$(x+1+m)^{(m)} = (x+1+m)(x+m)\cdots(x+2) \tag{6.1.34}$$

$$(x+m)^{(m)} = (x+m)(x+m-1)\cdots(x+1) \tag{6.1.35}$$

so that, substituting in (6.1.33) and factoring,

$$\Delta x^{(-m)} = \frac{1}{(x+m)\cdots(x+2)}\left(\frac{1}{x+1+m} - \frac{1}{x+1}\right) \tag{6.1.36}$$

$$= \frac{-m}{(x+1+m)(x+m)\cdots(x+1)} \tag{6.1.37}$$

which is precisely (6.1.31). This completes the proof that the factorial sum theorem (6.1.16) holds for negative factorials. Incidentally, we have proved that (6.1.29) holds for negative factorials as well.

ILLUSTRATIVE PROBLEM 6.1-3   Find

$$\sum_{k=1}^{n} \frac{1}{k(k+1)}.$$

Solution   We convert the problem to $\sum_{k=1}^{n}(k-1)^{(-2)}$. To make use of the factorial sum theorem, we let $x=k-1$. Then when $k=1$, $x=0$, and when $k=n$, $x=n-1$, so

$$\sum_{k=1}^{n} (k-1)^{(-2)} = \sum_{x=0}^{n-1} x^{(-2)} = \frac{1}{-1}(n^{(-1)} - 0^{(-1)})$$

$$= -\left(\frac{1}{n+1} - \frac{1}{1}\right) = \frac{n}{n+1}$$

6.1.4 Exponentials

Recalling that

$$\Delta a^{cx} = a^{cx}(a^c - 1) \tag{6.1.38}$$

we find immediately that

$$\sum_{x=x_1}^{x_2-1} a^{cx} = \frac{1}{a^c - 1} \sum_{x_1}^{x_2-1} \Delta a^{cx} \tag{6.1.39}$$

leading to the following theorem:

EXPONENTIAL SUM THEOREM   For $ac \neq 0$,

$$\sum_{x=x_1}^{x=x_2-1} a^{cx} = \frac{1}{a^c - 1}(a^{cx_2} - a^{cx_1}) \tag{6.1.40}$$

ILLUSTRATIVE PROBLEM 6.1-4   Find $3^4 + 3^6 + 3^8 + \cdots + 3^{30}$.

Solution   Here $a=3$, $c=2$, $x_1=2$, $x_2-1=15$, and the sum is

$$\frac{1}{3^2 - 1}(3^{32} - 3^4) = \frac{1}{8}(3^{32} - 3^4)$$

PROBLEM SECTION 6.1

A. Verify the following using the sum-difference theorem. The sums are from $k=1$ to $k=n$.

(1) $\Sigma k = n(n+1)/2$

(2) $\Sigma k^3 = n^2(n+1)^2/4$

(3) $\Sigma k^4 = n(n+1)(6n^3 + 9n^2 + n - 1)/30$

(4) $\Sigma k^5 = n^2(2n^4 + 6n^3 + 5n^2 - 1)/12$

(5) $\Sigma k(k+1) = n(n+1)(n+2)/3$

(6) $\Sigma(2k-1)^3 = n^2$

(7) $\Sigma \dfrac{1}{k(k+1)(k+2)} = \dfrac{n(n+3)}{4(n+1)(n+2)}$

(8) $\Sigma \dfrac{1}{k(k+1)(k+2)(k+3)} = \dfrac{n(n^2+6n+11)}{18(n+1)(n+2)(n+3)}$

B.

(9) Find $1 \times 2 \times 3 + 2 \times 3 \times 4 + \cdots + (n-2)(n-1)n$.

(10) Find $3 \times 5 \times 7 + 5 \times 7 \times 9 + \cdots$ to n terms.

(11) Find the sum of n terms of the series whose terms are generated by $n^3 + 7n$.

(12) Find the sum of n terms of the series

$$\frac{1}{1 \times 4 \times 7} + \frac{1}{4 \times 7 \times 10} + \frac{1}{7 \times 10 \times 13} + \cdots .$$

C. Prove by the methods of finite differences:

(13) $a + (a+d) + \cdots + (a + [n-1]d) = n(2a + [n-1]d)/2$

(14) $a + ax + \cdots + ar^{n-1} = a(r^n - 1)/(r-1)$

6.2 OTHER SERIES

6.2.1 Powers of Trigonometric Functions

A series like

$$\sin^p \theta + \sin^{p+1} \theta + \cdots + \sin^q \theta \tag{6.2.1}$$

is easily handled by the exponential sum theorem. Using (6.1.40) with $a = \sin\theta$, $c = 1$, $x_1 = p$ and $x_2 - 1 = q$, we have

$$\sum_{x=p}^{q+1} \sin^x \theta = \frac{1}{\sin\theta - 1} \left[(\sin\theta)^{q+1} - (\sin\theta)^p \right] \tag{6.2.2}$$

where $\sin\theta \neq 1$. Of course, the problem is also handled easily by the usual method for a geometric series. In fact, if the increment in x is not 1, the geometric formula must be used.

6.2.2 Trigonometric Series with Incremented Arguments

We have, for $a \neq 0$,

$$\Delta \sin(ay + b) = 2 \sin\tfrac{1}{2}a \cos\left[a\left(y + \tfrac{1}{2}\right) + b \right] \tag{6.2.3}$$

from which

$$\cos\left[a\left(y+\tfrac{1}{2}\right)+b\right] = \frac{\Delta \sin(ay+b)}{2\sin\tfrac{1}{2}a} \qquad (6.2.4)$$

Then

$$\sum_{y=y_1}^{y=y_2-1} \cos\left[a\left(y+\tfrac{1}{2}\right)+b\right] = \frac{1}{2\sin\tfrac{1}{2}a} \sum_{y_1}^{y_2-1} \Delta \sin(ax+b) \qquad (6.2.5)$$

$$= \frac{1}{2\sin\tfrac{1}{2}a}\left[\sin(ay_2+b) - \sin(ay_1+b)\right] \quad (6.2.6)$$

Now, let $x = y + \tfrac{1}{2}$. Then

$$\sum_{x=y_1+1/2}^{y_2-1/2} \cos(ax+b) = \frac{1}{2\sin\tfrac{1}{2}a}\left[\sin(ay_2+b) - \sin(ay_1+b)\right] \quad (6.2.7)$$

In order to make the results more useful, let $x_1 = y_1 + \tfrac{1}{2}$ and $x_2 = y_2 - \tfrac{1}{2}$. Then we have

$$\sum_{x=x_1}^{x_2} \cos(ax+b) = \frac{1}{2\sin\tfrac{1}{2}a}\left\{\sin\left[a\left(x_2+\tfrac{1}{2}\right)+b\right] - \sin\left[a\left(x_1-\tfrac{1}{2}\right)+b\right]\right\}$$

$$(6.2.8)$$

Using the formula

$$\sin A - \sin B = 2\cos\tfrac{1}{2}(A+B)\sin\tfrac{1}{2}(A-B) \qquad (6.2.9)$$

this simplifies somewhat to the following theorem.

COSINE SUM THEOREM   For $a \neq 0$,

$$\sum_{x=x_1}^{x_2} \cos(ax+b) = \frac{\cos\tfrac{1}{2}\left[a(x_2+x_1)+2b\right]\sin\tfrac{1}{2}a\left[x_2-x_1+1\right]}{\sin\tfrac{1}{2}a} \qquad (6.2.10)$$

The proof of the following theorem is left as an exercise:

SINE SUM THEOREM   For $a \neq 0$,

$$\sum_{x=x_1}^{x_2} \sin(ax+b) = \frac{\sin\tfrac{1}{2}\left[a(x_2+x_1)+2b\right]\sin\tfrac{1}{2}a\left[x_2-x_1+1\right]}{\sin\tfrac{1}{2}a} \qquad (6.2.11)$$

ILLUSTRATIVE PROBLEM 6.2-1   Using the proper formula, find

$$\cos(-1.205) + \cos(-0.635) + \cos(-0.065) + \cos(0.505)$$

Solution A difference table shows that the arguments are linearly ordered with $a = 0.57$, $b = -2.63$, and $x = 2.5$, 3.5, 4.5 and 5.5. Since x is stepped by 1 each time, (6.2.10) can be used. Then the sum is given by

$$\frac{\cos\tfrac{1}{2}[0.57(5.5+2.5)+2(-2.63)]\sin\tfrac{1}{2}[0.57(5.5-2.5+1)]}{\sin\tfrac{1}{2}(0.57)}$$

which is 3.03583.

We did this as an illustration, and there is no advantage for the sum of four cosines. However, the work would not be harder for the sum of 100 million terms.

ILLUSTRATIVE PROBLEM 6.2-2   Prove the important

THEOREM

$$\sum_{x=0}^{2N-1} \cos\frac{\pi k x}{N}\cos\frac{\pi m x}{N} = N\delta(m,k) \qquad \text{for} \quad 0 < m+k < 2N,$$

where $\delta(m,k)$ is the *Kronecker delta function* which equals 0 if $m \neq k$ and equals 1 if $m = k$.

Case 1: $m \neq k$. To use (6.2.10), we must express the product of cosines as a sum. We use the identity

$$\cos A\cos B = \tfrac{1}{2}\big[\cos(A+B)+\cos(A-B)\big]$$

with $A = \pi k x/N$ and $B = \pi m x/N$. Then

$$\sum_{x=0}^{2N-1} \cos\frac{\pi k x}{N}\cos\frac{\pi m x}{N}$$

$$= \frac{1}{2}\left[\sum_{x=0}^{2N-1}\cos\frac{\pi(k+m)x}{N} + \sum_{x=0}^{2N-1}\cos\frac{\pi(k-m)x}{N}\right]$$

Using (6.2.10) for the first of the two summations, with $x_1 = 0$, $x_2 = 2N-1$, $a = \pi(k+m)/N$, $b = 0$, we have

$$\sum_{x=0}^{2N-1} \cos\frac{\pi(k+m)x}{N}$$

$$= \frac{\cos\big[\tfrac{1}{2}(\pi(k+m)/N)(2N-1)\big]\sin\big[\tfrac{1}{2}(\pi(k+m)/N)(2N)\big]}{\sin\big[\pi(k+m)/2N\big]}$$

In the numerator, the sine has an argument of $\pi(k+m)$, i.e., π times an integer. Therefore, the summation vanishes.

Using (6.2.10) for the second of the two summations with $x_1 = 0$, $x_2 = 2N - 1$, $a = \pi(k-m)x/N$, the same result ensues. Therefore, for $m \neq k$, the entire summation is 0.

Case 2: $m = k$. The summation can be rewritten as

$$\sum_{x=0}^{2N-1} \cos^2 \frac{\pi k x}{N}$$

We use the trigonometric identity

$$\cos^2 A = \tfrac{1}{2}(1 + \cos 2A)$$

with $A = \pi k x / N$. Then

$$\sum_{x=0}^{2N-1} \cos^2 \frac{\pi k x}{N} = \frac{1}{2} \sum \left(1 + \cos \frac{2\pi k x}{N} \right)$$

$$= \frac{1}{2} \sum_{x=0}^{2N-1} 1 + \frac{1}{2} \sum_{x=0}^{2N-1} \cos \frac{2\pi k x}{N}$$

The first of the two summations is equal to $2N$, and the second, for reasons explained in case 1, is equal to 0. This completes the proof. $\square$

This theorem will be needed in the chapter on Fourier series.

To some extent, the method of thinking employed for sines and cosines can sometimes be applied to other functions.

ILLUSTRATIVE PROBLEM 6.2-3   Find $\sum_{k=0}^{n-1} \csc^2(2^k x)$.

Solution   We use

$$\Delta \cot(2^t x) = -\csc(2^{t+1}x)$$

and let $t + 1 = k$. Then

$$\csc(2^k x) = -\Delta \cot(2^{k-1}x)$$

$$\sum_{k=0}^{n-1} \csc(2^k x) = -\left[\cot(2^{n-1}x) - \cot(2^{-1}x) \right]$$

$$= \cot\tfrac{x}{2} - \cot(2^{n-1}x)$$

6.2.3 Logarithms

We have proved that

$$\Delta \ln(ay + b) = \ln\left(1 + \frac{a}{ay + b}\right) \qquad (6.2.12)$$

At first glance, it might seem that it would be easy to develop a sum formula for logarithms. We shall go through the steps naively, develop a spurious formula, then tell why it is incorrect.

Summing both members of (6.2.12), we have

$$\sum_{y=y_1}^{y_2-1} \ln\left(1 + \frac{a}{ay + b}\right) = \sum_{y=y_1}^{y_2-1} \Delta \ln(ay + b) \qquad (6.2.13)$$

$$= \ln \frac{ay_2 + b}{ay_1 + b} \qquad (6.2.14)$$

Let

$$x = 1 + \frac{a}{ay + b} \qquad (6.2.15)$$

so that $x_1 = 1 + a/(ay_1 + b)$, and $x_2 = a/(ay_2 + b)$. Then

$$\sum_{x_1}^{x_2} \ln x = \ln \frac{ay_2 + b}{ay_1 + b} \qquad (6.2.16)$$

We now solve for $ay + b$:

$$ay_1 + b = a/(x_1 - 1) \qquad (6.2.17)$$
$$ay_2 + b = a/x_2 \qquad (6.2.18)$$

Substituting in (6.2.16),

$$\sum_{x_1}^{x_2} \ln x = \ln\left[\frac{a}{x_2} \frac{x_1 - 1}{a}\right] = \ln\left(\frac{x_1 - 1}{x_2}\right) \qquad (6.2.19)$$

Test with $x_1 = 2$ and $x_2 = 4$:

$$\ln 2 + \ln 3 + \ln 4 = \ln\tfrac{1}{4} \qquad (6.2.20)$$

which is nonsense.

The fallacy is by no means obvious. To illuminate it, consider

$$\sum_{x=1}^{3} (2x + 3) = 5 + 7 + 9 = 21 \qquad (6.2.21)$$

Now, let $y = 2x + 3$. When $x = 1$, $y = 5$; when $x = 3$, $y = 9$. Thus

$$\sum_{y=5}^{9} y = 5 + 6 + 7 + 8 + 9 = 35 \qquad (6.2.22)$$

This should make it clear that substitution must be done carefully. The reason for the difficulty is that the dummy in the use of Σ arbitrarily increments by 1, and the replaced dummy must also increment by 1 the same number of times. Observe that in (6.2.21) there are *three* contributions, but in (6.2.22) there are *five*.

Now consider

$$\sum_{x=1}^{3} (x+3) = 4 + 5 + 6 = 15 \qquad (6.2.23)$$

and let $y = x + 3$; then

$$\sum_{y=4}^{6} y = 4 + 5 + 6 = 15 \qquad (6.2.24)$$

which is perfectly satisfactory. The moral is that it is permissible to make a substitution of the form

$$y = x + b \qquad (6.2.25)$$

where b is a constant, but no substitution can be made which does not preserve the size and number of steps.

Returning to the logarithm problem, we can accept the derivation up to (6.2.14). Now we consider the special case $a = 1$ and let $x = y + b$. Then

$$\sum_{x=x_1}^{x_2} \ln\left(1 + \frac{1}{x}\right) = \ln \frac{x_2 + 1}{x_1}, \qquad (6.2.26)$$

which is true but trivial.

Comment   Because of the arbitrary unit increment, the Σ operator may, in the long run, not be the most suitable for computer science. It may be convenient to define another summation operator such that the steps (increments) can be taken as in a DO loop.

6.2.4   Hyperbolic Functions

$$\Delta \cosh(ay + b) = 2\sinh\left[a\left(y + \tfrac{1}{2}\right) + b\right]\sinh\tfrac{1}{2}a \qquad (6.2.27)$$

Summing both members and rearranging,

$$\sum_{y=y_1}^{y_2-1} \sinh\left[a\left(y + \tfrac{1}{2}\right) + b\right] = \frac{1}{2\sinh\tfrac{1}{2}a} \sum_{y_1}^{y_2-1} \cosh(ay + b) \qquad (6.2.28)$$

$$= \frac{1}{2\sinh\tfrac{1}{2}a}\left[\cosh(ay_2 + b) - \cosh(ay_1 + b)\right] \qquad (6.2.29)$$

Let

$$x = y + \tfrac{1}{2} \qquad (6.2.30)$$

then

$$\sum_{x=y_1+1/2}^{y_2-1/2} \sinh(ax+b) = \frac{1}{2\sinh\frac{1}{2}a} \left[\cosh(ay_2+b) - \cosh(ay_1+b) \right] \quad (6.2.31)$$

Let

$$x_1 = y_1 + \tfrac{1}{2} \quad \text{and} \quad x_2 = y_2 - \tfrac{1}{2} \qquad (6.2.32)$$

Then we have the

HYPERBOLIC SINE SUM THEOREM

$$\sum_{x=x_1}^{x_2} \sinh(ax+b)$$

$$= \frac{\cosh\left[a\left(x_2 + \tfrac{1}{2}\right) + b \right] - \cosh\left[a\left(x_1 - \tfrac{1}{2}\right) + b \right]}{2\sinh\frac{1}{2}a} \qquad (6.2.33)$$

The following theorem is left as an exercise.

HYPERBOLIC COSINE SUM THEOREM

$$\sum_{x=x_1}^{x_2} \cosh(ax+b)$$

$$= \frac{\sinh\left[a\left(x_2 + \tfrac{1}{2}\right) + b \right] - \sinh\left[a\left(x_1 - \tfrac{1}{2}\right) + b \right]}{2\sinh\frac{1}{2}a} \qquad (6.2.34)$$

6.2.5 Combinatorials

The following theorem is left as an exercise.

COMBINATORIAL SUM THEOREM   For integral x and r,

$$\sum_{x=x_1}^{x_2} \binom{x}{r} = \binom{x_2+1}{r+1} - \binom{x_1}{r+1} \qquad (6.2.35)$$

Hint: Use $\binom{n}{k} = \binom{n-1}{k} + \binom{n-1}{k-1}$.

PROBLEM SECTION 6.2

A. These exercises are designed for practice in the use of the formulas. Using the proper formula, evaluate the sums. Check by direct addition. Our answers may not be precisely the same as yours, depending upon the machine used.

(1) $(\sin 2)^{3.1} + (\sin 2)^{4.1} + (\sin 2)^{5.1} + (\sin 2)^{6.1}$
Hint: This is a geometric series.

(2) $(\sin 2)^{3.2} + (\sin 2)^{5.2} + (\sin 2)^{7.2} + (\sin 2)^{9.2}$
Hint: This is a geometric series.

(3) $\cos 5.3 + \cos 6.8 + \cos 8.3 + \cos 9.8$
Hint: Write the arguments in the form $ax + b$.

(4) $\sin(-3.7) + \sin(-1.7) + \sin(0.3) + \sin(2.3) + \sin(4.3)$
Hint: Write the arguments in the form $ax + b$.

(5) $\sinh 0.125 + \sinh 0.355 + \sinh 0.585 + \sinh 0.815$

(6) $\cosh 1.08 + \cosh 1.12 + \cosh 1.20 + \cosh 1.62$

(7) Find the sum to n terms:

$\cos a - \cos(a + b) + \cos(a + 2b) - \cos(a + 3b) + \cdots$.
Hint: Compare with $\cos(a + k[b + \pi])$.

(8) Find the sum to n terms:

$\sin a + \cos(a + b) - \sin(a + 2b) - \cos(a + 3b) + \sin(a + 4b) + \cdots$.
Hint: Compare with $\sin(a + k[b + \frac{1}{2}\pi])$.

B. Prove the following theorems, which are used in the theory of Fourier series:

(9) $\displaystyle\sum_{x=0}^{2N-1} \cos\frac{\pi x}{N} = 0$

(10) $\displaystyle\sum_{x=0}^{2N-1} \sin\frac{\pi x}{N} = 0$

(11) $\displaystyle\sum_{x=0}^{2N-1} \cos\frac{\pi k x}{N} \sin\frac{\pi m x}{N} = 0$

(12) $\displaystyle\sum_{x=0}^{2N-1} \sin\frac{\pi k x}{N} \sin\frac{\pi m x}{N} = N\delta(m,k)$ for $0 < m + k < 2N$
Hint: Use $\cos 2A = 1 - 2\sin^2 A$.

C. Prove:
(13) The sine sum theorem
(14) The hyperbolic cosine sum theorem
(15) The combinatorial sum theorem

6.3 SUMMATION BY PARTS

6.3.1 The Basic Theorem

Let u and v be functions of x. Then

$$\Delta(uv) = u\,\Delta v + v_+\,\Delta u \tag{6.3.1}$$

where v_+ is $v(x+1)$, assuming that $h=1$. Summing, we have

$$\sum_{x=a}^{b-1} \Delta(uv) = \sum_{x=a}^{b-1} u\,\Delta v + \sum_{x=a}^{b-1} v(x+1)\,\Delta u \qquad (6.3.2)$$

Rearranging slightly and letting $x_1 = a$, $x_2 = b-1$, we have the

---

PARTS THEOREM

$$\sum_{x=x_1}^{x_2} u(x)\,\Delta v(x) = uv\big|_{x_1}^{x_2+1} - \sum_{x=x_1}^{x_2} v(x+1)\,\Delta u(x) \qquad (6.3.3)$$

The calculation for uv is done in the customary way:

$$uv\big|_{x_1}^{x_2+1} = u(x_2+1)v(x_2+1) - u(x_1)v(x_1) \qquad (6.3.4)$$

6.3.2 Indefinite Summands

It has probably occurred to the reader that in using the parts theorem there may be a problem finding v after Δv has been selected. We pause to illustrate the difficulty and one way of reducing it.

ILLUSTRATIVE PROBLEM 6.3-1   Given $\Delta v = \sin(ax+b)$, find $v(x)$.

Solution   First, we shall guess that

$$v = \cos(t_1 x + t_2) \qquad (6.3.5)$$

in the hope this will give us the proper Δv. Then

$$\Delta v = \cos[t_1(x+1)+t_2] - \cos[t_1 x + t_2] \qquad (6.3.6)$$

We use the trigonometric identity

$$\cos A - \cos B = -2\sin\tfrac{1}{2}(A+B)\sin\tfrac{1}{2}(A-B) \qquad (6.3.7)$$

and obtain, from (6.3.6),

$$\Delta v = -2\sin\left(t_1 x + \tfrac{1}{2}t_1 + t_2\right)\sin\tfrac{1}{2}t_1 \qquad (6.3.8)$$

We compare this with

$$\Delta v = \sin(ax+b) \qquad (6.3.9)$$

If (6.3.8) and (6.3.9) are indeed equal, then

$$\begin{cases} -2\sin\tfrac{1}{2}t_1 = 1 \\ \qquad t_1 = a \\ \tfrac{1}{2}t_1 + t_2 = b \end{cases} \qquad (6.3.10)$$

Unfortunately, this is not possible for arbitrary a and b. We start again with a new supposition:

$$v = t_0 \cos(t_1 x + t_2) \tag{6.3.11}$$

which leads to

$$\Delta v = -2t_0 \sin\left(t_1 x + \tfrac{1}{2}t_1 + t_2\right)\sin\tfrac{1}{2}t_1 \tag{6.3.12}$$

Comparing (6.3.12) and (6.3.9), we have

$$\begin{cases} -2t_0 \sin\tfrac{1}{2}t_1 = 1 \\ \qquad\quad t_1 = a \\ \quad \tfrac{1}{2}t_1 + t_2 = b \end{cases} \tag{6.3.13}$$

These are easily solved to produce

$$\begin{cases} t_1 = a \\ t_2 = b - \tfrac{1}{2}a \\ t_0 = -\dfrac{1}{2\sin\tfrac{1}{2}a} \end{cases} \tag{6.3.14}$$

from which

$$v = -\frac{\cos\left(ax + b - \tfrac{1}{2}a\right)}{2\sin\tfrac{1}{2}a} \tag{6.3.15}$$

We shall call this an *indefinite summand* of $\sin(ax + b)$, because the addition of a constant will not affect Δv. To check our answer, we difference (6.3.15):

$$\Delta v = -\frac{1}{2\sin\tfrac{1}{2}a}\left[\cos\left(ax + b + \tfrac{1}{2}a\right) - \cos\left(ax + b - \tfrac{1}{2}a\right)\right] \tag{6.3.16}$$

$$= -\frac{1}{2\sin\tfrac{1}{2}a}\left[-2\sin(ax + b)\sin\tfrac{1}{2}a\right] \tag{6.3.17}$$

which leads to the proper value of Δv.

Table A.7 in the Appendix lists some indefinite summands useful in the application of the parts theorem.

6.3.3 Applications

ILLUSTRATIVE PROBLEM 6.3-2   Find

$$\sum_{k=0}^{m-1} kx^k$$

Solution   Let $u = k$ and $\Delta v = x^k$. Then $\Delta u = 1$ and $v = (x-1)^{-1}x^k$. Summing by parts,

$$\sum_{k=0}^{m-1} kx^k = \frac{kx^k}{x-1} \Big|_0^m - \sum_{k=0}^{m-1} \frac{1}{x-1} \cdot x^{k+1} \cdot 1$$

$$= \frac{mx^m}{x-1} - \frac{1}{x-1} \sum_{k=0}^{m-1} x^{k+1}$$

$$= \frac{mx^m}{x-1} - \frac{x}{x-1} \sum_{k=0}^{m-1} x^k$$

The remaining summation in the right member is simply $1 + x + x^2 + \cdots + x^{m-1}$, a geometric series whose sum is $(1-x^m)/(1-x)$. Therefore

$$\sum_{k=0}^{m-1} kx^k = \frac{mx^m}{x-1} + \frac{x(1-x^m)}{(1-x)^2}$$

ILLUSTRATIVE PROBLEM 6.3-3   Find

$$\sum_{k=0}^{m-1} k^2 x^k$$

Solution   Let $u = k^2$ and $\Delta v = x^k$. Then $\Delta u = 2k+1$, and $v = x^k/(x-1)$. Summing by parts,

$$\sum_{k=0}^{m-1} k^2 x^k = \frac{k^2 x^k}{x-1} \Big|_0^m - \sum_{k=0}^{m-1} \frac{x^{k+1}}{x-1}(2k+1)$$

$$= \frac{k^2 x^k}{x-1} \Big|_0^m - 2 \sum_{k=0}^{m-1} \frac{kx^{k+1}}{x-1} - \sum_{k=0}^{m-1} \frac{x^{k+1}}{x-1}$$

The first summation in the right member can be written as

$$\frac{x}{x-1} \sum_{k=0}^{m-1} kx^k$$

which was evaluated in Illustrative Problem 6.3-2. The second summation in the right member is

$$\frac{x}{x-1} \sum_{k=0}^{m-1} x^k$$

which is a geometric series. After the substitutions, the result is

$$\frac{m^2 x^m}{x-1} + \frac{x(1-x^m)}{(x-1)^2} - \frac{2x[mx^m(x-1) + x(1-x^m)]}{(x-1)^3}$$

Complicated as this is, it is less error-laden, for computing purposes, than the original problem.

ILLUSTRATIVE PROBLEM 6.3-4   Find

$$I = \sum_{k=1}^{n-1} k \sin(2ak + b)$$

Solution   Let $u = k$. Then $\Delta v = \sin(2ak + b)$, $\Delta u = 1$, and

$$v = -\cos\left[2a\left(k - \tfrac{1}{2}\right) + b\right]/\sin a.$$

Thus

$$I = \frac{-k\cos\left[2a\left(k - \tfrac{1}{2}\right) + b\right]}{\sin a}\Bigg|_{1}^{n} - \sum_{k=1}^{n-1} - \frac{\cos\left[2a\left(k + \tfrac{1}{2}\right) + b\right]}{\sin a}$$

$$= \frac{-n\cos\left[2a\left(n - \tfrac{1}{2}\right) + b\right]}{\sin a} + \frac{\cos[a + b]}{\sin a}$$

$$+ \frac{1}{\sin a}\sum_{k=1}^{n-1}\cos[2ak + (a + b)]$$

Referring to the cosine sum theorem and substituting $2a$ for a, k for x and $a + b$ for b, we have

$$I = \frac{-n\cos(2an - a + b) + \cos(a + b)}{\sin a} + \frac{\sin(2an + b) - \sin(2a + b)}{2\sin^2 a}$$

PROBLEM SECTION 6.3

A.

(1) Find an indefinite sum of $ax^3 + b$.

(2) Find an indefinite sum of $ax^4 + b$.

(3) Find, to n terms, $\sin^2 q + \sin^2(q + r) + \sin^2(q + 2r) + \cdots$.
 Hint: Use $\sin^2 A = \tfrac{1}{2}(1 - \cos 2A)$.

(4) Find, to n terms, $\cos^3 t + \cos^3 3t + \cos^3 5t + \cdots$.
 Hint: Use $\cos^3 A = \tfrac{1}{4}(3\cos A + \cos 3A)$.

(5) Find, to n terms,

$$\tan\frac{a}{2}\sec a + \tan\frac{a}{2^2}\sec\frac{a}{2} + \tan\frac{a}{2^3}\sec\frac{a}{2^2} + \cdots.$$

(6) Find, to n terms, $\csc p \csc 3p + \csc 3p \csc 5p + \csc 5p \csc 7p + \cdots$.

(7) Find, to $2n$ terms, $\sin ns + \sin(n - 1)s + \sin(n - 2)s + \cdots$.

(8) Find, to $n - 1$ terms, $\sin \pi/n + \sin 2\pi/n + \sin 3\pi/n + \cdots$.

B.

(9) Find $\displaystyle\sum_{x=0}^{n-1} xa^x$.

(10) Find $\displaystyle\sum_{x=0}^{n-1} (x + 2)(x + 3)2^x$.

(11) Find $\displaystyle\sum_{x=0}^{n-1} x \sinh x$.

(12) Find $\displaystyle\sum_{x=0}^{n-1} 2^x \sin x$.

C.

    (13) Find, to n terms, $2\times 2 + 4\times 4 + 7\times 8 + 11\times 16 + 16\times 32 + \cdots$.

    (14) Find, to n terms, $1\times 3 + 3\times 3^2 + 5\times 3^3 + 7\times 3^4 + \cdots$.

    (15) Fing, to n terms, $1\times 3 + 4\times 7 + 9\times 13 + 16\times 21 + \cdots$.

    (16) Find, to n terms, $1 + 3x + 6x^2 + 10x^3 + \cdots$.

6.4 HOMOGENEOUS LINEAR DIFFERENCE EQUATIONS

6.4.1 Background

We interrupt our discussion of summation to devote two sections to the exploration of difference equations. These equations will assume a more and more important role as we continue our treatment of the problems of computer science.

An example of a *first-order difference equation* is

$$\Delta f(k) = a^k (a-1) \tag{6.4.1}$$

which, we know, has a *particular solution*

$$f(k) = a^k \tag{6.4.2}$$

However, this is not the only solution. We may add any function $g_1(k)$:

$$f(k) = a^k + g_1(k) \tag{6.4.3}$$

provided that $\Delta g_1(k) = 0$. For example, $g_1(k)$ may be equal to $\arctan(x^2 - 7\ln x)$ provided that x is not a function of k; or $g_1(k)$ may be $\sin k\pi$ if k is an integer; or $g_1(k)$ may be 12.7. What we are saying is that (6.4.3) is the *general* solution of (6.4.1) and includes precisely one *periodic arbitrary constant* (pac). Note that the pac may not really be a constant; it is merely independent of k.

The formula (6.4.1) can be rewritten as

$$f(k+1) - f(k) = a^k (a-1) \tag{6.4.4}$$

It is recognizable as *first-order* because $(k+1) - (k) = 1$.

An example of a *second-order difference equation* is

$$\Delta^2 f(k) = 2^k \tag{6.4.5}$$

We can write this as

$$\Delta\big[\Delta f(k)\big] = 2^k \tag{6.4.6}$$

which is partly solved by Appendix Table A.7:

$$\Delta f(k) = 2^k + g_1(k) \tag{6.4.7}$$

Using the table once again

$$f(k) = 2^k + k g_1(k) + g_2(k) \tag{6.4.8}$$

where g_1 and g_2 are periodic arbitrary constants.

Note that the general solution of a second-order difference equation has *two* pac. In general, the complete solution of an nth-order difference equation must have n pac.

Now, note that (6.4.5) can be written as

$$f(k+2) - 2f(k+1) + f(k) = 2^k \tag{6.4.9}$$

It is recognizable as *second-order* because the difference of the arguments, $(k+2)-(k)$, is equal to 2.

The general form of a linear difference equation is

$$a_n(k)f(k+n) + a_{n-1}f(k+n-1) + \cdots + a_0(k)f(k) = h(x) \tag{6.4.10}$$

It is *linear* because the fs are all to power 1. The following would be *non-linear*:

$$3\left[f(k+1)\right]^2 - 5\left[f(k)\right]^{2/3} = h(x) \tag{6.4.11}$$

because of the exponents.

In this section, we consider equations like (6.4.10) where $h(k)=0$. Such equations are called *homogeneous*. Two cases are especially easy to solve: case I, where the homogeneous equation is first-order and the coefficients may or may not be constant with reference to k; and case II, where the homogeneous equation is of any order but the coefficients are constant. There are other equations (case III) which do not seem to fit into either of these two categories but which, under a suitable transformation, change to case I or case II.

6.4.2   Case I: First-Order Homogeneous Linear Equations

Let the equation be

$$a(k)f(k+1) + b(k)f(k) = 0 \tag{6.4.12}$$

For $a(k) \neq 0$,

$$f(k+1) = c(k)f(k) \tag{6.4.13}$$

where $c(k) = -b(k)/a(k)$.

Starting with a specific value of r for which (6.4.13) is true, we see that

$$f(r+1)=c(r)f(r) \qquad (6.4.14)$$

$$f(r+2)=c(r+1)f(r+1) \qquad (6.4.15)$$

$$f(r+3)=c(r+2)f(r+2) \qquad (6.4.16)$$

$$\vdots$$

$$f(k-1)=c(k-2)f(k-2) \qquad (6.4.17)$$

$$f(k)=c(k-1)f(k-1) \qquad (6.4.18)$$

Multiplying all the equations and canceling, we have

$$f(k)=\left[c(r)c(r+1)c(r+2)\cdots c(k-2)c(k-1)\right]f(r), \quad (6.4.19)$$

which may be abbreviated as in the following theorem.

THEOREM: FIRST-ORDER HOMOGENEOUS LINEAR EQUATION If $a(k)f(k+1)+b(k)f(k)=0$, then

$$f(k)=A\left[\prod_{t=r}^{k-1}\left(-\frac{b(t)}{a(t)}\right)\right]f(r) \qquad (6.4.20)$$

where A is a pac.

As usual, the symbol $\prod_{t=r}^{k-1}$ means the product from $t=r$ to $t=k-1$ by increments of 1. For convenience, we choose $r=0$ if possible.

ILLUSTRATIVE PROBLEM 6.4-1   Solve $f(k+1)-e^{2k}f(k)=0$.

Solution $a(k)=1$, $b(k)=-e^{2k}$. Since the use of $r=0$ does not cause any difficulties (such as a loss of definition), we choose it as the lower limit of the product. Then, by the theorem,

$$f(k)=A\left[\prod_{t=0}^{k-1}e^{2t}\right]f(0) \qquad (6.4.21)$$

To evaluate the product, let

$$y=\prod_{t=0}^{k-1}e^{2t} \qquad (6.4.22)$$

Then

$$y=e^0e^2e^4\cdots e^{2(k-1)} \qquad (6.4.23)$$

$$\ln y=0+2+4+\cdots+(2k-2) \qquad (6.4.24)$$

The right member of (6.2.24) is an arithmetic series with sum $(k-1)k$. Therefore

$$y = e^{k(k-1)} \qquad (6.4.25)$$

Combining $Af(0)$ into a single pac, assuming $f(0)$ exists, we have the solution:

$$f(k) = Be^{k(k-1)} \qquad (6.4.26)$$

where B is a pac.

ILLUSTRATIVE PROBLEM 6.4-2   Solve Illustrative Problem 6.4-1, given the *boundary condition* $f(0) = \cos \pi x$, where x is not a function of k.

Solution   Substituting in (6.4.26) with $k = 0$,

$$\cos \pi x = B \qquad (6.4.27)$$

$$f(x) = (\cos \pi x)e^{k(k-1)} \qquad (6.4.28)$$

This is a *particular solution* satisfying the required boundary condition.

6.4.3   Case IIA: Linear Equations with Constant Coefficients and Distinct Solutions

We are interested in equations of the form

$$a_n f(k+n) + a_{n-1} f(k+n-1) + \cdots + a_{n-r} f(k+n-r) = 0 \quad (6.4.29)$$

where the as are independent of k. To solve this, we assume that

$$f(k) = t^k \qquad (6.4.30)$$

where t is a constant. (The idea is taken from the theory of differential equations.) Substituting in (6.4.29),

$$a_n t^{k+n} + a_{n-1} t^{k+n-1} + \cdots + a_{n-r} t^{k+n-r} = 0 \qquad (6.4.31)$$

Assuming that t^{k+n-r} does not vanish, we divide both members of (6.4.31) by this quantity and obtain the algebraic equation

$$a_n t^r + a_{n-1} t^{r-1} + \cdots + a_{n-r} = 0 \qquad (6.4.32)$$

which can always be solved for t. This equation is called the *characteristic equation* of (6.4.29). If there are r distinct solutions $t_1, t_2, \ldots, t_r$ for the characteristic equation, then there are r distinct solutions for $f(k)$. Each of these distinct solutions multiplied by a pac is obviously a solution as well, so that the general solution of (6.4.29) is

$$f(k) = A_1 t_1^k + A_2 t_2^k + \cdots + A_r t_r^k \qquad (6.4.33)$$

Since there are r pacs, this is a general solution.

ILLUSTRATIVE PROBLEM 6.4-3   Solve

$$f(k+2)-f(k+1)-6f(k)=0 \tag{6.4.34}$$

Solution   Let

$$f(k)=t^k \tag{6.4.35}$$

Then

$$t^{k+2}-t^{k+1}-6t^k=0 \tag{6.4.36}$$

Dividing by $t^k \neq 0$,

$$t^2-t-6=0 \tag{6.4.37}$$

$$(t-3)(t+2)=0 \tag{6.4.38}$$

$$t=3 \quad \text{or} \quad t=-2 \tag{6.4.39}$$

$$f(k)=A\,3^k+B\,(-2)^k \tag{6.4.40}$$

This is a general solution because there are two pac and (6.4.34) is second-order. Some algebraic equations may require the use of formulas or even of an iterative procedure on the computer.

ILLUSTRATIVE PROBLEM 6.4-4   Solve

$$3f(k+2)+2cf(k+1)+\pi xf(k)=0 \tag{6.4.41}$$

and find the particular solution for the boundary condition $f(0)=1, f(1)=2$.

Solution   Let

$$f(k)=t^k \tag{6.4.42}$$

Then

$$3t^{k+2}+2ct^{k+1}+\pi xt^k=0 \tag{6.4.43}$$

Dividing by $t^k \neq 0$,

$$3t^2+2ct+\pi x=0 \tag{6.4.44}$$

Using the quadratic formula, we have

$$t=\frac{-2c \pm \sqrt{4c^2-12\pi x}}{6} \tag{6.4.45}$$

so that

$$f(x)=A\left(\frac{-2+\sqrt{4c^2-12\pi x}}{6}\right)^k+B\left(\frac{-2-\sqrt{4c^2-12\pi x}}{6}\right)^k \tag{6.4.46}$$

where A and B are pac. At $k=0$ and 1, we have

$$f(0)=1=A+B \tag{6.4.47}$$

$$f(1)=2=\left(\frac{-2+\sqrt{4c^2-12\pi x}}{6}\right)A+\left(\frac{-2-\sqrt{4c^2-12\pi x}}{6}\right)B \tag{6.4.48}$$

This pair of simultaneous equations can be solved in a variety of ways. Solving by determinants, the result is

$$f(k)=\left(\frac{14+\sqrt6}{2\sqrt6}\right)\left(\frac{-2+\sqrt{4c^2-12\pi x}}{6}\right)^k-\left(\frac{14-\sqrt6}{2\sqrt6}\right)\left(\frac{-2-\sqrt{4c^2-12\pi x}}{6}\right)^k$$

$$\tag{6.4.49}$$

One other important situation arises when the solution of the characteristic equation yields (pairs of) complex roots: complex conjugates such as $a+bi$ and $a-bi$.

Suppose a pair of roots of the characteristic equation is represented by

$$t=a\pm bi. \tag{6.4.50}$$

Let

$$r=\sqrt{a^2+b^2} \tag{6.4.51}$$

Then

$$a\pm bi=r\left(\frac{a}{r}\pm i\frac{b}{r}\right) \tag{6.4.52}$$

Since $|a|<r$ and $|b|<r$, let

$$a/r=\cos\theta \quad \text{and} \quad b/r=\sin\theta \tag{6.4.53}$$

where $\tan\theta=b/a$. Then (6.4.52) becomes

$$a\pm bi=r(\cos\theta\pm i\sin\theta) \tag{6.4.54}$$

and by De Moivre's theorem,

$$(a\pm bi)^k=r^k(\cos k\theta\pm i\sin k\theta) \tag{6.4.55}$$

From (6.4.50), a pair of roots of the functional equation leads to

$$f(k)=A(a+bi)^k+B(a-bi)^k \tag{6.4.56}$$

Substituting from (6.4.55),

$$f(k)=Ar^k(\cos k\theta+i\sin k\theta)+Br^k(\cos k\theta-i\sin k\theta) \tag{6.4.57}$$

Factoring r^k and combining constants, we have the following:

COMPLEX-ROOTS THEOREM   If two roots of a characteristic equation are $a \pm bi$, then the corresponding roots of the general equation can be expressed as

$$r^k(A \cos k\theta + B \sin k\theta) \qquad (6.4.58)$$

where $r = \sqrt{a^2 + b^2}$ and $\tan \theta = b/a$.

ILLUSTRATIVE PROBLEM 6.4-5   Solve

$$f(k+2) - f(k+1) + f(k) = 0 \qquad (6.4.59)$$

Solution   The characteristic equation is

$$t^2 - t + 1 = 0 \qquad (6.4.60)$$

$$t = \frac{1 \pm i\sqrt{3}}{2} \qquad (6.4.61)$$

Here

$$a = \tfrac{1}{2}$$
$$b = \sqrt{3}\,/2$$
$$\tan^{-1}\theta = \sqrt{3} \qquad (6.4.62)$$
$$\theta = \pi/3$$

Therefore

$$f(k) = A \cos \frac{\pi k}{3} + B \sin \frac{\pi k}{3} \qquad (6.4.63)$$

These problems should always be checked. Frequently, the checking is much more tedious than the solution.

6.4.4   Case IIB. Linear Equations with Constant Coefficients and Repeated Roots

We shall proceed naively in the next problem to illuminate the difficulty, then backtrack and solve it correctly.

ILLUSTRATIVE PROBLEM 6.4-6   Solve

$$f(k+2) - 2f(k+1) + f(k) = 0 \qquad (6.4.64)$$

Solution   The characteristic equation is

$$t^{k+1} - 2t^{k+1} + t^k = 0 \qquad (6.4.65)$$

which, after the usual factoring, becomes

$$t^2 - 2t + 1 = 0 \tag{6.4.66}$$

so that it might appear that

$$t = A(1)^k + B(1)^k \tag{6.4.67}$$

Unfortunately, this is equivalent to

$$t = A \tag{6.4.68}$$

which cannot be a general solution, because (6.4.64) requires *two* periodic arbitrary constants. There must be another part to the general solution. Borrowing again from the theory of differential equations, we assert (and shall prove) that if x is a double root of the characteristic equation, the general solution contains

$$Ax^k + Bkx^k \tag{6.4.69}$$

In this case, the general solution is

$$f(k) = A(1)^k + Bk(1)^k = A + Bk \tag{6.4.70}$$

The theorem we are about to prove is the

REPEATED-ROOTS THEOREM   Given that
$$a_n f(k+n) + a_{n-1} f(k+n-1) + \cdots + a_0 f(k) = 0 \tag{6.4.71}$$

where the *a*s are independent of k, and that the characteristic equation has an *m*tuple root t (a root repeated m times), then the general solution includes
$$(A_0 + A_1 k + A_2 k^2 + \cdots + A_{m-1} k^{m-1}) t^k \tag{6.4.72}$$

Proof   We write (6.4.71) as

$$\sum_{j=0}^{n} a_j f(k+j) = 0 \tag{6.4.73}$$

Substituting

$$f(k) = t^k \tag{6.4.74}$$

we have

$$\sum_{j=0}^{n} a_j t^{k+j} = 0 \tag{6.4.75}$$

Dividing by t^k in the usual fashion,

$$\sum_{j=0}^{n} a_j t^j = 0 \tag{6.4.76}$$

We wish to prove that (6.4.72) satisfies (6.4.73). We shall need the following theorem from the theory of equations.

> MULTIPLE-ROOT THEOREM   If a polynomial, $f(t)$, has an $m$tuple root t, then $f(t) = f'(t) = \cdots = f^{(m-1)}(t) = 0$.

Before displaying the proof, we offer an example. Suppose

$$f(t) = (t-3)^2(t-4) \tag{6.4.77}$$

so that 3 is a double root. We rewrite this as

$$f(t) = g(t)h(t) \tag{6.4.78}$$

where $g(t) = (t-3)^2$ and $h(t) = (t-4)$. Note that $g(3) = 0$, $g'(t) = 2(t-3)$, so that $g'(3) = 0$, $g''(t) = 2$; thus $g''(3) \neq 0$. Also $h(3) = -1$, $h'(t) = 1$ for all t. We differentiate twice:

$$f'(t) = g(t)h'(t) + g'(t)h(t) \tag{6.4.79}$$

$$f''(t) = g(t)h''(t) + 2g'(t)h'(t) + g''(t)h(t) \tag{6.4.80}$$

When $t = 3$, (6.4.79) vanishes but (6.4.80) does not. Geometrically, this means that the graph is tangent to the t-axis at $t = 3$.

The general proof follows: Let

$$f(t) = g(t)h(t) \tag{6.4.81}$$

where

$$g(t) = (t - t_1)^m \tag{6.4.82}$$

We note that

$$\begin{cases} g(t) = g'(t) = g''(t) = \cdots = g^{(m-1)}(t) = 0 \\ g^{(m)}(t) \neq 0 \end{cases} \tag{6.4.83}$$

Differentiating, we have

$$f(t) = g(t)h(t) \tag{6.4.84}$$

$$f'(t) = g(t)h'(t) + g'(t)h(t) \tag{6.4.85}$$

$$f''(t) = g(t)h''(t) + 2g'(t)h'(t) + g''(t)h(t) \tag{6.4.86}$$

$$f'''(t) = g(t)h'''(t) + 3g'(t)h''(t) + 3g''(t)h'(t) + g'''(t)h(t) \tag{6.4.87}$$

$$\vdots$$

$$f^{(m-1)}(t) = g(t)h^{(m-1)}(t) + \binom{m-1}{1} g'(t)h^{(m-2)}(t)$$

$$+ \binom{m-1}{2} g''(t)h^{(m-3)}(t) + \cdots + g^{(m-1)}(t)h(t) \tag{6.4.88}$$

and, therefore, using (6.4.83):

$$f(t) = f'(t) = \cdots = f^{(m-1)}(t) = 0 \tag{6.4.89}$$

This proves the multiple-root theorem. To complete the proof of the repeated-roots theorem, we also need:

$$f^{(m)}(t) = g(t)h^{(m)}(t) + \cdots + g^{(m)}(t)h(t) \tag{6.4.90}$$

where $g^{(m)}(t)$ and $h(t)$ do not vanish.

Before proceeding, we must digress for another moment to discuss the differentiation of a summation. Suppose the summation is

$$f(t) = \sum_{k=0}^{4} p_k t^k = p_0 + p_1 t + p_2 t^2 + p_3 t^3 + p_4 t^4 \tag{6.4.91}$$

Then

$$f'(t) = \sum_{k=0}^{4} k p_k t^{k-1} = p_1 + 2p_2 t + 3p_3 t^2 + 4p_4 t^3 \tag{6.4.92}$$

$$f''(t) = \sum_{k=0}^{4} k^{(2)} p_k t^{k-2} = 2^{(2)} p_2 + 3^{(2)} p_3 t + 4^{(2)} p_4 t^2 \tag{6.4.93}$$

By the multiple-root theorem, we have, when t_1 is an $m$tuple root,

$$f(t_1) = \sum_{k=0}^{n} p_k t_1^k = 0 \tag{6.4.94}$$

$$f'(t_1) = \sum_{k=0}^{n} k p_k t_1^{k-1} = 0 \tag{6.4.95}$$

$$f''(t_1) = \sum_{k=0}^{n} k^{(2)} p_k t_1^{k-2} = 0 \tag{6.4.96}$$

$$\vdots$$

$$f^{(m-1)}(t_1) = \sum_{k=0}^{n} k^{(m-1)} p_k t_1^{k-m+1} = 0 \tag{6.4.97}$$

It is certainly true that for arbitrary constants $d_1, \ldots, d_{m-1}$) and m tuple root t_1,

$$d_1 f(t_1) + d_2 t_1 f'(t_1) + d_3 t_1^2 f''(t_1) + \cdots + d_{m-1} t_1^{m-1} f^{(m-1)}(t_1) = 0 \quad (6.4.98)$$

since each term is 0.

We have, after replacing k by j in (6.4.94)—(6.4.97),

$$\sum_{j=0}^{n} a_j \left(d_1 + d_2 j + d_3 j^{(2)} + \cdots + d_{m-1} j^{(m-1)} \right) t_1^j = 0 \quad (6.4.99)$$

Expressing $j^{(2)}$ as $j^2 - j$, $j^{(3)}$ as $j^3 - 3j^2 + 2j$, etc., and collecting terms, we have

$$\sum_{j=0}^{n} a_j \left(A_0 + A_1 j + \cdots + A_{m-1} j^{m-1} \right) t_1^j = 0 \quad (6.4.100)$$

which by comparison with (6.4.76) completes the proof of the repeated-roots theorem. $\square$

ILLUSTRATIVE PROBLEM 6.4-7   Solve

$$f(k+3) - 7f(k+2) + 16f(k+1) - f(k) = 0$$

Solution   The characteristic equation, after reduction, is

$$t^3 - 7k^2 + 16t - 1 = 0$$

The roots are $2, 2, 3$. The answer is, therefore,

$$f(k) = (A + Bk) 2^k + C 3^k$$

where A, B and C are periodic arbitrary constants.

ILLUSTRATIVE PROBLEM 6.4-8   Solve

$$f(k+6) - 4f(k+5) + f(k+4) + 10f(k+3) - 4f(k+1) - 8f(k) = 0$$

Solution   The characteristic equation, after reduction, is

$$(t-2)^3 (t+1)^2 = 0$$

with roots $2, 2, 2, -1, -1$. The general solution is therefore

$$f(k) = (A + Bk + Ck^2) 2^k + (D + Ek)(-1)^k$$

where A, B, C, D and E are periodic arbitrary constants.

6.4.5   Case III: Convertible Problems

We shall illustrate by examples.

ILLUSTRATIVE PROBLEM 6.4-9   Solve

$$[f(k+3)]^2 - 7[f(k+2)]^2 + 16[f(k+1)]^2 - [f(k)]^2 = 0$$

Solution   This is exactly the same as Illustrative Problem 6.4-7 once we make the substitution

$$g(k) = [f(k)]^2$$

ILLUSTRATIVE PROBLEM 6.4-10   Solve

$$\sqrt{f(k+2)} - 2\sqrt{f(k+1)} + \sqrt{f(k)} = 0$$

Solution   This is exactly the same as Illustrative Problem 6.4-6 after the substitution

$$g(k) = \sqrt{f(k)}$$

ILLUSTRATIVE PROBLEM 6.4-11   Solve

$$[f(n+1)]^2 - 2f(n) = 0$$

Solution

$$[f(n+1)]^2 = 2f(n), \; 2\ln f(n+1) = \ln f(n) + \ln 2.$$

Let

$$g(n) = \ln f(n)$$

Then

$$2g(n+1) - g(n) = \ln 2$$

Unfortunately, this is not homogeneous, but it is easily solved. The method of solution will be discussed in Section 6.5.

ILLUSTRATIVE PROBLEM 6.4-12   Solve

$$2f(n) - 2f(n-1) + n(n-1)f(n-2) = 0$$

Solution   As usual, we prefer to make the lowest argument equal to n, so we replace n by $n+2$. Then

$$2f(n+2) - 2(n+2)f(n+1) + (n+2)(n+1)f(n) = 0$$

This is a little tricky, but it is worth going through because we shall need the technique later. We divide by $(n+2)!$:

$$2\frac{f(n+2)}{(n+2)!} - 2\frac{f(n+1)}{(n+1)!} + \frac{f(n)}{n!} = 0$$

Now let $g(n) = f(n)/n!$. Then

$$2g(n+2) - 2g(n+1) + g(n) = 0$$

from which

$$g(n) = \left(\frac{\sqrt{2}}{2}\right)^n \left(A\cos\frac{n\pi}{4} + B\sin\frac{n\pi}{4}\right)$$

and

$$f(n) = n!\left(\frac{\sqrt{2}}{2}\right)^n \left(A\cos\frac{n\pi}{4} + B\sin\frac{n\pi}{4}\right)$$

PROBLEM SECTION 6.4

A. Solve for $f(k)$. The symbols a, b and c refer to functions independent of k. Your answers may differ in form and still be correct.

  (1) $f(k+1) - ca^{2k}f(k) = 0$

  (2) $f(k+1) - c^k a^{k+1}f(k) = 0$

  (3) $f(k+1) - k^2 f(k) = 0$

  (4) $f(k+1) - \dfrac{k}{k+1}f(k) = 0$

  (5) $f(k+1) - \pi^{2k+1}f(k) = 0$

  (6) $x^{2k+1}f(k+1) - f(k) = 0$

B.

  (7) $f(k+2) - f(k+1) - 2f(k) = 0$

  (8) $f(k+2) - 5f(k+1) + 4f(k) = 0$

  (9) $f(k+2) + f(x) = 0$

  (10) $f(k+3) + f(k) = 0$

C.

  (11) $f(k)f(k+2) = [f(k+1)]^2$

  (12) $[f(k+3)]^2 - 3[f(k+2)]^2 + 3[f(k+1)]^2 - [f(k)]^2 = 0$

  (13) $f(k+4)[f(k+2)]^6 f(k) = [f(k+1)]^4[f(k+3)]^4$

  (14) $f(k)[f(k+2)]^3 = [f(k+1)]^3 f(k+3)$

6.5 NON-HOMOGENEOUS LINEAR DIFFERENCE EQUATIONS

6.5.1 The Method of Superposition

Consider the homogeneous linear difference equation

$$f(k+1)-f(k)=0 \qquad\qquad (6.5.1)$$

The solution is

$$f(k)=A \qquad\qquad (6.5.2)$$

where A is a periodic arbitrary constant. We remind you that A could be something like 75.4; or arctan (x^2+1), where x is independent of k; or $\ln\cos(2\pi k+e^x)$, where x is independent of k; or any other value such that $\Delta A = 0$.

Now compare

$$f(k+1)-f(k)=k^2 \qquad\qquad (6.5.3)$$

Trial and error shows that

$$f(k)=\tfrac{1}{3}k^3-\tfrac{1}{2}k^2+\tfrac{1}{6}k \qquad\qquad (6.5.4)$$

is a satisfactory *particular solution* of (6.5.3).

However, (6.5.4) is not satisfactory as a general solution. A general solution of an nth-order difference equation has to (1) satisfy the equation and (2) have n periodic arbitrary constants. Equation (6.5.1) is called the *auxiliary equation* of (6.5.3). It is formed from that equation by assuming that the right member is 0. The solution of the auxiliary equation is called the *complementary solution*. Thus, A is the complementary solution of (6.5.3). The sum of (6.5.2) and (6.5.4) is

$$f(k)=\tfrac{1}{3}k^3-\tfrac{1}{2}k^2+\tfrac{1}{6}k+A \qquad\qquad (6.5.5)$$

which satisfies both the requirements mentioned above and is, therefore, the general solution of (6.5.3).

The idea (borrowed from the theory of differential equations) of adding a complementary and a particular solution to obtain a general solution is called the *method of superposition*. The more difficult task in our kind of problem is usually finding the particular solution.

As in the theory of differential equations, there are a great many tricks in the solution of difference equations. We shall consider only a few.

6.5.2 Guessing a Particular Solution

If the right member of a linear difference equation is not too complicated, it is sometimes not too difficult to guess the form of a particular solution. The following are well-known candidates. In the table, *soc* means *sine or cosine*.

Right Member	Guess
nth-degree polynomial	$c_0 + c_1 k + \cdots + c_n k^n$
$k^n a^k$	$(c_0 + c_1 k + \cdots + c_n k^n) a^k$
soc bk	$c_1 \sin bk + c_2 \cos bk$
a^k soc bk	$(c_1 \sin bk + c_2 \cos bk) a^k$

ILLUSTRATIVE PROBLEM 6.5-1 Solve

$$f(k+1) - 2f(k) = 4\sin 3k \tag{A}$$

Solution The complementary solution is easy:

$$[f(k)]_c = A\,2^k,$$

where A is a pac. To find a particular solution, try

$$[f(k)]_p = c_1 \sin 3k + c_2 \cos 3k$$

Substituting in (A),

$$c_1 \sin(3k+3) + c_2 \cos(3k+3) - 2c_1 \sin 3k - 2c_2 \cos 3k = 4\sin 3k$$

Expanding $\sin(3k+3)$ and $\cos(3k+3)$ in the usual fashion,

$$\sin 3k\,[c_1 \cos 3 - c_2 \sin 3 - 2c_1] + \cos 3k\,[c_1 \sin 3 + c_2 \cos 3 - 2c_2] = 4\sin 3k$$

Comparing coefficients, we have

$$\begin{cases} (\cos 3 - 2)c_1 + (-\sin 3)c_2 = 4 \\ (\sin 3)c_1 + (\cos 3 - 2)c_2 = 0 \end{cases}$$

from which

$$\begin{cases} c_1 = \dfrac{4(2 - \cos 3)}{4\cos 3 - 5} \\ c_2 = \dfrac{4\sin 3}{4\cos 3 - 5} \end{cases}$$

so that the general solution of (A) is

$$f(k) = A\,2^k + \left[\frac{4(2-\cos 3)}{4\cos 3 - 5}\right]\sin 3k + \left[\frac{4\sin 3}{4\cos 3 - 5}\right]\cos 3k \tag{B}$$

ILLUSTRATIVE PROBLEM 6.5-2 In Illustrative Problem 6.5-1, we are given the boundary condition

$$f(0) = 0 \tag{C}$$

Find the particular solution which satisfies (C).

Solution   Substituting in (B), we find

$$A = \frac{-4\sin 3}{4\cos 3 - 5}$$

so that

$$f(k) = \frac{4}{4\cos 3 - 5}\left[(2 - \cos 3)\sin 3k + \sin 3\cos 3k - 2^k \sin 3\right]$$

6.5.3   First-Order Equations

We have already proved that if

$$f(k+1) - c(k)f(k) = 0 \qquad\qquad (6.5.6)$$

then

$$f(k) = A_1 f(r) \prod_{t=r}^{k-1} c(t) \qquad\qquad (6.5.7)$$

where A_1 is a pac. Now we shall turn our attention to

$$f(k+1) - c(k)f(k) = g(k) \qquad\qquad (6.5.8)$$

assuming it is not easily solved by guessing, as in Section 6.5.2. If we return to (6.5.6) and divide through by

$$P(k) = \prod_{t=r}^{k} c(t)$$

we find

$$\frac{f(k+1)}{P(k)} - \frac{c(k)f(k)}{P(k)} = 0 \qquad\qquad (6.5.9)$$

A little reflection shows that

$$P(k) = P(k-1)\cdot c(k) \qquad\qquad (6.5.10)$$

so that (6.5.9) becomes

$$\frac{f(k+1)}{P(k)} - \frac{f(k)}{P(k-1)} = 0 \qquad\qquad (6.5.11)$$

If we now let

$$z(k) = f(k)/P(k-1) \tag{6.5.12}$$

the left member of (6.5.11) becomes

$$\Delta z(k) = 0 \tag{6.5.13}$$

i.e., it is an exact difference. We now return to (6.5.8) to see what happens:

$$\frac{f(k+1)}{P(k)} - \frac{f(k)}{P(k-1)} = \frac{g(k)}{P(k)} \tag{6.5.14}$$

If we are lucky, this may be easy to solve.

ILLUSTRATIVE PROBLEM 6.5-3   Solve

$$f(k+1) - kf(k) = 1 \tag{D}$$

Solution   Choosing $r = 1$,

$$P(k) = \prod_{t=1}^{k} t = k!$$

Dividing (D) by $k!$, we have

$$\frac{f(k+1)}{k!} - \frac{f(k)}{(k-1)!} = \frac{1}{k!}$$

We substitute $z(k) = f(k)/(k-1)!$. Then

$$z(k+1) - z(k) = \frac{1}{k!}$$

$$\Delta z(k) = \frac{1}{k!}$$

$$\sum_{t=1}^{k-1} \Delta z(t) = \sum_{t=1}^{k-1} \frac{1}{t!}$$

$$z(k) - z(1) = \sum_{t=1}^{k-1} \frac{1}{t!}$$

$$f(k) = \left[\sum_{t=1}^{k-1} \frac{1}{t!} + f(1) \right] (k-1)!$$

This does not look very comfortable, but is easy to compute. Sometimes we're luckier than other times.

PROBLEM SECTION 6.5

A. Because there are countless particular solutions, this part of your solution
   may be correct yet not agree with ours. Solve:

(1) $f(x+2)-6f(x+1)+4f(x)=0$

(2) $f(k+2)-f(k+1)-6f(k)=k$

(3) $f(k+2)-5f(k+1)+6f(k)=5^k$

(4) $f(k+2)-3f(k+1)-4f(k)=3^k$

(5) $f(k+3)-5f(k+2)+8f(k+1)-4f(k)=k2^k$

(6) $f(k+2)+n^2f(k)=\cos mk, \ m\neq\pi/2$

B.

(7) $f(k+1)-\dfrac{f(k)}{k+1}=\dfrac{k-1}{k(k+1)}$

(8) $f(k+1)-e^k e^{k!}f(k)=ke^k e^{k!}+k+1$

(9) $k^2f(k+1)-(k+1)^2f(k)=k(k+1)$

(10) $f(k+1)-e^{2k}f(k)=3k^2e^{k+k^2}$

C.

(11) $k(k+1)f(k+2)-2k(k+2)f(k+1)+(k+1)(k+2)f(k)=k(k+1)(k+2)$
   Hint: Let $g(k)=f(k)/k$.

(12) $(2k-1)f(k+2)-(8k-2)f(k+1)+(6k+3)f(k)=0$
   Hint: Try $f(k)=k$ as a particular solution.

(13) $\dfrac{f(k+1)}{f(k)}=e[f(k+1)]^{-1/k}$
   Hint: Let $g(k)=\ln f(k)$.

(14) $f(k+2)-(2k+1)f(k+1)+k^2f(k)=0$
   Hint: Let $g(k)=f(k+1)-kf(k)$.

Chapter 7

The Taylor Series

If Chapter 6 was not done completely, omit the material from Section 7.2.3 on.

In the course of his work, a mathematician, statistician, scientist, social scientist or business analyst encounters formulas like the following:

$$y = Ae^{-bt}\cos(\omega t - \varphi) \qquad \text{(damped oscillations—electronics)}$$

$$h_1 = \frac{\ln\dfrac{1-a}{b}}{\ln\dfrac{p_2(1-p_1)}{p_1(1-p_2)}} \qquad \text{(sequential sampling—statistics)}$$

$$v^2 = \frac{gl}{2\pi}\tanh\frac{2\pi h}{l} \qquad \text{(wave velocity—hydrodynamics)}$$

$$L = \sqrt{\frac{2Q(B+S)}{VC}} \qquad \text{(economic lot size—business)}$$

$$S = k\ln R \qquad \text{(Fechner's law—psychology)}$$

In evaluating such formulas, the computer scientist almost always uses the software subroutines for e^x, $\sin x$, $\ln x$, and other common functions. How does the computer provide them? Does it contain a set of tables? No, it does not. According to an IBM 360 manual, square root is done by the Newton-Raphson method (already explained); logarithms, sine, cosine and double-precision exponentiation are done by Chebyshev series (to be explained); arctangent, hyperbolic tangent and single-precision exponentiation are done by the rational-fraction method or an equivalent (to be explained). In each case, the result is attained quickly and efficiently.

In this chapter, we shall briefly explore the use of Taylor series to represent functions. It may occasion some surprise that we are bothering to do this, since *none* of the built-in subroutines is for Taylor series. However, much of the theory of computing is inextricably bound up with Taylor series. Furthermore, our aims are not always the same as those of computer manufacturers. The manufacturer needs to conserve space (bytes) and microseconds to present an efficient, fast subroutine—even if it takes a year to develop the appropriate algorithm. In our work, day to day, if we need to express a function as a series, we are often willing to trade off the machine time and space for our own valuable hours (or months) of labor.

7.1 THE TAYLOR SERIES (PART I)

7.1.1 Taylor-Series Approximations

Any function that is differentiable *ad infinitum* can theoretically be approximated by a polynomial. A typical function of this kind is $y = f(x) = \sin x$, for which

$$f(x) = \sin x$$
$$f'(x) = \cos x$$
$$f''(x) = -\sin x$$
$$f'''(x) = -\cos x$$
$$f^{iv}(x) = \sin x$$
$$\vdots \qquad\qquad\qquad (7.1.1)$$

We shall eventually review how the approximations are obtained, but at this point we shall jump ahead to show what the result is for $f(x) = \sin x$. The first approximation is

$$y = x \qquad\qquad (7.1.2)$$

This is shown as a straight line in Figure 7.1-1. According to the graph, this simple approximation fits only at the origin and the error increases as we leave the vicinity of the origin.

The second approximation turns out to be

$$y = x - \tfrac{1}{6}x^3 \qquad\qquad (7.1.3)$$

which is a little better. At $x = 1$, this curve separates sharply downwards. The third approximation,

$$y = x - \tfrac{1}{6}x^3 + \tfrac{1}{120}x^5 \qquad\qquad (7.1.4)$$

separates less sharply upward at about $x = 1.5$. The fourth approximation,

$$y = x - \tfrac{1}{6}x^3 + \tfrac{1}{120}x^5 - \tfrac{1}{5040}x^7 \qquad\qquad (7.1.5)$$

seems to fit rather well up to about 2.5, where it separates downwards.

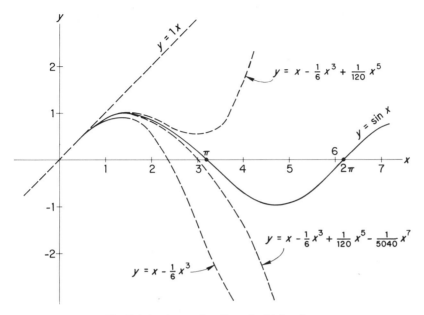

Fig. 7.1-1 Approximations to $f(x) = \sin x$.

None of the polynomial curves is exactly right—even where one seems (visually) to coincide with the sine curve—because the sine function is *not* a polynomial at all. However, by making the polynomial long enough, we can, in theory, bend it up and down to get as close as we need to. In practice, the propagation of errors is such that long polynomials are almost always very inaccurate. For the present (and in Section 7.2) we shall concentrate on the easiest way to obtain a polynomial expansion of a function which possesses a complete array of continuous derivatives at the point under consideration.

7.1.2 $\sin x$

Let us now find a Maclaurin series for $\sin x$. We remind you that computer manufacturers do not use Taylor series for their built-in subprograms, because other types of series converge faster and more accurately; but a Taylor series is perfectly satisfactory for many applications.

In the illustrative problem which follows, we shall review the formation of a Maclaurin series and write the computer program in three ways:

1. Simple recursion with forwards addition
2. Simple recursion with backwards addition
3. The method of nesting (Horner's method)

The first of these is easiest to program but usually least accurate. The other two are similar in roundoff characteristics and are sometimes used to check each other in very important problems. We shall postpone the (difficult) error analysis to the end of Section 7.2.

ILLUSTRATIVE PROBLEM 7.1-1   (a) Write a Maclaurin series for $\sin x$. (b) Find $\sin x$ from 0.1 to 1.6 radians, incrementing by 0.1 radians. (c) Find $\sin x$ from 1 to 10 radians, incrementing by 1 radian.

Solution

(a)
$$f(x) = \sin x, \qquad f(0) = 0$$
$$f'(x) = \cos x, \qquad f'(0) = 1$$
$$f''(x) = -\sin x, \qquad f''(0) = 0$$
$$f'''(x) = -\cos x, \qquad f'''(0) = -1$$
$$f^{\text{iv}}(x) = \sin x, \qquad f^{\text{iv}}(0) = 0$$
$$\vdots \qquad\qquad \vdots \tag{7.1.6}$$

$$f(x) = f(0) + \frac{f'(0)}{1!}x + \frac{f''(0)}{2!}x^2 + \cdots \tag{7.1.7}$$

$$\sin x = 0 + \frac{1}{1!}x + \frac{0}{2!}x^2 + \frac{-1}{3!}x^3 + \frac{0}{4!}x^4 + \cdots \tag{7.1.8}$$

$$\sin x = x - \frac{1}{3!}x^3 + \frac{1}{5!}x^5 - \frac{1}{7!}x^7 + \cdots \tag{7.1.9}$$

Note that (7.1.9) is an alternating series whether x is positive or negative. Therefore, for all values of x, the error of truncation is given by

$$E_n \leqslant |u_{n+1}| = \left| \frac{x^{2n+1}}{(2n+1)!} \right| \tag{7.1.10}$$

Before proceeding to the calculation, let us discuss the three methods for programming the sum.

Recursion (backwards and forwards). If we number the terms, in (7.1.9) beginning with $k = 1$, the general term is

$$u_k = (-1)^{k-1} \frac{x^{2k-1}}{(2k-1)!} \tag{7.1.11}$$

and

$$u_{k-1} = (-1)^{k-2} \frac{x^{2k-3}}{(2k-3)!} \tag{7.1.12}$$

Therefore

$$\frac{u_k}{u_{k-1}} = \frac{(-1)^{k-1}x^{2k-1}}{(2k-1)!} \bigg/ \frac{(-1)^{k-2}x^{2k-3}}{(2k-3)!} \tag{7.1.13}$$

$$\frac{u_k}{u_{k-1}} = \frac{(-1)^{k-1}x^{2k-1}(2k-3)!}{(-1)^{k-2}x^{2k-3}(2k-1)!} \tag{7.1.14}$$

Using

$$(2k-1)! = (2k-1)(2k-2)\cdot(2k-3)! \qquad (7.1.15)$$

$$\frac{x^{2k-1}}{x^{2k-3}} = x^{(2k-1)-(2k-3)} = x^2 \qquad (7.1.16)$$

we obtain the recursion relation

$$u_k = -\frac{x^2}{(2k-1)(2k-2)} u_{k-1} \qquad (7.1.17)$$

In programming (7.1.17), it is possible to add the terms forwards or backwards. If they are added forwards, only one storage address is needed for u_k, since addition can take place immediately. If they are added backwards, the successive values of u_k must be stored as a vector and the addition done as a separate step. We shall illustrate.

Nesting. Consider the first three terms of (7.1.9) as an example. Then

$$\sin x \sim x - \frac{x^3}{6} + \frac{x^5}{120} \qquad (7.1.18)$$

which can be rewritten as

$$\sin x \sim x\left[1 - x^2\left(\frac{1}{6} - \frac{x^2}{120}\right)\right]$$

The program for calculating $\sin x$ by three methods—forwards addition, backwards addition and nesting—is shown in Figure 7.1-2.

(b) The results of the three methods from $x = 0.1$ to 1.6 radians (a little more than $\pi/2$) are shown in Figure 7.1-3. These were done in single precision on an IBM 360, taking enough terms in each case so that the theoretical error of truncation was less than 10^{-8}. The errors are, therefore, entirely roundoff. For comparison, the error of the built-in IBM subprogram (single-precision) is also shown.

Comparing the three methods of procedure, we see that in general either backwards addition or nesting leads to less roundoff error than forwards addition, as promised in Chapter 1. Because roundoff errors are unpredictable, there are individual cases, e.g. $\sin 1.2$, where they vanished fortuitously in the forward method but not in the other two; however, this cannot be depended upon. Up to 0.8 radians (about $\pi/4$), using the values in the table, the average absolute errors in the different methods were:

forward:	12.0×10^{-8}
backward:	3.9×10^{-8}
nesting:	4.0×10^{-8}

```
      DOUBLE PRECISION TV(16),EF(16),EB(16),EN(16),A
      DIMENSION U(25),F(16),B(16),ZN(16),NK(16)
      TV(1)=0.09983341664682815
      TV(2)=0.1986693307950612
      TV(3)=0.2955202066613396
      TV(4)=0.3894183423086505
      TV(5)=0.4794255386042030
      TV(6)=0.5646424733950354
      TV(7)=0.6442176872376911
      TV(8)=0.7173560908995228
      TV(9)=0.7833269096274834
      TV(10)=0.8414709848078965
      TV(11)=0.8912073600614353
      TV(12)=0.9320390859672263
      TV(13)=0.9635581854171930
      TV(14)=0.9854497299884602
      TV(15)=0.9974949866040544
      TV(16)=0.9995736030415052
C CRITERION FOR CALCULATING TERMS IN RECURSION RELATION
      C=1.E-8
C LOOP TO FIND SIN X FROM 0.1 TO 1.6 RADIANS.
      DO 1 I=1,16
C INITALIZING
      DO 2 K=1,25
2     U(K)=0.0
      X=I
      X=X/10.0
      U(1)=X
      K=2
4     J=2*(2*K-1)*(K-1)
      D=J
      U(K)=-X**2*U(K-1)/D
      IF(ABS(U(K))-C)6,6,5
5     K=K+1
      GO TO 4
6     F(I)=0.0
      B(I)=0.0
      NK(I)=K-1
      KT=K-1
      DO 7 K=1,KT
7     F(I)=F(I)+U(K)
      L=KT
8     B(I)=B(I)+U(L)
      L=L-1
      IF(L)1,1,8
1     CONTINUE
C CALCULATION FOR NESTING
      DO 3 I=1,16
      X=I
      X=X/10.0
      Y=X**2
      K=NK(I)
      P=1.0
10    L=2*K
      D=(L-1)*(L-2)
      P=Y*P/D
      P=1.0-P
      IF(K-2)12,12,11
11    K=K-1
      GO TO 10
12    ZN(I)=P*X
3     CONTINUE
C CALCULATION OF ERRORS
      DO 13 I=1,16
      A=F(I)
      EF(I)=TV(I)-A
      A=B(I)
      EB(I)=TV(I)-A
      A=ZN(I)
```

```
13      EN(I)=TV(I)-A
        WRITE(3,100)
100     FORMAT(1H1,T2,'RADIANS',T15,'TV-FORWARD',T31,
        'TV-BACKWARD',T47,1'TV-NESTING',T63,'TV-IBM'//)
C RESULTS
        DO 14 I=1,16
        X=I
        X=X/10.0
        A=SIN(X)
        E=TV(I)-A
14      WRITE(3,102)X,EF(I),EB(I),EN(I),E
102     FORMAT(1H ,T4,F3.1,    1PD16.2,3D16.2)
        END
```

Fig. 7.1-2 $\sin x$ **computed forwards, backwards and by nesting.**

RADIANS	TV-FORWARD	TV-BACKWARD	TV-NESTING	TV-IBM
0.1	5.96D-08	0.0	0.0	5.96E-08
0.2	5.96D-08	0.0	0.0	5.96E-08
0.3	1.19D-07	5.96D-08	5.96D-08	5.96E-08
0.4	5.96D-08	0.0	0.0	0.0
0.5	0.0	0.0	-5.96D-08	0.0
0.6	1.19D-07	0.0	0.0	0.0
0.7	1.79D-07	0.0	0.0	-5.96E-08
0.8	1.19D-07	0.0	5.96D-08	0.0
0.9	5.96D-08	-1.19D-07	-5.96D-08	-5.96E-08
1.0	0.0	-1.19D-07	-5.96D-08	-5.96E-08
1.1	1.19D-07	0.0	0.0	1.79E-07
1.2	0.0	-1.79D-07	-1.79D-07	0.0
1.3	0.0	-1.19D-07	0.0	5.96E-08
1.4	-1.19D-07	-2.38D-07	-2.98D-07	-5.96E-08
1.5	1.79D-07	-5.96D-08	-2.38D-07	-5.96E-08
1.6	5.96D-07	-1.79D-07	-2.98D-07	-5.96E-08

Fig. 7.1-3 **Output of the** $\sin x$ **program. TV = true value.**

From 0.9 to 1.6 radians, the average absolute errors, using the values in the table, were

forward:	15.3×10^{-8}
backward:	9.6×10^{-8}
nesting:	12.4×10^{-8}

The IBM single-precision subprogram (which is not a Taylor series) makes use of the periodicity of the sine function to reduce the argument to a value less than $\pi/4$. The average absolute error up to 0.8 radians (a little more than $\pi/4$) was 5.6×10^{-8}—more, in fact, than either backwards addition or nesting for the specific values tested.

This example shows that the important error in this kind of problem is roundoff, not truncation. In fact, taking more terms would have increased the error. Furthermore, it should be clear than none of the methods can guarantee accuracy to the number of digits printed on the output device.

(c) In working with a periodic function, the intelligent thing to do is to reduce the argument to as small a value as possible, as the IBM subprogram does. This reduces the number of terms needed to minimize the truncation error, and therefore, since the number of operations is reduced, it minimizes the roundoff error as well. If this is not done, the results are atrocious. For 10 radians, here are the errors for the three methods:

forwards addition: 19697×10^{-8}
backwards addition: 41149×10^{-8}
nesting: 19787×10^{-8}

This should emphasize the basic problem with the use of a Taylor series. If the argument is large, a great many terms must be used to reduce the truncation error to a satisfactory level, but in the process the roundoff error propagates and the results are thoroughly unreliable.

PROBLEM SECTION 7.1

A. Write five terms of a Maclaurin series. In each case, find the general term u_n and write a recursion relation for u_n in terms of u_{n-1}, counting the first term as term 1: (1) $\cos x$, (2) e^{-x} with $x > 0$, (3) $\ln(1+x)$ with $x > 0$, (4) $(\sin x)/x$ [*Hint:* $\lim_{x \to 0}(\sin x)/x = 1$].

B. Find four terms of a Maclaurin series for each of the following: (5) $e^{-x}\cos x$, (6) $\sec x$, (7) $x\cos x$, (8) $\arctan x$, (9) $e^{\arcsin x}$, (10) $\ln(1+x^2)$.

C. (a) Use four terms of the Maclaurin series developed in part A to calculate the following. (b) Estimate the theoretical error of truncation. (c) Find the absolute and relative errors. The values in parentheses are the true values, accurate to the number of places displayed. (No answers can be given for this section because the errors will depend upon the machine and the procedure.)
   (11) $\cos 0.25$ (0.9689124217)
   (12) $\cos 3.00$ (-0.9899924966)
   (13) $e^{-0.513}$ (0.5986967916)
   (14) $e^{-4.80}$ (0.0082297470 49)
   (15) $\ln 1.03$ (0.0295588022 4)
   (16) $\ln 19.93$ (0.1764711431)
   (17) $(\sin 0.58)/0.58$ (0.9448688566)
   (18) $(\sin 1.598)/1.598$ (0.6255506902)

7.2   THE TAYLOR SERIES (PART II)

7.2.1   Expansion of e^x

We can find an expansion of e^x about $x = b$ using the method already explained and illustrated:

$$f(x) = e^x, \qquad\qquad f(b) = e^b$$
$$f'(x) = e^x, \qquad\qquad f'(b) = e^b$$
$$f''(x) = e^x, \qquad\qquad f''(b) = e^b$$
$$\vdots$$

Then Taylor's theorem states that if the series is convergent,

$$f(x) = f(b) + \frac{f'(b)}{1!}(x-b) + \frac{f''(b)}{2!}(x-b)^2 + \cdots \qquad (7.2.1)$$

so that

$$e^x = e^b + \frac{e^b}{1!}(x-b) + \frac{e^b}{2!}(x-b)^2 + \cdots \qquad (7.2.2)$$

or

$$e^x = e^b \left[1 + \frac{(x-b)}{1!} + \frac{(x-b)^2}{2!} + \cdots \right] \qquad (7.2.3)$$

If $b = 0$, we obtain the Maclaurin series

$$e^x = 1 + x + \frac{x^2}{2!} + \frac{x^3}{3!} + \cdots \qquad (7.2.4)$$

7.2.2 The Error of Truncation (Lagrange Remainder)

We have already discussed the error of truncation, E_n, for a convergent alternating series. If the series does not alternate, there are several methods which can be used. We present the following theorem without proof.

LAGRANGE REMAINDER THEOREM   If $f(x)$ is expanded into a Taylor series about $x = b$, then

$$f(x) = f(b) + \frac{f'(b)}{1!}(x-b) + \frac{f''(b)}{2!} + \cdots + \frac{f^{(n-1)}(b)}{(n-1)!} + R_n \qquad (7.2.5)$$

and the *remainder*, R_n, after n terms, is given by

$$R_n = \frac{f^{(n)}(c)}{n!}(x-b)^n \qquad (7.2.6)$$

where $c \in (x,b)$.

In computing practice, we find an upper bound for $|f^{(n)}(c)|$ on the interval. We shall call it $f^{(n)}_{\max}$. Then

$$E_n \leqslant \left| \frac{f^{(n)}_{\max}}{n!}(x-b)^n \right| \qquad (7.2.7)$$

is an estimate of the error of truncation after n terms.

ILLUSTRATIVE PROBLEM 7.2-1   Find e^x for $x = 0.1, 0.2, \ldots, 0.9, 1, 2, \ldots, 10$.

Solution   The Taylor series is

$$e^x = e^b + \frac{e^b}{1!}(x-b) + \frac{e^b}{2!}(x-b)^2 + \cdots \tag{7.2.8}$$

with

$$R_n = \frac{f^{(n)}(c)}{n!}(x-b)^n, \qquad c \in (x,b) \tag{7.2.9}$$

For $b = 0$, (7.2.8) becomes the Maclaurin series

$$e^x = 1 + x + \frac{x^2}{2!} + \frac{x^3}{3!} + \cdots \tag{7.2.10}$$

and

$$R_n = \frac{e^c}{n!}x^n, \qquad c \in (0,x) \tag{7.2.11}$$

In order to continue, we must know something about e^x in the interval $(0,x)$. We can easily prove that $e < 3$. We use (7.2.10) with $x = -1$. Then

$$e^{-1} = 1 - 1 + \frac{1}{2!} - \frac{1}{3!} + \frac{1}{4!} - \cdots \tag{7.2.12}$$

Since this is an alternating series,

$$e^{-1} > \frac{1}{2} - \frac{1}{2} + \frac{1}{2!} - \frac{1}{3!} + \frac{1}{4!} \tag{7.2.13}$$

$$e^{-1} > \tfrac{1}{3} \tag{7.2.14}$$

$$e < 3 \tag{7.2.15}$$

We shall use 3 as an upper bound (obviously not the *least* upper bound). Then

$$E_n \leqslant \left| \frac{3^x}{n!}x^n \right| = \frac{3^x}{n!}x^n \tag{7.2.16}$$

the absolute-value signs being dropped because $x > 0$ in this problem. (If $x < 0$, the series is alternating, and it is easier to use the theorem for alternating series.)

As usual, we have a choice of methods for computing the sum: the recursion method with forwards addition, the recursion method with backwards addition, and nesting.

For recursion, assuming we start with $k = 1$, the general term for (7.2.10) is

$$u_k = x^{k-1}/(k-1)! \tag{7.2.17}$$

Then

$$u_{k-1} = x^{k-2}/(k-2)! \tag{7.2.18}$$

from which

$$u_k = \frac{x^{k-1}}{x^{k-2}} \frac{(k-2)!}{(k-1)!} u_{k-1} \qquad (7.2.19)$$

$$u_k = \frac{x}{k-1} u_{k-1} \qquad (7.2.20)$$

We can also calculate the maximum theoretical error of truncation by recursion. From (7.2.16),

$$E_n \leqslant \frac{3^x}{n!} x^n \qquad (7.2.21)$$

$$E_{n-1} \leqslant \frac{3^x}{(n-1)!} x^{n-1} \qquad (7.2.22)$$

from which

$$E_n \leqslant \frac{x}{n} E_{n-1} \qquad (7.2.23)$$

For *nesting*, the series (7.2.10) is rewritten as

$$e^x = 1 + x \left(1 + \frac{x}{2} \left(1 + \frac{x}{3} \left(1 + \frac{x}{4} (1 + \cdots) \cdots \right) \right) \right) \qquad (7.2.24)$$

Figure 7.2-1 shows the computer program to perform the necessary calculations. The computer was programmed, in single precision, to calculate successive values of u_k until the theoretical error of truncation was less than 10^{-8}. The errors, shown

```
      DOUBLE PRECISION TV(19),Q(19),R(19),S(19),T(19),W
      DIMENSION U(70),NK(19),E(19),B(19),ZN(19),F(19)
C INITIAL VALUES.
      TV(1)=1.105170918075648
      TV(2)=1.221402758160170
      TV(3)=1.349858807576003
      TV(4)=1.491824697641270
      TV(5)=1.648721270700128
      TV(6)=1.822118800390509
      TV(7)=2.013752707470477
      TV(8)=2.225540928492468
      TV(9)=2.459603111156950
      TV(10)=2.718281828459045
      TV(11)=7.389056098930650
      TV(12)=20.085536923187668
      TV(13)=54.598150033144239
      TV(14)=148.413159102577
      TV(15)=403.428793492735
      TV(16)=1096.633158428459
      TV(17)=2980.957987041728
      TV(18)=8103.083927575384
      TV(19)=22026.465794806717
C
      C=1.E-8
C
      DO 1 K=1,19
      F(K)=0.0
      B(K)=0.0
1     ZN(K)=0.0
```

Fig. 7.2-1 (Continued on next page)

```
C FOR I = 0.1 TO 0.9, THEN I = 1 TO 10.
      DO 12 I=1,19
      X=I
      IF(I-9)2,2,3
2     X=X/10.0
      GO TO 31
3     X=FLOAT(I-9)
C CALCULATION OF U(K) AND E(K), THEORETICAL TRUNCATION MAXIMUM ERROR.
31    U(1)=1.0
      U(2)=X
      K=3
4     ZK=FLOAT(K-1)
      U(K)=X*U(K-1)/ZK
      A=U(K)*3.0**X
      IF(ABS(A)-C)6,6,5
5     K=K+1
      GO TO 4
6     NK(I)=K-1
      E(I)=A
C CALCULATION OF F(I).
      KT=NK(I)
      DO 7 K=1,KT
7     F(I)=F(I)+U(K)
C CALCULATION OF B(I)
      DO 8 K=1,KT
      L=KT-K+1
8     B(I)=B(I)+U(L)
C CALCULATION OF ZN(I).
      L=NK(I)-1
      P=1.0
9     Y=L
      P=X*P/Y
      P=1.0+P
      IF(L-2)11,11,10
10    L=L-1
      GO TO 9
11    P=X*P
      ZN(I)=1.0+P
C CALCULATION OF ERRORS.
      W=EXP(X)
      Q(I)=TV(I)-W
      W=F(I)
      R(I)=TV(I)-W
      W=B(I)
      S(I)=TV(I)-W
      W=ZN(I)
12    T(I)=TV(I)-W
C PRINT-OUT OF RESULTS.
      WRITE(3,100)
100   FORMAT(1H1,T49,'EXP(X) BY FOUR METHODS')
      WRITE(3,104)
104   FORMAT(1H0,T6,'ARG',T16,'TRUE VALUE',T32,'TV-IBM',T40,'TV-FORWARD'
     1,T51,'TV-BACKWARD',T64,'TV-NESTING',T75,'TRUNC.ERROR')
      DO 18 I=1,19
      X=I
      IF(I-9)16,16,17
16    X=X/10.0
      GO TO 18
17    X=FLOAT(I-9)
18    WRITE(3,103)X,TV(I),  Q(I),R(I),S(I),T(I),E(I)
103   FORMAT(1H ,T5,F4.1,3X,1PE14.7,5E12.2)
      STOP
      END
```

Fig. 7.2-1 Computer program to calculate e^x by two recursion methods and nesting. The errors are shown in Figure 7.2-2.

EXP(X) BY FOUR METHODS

ARG	TRUE VALUE	TV-IBM	TV-FORWARD	TV-BACKWARD	TV-NESTING	TRUNC. ERROR
0.1	1.1051702D 00	-9.54D-07	1.91D-06	0.0	0.0	1.55E-09
0.2	1.2214022D 00	-9.54D-07	1.91D-06	0.0	0.0	3.16E-09
0.3	1.3498583D 00	0.0	2.86D-06	0.0	0.0	2.26E-09
0.4	1.4918242D 00	-9.54D-07	2.86D-06	0.0	0.0	1.12E-09
0.5	1.6487207D 00	-9.54D-07	1.91D-06	0.0	0.0	9.32E-09
0.6	1.8221188D 00	0.0	4.77D-06	9.54D-07	9.54D-07	3.22E-09
0.7	2.0137520D 00	-9.54D-07	1.91D-06	0.0	0.0	1.07E-09
0.8	2.2255402D 00	-9.54D-07	3.81D-06	0.0	9.54D-07	5.18E-09
0.9	2.4596024D 00	-9.54D-07	3.81D-06	0.0	9.54D-07	1.58E-09
1.0	2.7182817D 00	0.0	2.86D-06	0.0	0.0	6.26E-09
2.0	7.3890553D 00	-9.54D-07	5.72D-06	0.0	0.0	3.32E-09
3.0	2.0085526D 01	-1.53D-05	9.16D-05	0.0	0.0	5.53E-09
4.0	5.4598145D 01	0.0	1.37D-04	3.05D-05	0.0	5.88E-09
5.0	1.4841315D 02	-1.53D-05	1.68D-04	7.63D-05	1.22D-04	5.12E-09
6.0	4.0342871D 02	0.0	2.69D-03	2.44D-04	2.44D-04	4.01E-09
7.0	1.0966331D 03	0.0	3.91D-03	1.95D-03	1.46D-03	2.95E-09
8.0	2.9809578D 03	-2.44D-04	8.06D-03	6.10D-03	2.69D-03	2.09E-09
9.0	8.1030820D 03	0.0	5.47D-02	1.95D-02	1.56D-02	7.18E-09
10.0	2.2026465D 04	0.0	7.81D-02	5.47D-02	5.47D-02	4.76E-09

Fig. 7.2-2   Output for the e^x program.

in Figure 7.2-2, are therefore all roundoff errors. We mention that wherever possible, the error of conversion was reduced by using fixed-point calculations (with integers), then converting to floating-point.

Careful study of the output in Figure 7.2-2 shows that for small values of x (up to $x=3$), the backwards addition and the nesting method were equally good. Above $x=3$, none of the methods is really very good, including the single-precision IBM subprogram (e.g. at $x=3$, 5 and 8). We remind you that the built-in subprogram is not a Taylor series.

ILLUSTRATIVE PROBLEM 7.2-2   Using a Maclaurin series, calculate $\arctan x$ for $x=0.1, 0.2, \ldots, 0.9$.

Solution   We leave details for the problem section. This series converges slowly for $x^2 < 1$. As usual, we programmed the computer to calculate successive terms until the theoretical truncation error was less than 10^{-8}. The results are shown in Figure 7.2-3.

ARCTAN X BY FOUR METHODS

ARG	TRUE VALUE	TV-IBM	TV-FORWARD	TV-BACKWARD	TV-NESTING
0.1	0.09966862	0.0	1.19D-07	5.96D-08	0.0
0.2	0.19739550	0.0	1.19D-07	0.0	0.0
0.3	0.29145676	5.96E-08	1.79D-07	0.0	0.0
0.4	0.38050634	-5.96E-08	1.79D-07	0.0	0.0
0.5	0.46364760	0.0	2.98D-07	0.0	0.0
0.6	0.54041946	0.0	2.38D-07	-5.96D-08	-5.96D-08
0.7	0.61072594	0.0	4.77D-07	-5.96D-08	0.0
0.8	0.67474091	0.0	7.15D-07	-5.96D-08	5.96D-08
0.9	0.73281509	0.0	1.67D-06	5.96D-08	0.0

Fig. 7.2-3   Output for the $\arctan x$ program.

7.2.3 Forward Recursion Error

In this subsection and the next, we shall develop estimates of forward recursion error and nesting recursion error. These formulas are not very useful, but are offered as an indication of the kind of calculation used to develop such formulas.

In the calculation of a truncated series, there are four kinds of error:

> *Input errors.*
> > Initial errors of the data
> > Errors of conversion
> *Procedural errors.*
> > Roundoff errors
> > Errors of truncation

We have already explained in Chapter 1 that these errors fall into two categories: the input errors, which are unavoidable; and the procedural errors, which can be reduced by careful procedure.

In forward recursion, the procedure consists of adding the new term

$$u_k = g(x,k)u_{k-1} \qquad\qquad (7.2.25)$$

to the old sum

$$S_k = S_{k-1} + u_k \qquad\qquad (7.2.26)$$

In most cases for useful series, $g(x,k)$ is of the form

$$g(x,k) = \frac{x^p}{d(k)}, \qquad p > 0 \qquad\qquad (7.2.27)$$

where $d(k)$ is a polynomial in k. For example, here are the ratios u_k/u_{k-1} for several Taylor series:

$\cos x$,	$-x^2/[(2k-2)(2k-3)]$
e^{-x},	$x/(k-1)$
$\sinh x$,	$x^2/[(2k-1)(2k-2)]$

One exception is

$\ln(1+x)$,	$-\left(\dfrac{k-1}{k}\right)x$

which is a very slowly converging series. We shall assume (7.2.26) in the following. Then u_k is calculated as shown in Figure 7.2-4.

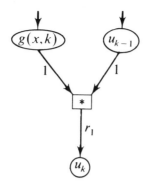

Fig. 7.2-4 Generating u_k.

In the following, we write $i_{u(k)}$ for i_{u_k} and $i_{u(k-1)}$ for $i_{u_{k-1}}$. Then

$$i_{u(k)} = i_g + i_{u(k-1)} + r_1 \tag{7.2.28}$$

We now need to know something about i_g. In (7.2.27), we assume the denominator is done (properly) in fixed point, so that there is no inherent error for the denominator. Also, we assume (properly) that enough terms are computed to render the truncation error negligible. All the error is then the input error of x and the roundoff. Then

$$\frac{\partial g}{\partial x} = \frac{px^{p-1}}{d(k)} \tag{7.2.29}$$

$$\frac{\delta g}{g} = \frac{px^{p-1}}{d(k)} \frac{d(k)}{x^p} \frac{\delta x}{x} x \tag{7.2.30}$$

$$i_g^I = pi_x^I = pi_x \tag{7.2.31}$$

To find i_g^O, we examine Figure 7.2-5 for the procedure, from which, since

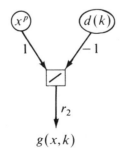

Fig. 7.2-5 Generating $g(x,k)$.

$d(k)$ is done in fixed point,

$$i_g^O = r_2 \tag{7.2.32}$$

Therefore

$$i_g = pi_x + r_2 \tag{7.2.33}$$

Substituting (7.2.33) in (7.2.28), we have

$$i_{u(k)} = pi_x + r_2 + i_{u(k-1)} + r_1 \tag{7.2.34}$$

which can be written as

$$i_{u(k)} - i_{u(k-1)} = pi_x + r_1 + r_2 \tag{7.2.35}$$
$$f(k) - f(k-1) = A \tag{7.2.36}$$

where A is constant with respect to k. Equation (7.2.36) is easily solved by standard methods already explained:

$$f(k) = (k-1)A + f(1) \tag{7.2.37}$$
$$\therefore \quad i_{u(k)} = (k-1)(pi_x + r_1 + r_2) + i_{u(1)} \tag{7.2.38}$$

We now return to (7.2.26) as shown in Figure 7.2-6.

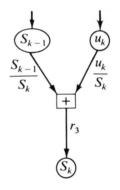

Fig. 7.2-6   Generating S_k.

In the following, we write i_k for i_{S_k} and i_{k-1} for $i_{S_{k-1}}$. Then

$$i_k = \frac{S_{k-1}}{S_k} i_{k-1} + \frac{u_k}{S_k} i_{u(k)} + r_3 \tag{7.2.39}$$

Multiplying by S_k and using (7.2.38) in the form

$$i_{u(k)} = (k-1)A + B = kA + (B - A) \tag{7.2.40}$$

we have

$$S_k i_k - S_{k-1} i_{k-1} = kAu_k + (B - A)u_k + S_k r_3 \tag{7.2.41}$$

We now sum each member from 1 to k:

$$\sum_{j=1}^{k} S_j i_j - \sum_{j=1}^{k} S_{j-1} i_{j-1} = A \sum_{j=1}^{k} ju_j + (B-A) \sum_{j=1}^{k} u_j + r_3 \sum_{j=1}^{k} S_j \quad (7.2.42)$$

Using the parts theorem for the first summation in the right member, with $u=j$, $\Delta u=1$, $\Delta v=u_j$, $v=S_{j-1}$, we have

$$\sum_{j=1}^{k} ju_j = \sum_{j=1}^{k} j\Delta S_{j-1} = jS_{j-1}|_1^{k+1} - \sum_1^{k} S_j \quad (7.2.43)$$

$$\sum_{j=1}^{k} ju_j = (k+1)S_k - \sum_{j=1}^{k} S_j \quad (7.2.44)$$

The left member of (7.2.42) is simply $S_k i_k$, and $\sum_{j=1}^{k} u_j = S_k$, so that (7.2.42) becomes

$$\epsilon_k = A\left[(k+1)S_k - \sum_{j=1}^{k} S_j\right] + (B-A)S_k + r_3 \sum_{j=1}^{k} S_j \quad (7.2.45)$$

$$\epsilon_k = \left[k(pi_x + r_1 + r_2) + i_{u(1)}\right]S_k + \left[r_3 - pi_x - r_1 - r_2\right]\sum_{j=1}^{k} S_j \quad (7.2.46)$$

Using $R_1 = \max(|i_x|, |i_{u(1)}|)$ and $R_2 = \max(|r_1|, |r_2|, |r_3|)$, we have

$$|\epsilon_k| \le \left[k(R_1 + 2R_2) + R_1\right]|S_k| + \left[pR_1 + 3R_2\right]\left|\sum_{j=1}^{k} S_j\right| \quad (7.2.47)$$

In general, the way Taylor series are used, the starting point is quite accurate and R_1 can be regarded as negligible. We then have an approximation as follows.

FORWARD-RECURSION ERROR THEOREM The maximum error ϵ of a sum S_k, derived by a forward recursion of the form

$$u_k = \frac{x^p}{d(k)} u_{k-1} \quad (7.2.48)$$

where $p > 0$ and $d(k)$ is a polynomial in k, can be estimated by

$$|\epsilon| \le 2kR|S_k| + 3R\left|\sum_{j=1}^{k} S_j\right| \quad (7.2.49)$$

where R is the roundoff error of the computer. The assumption is that the initial value has negligible error and that just enough terms have been taken to make the error of truncation negligible.

Examples   The Taylor series for $\sin x$ satisfies (7.2.48). The first six partial sums for $\sin 2$ (IBM single precision) are

k	S_k
1	2.0000000
2	0.6666666
3	0.9333333
4	0.9079365
5	0.9093474
6	0.9092961

Here $k=6$, $R=9.54\times 10^{-7}$, $S_k=0.9092961$, and $\Sigma_{j=1}^6 S_j = 6.3265799$. Then $|\epsilon| \leqslant 2.9\times 10^{-5}$. The actual error is 1.3×10^{-6}.

It is evident from (7.2.49) that as the number of terms increases, the estimate of $|\epsilon|$ also increases, which may or may not be realistic.

7.2.4 Nesting Recursion Error

Nesting serves the same purpose as recursion with backwards addition. However, it is more difficult to program. Nevertheless, it is almost mandatory for power series.

Suppose the power series is

$$S_4 = \sum_{k=0}^4 a_{4-k} x^k \tag{7.2.50}$$

$$S_4 = a_4 + a_3 x + a_2 x^2 + a_1 x^3 + a_0 x^4 \tag{7.2.51}$$

It can be nested as follows:

$$S_4 = a_4 + x\big(a_3 + x(a_2 + x(a_1 + a_0 x))\big) \tag{7.2.52}$$

obviously, (7.2.51) requires 14 multiplications, whereas (7.2.52) requires only 4. Both require 4 additions. Even without a formal analysis, it appears to be evident that (7.2.52) has to be great deal better for roundoff than (7.2.51).

More generally, if

$$S_n = \sum_{k=0}^n a_{n-k} x^k = a_n + a_{n-1} x + \cdots + a_0 x^n \tag{7.2.53}$$

$$S_n = a_n + x\big(a_{n-1} + x(a_{n-2} + x(\cdots + x(a_1 + a_0 x)\cdots))\big) \tag{7.2.54}$$

In programming (7.2.54), we can start with

$$P_0 = a_0 \tag{7.2.55}$$

then multiply by x and add a_1:

$$P_1 = xP_0 + a_1 \qquad (7.2.56)$$

and then multiply by x and add a_2:

$$P_2 = xP_1 + a_2 \qquad (7.2.57)$$

In general, the combination

$$\begin{cases} P_0 = a_0 & (7.2.58) \\ P_k = xP_{k-1} + a_k & \text{for} \quad k > 0 \qquad (7.2.59) \end{cases}$$

is a possible nesting recursion.

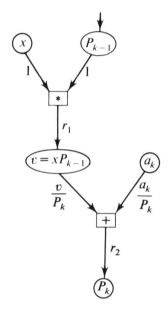

Fig. 7.2-7   Procedure for nesting.

The process chart for (7.2.59) is displayed in Figure 7.2-7. To simplify the problem, we shall assume that the inherent errors, i, of x and the coefficients a_k are equal to R_1 and that the roundoff characteristic of the computer is R_2. Then

$$i_x \sim i_{a(k)} \sim R_1 \qquad (7.2.60)$$

$$r_1 \sim r_2 \sim R_2 \qquad (7.2.61)$$

We shall write i_k for the inherent error of P_k, i_{k-1} for that of P_{k-1} and i_v for that of xP_{k-1}. From the process chart, we have

$$i_k = \frac{xP_{k-1}}{P_k} i_v + \frac{a_k}{P_k} R_1 + R_2 \qquad (7.2.62)$$

$$i_v = R_1 + i_{k-1} + R_2 \qquad (7.2.63)$$

from which

$$i_k = \frac{xP_{k-1}}{P_k}(R_1 + i_{k-1} + R_2) + \frac{a_k}{P_k}R_1 + R_2 \tag{7.2.64}$$

$$i_k = \frac{xP_{k-1}}{P_k}i_{k-1} + \frac{xP_{k-1}}{P_k}R_1 + \frac{xP_{k-1}}{P_k}R_2 + \frac{a_k}{P_k}R_1 + R_2 \tag{7.2.65}$$

We now use

$$a_k = P_k - xP_{k-1} \tag{7.2.66}$$

from (7.2.59), then multiply by P_k. After simplifying,

$$P_k i_k - xP_{k-1}i_{k-1} = (R_1 + R_2)P_k + R_2 xP_{k-1} \tag{7.2.67}$$

We shall seek a complementary solution in the usual way by considering

$$P_k i_k - xP_{k-1}i_{k-1} = 0 \tag{7.2.68}$$

Letting $r^k = P_k i_k$, we easily find that

$$(P_k i_k)_c = Cx^k \tag{7.2.69}$$

To find a general solution, assume that

$$P_k i_k = C(k)x^k \tag{7.2.70}$$

Substituting in (7.2.67), we have

$$C(k)x^k - C(k-1)x^k = (R_1 + R_2)P_k + R_2 xP_{k-1} \tag{7.2.71}$$

$$\Delta C(k-1) = (R_1 + R_2)\frac{P_k}{x^k} + R_2\frac{P_{k-1}}{x^{k-1}} \tag{7.2.72}$$

We sum from 1 to k:

$$\sum_{j=1}^{k}\Delta C(j-1) = (R_1 + R_2)\sum_{j=1}^{k}\frac{P_j}{x^j} + R_2\sum_{j=1}^{k}\frac{P_{j-1}}{x^{j-1}} \tag{7.2.73}$$

Using $P_0 = a_0$, $i_0 = R_1$ and (7.2.70), we have

$$\sum_{j=1}^{k}\Delta C(j-1) = C(k) - C(0) = C(k) - P_0 i_0 = C(k) - R_1 a_0 \tag{7.2.74}$$

Also,

$$R_2\sum_{j=1}^{k}\frac{P_{j-1}}{x^{j-1}} = R_2\sum_{j=0}^{k-1}\frac{P_j}{x^j} = R_2 a_0 + R_2\sum_{j=1}^{k-1}\frac{P_j}{x^j} \tag{7.2.75}$$

and

$$(R_1 + R_2) \sum_{j=1}^{k} \frac{P_j}{x^j} = (R_1 + R_2) \sum_{j=1}^{k-1} \frac{P_j}{x^j} + (R_1 + R_2) \frac{P_k}{x^k} \quad (7.2.76)$$

so that, returning to (7.2.73), we have

$$C(k) - R_1 a_0 = (R_1 + 2R_2) \sum_{j=1}^{k-1} \frac{P_j}{x^j} + R_2 a_0 + (R_1 + R_2) \frac{P_k}{x^k} \quad (7.2.77)$$

$$C(k) = (R_1 + 2R_2) \sum_{j=1}^{k-1} \frac{P_j}{x^j} + (R_1 + R_2) \left(a_0 + \frac{P_k}{x^k} \right) \quad (7.2.78)$$

Substituting in (7.2.70),

$$i_k = \frac{x^k}{P_k} \left[(R_1 + 2R_2) \sum_{j=1}^{k-1} \frac{P_j}{x^j} + (R_1 + R_2) \left(a_0 + \frac{P_k}{x^k} \right) \right] \quad (7.2.79)$$

from which the following rather unwieldy (but not difficult) formula follows.

NESTING-RECURSION ERROR THEOREM If a recursion is performed using the combination

$$\begin{cases} P_0 = a_0 \\ P_k = xP_{k-1} + a_k \end{cases}$$

for the power series

$$S = \sum_{k=0}^{n} a_{n-k} x^k$$

and if the inherent relative error of the input is R_1 and the roundoff error of the computer is R_2, then

$$|\epsilon| \leqslant |x^k| \left[(R_1 + 2R_2) \left| \sum_{j=1}^{k-1} \frac{P_j}{x^j} \right| + (R_1 + R_2) \left| a_0 + \frac{P_k}{x^k} \right| \right] \quad (7.2.80)$$

Example Consider

$$\sin y \sim y - \frac{y^3}{3!} + \frac{y^5}{5!} - \cdots - \frac{y^{15}}{15!} \quad (7.2.81)$$

Under ordinary circumstances, this would be programmed as

$$\sin y = y\left(1 - \frac{1}{2\times3}y^2\left(1 - \frac{1}{4\times5}y^2\left(1 - \frac{1}{6\times7}y^2(\cdots)\right)\right)\right) \quad (7.2.82)$$

but for our purposes, we shall consider it as

$$\sin y = y\left[1 + x\left[-\frac{1}{3!} + x\left(\frac{1}{5!} + x\left(-\frac{1}{7!}\right.\right.\right.\right.$$

$$\left.\left.\left.\left. + x\left(\cdots + x\left(\frac{1}{13!} + x\cdot\frac{-1}{15!}\right)\cdots\right)\cdots\right)\right]\right] \quad (7.2.83)$$

where $x = y^2$. In Table 7.2-1 we outline the calculation for $\sin 2$, so that $y = 2$ and $x = 4$. The results have been rounded for clarity.

Table 7.2-1

k	a_k	P_k	P_k/x^k
0	$-1/15!$	-7.647×10^{-13}	-7.647×10^{-13}
1	$1/13!$	$+1.575\times10^{-10}$	$+3.938\times10^{-11}$
2	$-1/11!$	-2.442×10^{-8}	-1.526×10^{-9}
3	$1/9!$	$+2.658\times10^{-6}$	$+4.153\times10^{-8}$
4	$-1/7!$	-1.878×10^{-4}	-7.335×10^{-7}
5	$1/5!$	$+7.582\times10^{-3}$	$+7.405\times10^{-6}$
6	$-1/3!$	-1.363×10^{-1}	-3.329×10^{-5}
7	$1/1!$	$+4.546\times10^{-1}$	$+2.775\times10^{-5}$

In the calculation of $\sin 2$, R_1 and R_2 were entirely roundoff, so that we could assume, for the IBM 360, single-precision, that each was $\leqslant 9.54\times10^{-7}$. Substituting in (7.2.80), we have

$$|\epsilon| \leqslant 4^7[3(2.657\times10^{-5}) + 2(-7.647\times10^{-13} + 6.937\times10^{-6})]R \quad (7.2.84)$$

$$|\epsilon| \leqslant 1.5\times10^{-6} \quad (7.2.85)$$

The actual error was 5.3×10^{-7}, not too far off.

PROBLEM SECTION 7.2

A. In each of the following: (a) Write a Taylor series to 4 terms. (b) Find the interval of absolute convergence. (c) Calculate the numerical value for the argument given. Of course there will be considerable error. (d) Estimate the error of truncation. (e) *Optional:* Estimate the total error. (f) Find the actual error. In this book we can give answers for only (a) and (b), because numerical errors are machine-dependent. In each problem, TV is the true value, correct to the number of places displayed.

(1) $\sin x$ at $x = -1$ (TV = -0.8414709848)

(2) $\cos x$ at $x = -0.85$ (TV = 0.6599831459)

(3) $\arcsin x$ at $x = 0.4$ (TV = 0.4115168461)

(4) $\arctan x$ at $x = -0.7$ (TV = -0.6107259644)

(5) $\sinh x$ at $x = 2$ (TV = 3.626860408)

(6) $\cosh x$ at $x = 4$ (TV = 27.289917197)

(7) $e^{\sin x}$ at $x = 1.6$ (TV = 2.7171233008)

(8) $e^{\cos x}$ at $x = 0.9$ (TV = 1.8619233267)

(9) 2^x at $x = 0.8$ (TV = 1.7411011)

(10) $e^{\arcsin x}$ at $x = 0.6$ (TV = 1.9031332302)

B. Same instructions as A.

(11) e^{-x^2} at $x = 0.9$, but expand about 1

$$(TV = 0.4448580662)$$

(12) $\ln x$ at $x = 1.7$, but expand about 1 [TV = 0.5306282511]

Chapter 8

Series of Powers and Fractions

In an abbreviated course, omit Section 8.1.

In this chapter, we pause to consider sums of the form

$$\sum_{k=1}^{\infty} k^p, \qquad p > 0 \tag{8.1}$$

and of the form

$$\sum_{k=1}^{\infty} \frac{P(k)}{Q(k)} \tag{8.2}$$

where $P(k)$ and $Q(k)$ are polynomials in k. These are sufficiently common in computing practice to warrant the special attention.

8.1 SERIES OF POWERS

8.1.1 Background

The problem of finding a formula for sums of the type $\sum_{k=1}^{n} k^p$, where p is a positive integer, has always been of great interest. In 1713, the *Ars Conjectandi* by Jacques Bernoulli was published posthumously. In it, he developed a solution of the form

$$\sum_{k=1}^{n} k^p = a_0 n^{p+1} + a_1 n^p + \cdots + a_{p+1} n^0 \tag{8.1.1}$$

and explained a rather involved method for finding the value of each coefficient.

From our vantage point, we can see that (8.1.1) can be evaluated easily, using the sum-difference theorem

$$\sum_{k=a}^{b-1} \Delta f(k) = f(b) - f(a) \qquad (8.1.2)$$

provided that we can find a difference such that

$$\Delta f(k) = f(k+1) - f(k) = nk^{n-1} \qquad (8.1.3)$$

8.1.2 The Bernoulli Polynomial

It is fairly obvious that $f(k)$ in (8.1.3) must be some kind of polynomial in k. We shall call it the *first-order Bernoulli polynomial, $B_n(k)$*.

By patient trial and error, we can get an idea of the form of the first few Bernoulli polynomials.

$B_1(k)$. We want a polynomial such that

$$\Delta B_1(k) = 1k^0 = 1 \qquad (8.1.4)$$

We let $B_1(k) = ak + b$. Then

$$\Delta B_1(k) = [a(k+1) + b] - [ak + b] = a \qquad (8.1.5)$$

By comparing coefficients,

$$a = 1 \qquad (8.1.6)$$

$$B_1(k) = k + b \qquad (8.1.7)$$

where b can be any constant and still satisfy (8.1.4).

$B_2(k)$. We want a polynomial such that

$$\Delta B_2(k) = 2k^1 = 2k \qquad (8.1.8)$$

Let $B_2(k) = ak^2 + bk + c$. Then, by comparing coefficients in

$$\Delta B_2(k) = [a(k+1)^2 + b(k+1) + c] - [ak^2 + bk + c] \qquad (8.1.9)$$

we find that $a = 1$ and $b = -1$, so that

$$B_2(k) = k^2 - k + c \qquad (8.1.10)$$

where c can be any constant and still satisfy (8.1.8).

$B_3(k)$. We want a polynomial such that

$$\Delta B_3(k) = 3k^2 \tag{8.1.11}$$

We let $B_3(k) = ak^3 + bk^2 + ck + d$, and substitute for $\Delta B_3(k)$. After simplifying, we find that

$$a = 1, \quad b = -\tfrac{3}{2}, \qquad c = \tfrac{1}{2} \tag{8.1.12}$$

$$B_2(k) = k^3 - \tfrac{3}{2}k^2 + \tfrac{1}{2}k + d \tag{8.1.13}$$

where d can be any constant.

From our point of view, it might be most convenient to let the constants in (8.1.7), (8.1.10) and (8.1.13) vanish, but for other reasons mathematicians impose another condition:

$$\int_0^1 B_n(k)\,dk = 0 \tag{8.1.14}$$

For $B_1(k)$, then,

$$\int_0^1 (k + b)\,dk = \tfrac{1}{2}k^2 + bk\big|_0^1 = \tfrac{1}{2} + b = 0 \tag{8.1.15}$$

$$b = -\tfrac{1}{2} \tag{8.1.16}$$

The first few first-order Bernoulli polynomials based on (8.1.14) are given in the Appendix, Table A.19.

8.1.3 mth-Order Bernoulli Polynomials

The original definition of the mth-order Bernoulli polynomial was based upon the identity

$$\frac{t^m e^{kt}}{(e^t - 1)^m} \equiv \sum_{n=0}^{\infty} \frac{t^n}{n!} B_n^{(m)}(k) \tag{8.1.17}$$

For the *first*-order polynomial, the superscript m is usually omitted, so that it can be defined as follows:

FIRST-ORDER BERNOULLI POLYNOMIAL, $B_n(k)$ The polynomial is defined by the coefficients B_n in the following Maclaurin series:

$$\frac{te^{kt}}{e^t - 1} = \sum_{n=0}^{\infty} \frac{t^n}{n!} B_n(k) \tag{8.1.18}$$

The series converges uniformly for $|t| < 2\pi$.

Although the expansion is not difficult, it will be more convenient for us to develop an easier algorithm. Note that for $n=0$, the right member of (8.1.18) starts with

$$\frac{t^0}{0!} B_0(k) = B_0(k) \qquad (8.1.19)$$

The expansion of the left member starts with a *1*. Therefore,

$$B_0(k) = 1 \qquad (8.1.20)$$

Now we rewrite (8.1.18) as

$$\frac{te^{kt}}{e^t - 1} = 1 + \sum_{n=1}^{\infty} \frac{t^n}{n!} B_n(k) \qquad (8.1.21)$$

and differentiate with respect to k:

$$\frac{t^2 e^{kt}}{e^t - 1} = \sum_{n=1}^{\infty} \frac{t^n}{n!} B_n'(k) \qquad (8.1.22)$$

Dividing by t and expressing $n!$ as $n(n-1)!$, we have

$$\frac{te^{kt}}{e^t - 1} = \frac{1}{n} \sum_{n=1}^{\infty} \frac{t^{n-1}}{(n-1)!} B_n'(k) \qquad (8.1.23)$$

Looking back at (8.1.18), we wish to make a comparison. We change the dummy from n to $m-1$. When $n=0$, $m=1$, so that (8.1.18) becomes

$$\frac{te^{kt}}{e^t - 1} = \sum_{m=1}^{\infty} \frac{t^{m-1}}{(m-1)!} B_{m-1}(k) \qquad (8.1.24)$$

Since the left members of (8.1.23) and (8.1.24) are identical, the right members must be identical as well. We change the dummy in (8.1.24) from m to n. Then

$$\sum_{n=1}^{\infty} \frac{t^{n-1}}{(n-1)!} B_{n-1}(k) = \frac{1}{n} \sum_{n=1}^{\infty} \frac{t^{n-1}}{(n-1)!} B_n'(k) \qquad (8.1.25)$$

from which

$$B_n'(k) = n B_{n-1}(k) \qquad (8.1.26)$$

Let us see whether (8.1.26) and (8.1.14) are compatible. We use $m = n-1$:

$$\int_0^1 B_m(k)\,dk = \int_0^1 \frac{1}{m+1} B'_{m+1}(k)\,dk$$

$$= \frac{B_{m+1}(k)}{m+1}\bigg|_0^1 = \frac{B_{m+1}(1) - B_{m+1}(0)}{m+1} \qquad (8.1.27)$$

But, for any subscript, we want $B(1) = B(0) = 1$, because

$$\Delta B_n(0) = n\,0^{n-1} = 0 \qquad (8.1.28)$$

Then (8.1.26) and (8.1.14) are compatible.

We haven't really proved anything, but it can be shown that the original definition is equivalent to the following recursive definition. We shall limit ourselves to a proof (in the next subsection) that with the following definition, (8.1.3) is satisfied—which is all we really wanted.

RECURSIVE DEFINITION: BERNOULLI POLYNOMIAL

$$B_0(k) = 1, \qquad B'_n(k) = nB_{n-1}(k) \quad \text{and} \quad \int_0^1 B_n(k)\,dk = 0 \qquad (8.1.29)$$

ILLUSTRATIVE PROBLEM 8.1-1   Find (a) $B_1(k)$, (b) $B_2(k)$, (c) $B_3(k)$.

Solution
(a)

$$B'_1(k) = 1B_0(k) = 1 \qquad (8.1.30)$$

$$\therefore \quad B_1(k) = k + C \qquad (8.1.31)$$

$$\int_0^1 B_1(k)\,dk = \frac{1}{2}k^2 + Ck\bigg|_0^1 = \frac{1}{2} + C = 0 \qquad (8.1.32)$$

$$B_1(k) = k - \frac{1}{2} \qquad (8.1.33)$$

(b)

$$B'_2(k) = 2B_1(k) = 2k - 1 \qquad (8.1.34)$$

$$\therefore \quad B_2(k) = k^2 - k + C \qquad (8.1.35)$$

$$\int_0^1 B_2(k)\,dk = \frac{k^3}{3} - \frac{k^2}{2} + Ck\bigg|_0^1 = -\frac{1}{6} + C = 0 \qquad (8.1.36)$$

$$B_2(k) = k^2 - k + \frac{1}{6} \qquad (8.1.37)$$

(c)

$$B_3'(k) = 3B_2(k) = 3k^2 - 3k + \frac{1}{2} \tag{8.1.38}$$

$$\therefore \quad B_3(k) = k^3 - \frac{3}{2}k^2 + \frac{1}{2}k + C \tag{8.1.39}$$

$$\int_0^1 B_3(k)\,dk = \frac{1}{4}k^4 - \frac{1}{2}k^3 + \frac{1}{4}k^2 + Ck\Big|_0^1 = C = 0 \tag{8.1.40}$$

$$B_3(k) = k^3 - \frac{3}{2}k^2 + \frac{1}{2}k \tag{8.1.41}$$

8.1.4 Differencing the Bernoulli Polynomial

We shall now prove that the recursive definition (8.1.29) does indeed give us the difference we want.

BERNOULLI DIFFERENCE THEOREM   For $h = 1$,

$$\Delta B_n(k) = nk^{n-1}, \qquad n = 1, 2, 3, \ldots \tag{8.1.42}$$

Proof We know by actual trial that

$$\Delta B_1(k) = \left(k + 1 - \tfrac{1}{2}\right) - \left(k - \tfrac{1}{2}\right) = 1 \tag{8.1.43}$$

so that (8.1.42) is true for $n = 1$. Assume that it is true for $n = m \geqslant 1$, i.e., that

$$\Delta B_m(k) = B_m(k+1) - B_m(k) = mk^{m-1} \tag{8.1.44}$$

From the recursive definition, we have

$$B_{m+1}'(k+1) = (m+1)B_m(k+1) \tag{8.1.45}$$

$$B_{m+1}'(k) = (m+1)B_m(k) \tag{8.1.46}$$

Subtracting (8.1.46) from (8.1.45),

$$B_{m+1}'(k+1) - B_{m+1}'(k) = (m+1)\big[B_m(k+1) - B_m(k)\big] \tag{8.1.47}$$

Using (8.1.44),

$$B_{m+1}'(k+1) - B_{m+1}'(k) = (m+1)mk^{m-1} \tag{8.1.48}$$

Integrating both members,

$$B_{m+1}(k+1) - B_{m+1}(k) = (m+1)k^m + C \tag{8.1.49}$$

To find C, let $k=0$ and $m=1$. Then, using (8.1.37),

$$B_1(1) - B_1(0) = C = 0 \tag{8.1.50}$$

which completes the proof. $\square$

8.1.5 Summing a Finite Power Series

The Bernoulli difference theorem affords an immediate method for finding

$$S_n = \sum_{k=0}^{n} k^p = \sum_{k=1}^{n} k^p, \qquad p = 1, 2, 3, \dots \tag{8.1.51}$$

We use

$$\Delta B_{p+1}(k) = (p+1)k^p \tag{8.1.52}$$

Then

$$S_n = \sum_{k=0}^{n} \frac{1}{p+1} \Delta B_{p+1}(k) \tag{8.1.53}$$

Applying the sum-difference theorem, we have

BERNOULLI POWER-SUM THEOREM

$$\sum_{k=0}^{n} k^p = \frac{1}{p+1}[B_{p+1}(n+1) - B_{p+1}(0)] \tag{8.1.54}$$

ILLUSTRATIVE PROBLEM 8.1-2   Find

$$\sum_{k=0}^{8} k^7$$

Solution

$$S = \frac{1}{8}[B_8(9) - B_8(0)] \tag{8.1.55}$$

From the Appendix, Table A.19, after nesting, the result is

$$B_8(k) - B_8(0) = \frac{1}{3}k^2\big(2 + k^2\big(-7 + k^2(14 + k(-12 + 3k)))\big) \tag{8.1.57}$$

For $k=9$, the result is 26,379,648, and $S = 3,297,456$. By brute force, this little problem requires 42 multiplications and 7 additions. By the theorem, it requires 6 multiplications, 4 additions and 1 division. Assuming the calculations are done properly, in fixed point, there is no error either way.

PROBLEM SECTION 8.1

A. By using the recursive definition, find the following Bernoulli polynomials. The answers are in Table A.19.

(1) B_4, (2) B_5, (3) B_6, (4) B_7, (5) B_8, (6) B_9

B. Find the sums, using the power sum theorem.

(7) $\sum\limits_{k=1}^{25} k^2$, (8) $\sum\limits_{k=1}^{20} k^3$, (9) $\sum\limits_{k=1}^{50} k^4$, (10) $\sum\limits_{k=1}^{n} k^5$

C.
(11) The original definition for a Bernoulli polynomial of order 1 is given by

$$\frac{te^{xt}}{e^t-1} = \sum_{k=0}^{\infty} \frac{t^k}{k!} B_k(x) \tag{8.1.58}$$

By substituting $1-x$ for x, prove that

$$B_k(1-x) = (-1)^k B_k(x) \tag{8.1.59}$$

(12) The original definition for a Bernoulli polynomial in order n, $B_k^{(n)}(x)$, is

$$\frac{t^n e^{xt}}{(e^t-1)^n} = \sum_{k=0}^{\infty} \frac{t^k}{k!} B_k^{(n)}(x) \tag{8.1.60}$$

Prove the *complementary-argument theorem*:

$$B_k^{(n)}(n-x) = (-1)^k B_k^{(n)}(x) \tag{8.1.61}$$

(13) Using (8.1.59), show that $B_{2k+1}(\tfrac{1}{2}) = 0$.
(14) Prove by induction that $B_n(x)$ is a polynomial in x.

8.2 SERIES OF FRACTIONS

8.2.1 Background

We have pointed out several times that the brute-force methods of summation invite heavy roundoff errors. Sometimes the clever application of theory will avoid the errors by supplying a formula, or even an exact answer, for the sum.

8.2.2   Infinite Sums of Fractions

Using summation by parts, it can be shown that

$$\sum_{k=1}^{n} \frac{2k+3}{k(k+1)} \frac{1}{3^k} = 1 - \frac{1}{n+1} \frac{1}{3^n} \tag{8.2.1}$$

It is obvious that as $n \to \infty$, the sum approaches 1:

$$\sum_{k=1}^{\infty} \frac{2k+3}{k(k+1)} \frac{1}{3^k} = 1 \tag{8.2.2}$$

It is not easy to develop (8.2.1), but once we have it, the transition to (8.2.2) is painless and the answer is exact.

We shall need a formula for the sum of any series of the form

$$S' = \sum_{k=a}^{\infty} \frac{1}{(k+c)^{(N)}} \tag{8.2.3}$$

where $N > 1$. Using the definition for negative factorial functions,

$$y^{(-m)} = \frac{1}{(y+m)^{(m)}} \tag{8.2.4}$$

we write (8.2.3) in the form

$$S' = \sum_{k=a}^{\infty} (k-N+c)^{(-N)} \tag{8.2.5}$$

In order to use the factorial sum theorem, we let

$$y = k - N + c \tag{8.2.6}$$

and temporarily let the upper limit in (8.2.5) be $b-1$. When $k=a$, $y = a - N + c$. When $k = b-1$, $y = b-1-N+c$. Substituting in (8.2.5), we have

$$S' = \sum_{y=a-N+c}^{y=b-1-N+c} y^{(-N)} = \frac{1}{-N+1} \Big[(b-1-N+c)^{(-N+1)}$$

$$- (a-N+c)^{(-N+1)} \Big] \tag{8.2.7}$$

As b goes to infinity (with $N > 1$), this becomes

$$S' = \frac{1}{-N+1} \Big[-(a-N+c)^{(-N+1)} \Big] \tag{8.2.8}$$

Using the definition of negative factorial functions once again, we obtain the following theorem:

THEOREM: INFINITE SERIES OF FACTORIAL FUNCTIONS   For N an integer >1,

$$\sum_{k=a}^{\infty} \frac{1}{(k+c)^{(N)}} = \frac{1}{N-1} \frac{1}{(a+c-1)^{(N-1)}} \qquad (8.2.9)$$

Corollary 1

$$\sum_{k=1}^{\infty} \frac{1}{(k+c)^{(N)}} = \frac{1}{(N-1)} \frac{1}{c^{(N-1)}}$$

Corollary 2

$$\sum_{k=1}^{\infty} \frac{1}{(k+N)^{(N)}} = \frac{1}{(N-1)} \frac{1}{N!}$$

ILLUSTRATIVE PROBLEM 8.2-1   Find

$$\sum_{k=2}^{\infty} \frac{1}{(k+1)k(k-1)}$$

Solution   The denominator is $(k+1)^{(3)}$, so that $a=2$, $c=1$ and $N=3$. Applying (8.2.9), the sum is $\frac{1}{2}(1/2!) = \frac{1}{4}$.

8.2.3   The Comparison Method (Kummer's Method)

A very important series is the Riemann zeta-function

$$\zeta(n) = \sum_{k=1}^{\infty} \frac{1}{k^n} \qquad (8.2.10)$$

A short table of values is given in the Appendix as Table A.12. It can be seen that the zeta-function is known precisely for *even* values of n. (We shall discuss this in the next chapter.) We shall now attack the problem of finding $\zeta(3)$, for which the true value, correct to ten significant figures, is 1.2020 56903. We shall pretend that we do not know the true value and write a very short program to calculate it by brute force.

The results are shown in Table 8.2-1. Note that S_9 and S_{10} agree to *four* significant figures, but we happen to know that they are correct to only *three* after rounding. (This shows, once again, that the agreement of successive values is not necessarily an indication of the correct answer.)

We could take more terms, but the propagation of roundoff errors makes the results less than credible. We must seek a method which speeds the convergence of the partial sums, thus minimizing the roundoff error.

Table 8.2-1 $\zeta(3)$ by Brute Force			
k	$U(k)$	$S(k)$	$E(k)$
1	1.0000000000	1.0000000000	2.02D-01
2	0.1250000000	1.1250000000	7.71D-02
3	0.0370370370	1.1620370370	4.00D-02
4	0.0156250000	1.1776620370	2.44D-02
5	0.0080000000	1.1856620370	1.64D-02
6	0.0046296296	1.1902916667	1.18D-02
7	0.0029154519	1.1932071186	8.85D-03
8	0.0019531250	1.1951602436	6.90D-03
9	0.0013717421	1.1965319857	5.52D-03
10	0.0010000000	1.1975319857	4.52D-03

We search for a "similar" series for which the sum is known. We shall call this S', the *comparison series*. Then it is always true that

$$S = S' - (S' - S) \tag{8.2.11}$$

If $S' - S$ converges more rapidly than S, then we can make a more efficient program.

In this case, for reasons which will soon be apparent, we can use the series in Illustrative Problem 8.2-1 as a comparison series.

The fact that this series begins at 2 instead of 1 is a little bothersome, but we can rewrite $\zeta(3)$ as

$$\zeta(3) = 1 + \sum_{k=2}^{\infty} \frac{1}{k^3} \tag{8.2.12}$$

Now, let S' be the series in Problem 8.2-1, and S the series in the right member of (8.2.12). Then

$$S' - S = \sum_{k=2}^{\infty} \left[\frac{1}{(k+1)k(k-1)} - \frac{1}{k^3} \right] = \sum_{k=2}^{\infty} \frac{1}{(k+1)(k-1)k^3} \tag{8.2.13}$$

Putting the pieces toegther, we have, from (8.2.11),

$$\zeta(3) = \frac{1}{4} + 1 - \sum_{k=2}^{\infty} \frac{1}{(k+1)(k-1)k^3} \tag{8.2.14}$$

The result is shown in Table 8.2-2.

	Table 8.2-2 $\zeta(3)$ by (8.2.14)		
k	$U(k)$	$S(k)$	$E(k)$
2	0.0416666667	1.2083333333	-6.28D-03
3	0.0046296296	1.2037037037	-1.65D-03
4	0.0010416667	1.2026620370	-6.05D-04
5	0.0003333333	1.2023287037	-2.72D-04
6	0.0001322751	1.2021964286	-1.40D-04
7	0.0000607386	1.2021356900	-7.88D-05
8	0.0000310020	1.2021046880	-4.78D-05
9	0.0000171468	1.2020875412	-3.06D-05
10	0.0000101010	1.2020774402	-2.05D-05

It can be seen that the device improved matters considerably at the expense of a little planning. The summation in (8.2.14) converges faster than the original because the fraction in it is like $1/k^5$, whereas the original was $1/k^3$. We shall now continue the improvement.

ILLUSTRATIVE PROBLEM 8.2-2 In this section, we speeded the calculation of $\zeta(3)$ by developing (8.2.13). Speed it some more.

Solution The fraction in (8.2.13) is like $1/k^5$. We seek a comparison series similar to this so that the *difference*, $S' - S$, will be still faster. We could, for example, choose

$$\frac{1}{k^{(5)}} = \frac{1}{k(k-1)(k-2)(k-3)(k-4)} \qquad (8.2.15)$$

From the standpoint of algebraic manipulation, it is more convenient to use

$$\frac{1}{(k+2)^{(5)}} = \frac{1}{(k+2)(k+1)k(k-1)(k-2)} \qquad (8.2.16)$$

for which [using (8.2.9)] the sum is $\frac{1}{4}(1/4^{(4)}) = \frac{1}{96}$, beginning with $k = a = 3$. We now rewrite (8.2.14):

$$\zeta(3) = \frac{1}{4} + 1 - \frac{1}{3 \times 1 \times 8} - \sum_{k=3}^{\infty} \frac{1}{(k+1)(k-1)k^3} \qquad (8.2.17)$$

and consider the summation in the right member of (8.2.17) as S, with a similar summation from (8.2.16) as S'. Then

$$S' - S = \sum_{k=3}^{\infty} \frac{1}{(k+1)(k-1)k^3} - \frac{1}{(k+3)(k+1)k(k-1)(k-2)} \qquad (8.2.18)$$

$$= \sum_{k=3}^{\infty} \frac{-4}{(k^2-4)(k^2-1)k^3} \qquad (8.2.19)$$

which, as you see, converges like $1/k^7$. We rewrite (8.2.17):

$$\zeta(3) = \frac{1}{4} + 1 - \frac{1}{24} - \frac{1}{96} + 4\sum_{k=3}^{\infty} \frac{1}{(k^2-4)(k^2-1)k^3} \qquad (8.2.20)$$

The results are shown in Table 8.2-3.

Table 8.2-3 $\zeta(3)$ **using (8.2.20)**

k	$U(k)$	$S(k)$	$E(k)$
3	0.0037031037	1.2016197346	4.37D-04
4	0.0003472222	1.2019669568	8.99D-05
5	0.0000634921	1.2020304489	2.65D-05
6	0.0000165344	1.2020469833	9.92D-06
7	0.0000053990	1.2020523822	4.52D-06
8	0.0000020668	1.2020544490	2.45D-06
9	0.0000008907	1.2020553398	1.56D-06
10	0.0000004209	1.2020557607	1.14D-06

It is evident that the convergence is relatively rapid. After only 8 computations, the result is correct to the nearest millionth, whereas the original computation was only correct to the nearest hundredth after 10 computations.

8.2.4 Using $\zeta(n)$ as a Comparison Series

In the preceding, we used the computation of $\zeta(3)$ as an illustration. However, the result would have been of no use to us if we had not known the correct answer in advance. Let us think about the error. In (8.2.20), we wrote

$$\zeta(3) = \frac{1}{4} + 1 - \frac{1}{24} - \frac{1}{96} + 4\sum_{k=3}^{\infty} \frac{1}{(k^2-4)(k^2-1)k^3} \qquad (8.2.21)$$

but in the computer program we actually used

$$\zeta(3) \sim \frac{1}{4} + 1 - \frac{1}{24} - \frac{1}{96} + 4\sum_{k=3}^{10} \frac{1}{(k^2-4)(k^2-1)k^3} \qquad (8.2.22)$$

Evidently, there is an error of truncation between (8.2.21) and (8.2.22):

$$E_T \leqslant \left| \sum_{k=11}^{\infty} \frac{4}{(k^2-4)(k^2-1)k^3} \right| \qquad (8.2.23)$$

If we did not know the correct answer, it would have been necessary to estimate this error as well as the roundoff errors. We shall shirk this because values of the Riemann zeta-function are known precisely to many places. Instead, we shall use the zeta-function as a very useful comparison series, often more useful than the factorial function series, and shall calculate the error properly.

ILLUSTRATIVE PROBLEM 8.2-3   Find

$$\sum_{k=1}^{\infty} \frac{k}{(k^3+5)}$$

correct to the nearest millionth.

Solution   We remind you once more (we hope unnecessarily) that the easiest thing to do is to calculate the partial sums by brute force, stopping when two adjacent partial sums agree to the required number of places. The only satisfying fact about this procedure is that one can be virtually sure that the answer is wrong. In this problem, for which we do not know the answer, the use of brute force produced the column so labeled in Table 8.2-4. Note that after 20 partial sums, the result has not yet settled down in the hundredths place.

The fraction in the summation converges as k/k^3 or $1/k^2$ as k approaches infinity. We can try

$$S' = \sum_{k=1}^{\infty} \frac{1}{k(k+1)} = 1 \tag{8.2.24}$$

as a comparison series. Then

$$S' - S = \sum_{k=1}^{\infty} \frac{-k^2+5}{k(k+1)(k^3+5)} \tag{8.2.25}$$

$$S = S' - (S'-S) = 1 + \sum_{k=1}^{\infty} \frac{k^2-5}{(k^3+5)k(k+1)} \tag{8.2.26}$$

This doesn't look very good. The summation in (8.2.26) converges as $k^2/k^5 = 1/k^3$, which is only a little better than before.

We now try instead the appropriate Riemann zeta-function, i.e., $\zeta(2)$:

$$S' = \zeta(2) = \sum_{k=1}^{\infty} \frac{1}{k^2} \sim 1.64493\,4067 \tag{8.2.27}$$

$$S' - S = \sum_{k=1}^{\infty} \left(\frac{1}{k^2} - \frac{1}{k^3+5} \right) = 5 \sum_{k=1}^{\infty} \frac{1}{k^2(k^3+5)} \tag{8.2.28}$$

This converges as $1/k^5$. Then

$$S = S' - (S'-S) = 1.64493\,4067 - 5 \sum_{k=1}^{\infty} \frac{1}{k^2(k^3+5)} \tag{8.2.29}$$

Table 8.2-4 Computing $\sum_{k=1}^{\infty} k/(k^3+5)$

k	BRUTE FORCE	(8.2.29)	(8.2.34)	(8.2.39)
1	0.16666666666666666	0.816003672281902	0.6269642512003577	0.72888881937662772
2	0.32051282051282205	0.715446521074344 1	0.6870604050465114	0.69132809761243 41
3	0.41426282051282205	0.698085409963233 0	0.6902754256226429	0.69073272343166 89
4	0.47223383500055741	0.693556424455986 7	0.6906295261539695	0.69070508069786 03
5	0.51069537346711 26	0.692017962917525 2	0.6906909079107 6349	0.69070261915933 992
6	0.53784469473408 09	0.691389506406715 7	0.6907053386813515	0.69070228240929 76
7	0.55795963726281 65	0.691096285670145 3	0.6907096130361119	0.69070222010092 05
8	0.57343325077129 8	0.690945173484458 7	0.6907110887410501	0.69070220568973 89
9	0.58569510545860 12	0.690861074853584 4	0.6907116655492176	0.69070220173357 91
10	0.59564535421482 01	0.690811323609803 3	0.6907119143054363	0.69070220048980 25
11	0.60387888714895 19	0.690780393734017 7	0.6907120304958034	0.69070220005332 31
12	0.61080329568905 57	0.690760357829677 2	0.6907120884700635	0.69070219988557 47
13	0.61670701957643 08	0.690746921953738 7	0.69071211904784 20	0.69070219981598 41
14	0.62179978058043 23	0.690737642141413 6	0.690712119c478420	0.69070219978517 50
15	0.62623765040291 75	0.690731067519454 4	0.6907121214918703	0.69070219977707 474
16	0.63013913784500 48	0.690726304961541 7	0.6907168898634524	0.69070219976364 91
17	0.63359827555626 95	0.690722787060349 8	0.6907168743072674	0.69070219976000 76
18	0.63667960346855 32	0.690720143219547 1	0.6907168608337626	0.69070219975806 42
19	0.63944766687366 185	0.690718125385119 4	0.6907167249092718	0.69070219975699 13
20	0.64194610721257 10	0.690716563861071 9	0.6907165662364 4975	0.69070219975638 38

The result of the computation is shown in Table 8.1-4 in column 3. Note that it has "settled down" to 10^{-4} by the twelfth partial sum. (This doesn't promise that it is correct, but if the sums do *not* settle down, one can be sure that they are not correct.)

This convergence is still too slow. We return to (8.2.29) and consider the summation in the right member, which we shall call S_1. Perhaps we can improve it. We let

$$S_1' = \zeta(5) \sim 1.0369277755 \tag{8.2.30}$$

$$S_1' - S = \sum_{k=1}^{\infty} \left[\frac{1}{k^5} - \frac{1}{k^2(k^3+5)} \right] = \sum_{k=1}^{\infty} \frac{5}{k^5(k^3+5)} \tag{8.2.31}$$

$$S_1 = S_1' - (S_1' - S) = 1.03692\ 7755 - 5 \sum_{k=1}^{\infty} \frac{1}{k^5(k^3+5)} \tag{8.2.32}$$

Substituting in (8.2.29), we have

$$S = 1.64493\ 7067 - 5 \left[1.03692\ 7755 - 5 \sum_{k=1}^{\infty} \frac{1}{k^5(k^3+5)} \right] \tag{8.2.33}$$

$$S = -3.539704708 + 25 \sum_{k=1}^{\infty} \frac{1}{k^5(k^3+5)} \tag{8.2.34}$$

The results of the calculation are also shown in Table 8.2-4 in column 4. Note that the partial sums have settled down to 10^{-5} by the seventh partial sum.

This is much better, but still not quite fast enough to satisfy us about the millionths place. We return to (8.2.34) and consider the summation in the right member. It converges as $1/k^8$. Therefore, we use

$$S_2' = \zeta(8) \sim 1.00407\ 7356 \tag{8.2.35}$$

$$S_2' - S_2 = \sum_{k=1}^{\infty} \frac{5}{k^8(k^3+5)} \tag{8.2.36}$$

$$S_2 = 1.00407\ 7356 - 5 \sum_{k=1}^{\infty} \frac{1}{k^8(k^3+5)} \tag{8.2.37}$$

Substituting in (8.2.34),

$$S = -3.53970\ 4708 + 25 \left[1.00407\ 7356 - 5 \sum_{k=1}^{\infty} \frac{1}{k^8(k^3+5)} \right] \tag{8.2.38}$$

$$= 21.56222\ 919 - 125 \sum_{k=1}^{\infty} \frac{1}{k^8(k^3+5)} \tag{8.2.39}$$

By looking in the last column of the table, you see that the *fifth* partial sum appears to have settled to 10^{-6}.

Can we be sure that $S=0.690710$ is correct to the nearest millionth? Certainly not. We must estimate the error. In the following, we do so. Our plan is to find the error of the summation in (8.2.39), remembering that the error is to be multiplied by 125. We need an error less than $\frac{1}{125}$ of a millionth.

8.2.5 Calculating the Error

In (8.2.39) we have

$$k^8(k^3+5)>k^{11} \tag{8.2.40}$$

Therefore

$$\Sigma \frac{1}{k^8(k^3+5)} < \Sigma \frac{1}{k^{11}} \tag{8.2.41}$$

Since we summed this series up to $k=7$, the error of truncation is

$$\sum_{k=8}^{\infty} \frac{1}{k^8(k^3+5)} < \sum_{k=8}^{\infty} \frac{1}{k^{11}} \tag{8.2.42}$$

To evaluate the right member, we write

$$E_T \leqslant \left| \sum_{k=1}^{\infty} \frac{1}{k^{11}} - \sum_{k=1}^{7} \frac{1}{k^{11}} \right| \tag{8.2.43}$$

The first summation is simply $\zeta(11)$, which can be found in the Appendix, Table A.12. The second summation is in Table A.10. The error of truncation is therefore approximately $10^{-9} \times 125$, well within the limit requested.

A program using forwards addition (to magnify roundoff error) is shown in Figure 8.2-1. However, if we program more carefully, the roundoff error is not too bad. In (8.2.39), the calculation of the denominator should be done in fixed point. Then it must be converted to floating point and inverted. This should be stored as u_k. The contributions should be added backwards. Let us see what the error looks like.

Let $a_k = k^8(k^3+5)$. Then the relative error of conversion to floating point depends upon the machine. We shall call it r_1. To find u_k, we must find the reciprocal of a_k (Figure 8.2-2).

```
      DOUBLE PRECISION B(20),F1(20),F2(20),F3(20),C(8),A,D,S,ZK
      C(2)=1.644934066848226
      C(5)=1.036927755143370
      C(8)=1.004077356197944
C CALCULATION BY BRUTE FORCE.
      S=0.0
      DO 1 K=1,20
      ZK=K
      D=K**3+5
      D=ZK/D
      S=S+D
1     B(K)=S
C CALCULATION BY FORMULA 29
      S=0.0
      DO 2 K=1,20
      D=K**2*(K**3+5)
      D=1.0/D
      S=S+D
2     F1(K)=C(2)-5.0*S
C CALCULATION BY FORMULA 34.
      S=0.0
      A=C(2)-5.0*C(5)
      DO 3 K=1,20
      D=K**5*(K**3+5)
      D=1.0/D
      S=S+D
3     F2(K)=A+25.0*S
C CALCULATION BY FORMULA 39.
      S=0.0
      A=A+25.0*C(8)
      DO 4 K=1,20
      ZK=K
      D=ZK**8*(ZK**3+5.0)
      D=1.0/D
      S=S+D
4     F3(K)=A-125.0*S
      WRITE(3,100)
100   FORMAT(1H1,T32,'COMPARISON OF FOUR WAYS TO COMPUTE THE SUM OF K/(K
     1**3+5)'/)
      WRITE(3,101)
101   FORMAT(1H ,3X,'K',T23,'BRUTE FORCE',T51,'FORMULA 29',T78,'FORMULA
     134',T105,'FORMULA 39')
      DO 5 K=1,20
5     WRITE(3,102)K,B(K),F1(K),F2(K),F3(K)
102   FORMAT(1H ,3X,I2,4F27.16)
      END
```

Fig. 8.2-1 Program for computing $\sum\limits_{1}^{\infty} k/(k^3+5)$

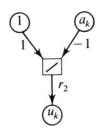

Fig. 8.2-2 Process chart for $1/a_k$.

Then, remembering that $i_1 = 0$,

$$i_{u(k)} = i_1 - i_{a(k)} + r_2 \tag{8.2.44}$$

$$i_{u(k)} = -r_1 + r_2 \tag{8.2.45}$$

$$|e_{u(k)}| \leqslant 2r|u_k| \tag{8.2.46}$$

where $r = \max(|r_1|, |r_2|)$. Using the backwards-sum error theorem (1.2.42), the error, including roundoff and conversion but not truncation, is given by

$$E \leqslant \left| \sum_{k=1}^{n} 2ru_k + r \sum_{k=1}^{n-1} ku_k + r(n-1)u_n \right| \tag{8.2.47}$$

which simplifies slightly to

$$E \leqslant \left| (n+1)ru_n + r \sum_{k=1}^{n-1} (2+k)u_k \right| \tag{8.2.48}$$

For $n = 7$, with $r = 2.2 \times 10^{-16}$ for the IBM 360, this amounts to

$$8 \times 2.2 \times 10^{-16} \times 4.98 \times 10^{-10} + 2.2 \times 10^{-16} \times \sum_{1}^{6} (2+k)u_k \tag{8.2.49}$$

The first contribution is negligible (about 8.8×10^{-25}). The rest of it comes to about 1.10×10^{-16}.

If programs are planned to use integer multiplication instead of floating-point multiplication, speed may be lost but there is a tremendous gain in accuracy. We do not continue the analysis, because it is obvious that we are well within the precision required.

8.2.6   Other Comparison Series

Appendix Table A.13 displays some other series which may serve as comparison series.

We also call attention to a very interesting method of summation by *polygamma functions*, explained and illustrated in Abramowitz and Stegun, pp. 264–265.

PROBLEM SECTION 8.2

A. In each case, develop an efficient method for finding the numerical values of the Riemann zeta-functions. The true values are given in Table A.12.
   (1) $\zeta(4)$,   (2) $\zeta(5)$,   (3) $\zeta(6)$,   (4) $\zeta(7)$

B. In each case, use $\zeta(n)$ as a comparison series and write a formula involving a series which converges at least as fast as $1/k^7$:

(5) $\displaystyle\sum_{k=1}^{\infty} \frac{1}{k^2+1}$

(7) $\displaystyle\sum_{k=1}^{\infty} \frac{1}{k^2(k^2+1)}$

(6) $\displaystyle\sum_{k=1}^{\infty} \frac{k}{k^4+1}$

(8) $\displaystyle\sum_{k=1}^{\infty} \frac{k}{(k^2+1)^2}$

C. In each case: (a) Use an infinite factorial series and write a formula involving a series which converges at least as fast as $1/k^5$. (b) Now try a Riemann function and compare for ease. Our answers may not be like yours because of different choices. (c) Estimate the error of truncation after 10 terms for each method.

(9) $\displaystyle\sum_{k=1}^{\infty} \frac{k^2}{k^6+3}$

(10) $\displaystyle\sum_{k=1}^{\infty} \frac{(k^2+5)}{k(k^6-8)}$

Chapter 9

The Fourier Series

In an abbreviated course, omit Section 9.1.

We have pointed out that computer scientists rely heavily on series. In Chapter 7, we outlined the representation of a known function by a series of the *point* type: the Taylor series. In this chapter, we shall extend the discussion to representation by a series of the *interval* type, the best known being the Fourier series. The infinite Fourier series is valuable for theoretical reasons, whereas the finite Fourier series finds extremely heavy use in engineering and some use in commerce—e.g., in forecasting the sale of seasonal items.

Part of the job of a competent computer scientist is to decide which type of representation is most appropriate for a specific situation. Appropriateness may be in terms of convenience, programming time, computing time or error characteristics. Chapters 7, 9, 10, and 11 are designed to give you enough information to reach a decision.

The first half of this chapter deals with the representation of a function by an infinite Fourier series. The most important outcome of this section is a display of the pleasant error characteristics of a Fourier series.

The second half of the chapter deals with *harmonic analysis*, the fitting of a finite Fourier series to a known function or to experimental data.

In this chapter, it will be necessary to refer to a few theorems from advanced calculus. In some cases, the complete proofs are beyond the scope (and space) permissible in this book. References are to I. S. Sokolnikoff, *Advanced Calculus* (McGraw-Hill, 1939), but the theorems are standard and may be found in other books on advanced calculus.

9.1 THE INFINITE FOURIER SERIES

9.1.1 Background

About the middle of the 18th century, physicists and mathematicians such as d'Alembert, Euler and Bernoulli became interested in the theory of vibrations. For example, if a violin string is stretched between two fixed

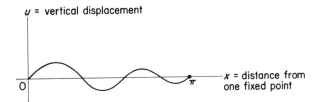

Fig. 9.1-1   Waveform of a plucked violin string.

points, 0 and π, the shape of the string at any time t after it is plucked (Figure 9.1-1) is expressible as the displacement u from the rest position. In 1753, Daniel Bernoulli gave a solution in the form

$$u = b_1 \sin x \cos ct + b_2 \sin 2x \cos 2ct + \cdots \qquad (9.1.1)$$

In 1807, the physicist J. B. J. Fourier, involved in the study of the conduction of heat, announced that it was possible to express an "arbitrary function" $f(x)$ in the following form:

$$f(x) = c + \sum_{k=1}^{\infty} (a_k \cos kx + b_k \sin kx) \qquad (9.1.2)$$

This formulation was not readily acceptable to mathematicians; in fact, it is *not* true for an "arbitrary function". However, in 1829, Dirichlet managed to establish one of the first theorems to display *sufficient* conditions under which a series like that in the right member of (9.1.2) would converge properly.

Our interest in the Fourier series (which we shall define precisely) stems from at least three properties:

1. It is indispensable for representing periodic phenomena.
2. It is indispensable for representing functions with discontinuities.
3. It has, from the viewpoint of the computer scientist, better error characteristics over an interval than a Taylor series.

We shall illustrate the three properties in this section.

9.1.2   Some Informal Definitions

Consider Figure 9.1-2. We are interested in the interval $[a,b]$ of x. It should be obvious that this function cannot be expressed as a Taylor series, since (1) it is discontinuous at a, p, q, s and b, and (2) it does not have a derivative at r. The Taylor series requires an infinitely *smooth* function. (We shall define "smooth" formally.) In the graph, the hollow circles at A, B, D, E, G, I, K and L represent the ends of *open* intervals, and the filled circles at M, C, F, J represent the values of y assigned at the corresponding values of x.

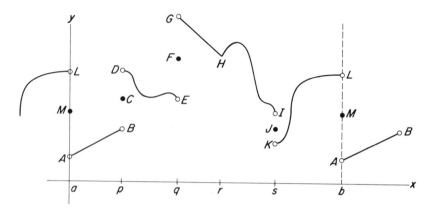

Fig. 9.1-2   A function with discontinuities.

Going from left to right, starting at A, we see that the graph represents a function, since for every value of x there is one and only one value of y. At point B, there is a *jump discontinuity*. Observe that as we proceed from A to B, $f(x)$ approaches but does not reach a limit which we shall call $f(p-)$. This is the ordinate up to B. The symbolism is intended to convey the thought that x is approaching p from "below", i.e., from the left.

Similarly, as we proceed from E to D, $f(x)$ approaches but does not reach a limit which we shall call $f(p+)$. This means that x is approaching p from "above", i.e., from the right.

This gives us two candidates for the value of $f(x)$ as x approaches p, one from the left and one from the right. We arbitrarily define $f(x)$ when $x=p$ as

$$f(p)=\tfrac{1}{2}\left[f(p-)+f(p+)\right] \tag{9.1.3}$$

Similarly, we define $f(x)$ at $x=q$ as

$$f(q)=\tfrac{1}{2}\left[f(q-)+f(q+)\right] \tag{9.1.4}$$

Assume that the function for $x\in[a,b]$ is repeated *periodically* in both directions. We pause to define periodicity:

PERIODIC FUNCTION   A function $f(x)$ is periodic with period p iff $f(x)=f(x+np)$ for all integral n, where p is the smallest number that makes this true.

For example, the sine and cosine functions are periodic with period 2π, and the tangent function is periodic with period π.

Then, when $x=a$ or $x=b$, we define

$$f(a)=f(b)=\tfrac{1}{2}\left[f(b-)+f(a+)\right] \tag{9.1.5}$$

This is the assigned value of y at point M.

Now let us turn our attention to the problem of smoothness. (This is defined in various ways in different textbooks.) We shall say that a function is *smooth* on an interval if it has a continuous derivative on the interval. The function in the figure is obviously not smooth as a whole, but it is smooth on the open intervals (a,p), (p,q), (q,r), (r,s), (s,b). If a function is smooth on an interval except at a *finite* number of points, we shall say that it is *sectionally smooth* or *piecewise smooth*.

At H, note that the function is continuous, but the derivative is not. If a function is piecewise smooth and has continuous second derivatives, we shall call it *piecewise very smooth*.

We shall also need the notion of *orthogonal functions* in this chapter and the next. There are two definitions—one for integrals, and one for sums. At the moment, we shall need only the first of these.

ORTHOGONAL FUNCTIONS (BY INTEGRALS) On the interval (a,b), $f(x)$ and $g(x)$ are orthogonal to each other relative to the weighting function $w(x) \geqslant 0$ iff

$$\int_a^b w(x) f(x) g(x) dx = 0 \qquad (9.1.6)$$

ILLUSTRATIVE PROBLEM 9.1-1 Show that on the interval $(-\pi, \pi)$, $\sin mx$ and $\sin nx$ are orthogonal relative to the weighting function $w(x) = 1$, where m and n are unequal integers.

Solution

$$J = \int_{-\pi}^{\pi} \sin mx \sin nx \, dx \qquad (9.1.7)$$

From Appendix Table A.8, formula (17),

$$\sin mx \sin nx = \tfrac{1}{2}\cos(m-n)x - \tfrac{1}{2}\cos(m+n)x \qquad (9.1.8)$$

Integrating, we obtain

$$J = \frac{1}{2}\frac{\sin(m-n)x}{m-n} - \frac{1}{2}\frac{\sin(m+n)x}{m+n}\bigg|_{-\pi}^{\pi} = 0 \qquad (9.1.9)$$

Note that if $m = n$, (9.1.7) becomes

$$K = \int_{-\pi}^{\pi} \sin^2 x \, dx \qquad (9.1.10)$$

$$= \int_{-\pi}^{\pi} \frac{1 - \cos 2x}{2} dx = \pi \qquad (9.1.11)$$

We shall need the following results, all easy to prove using formulas (17)–(19) from Table A.8:

	$m \neq n$	$m = n$	
$\int_{-\pi}^{\pi} \sin mx \sin nx \, dx =$	0	π	(9.1.12)
$\int_{-\pi}^{\pi} \cos mx \cos nx \, dx =$	0	π	(9.1.13)
$\int_{-\pi}^{\pi} \sin mx \cos nx \, dx =$	0	0	(9.1.14)

We shall also need

$$\int_{-\pi}^{\pi} \sin nx\, dx = 0 \tag{9.1.15}$$

$$\int_{-\pi}^{\pi} \cos nx\, dx = 0 \tag{9.1.16}$$

We note in passing that (9.2.12) can be expressed more compactly as

$$\int_{-\pi}^{\pi} \sin mx \sin nx\, dx = \pi \delta_n^m \tag{9.1.17}$$

The symbol δ_n^m is called the *Kronecker delta* and is defined as 1 if $m = n$, 0 otherwise.

9.1.3 Least-Squares Representation

Referring to Figure 7.2-4, you see that a Taylor series is extremely accurate near the point about which the function was expanded, but the error of truncation becomes larger and larger as we leave that point. In many (or most) cases, we are interested in the values of a function throughout a specific interval, and we would like the errors on the interval to be (1) no greater than a fixed known quantity, and (2) as small as possible.

Suppose the approximating function for $f(x)$ is $\tilde{f}(x)$. Then the error at every point is

$$E(x) = |f(x) - \tilde{f}(x)| \tag{9.1.18}$$

The total error over the interval $[-\pi, \pi]$ is then

$$E = \int_{-\pi}^{\pi} |f(x) - \tilde{f}(x)|\, dx \tag{9.1.19}$$

This is shown as the shaded area in Figure 9.1-3.

We would like to minimize (9.1.19). Unfortunately, integrals with absolute-value integrands are difficult to work with. Instead, we shall use an integral which serves the same purpose:

$$E = \int_{-\pi}^{\pi} \left[f(x) - \tilde{f}(x) \right]^2 dx \tag{9.1.20}$$

In other words, we avoid the negative integrands by squaring the errors. The result of minimizing (9.1.20) is called a *least-squares representation* of $f(x)$.

We now inquire as follows: if $\tilde{f}(x)$ in (9.1.20) has the form (9.1.2) suggested by Fourier, except that we use $\frac{1}{2}a_0$ in place of c, then what must be the values of the coefficients $a_0, a_1, \ldots, b_1, b_2, \ldots$ to minimize E?

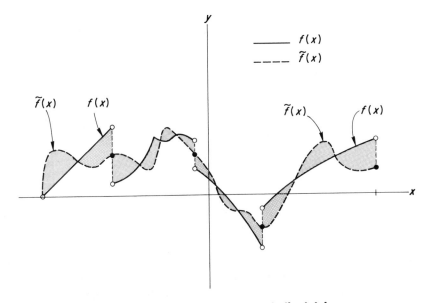

Fig. 9.1-3 The shaded area represents the total error.

We investigate

$$E = \int_{-\pi}^{\pi} \left[f(x) - \tfrac{1}{2}a_0 - \sum_{k=1}^{n} (a_k \cos kx + b_k \sin kx) \right]^2 dx \quad (9.1.21)$$

In the usual way, we shall find the partial derivatives of E with respect to a_0, a_k ($k>0$) and b_k ($k>0$), and then set these derivatives equal to zero. Since the derivatives and the integrand are obviously continuous on the interval, we can differentiate under the integral sign. The results are as follows [we are using I for the integrand in (9.1.21)]:

$$\frac{\partial E}{\partial a_0} = -\int_{-\pi}^{\pi} I \, dx \quad (9.1.22)$$

$$\frac{\partial E}{\partial a_k} = -2\int_{-\pi}^{\pi} I \cos kx \, dx \quad (9.1.23)$$

$$\frac{\partial E}{\partial b_k} = -2\int_{-\pi}^{\pi} I \sin kx \, dx \quad (9.1.24)$$

Expressing each integral as a sum of three integrals and using (9.1.12)–(9.1.14), we obtain

$$\frac{\partial E}{\partial a_0} = -\int_{-\pi}^{\pi} f(x) \, dx + \pi a_0 \quad (9.1.25)$$

$$\frac{\partial E}{\partial a_k} = 2\int_{-\pi}^{\pi} f(x) \cos kx \, dx + 2\pi a_k \quad (9.1.26)$$

$$\frac{\partial E}{\partial b_k} = -2\int_{-\pi}^{\pi} f(x) \sin kx \, dx + 2\pi b_k \quad (9.1.27)$$

A *necessary* condition for minimizing E is that the three partial derivatives vanish. This leads to a definition of the Fourier coefficients:

FOURIER COEFFICIENTS $(-\pi$ TO $\pi)$

$$a_0 = \frac{1}{\pi} \int_{-\pi}^{\pi} f(x)\,dx \tag{9.1.28}$$

$$a_k = \frac{1}{\pi} \int_{-\pi}^{\pi} f(x) \cos kx\,dx, \qquad k > 0 \tag{9.1.29}$$

$$b_k = \frac{1}{\pi} \int_{-\pi}^{\pi} f(x) \sin kx\,dx, \qquad k > 0 \tag{9.1.30}$$

To assure us that the use of these formulas really provides a minimum for E rather than a saddle surface or a maximum, the second derivatives should be checked. We shall omit the details (see Sokolnikoff, pp. 380–382). They do check.

We now define a Fourier series for a function periodic on the interval $(-\pi$ to $\pi)$:

FOURIER SERIES FOR $f(x)$ ON $(-\pi,\pi)$ The Fourier series of a function periodic from $-\pi$ to π is the series developed formally as

$$\frac{a_0}{2} + \sum_{k=1}^{\infty} (a_k \cos kx + b_k \sin kx) \tag{9.1.31}$$

where a_0, a_k and b_k are given by (9.1.28)–(9.1.30).

We observe that from a mathematical point of view this is a strange definition, because there is nothing in (9.1.31) requiring that the series actually converge to $f(x)$. In fact, even if $f(x)$ is integrable on the interval, the Fourier series may not converge or, if it does converge, may not converge to $f(x)$. (This is what is meant by "formally".) More restrictive conditions are needed. What we have shown is that *if* the Fourier series converges to $f(x)$, then the appropriate choices for a_0, a_k and b_k are the Fourier coefficients (9.1.28)–(9.1.30).

9.1.4   Change of Scale

By a simple change of scale, any interval $[a,b]$ can be changed to $[-\pi,\pi]$. For convenience, we shall redefine the Fourier coefficients for a general interval.

FOURIER COEFFICIENTS   For a function periodic on the interval $[a, b]$ of length $2L$, the Fourier coefficients are given by

$$a_0 = \frac{1}{L} \int_a^b f(x)\, dx \tag{9.1.32}$$

$$a_k = \frac{1}{L} \int_a^b f(x) \cos\frac{k\pi x}{L}\, dx \tag{9.1.33}$$

$$b_k = \frac{1}{L} \int_a^b f(x) \sin\frac{k\pi x}{L}\, dx \tag{9.1.34}$$

FOURIER SERIES   For a function $f(x)$ with a period of $2L$ from a to b, the Fourier series is defined as

$$\frac{a_0}{2} + \frac{1}{L} \sum_{k=1}^{\infty} \left(a_k \cos\frac{k\pi x}{L} + b_k \sin\frac{k\pi x}{L} \right) \tag{9.1.35}$$

These formulas are derived by replacing x with $\pi x / L$ throughout the derivation. Note that if $L = \pi$, (9.1.32)–(9.1.34) reduce to (9.1.28)–(9.1.30), whereas (9.1.35) reduces to (9.1.31).

9.1.5   The Fourier Sine Series

We start with a very simple problem.

ILLUSTRATIVE PROBLEM 9.1-1   Expand $f(x) = x$ as a Fourier series on the interval $(-\pi, \pi)$.

Solution   The first step is a sketch (Figure 9.1-4). Note that we delete the endpoints, and that we assume that the function is periodic on the interval. In order to define the function properly at the endpoints, we use the average of the limits at the undefined endpoints. In this case, $f(-3\pi) = f(-\pi) = f(\pi) = \cdots = 0$.
Now we calculate the Fourier coeffieients, using (9.1.28)–(9.1.30):

$$a_0 = \frac{1}{\pi} \int_{-\pi}^{\pi} x\, dx = 0 \tag{9.1.36}$$

$$a_k = \frac{1}{\pi} \int_{-\pi}^{\pi} x \cos kx\, dx = 0, \qquad k > 0 \tag{9.1.37}$$

$$b_k = \frac{1}{\pi} \int_{-\pi}^{\pi} x \sin kx\, dx = -\frac{2}{k} \cos k\pi, \qquad k > 0 \tag{9.1.38}$$

Substituting in (9.1.31) and simplifying,

$$f(x) = x = 2\left[\sin x - \frac{\sin 2x}{2} + \frac{\sin 3x}{3} - \cdots \right] = 2 \sum_{k=1}^{\infty} \frac{\sin kx}{k} (-1)^{k-1} \tag{9.1.39}$$

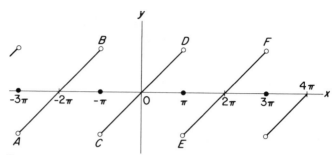

Fig. 9.1-4 $f(x)$ on $(-\pi, \pi)$ **represented as a periodic function.**

which completes the solution. Note that at $x = \pi$ or $-\pi$, $f(x) = x = 0$ because of the assumption we were forced to make. This is a reminder not to use the Fourier series result at a point of discontinuity; it may be an average.

Figure 9.1-5 shows the way the partial sums of this Fourier series behaves for various values of k. In particular, note how *the approximating curves produce alternate positive and negative errors of approximately equal size.*

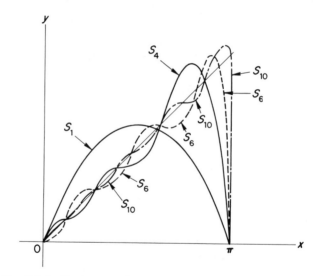

Fig. 9.1-5   Various partial sums for Illustrative Problem 9.1-1.

Why did all the cosine coefficients $a_0, \ldots, a_k$ vanish? The reason is that $f(x) = x$ is an *odd function*, i.e., $f(-x) = -f(x)$. Consider any odd function and investigate these coefficients:

$$a_0 = \frac{1}{\pi} \int_{-\pi}^{\pi} f(x) \, dx \tag{9.1.40}$$

$$a_0 = \frac{1}{\pi} \int_{-\pi}^{0} f(x) \, dx + \frac{1}{\pi} \int_{0}^{\pi} f(x) \, dx \tag{9.1.41}$$

In (9.1.41), let $y = -x$ in the first integral. Then $f(x) = -f(y)$. Therefore,

$$\frac{1}{\pi}\int_\pi^0 -f(y)(-dy) = \frac{1}{\pi}\int_\pi^0 f(y)\,dy = -\frac{1}{\pi}\int_0^\pi f(y)\,dy \qquad (9.1.42)$$

Since y is a dummy, we can replace it by x and substitute in (9.1.41):

$$a_0 = -\frac{1}{\pi}\int_0^\pi f(x)\,dx + \frac{1}{\pi}\int_0^\pi f(x)\,dx = 0 \qquad (9.1.43)$$

Similarly, in (9.1.29), $a_k = 0$ if $f(x)$ is odd, since $\cos kx$ is an *even* function, making the integrand odd. The reader is urged to make the substitutions, remembering that $\cos(-\theta) = \cos\theta$.

The practical significance of this result is that *if $f(x)$ is an odd function, there is no need to calculate a_0 or a_k.*

ILLUSTRATIVE PROBLEM 9.1-2   Expand

$$f(x) = \begin{cases} 0 & (-\pi, 0) \\ \pi & (0, \pi) \end{cases}$$

as a Fourier series on the interval $(-\pi, \pi)$.

Solution   (Figure 9.1-6.) Applying the formulas, we have

$$f(x) = 2\sum_{k=1}^\infty \frac{\sin kx}{k} \qquad (9.1.44)$$

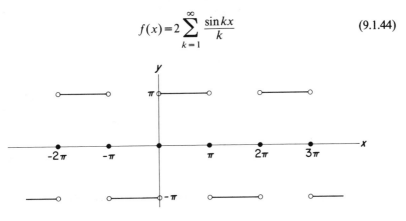

Fig. 9.1-6 Illustrative Problem 9.1-2.

Figure 9.1-7 shows how the partial sums approximate the square wave of Figure 9.1-6. Note, once again, how successive errors are approximately equal and opposite in sign.

ILLUSTRATIVE PROBLEM 9.1-3   Starting with

$$f(x) = \begin{cases} -1, & -2 < x < 0 \\ +1, & 0 < x < 2 \end{cases}$$

develop a Fourier series.

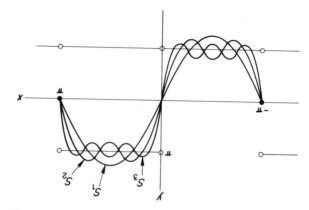

Fig. 9.1-7   Partial sums for Illustrative Problem 9.1-2.

Solution   (Figure   9.1-8.)   We   assume   a   period $=4$ and   use   the   definitions (9.1.32)–(9.1.35):

$$2L=4, \qquad L=2 \tag{9.1.45}$$

Since it is an odd function, as shown by the diagram, we know it is a sine series. Therefore, we need not calculate a_0 and a_k. We have

$$b_k = \frac{1}{L}\int_{-L}^{L} f(x)\sin\frac{k\pi x}{L}\,dx \tag{9.1.46}$$

$$= \frac{1}{2}\int_{-2}^{2} 1\sin\frac{k\pi x}{2}\,dx \tag{9.1.47}$$

$$= \frac{2}{k\pi}(1-\cos k\pi) \tag{9.1.48}$$

Observe that

$$b_k = \frac{2}{L}\int_{0}^{L} f(x)\sin\frac{k\pi x}{L}\,dx \tag{9.1.49}$$

would have given the same result because of the symmetry of the integrand in

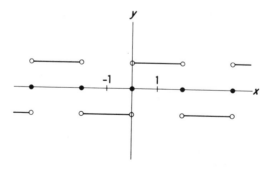

Fig. 9.1-8   Illustrative Problem 9.1-3.

(9.1.47). The computation of Fourier coefficients can often be simplified by taking advantage of symmetry.

Using (9.1.48), we have $b_1 = 4/\pi$, $b_2 = 0$, $b_3 = \frac{4}{3}\pi, \dots$, from which

$$f(x) = \sum_{k=1}^{\infty} \frac{4}{k\pi}(1 - \cos k\pi) \sin \frac{k\pi x}{2} \tag{9.1.50}$$

It happens that there are some interesting bonuses in the study of Fourier series, especially in the use of special cases. In (9.1.50), let $x = 1$. Then, from the original definition of $f(x)$ in the problem, when $x = 1$, $f(x) = 1$. Substituting into (9.1.50),

$$1 = \frac{4}{\pi}\left(1 - \frac{1}{3} + \frac{1}{5} - \frac{1}{7} + \cdots\right) \tag{9.1.51}$$

from which

$$\sum_{k=1}^{\infty} (-1)^{k-1} \frac{1}{2k-1} = \frac{\pi}{4} \tag{9.1.52}$$

a fascinating result. Unfortunately, one cannot calculate π from this relationship because the series is too slow. As you see, it converges with the speed of $1/k$.

9.1.7 Fourier Cosine Series

If $f(x)$ is an even function, then in (9.1.30) or (9.1.34), since $\sin kx$ is odd, the product $f(x)\sin kx$ will be odd. Therefore a Fourier series for *even* $f(x)$ has $b_k = 0$. This results in a Fourier cosine series.

ILLUSTRATIVE PROBLEM 9.1-4   Find a Fourier series for $f(x) = 1 - x$ for $x \in (0, 1)$, (a) as a cosine series and (b) as a sine series.

Solution   (a) We shall assume that $f(x)$ is an even function on the interval $(-1, 1)$ as shown in Figure 9.1-9. Using (9.1.32) and (9.1.33) with $2L = 2$, we have

$$a_0 = \int_{-1}^{1} (1 - x)\,dx = 1 \tag{9.1.53}$$

$$a_k = \int_{-1}^{1} (1 - x)\cos k\pi x\,dx = \begin{cases} 0 & \text{for } k \text{ even,} \\ 4/(n^2\pi^2) & \text{for } k \text{ odd.} \end{cases} \tag{9.1.54}$$

Therefore

$$f(x) = 1 - x = \frac{1}{2} + \frac{4}{\pi^2} \sum_{k=1}^{\infty} \frac{\cos(2k-1)\pi x}{(2k-1)^2} \tag{9.1.55}$$

Note that this function is continuous everywhere, so that there are no problems with discontinuities.

(b) If we assume $f(x)$ to be an odd function on $(-1, 1)$, we obtain the sketch in

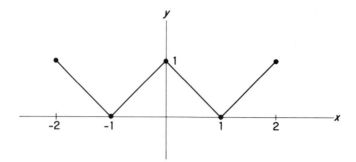

Fig. 9.1-9   Illustrative Problem 9.1-4(a).

Figure 9.1-10. Then

$$b_k = \int_{-1}^{1} (1-x) \sin k\pi x \, dx = \frac{2}{k\pi} \tag{9.1.56}$$

and

$$f(x) = 1 - x = \frac{2}{\pi} \sum_{k=1}^{\infty} \frac{\sin k\pi x}{k} \tag{9.1.57}$$

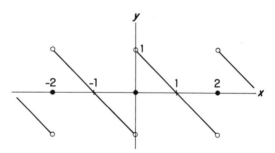

Fig. 9.1-10   Illustrative Problem 9.1-4(b).

Is there more than one Fourier series for a function? A theorem, the *uniqueness theorem*, which we shall not prove, states that on a specified interval there is one and only one Fourier series for a specified function. If the interval or the function is changed, another Fourier series may be obtained. In Illustrative Problem 9.1-4, we changed the function.

We should mention that the *converse* of the uniqueness theorem is not true, i.e., a Fourier series may be a valid representation of more than one function. This will happen if, for example, a finite number of points in $f(x)$ are changed. Figure 9.1-11 shows the function

$$f(x) = \begin{cases} 1-x, & (0, 0.5) \\ 2, & 0.5 \\ 1-x, & (0.5, 1) \\ -0.5, & 1 \\ 1-x, & (1, 2) \end{cases} \tag{9.1.58}$$

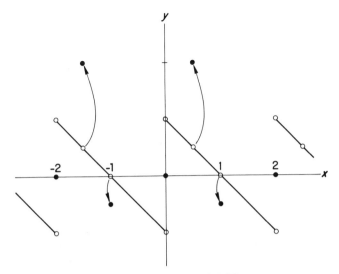

Fig. 9.1-11 Equation (9.1.58).

The integrals are precisely the same as in Illustrative Problem 9.1-4. Therefore the Fourier series will be precisely the same for Figure 9.1-10.

9.1.8 The Riemann Zeta-Function $\zeta(n)$

We remind you of the Riemann zeta-function which was used as a comparison series in the previous chapter:

$$\zeta(n) = \sum_{k=1}^{\infty} k^{-n} \tag{9.1.59}$$

For *even* values of n, this function can be expressed in closed form.

ILLUSTRATIVE PROBLEM 9.1-5 Find the Fourier series for $f(x) = x^2/4$ on the interval $(-\pi, \pi)$.

Solution Since $f(x)$ is obviously even, $b_k = 0$. By applying (9.1.28) and (9.1.29),

$$a_0 = \pi^2/6 \tag{9.1.60}$$

$$a_k = (-1)^k/k^2 \tag{9.1.61}$$

Then the Fourier series is

$$f(x) = \frac{\pi^2}{12} - \sum_{k=1}^{\infty} \frac{(-1)^{k-1}\cos kx}{k^2} \tag{9.1.62}$$

Now we look for bonuses. Let $x = 0$. Then

$$\frac{\pi^2}{12} = \frac{1}{1^2} - \frac{1}{2^2} + \frac{1}{3^2} - \frac{1}{4^2} + \cdots \tag{9.1.63}$$

Let $x = \pi$; then (9.1.62) becomes

$$\frac{\pi^2}{6} = \zeta(2) = \frac{1}{1^2} + \frac{1}{2^2} + \frac{1}{3^3} + \cdots \tag{9.1.64}$$

Adding (9.1.63) and (9.1.64), we obtain

$$\frac{\pi^2}{8} = \frac{1}{1^2} + \frac{1}{3^2} + \frac{1}{5^2} + \cdots \tag{9.1.65}$$

The following theorem, important in its own right, is sometimes useful in extending results of previous calculations, as we shall demonstrate in Illustrative Problem 9.1-6.

THEOREM: PARSEVAL'S EQUATION   For a valid Fourier series on $(-\pi, \pi)$,

$$\frac{1}{\pi} \int_{-\pi}^{\pi} [f(x)]^2 \, dx = \frac{a_0^2}{2} + \sum_{k=1}^{\infty} (a_k^2 + b_k^2) \tag{9.1.66}$$

Proof   We shall prove only the case where the series in the right member is uniformly convergent. We are given that

$$f(x) = \tfrac{1}{2} a_0 + \sum_{k=1}^{\infty} (a_k \cos kx + b_k \sin kx) \tag{9.1.67}$$

is valid. Multiplying both members by $f(x)$, then integrating, we get

$$\int_{-\pi}^{\pi} [f(x)]^2 \, dx = \frac{1}{2} \int_{-\pi}^{\pi} f(x) \, dx$$
$$+ \sum_{k=1}^{\infty} \left(a_k \int_{-\pi}^{\pi} f(x) \cos kx \, dx + b_k \int_{-\pi}^{\pi} f(x) \sin kx \, dx \right) \tag{9.1.68}$$

The term-by-term integration, amounting to an interchange of Σ and $\int$, is justified by the uniform convergence. Substituting for the integrals from (9.1.28)–(9.1.30), we have

$$\int_{-\pi}^{\pi} [f(x)]^2 \, dx = \tfrac{1}{2} \pi a_0 + \sum_{k=1}^{\infty} (a_k \pi a_k + b_k \pi b_k) \tag{9.1.69}$$

which completes the proof. $\square$

ILLUSTRATIVE PROBLEM 9.1-6   Apply (9.1.66) to Illustrative Problem 9.1-5.

Solution   We use

$$f(x) = x^2/4 \quad \Rightarrow \quad [f(x)]^2 = x^4/16 \tag{9.1.70}$$

$$a_0 = \pi^2/6 \quad \Rightarrow \quad \tfrac{1}{2}a_0^2 = \pi^4/72 \tag{9.1.71}$$

$$a_k = (-1)^k/k^2 \quad \Rightarrow \quad a_k^2 = 1/k^4 \tag{9.1.72}$$

$$b_k = 0 \tag{9.1.73}$$

Then, substituting in (9.1.66),

$$\frac{1}{\pi} \int_{-\pi}^{\pi} \frac{x^4}{16} dx = \frac{\pi^4}{72} + \sum_{k=1}^{\infty} \frac{1}{k^4} \tag{9.1.74}$$

from which, after some arithmetic,

$$\zeta(4) = \pi^4/90 \tag{9.1.75}$$

An easy corollary of (9.1.66) is the following theorem.

THEOREM: PARSEVAL'S EQUATION, EXTENDED   For two Fourier series $f(x)$ and $g(x)$,

$$\frac{1}{\pi} \int_{-\pi}^{\pi} f(x) g(x) dx = \tfrac{1}{2} a_0 a_0' + \sum_{k=1}^{\infty} (a_k a_k' + b_k b_k') \tag{9.1.76}$$

Proof   Use (9.1.66) on $f(x) + g(x)$, then on $f(x) - g(x)$; then subtract. □

9.1.9   Full Fourier Series

Where a function is neither odd nor even, both sines and cosines appear in the Fourier series.

ILLUSTRATIVE PROBLEM 9.1-7   Find a Fourier series on the interval $(-\pi, \pi)$ for the function

$$f(x) = \begin{cases} 0, & (-\pi, 0] \\ 1 - \cos 2x, & [0, \pi) \end{cases} \tag{9.1.77}$$

Solution   (Figure 9.1-12.)

$$a_0 = \frac{1}{\pi} \int_0^{\pi} (1 - \cos 2x) \, dx = 1 \tag{9.1.78}$$

$$a_k = \frac{1}{\pi} \int_0^{\pi} (1 - \cos 2x) \cos kx \, dx \tag{9.1.79}$$

$$= \frac{1}{\pi} \int_0^{\pi} \cos kx \, dx - \frac{1}{\pi} \int_0^{\pi} \cos 2x \cos kx \, dx \tag{9.1.80}$$

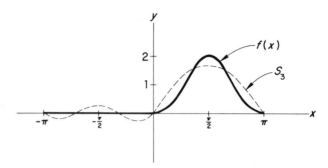

Fig. 9.1-12   Illustrative Problem 9.1-7.

The first integral in (9.1.80) obviously vanishes. In the second integral, we use

$$\cos 2x \cos kx = \tfrac{1}{2}\cos(k-2)x + \tfrac{1}{2}\cos(k+2)x \tag{9.1.81}$$

Then

$$\int_0^\pi \cos 2x \cos kx\, dx = \frac{1}{2}\left[\int_0^\pi \cos(k-2)x\, dx + \int_0^\pi \cos(k+2)x\, dx\right] \tag{9.1.82}$$

If $k=2$,

$$\int_0^\pi \cos 2x \cos kx\, dx = \frac{1}{2}\left[\int_0^\pi \cos 0\, dx + \int_0^\pi \cos 4x\, dx\right] \tag{9.1.83}$$

$$= \frac{\pi}{2} + 0 \tag{9.1.84}$$

Therefore $a_2 = -\tfrac{1}{2}$.

If $k \neq 2$, from (9.1.82),

$$\int_0^\pi \cos 2x \cos kx\, dx = \frac{1}{2}\left[\frac{\sin(k-2)x}{k-2} + \frac{\sin(k+2)x}{k+2}\right]\Bigg|_0^\pi \tag{9.1.85}$$

which is 0. Therefore, for $k \geqslant 1$, only a_2 does not vanish.
For the b_k,

$$b_k = \frac{1}{\pi}\int_0^\pi (1-\cos 2x)\sin kx\, dx \tag{9.1.86}$$

$$= \frac{1}{\pi}\left[\int_0^\pi \sin kx\, dx - \int_0^\pi \cos 2x \sin kx\, dx\right] \tag{9.1.87}$$

Using formula (19) in Appendix Table A.8, the second integral can be expanded, so that

$$b_k = \frac{1}{\pi}\left[\int_0^\pi \sin kx\, dx - \frac{1}{2}\int_0^\pi \sin(k+2)x\, dx - \frac{1}{2}\int_0^\pi \sin(k-2)x\, dx\right] \tag{9.1.88}$$

For *even* values of k, b_k vanishes. For *odd* values of k,

$$b_k = \frac{1}{\pi}\left[\frac{-\cos kx}{k} + \frac{1}{2}\frac{\cos(k+2)x}{k+2} + \frac{1}{2}\frac{\cos(k-2)x}{k-2}\right]_0^\pi \qquad (9.1.89)$$

For odd values of k, $\cos k\pi = -1$. Also $\cos 0 = 1$. Therefore,

$$b_k = \frac{1}{\pi}\left[\frac{2}{k} - \frac{1}{k+2} - \frac{1}{k-2}\right] = \frac{-8}{\pi(k+2)k(k-2)} \qquad (9.1.90)$$

Therefore the Fourier series is

$$\frac{1}{2} - \frac{1}{2}\cos 2x + \frac{8}{3\pi}\sin x - \frac{8}{15\pi}\sin 3x - \frac{8}{105\pi}\sin 5x - \cdots \qquad (9.1.91)$$

which can be written as

$$\frac{1}{2}(1-\cos 2x) - \frac{8}{\pi}\sum_{k=1}^{\infty}\frac{\sin(2k-1)x}{(2k-3)(2k-1)(2k+1)} \qquad (9.1.92)$$

In Figure 9.1-12, the dashed line represents the approximation using three terms. Once again, note the alternating maxima and minima in the approximating curve.

9.1.10 Errors

It is difficult, possible and unnecessary to estimate the error of a Fourier series; the true values are always known at the beginning, so that the *actual* errors can be computed for various truncations of the infinite series.

Consider the function of Illustrative Problem 9.1-7, as represented by (9.1.92). To find out about the errors, a test was made at intervals of 5° from $-180°$ to $+180°$. A part of the error table is shown in Figure 9.1-13, and the program is displayed in Figure 9.1-14.

The maximum absolute values of the actual errors for various numbers of terms are approximately as follows:

NUMBER OF TERMS	MAXIMUM ABSOLUTE ERROR
3	0.18
5	0.014
10	0.0015
15	0.00025
20	0.25
25	2.5
30	2.5

This, once again, emphasizes the fact that increasing the number of terms does *not* guarantee better results.

DEGREES	24 TERMS	25 TERMS	26 TERMS	27 TERMS	28 TERMS	29 TERMS	30 TERMS
0	0.0	0.0	0.0	0.0	0.0	0.0	0.0
5	1.31D-04	1.12D-04	9.14D-05	7.18D-05	5.32D-05	3.61D-05	2.09D-05
10	-3.85D-05	-1.05D-05	1.25D-05	2.92D-05	3.88D-05	4.17D-05	3.91D-05
15	-1.55D-05	-3.53D-05	-4.17D-05	-3.61D-05	-2.25D-05	-5.94D-06	8.87D-06
20	3.67D-05	3.67D-05	2.09D-05	-4.37D-07	-1.71D-05	-2.29D-05	-1.77D-05
25	-3.40D-05	-1.42D-05	1.03D-05	2.27D-05	1.77D-05	-2.21D-06	-1.17D-05
30	-1.76D-05	-1.04D-05	-2.27D-05	-1.18D-05	7.40D-06	1.60D-06	8.31D-06
35	2.03D-06	2.18D-05	1.14D-05	-1.02D-05	-1.51D-05	-1.10D-06	1.15D-05
40	-1.65D-05	-1.65D-05	7.75D-06	1.52D-05	-1.49D-06	-1.25D-05	-2.65D-06
45	2.10D-05	1.19D-06	-1.62D-05	-8.54D-07	1.27D-05	6.29D-07	-1.02D-05
50	-1.56D-05	-1.24D-05	8.16D-06	-1.22D-05	-2.60D-06	1.05D-05	-1.22D-06
55	4.01D-06	-1.58D-05	6.48D-06	8.37D-06	-1.02D-05	-1.22D-06	8.41D-06
60	8.02D-06	8.02D-06	-1.33D-05	5.52D-06	5.52D-06	-3.81D-07	3.95D-06
65	-1.53D-05	4.52D-06	6.66D-06	-1.11D-05	7.47D-06	-9.32D-06	-6.25D-06
70	1.51D-05	-1.29D-05	5.88D-06	2.11D-06	-7.50D-06	8.59D-06	-5.81D-06
75	-7.97D-06	-1.18D-05	-1.19D-05	9.04D-06	-4.56D-06	-1.24D-07	3.84D-06
80	-2.45D-06	-2.45D-06	5.95D-06	-7.99D-06	8.66D-06	-8.20D-06	6.89D-06
85	8.14D-06	-8.37D-06	5.72D-06	-3.44D-06	1.53D-06	3.93D-08	-1.30D-06
90	-1.49D-05	1.31D-05	-1.15D-05	1.02D-05	-9.05D-06	8.08D-06	-7.25D-06
95	8.14D-06	-8.37D-06	5.72D-06	-3.44D-06	1.53D-06	3.93D-08	-1.30D-06
100	-2.45D-06	-2.45D-06	5.95D-06	-7.99D-06	8.66D-06	-8.20D-06	6.89D-06
105	-7.97D-06	-1.18D-05	-1.19D-05	9.04D-06	-4.56D-06	-1.24D-07	3.84D-06
110	1.51D-05	-1.29D-05	5.88D-06	2.11D-06	-7.50D-06	8.59D-06	-5.81D-06
115	-1.53D-05	4.52D-06	6.66D-06	-1.11D-05	7.47D-06	-9.32D-06	-6.25D-06
120	8.02D-06	8.02D-06	-1.33D-05	5.52D-06	5.52D-06	-3.81D-07	3.95D-06
125	4.01D-06	-1.58D-05	6.48D-06	8.37D-06	-1.02D-05	-1.22D-06	8.41D-06
130	-1.56D-05	-1.24D-05	8.16D-06	-1.22D-05	-2.60D-06	1.05D-05	-1.22D-06
135	2.10D-05	1.19D-06	-1.62D-05	-8.54D-07	1.27D-05	6.29D-07	-1.02D-05
140	-1.65D-05	-1.65D-05	7.75D-06	1.52D-05	-1.49D-06	-1.25D-05	-2.65D-06
145	2.03D-06	2.18D-05	1.14D-05	-1.02D-05	-1.51D-05	-1.10D-06	1.15D-05
150	-1.76D-05	-1.04D-05	-2.27D-05	-1.18D-05	7.40D-06	1.60D-06	8.31D-06
155	-3.40D-05	-1.42D-05	1.03D-05	2.27D-05	1.77D-05	-2.21D-06	-1.17D-05
160	3.67D-05	3.67D-05	2.09D-05	-4.37D-07	-1.71D-05	-2.29D-05	-1.77D-05
165	-1.55D-05	-3.53D-05	-4.17D-05	-3.61D-05	-2.25D-05	-5.94D-06	8.87D-06
170	-3.85D-05	-1.05D-05	1.25D-05	2.92D-05	3.88D-05	4.17D-05	3.91D-05
175	1.31D-04	1.12D-04	9.14D-05	7.18D-05	5.32D-05	3.61D-05	2.09D-05

DEGREES	10 TERMS	11 TERMS	12 TERMS	13 TERMS	14 TERMS	15 TERMS	16 TERMS
0	0.0	0.0	0.0	0.0	0.0	0.0	0.0
5	-1.52D-03	-9.99D-04	-6.25D-04	-3.57D-04	-1.65D-04	-3.11D-05	6.09D-05
10	6.41D-04	7.33D-04	6.67D-04	5.29D-04	3.67D-04	2.13D-04	8.29D-05
15	8.31D-04	3.24D-04	-3.89D-05	-2.35D-04	-2.90D-04	-2.47D-04	-1.55D-04
20	-3.43D-04	-5.23D-04	-3.94D-04	-1.54D-04	5.36D-05	1.59D-04	1.59D-04
25	-6.23D-04	-1.47D-04	1.93D-04	2.65D-04	1.44D-04	-1.93D-05	-1.11D-04
30	1.34D-04	3.97D-04	2.09D-04	-6.84D-05	-1.74D-04	-9.18D-05	3.83D-05
35	5.10D-04	7.97D-05	-2.28D-04	-1.56D-04	5.41D-05	1.23D-04	3.14D-05
40	1.52D-05	-3.23D-04	-8.13D-05	1.59D-04	8.69D-05	-7.46D-05	-7.46D-05
45	-4.19D-04	-4.75D-04	2.18D-04	2.18D-05	-1.27D-04	-1.13D-05	8.06D-05
50	-1.26D-04	2.77D-04	-1.08D-05	-1.50D-04	4.86D-05	7.71D-05	-5.30D-05
55	3.31D-04	-2.94D-05	-1.86D-04	8.21D-05	6.37D-05	-8.49D-05	7.07D-06
60	2.07D-04	-2.48D-04	7.73D-05	7.73D-05	-1.05D-04	3.68D-05	3.68D-05
65	-2.40D-04	-1.80D-05	1.41D-04	-1.27D-04	4.54D-05	3.11D-05	-6.09D-05
70	-2.64D-04	2.30D-04	-1.23D-04	1.59D-05	5.25D-05	-7.32D-05	5.69D-05
75	1.46D-04	9.76D-06	-8.74D-05	1.09D-04	-9.49D-05	6.35D-05	-2.84D-05
80	2.97D-04	-2.20D-04	1.50D-04	-9.07D-05	4.48D-05	-1.13D-05	-1.13D-05
85	-4.89D-05	-3.10D-06	2.96D-05	-4.27D-05	4.69D-05	-4.72D-05	4.48D-05
90	-3.09D-04	2.17D-04	-1.58D-04	1.19D-04	-9.18D-05	7.22D-05	-5.79D-05
95	-4.89D-05	-3.10D-06	2.96D-05	-4.27D-05	4.69D-05	-4.72D-05	4.48D-05
100	-2.97D-04	-2.20D-04	1.50D-04	-9.07D-05	4.48D-05	-1.13D-05	-1.13D-05
105	1.46D-04	9.76D-06	-8.74D-05	1.09D-04	-9.49D-05	6.35D-05	-2.84D-05
110	-2.64D-04	2.30D-04	-1.23D-04	1.59D-05	5.25D-05	-7.32D-05	5.69D-05
115	-2.40D-04	-1.80D-05	1.41D-04	-1.27D-04	4.54D-05	3.11D-05	-6.09D-05
120	2.07D-04	-2.48D-04	7.73D-05	7.73D-05	-1.05D-04	3.68D-05	3.68D-05
125	3.31D-04	-2.94D-05	-1.86D-04	8.21D-05	6.37D-05	-8.49D-05	7.07D-06
130	-1.26D-04	2.77D-04	-1.08D-05	-1.50D-04	4.86D-05	7.71D-05	-5.30D-05
135	-4.19D-04	-4.75D-04	2.18D-04	2.18D-05	-1.27D-04	-1.13D-05	8.06D-05
140	1.52D-05	-3.23D-04	-8.13D-05	1.59D-04	8.69D-05	-7.46D-05	-7.46D-05
145	5.10D-04	7.97D-05	-2.28D-04	-1.56D-04	5.41D-05	1.23D-04	3.14D-05
150	1.34D-04	3.97D-04	2.09D-04	-6.84D-05	-1.74D-04	-9.18D-05	3.83D-05
155	-6.23D-04	-1.47D-04	1.93D-04	2.65D-04	1.44D-04	-1.93D-05	-1.11D-04
160	-3.43D-04	-5.23D-04	-3.94D-04	-1.54D-04	5.36D-05	1.59D-04	1.59D-04
165	8.31D-04	3.24D-04	-3.89D-05	-2.35D-04	-2.90D-04	-2.47D-04	-1.55D-04
170	6.41D-04	7.33D-04	6.67D-04	5.29D-04	3.67D-04	2.13D-04	8.29D-05
175	-1.52D-03	-9.99D-04	-6.25D-04	-3.57D-04	-1.65D-04	-3.11D-05	6.09D-05

Fig. 9.1-13   Part of the error-table output for Illustrative Problem 9.1-7.

```
       DOUBLE PRECISION S(72),E(14),FR(16),R,C,PI,SP,X,DEN,F(72),Y
C DEFINITIONS.
       C=0.01745329251994433D00
       PI=3.14159265358979D00
       SP = 8.D0/PI
C ALTERED TRUE VALUES.
       DO 1 I=1,36
       N=5*I-185
       X=N
       X=X*C*2.0D0
1          F(I)=0.5D0*(DCOS(X)-1.D0)
       DO 2 I=37,72
       N=5*I-185
       X=N
       X=X*C*2.0D0
2      F(I)=0.5D0*(1.D0-DCOS(X))
C INITIAL VALUES.
       DO 3 I=1,72
3      S(I)=0.0D0
       KT=3
C
4      N=-180
       I=1
       L=KT-1
C HEADING.
5      K1=KT+1
       K2=KT+2
       K3=KT+3
       K4=KT+4
       K5=KT+5
       K6=KT+6
       WRITE(3,100)KT,K1,K2,K3,K4,K5,K6
100    FORMAT(1H1,T5,'DEGREES',T14,7(3X,I2,1X,'TERMS'),/)
6      FR(L)=S(I)
       X=N
       X=X*C
       K6=KT+6
       DO 7 K=KT,K6
       J2=2*K-5
       J1=J2-2
       J3=J2+2
       DEN=J1*J2*J3
       Y=J2
       R=X*Y
7      FR(K)=FR(K-1)+DSIN(R)/DEN
C
       S(I)=FR(KT+6)
C
       K6=K+6
       DO 8 K=KT,K6
       J=K-2
8      E(J)=F(I)+SP*FR(K)
C PRINT A LINE.
       K1=KT-2
       K2=KT+4
       WRITE(3,101)N,(E(J),J=K1,K2)
101    FORMAT(1H ,T6,I4,T14,1PD11.2,6(D11.2))
C
       IF(I-36)9,10,11
9      I=I+1
       N=N+5
       GO TO 6
10     I=I+1
       N=N+5
       GO TO 5
11     IF(I-72)9,12,12
12     IF(KT-24)14,15,13
13     WRITE(3,102)
102    FORMAT(1H0,T25,'ERROR')
14      KT=KT+7
       GO TO 4
15     STOP
       END
```

Fig. 9.1-14 Program for Illustrative Problem 9.1-7.

PROBLEM SECTION 1

A. Sine series:
(1) Starting with

$$f(x) = \begin{cases} 0.1x, & 0 \leqslant x \leqslant 10 \\ 0.1(20-x), & 10 \leqslant x \leqslant 20 \end{cases}$$

write a Fourier series and show that

$$\sum_{k=1}^{\infty} \frac{1}{(2k-1)^2} = \frac{\pi^2}{8}$$

(2) Starting with $f(x) = 5$, $0 < x < 3$, write a Fourier series and show that

$$\sum_{k=1}^{\infty} \frac{1}{2k-1} \sin \frac{(2k-1)\pi}{3} = \frac{\pi}{4}$$

(3) Starting with $f(x) = 4x - x^2$, $0 < x < 4$, write a Fourier series and show that

$$\sum_{k=1}^{\infty} \frac{1}{(2k-1)^3} \sin \frac{(2k-1)\pi}{4} = \frac{3\pi^3}{128}$$

(4) Starting with

$$f(x) = \begin{cases} x, & -\pi \leqslant x < 0 \\ -x, & 0 \leqslant x < \pi \end{cases}$$

write a Fourier series and show that

$$\sum_{k=1}^{\infty} \frac{(-1)^k \sin kx}{k} = \frac{1}{2}\left[f(x) + \frac{\pi}{2} \right]$$

(5) Starting with $f(x) = x$, $0 \leqslant x < 2$, write a Fourier series and show that

$$\sum_{k=1}^{\infty} \frac{\sin k\pi x}{k} = \frac{\pi}{2}(1-x)$$

B. Cosine series:
(6) Starting with

$$f(x) = \begin{cases} 1, & 0 < x < \pi/2 \\ 0, & \pi/2 < x < \pi \end{cases}$$

write a Fourier series and show that

$$\sum_{k=1}^{\infty} (-1)^{k+1} \frac{\cos(2k-1)x}{2k-1} = \frac{\pi}{4}$$

(7) Starting with

$$f(x) = \begin{cases} -1, & 0 < x < 2 \\ +1, & 2 < x < 4 \end{cases}$$

write a Fourier series and show that

$$\sum_{k=1}^{\infty} (-1)^{k+1} \frac{\cos[(2k-1)\pi x/4]}{2k-1} = \frac{\pi}{4}$$

(8) Starting with

$$f(x) = \begin{cases} 0, & -\pi \leqslant x < -\pi/2 \\ \cos x, & -\pi/2 \leqslant x < \pi/2 \\ 0, & \pi/2 \leqslant x \leqslant \pi \end{cases}$$

write a Fourier series and show that for $x \in (-\pi/2, \pi/2)$.

$$\sum_{k=1}^{\infty} \frac{(-1)^k \cos 2kx}{1-k^2} = \frac{\pi}{2}\left(\cos x - \frac{1}{\pi}\right)$$

(9) Starting with $f(x) = x^2$, $-\pi \leqslant x < \pi$, write a Fourier series and show that

$$\sum_{k=1}^{\infty} \frac{(-1)^k \cos kx}{k^2} = \frac{x^2}{4} - \frac{\pi}{12}$$

(10) Starting with $f(x) = \sin 2x$ for $0 < x < \pi$, write a Fourier series and show that

$$\sum_{k=1}^{\infty} \frac{\cos(2k+1)x}{(2k-1)(2k+3)} = -\frac{\pi \sin 2x}{8}$$

C.

(11) Develop a Fourier sine series for $x(\pi - x)$ on $(0, \pi)$. Then, using Parseval's theorem, prove that $\zeta(6) = \pi^6/945$.

(12) Expand $f(x) = x^2$ as a Fourier series on $(0, 2\pi)$.

(13) If

$$f(x) = \begin{cases} -\pi, & (-\pi, 0) \\ x, & (0, \pi) \end{cases}$$

write a Fourier series for the interval $(-\pi, \pi)$, then show that

$$\sum_{k=1}^{\infty} \frac{1}{(2k-1)^2} = \frac{\pi^2}{8}$$

9.2 HARMONIC ANALYSIS

9.2.1 Background

In Section 9.1, we explored the representation of a function by an infinite Fourier series. The method used depended upon orthogonal functions defined by integrals. We know that, in fact, an infinite series must be truncated to permit programming for a computer. Instead, many engineering and commercial analyses prefer to use a finite Fourier series (FFS), usually with 6, 12, 24 or 60 terms. When this is done, the process is called *harmonic analysis*, by analogy with the sounds made by a musical instrument, in which the main wave is accompanied by harmonics or overtones.

9.2.2 Change of Scale

We shall confine ourselves to a standard interval: $[0, 2\pi]$. Any given interval can be translated and either stretched or shrunk to fit this arbitrary interval. For instance, suppose x is defined on $[t, u]$. Then to transform from x, defined on $[t, u]$, to θ, defined on $[0, 2\pi]$, we proceed as follows: Let

$$x = m\theta + b \tag{9.2.1}$$

Then

$$\begin{cases} t = 0m + b \\ u = 2\pi m + b \end{cases} \tag{9.2.2}$$

leading to the equation

$$x = \left(\frac{u - t}{2\pi} \right)\theta + t \tag{9.2.3}$$

ILLUSTRATIVE PROBLEM 9.2-1   If x is defined on $[-2.7, +4.3]$, what is the transformation equation relative to the standard interval $[0, 2\pi]$?

Solution

$$x = \frac{7}{2\pi}\theta - 2.7. \tag{9.2.4}$$

Suppose that the standard interval is divided into 12 subintervals. Then the corresponding values of θ will be $0°, 30°, 60°, \ldots, 330°$ or $0, \pi/6, 2\pi/6, 3\pi/6, \ldots, 11\pi/6$. If the standard interval is divided into 24 subintervals, the corresponding values of θ will be $0°$, $15°$, $30°, \ldots, 330°, 345°$ or $0, \pi/12, 2\pi/12, \ldots, 23\pi/12$.

In the first case, we can express the values as $\theta = r \cdot 2\pi/12 = r\pi/6$ where $r = 0, 1, \ldots, 11$; in the second case, $\theta = r \cdot 2\pi/24 = r\pi/12$, where $r =$

$0, 1, \ldots, 23$. In general, for $2N$ divisions, the values of θ are given by

$$\theta_r = \frac{r\pi}{N}, \qquad r = 0, 1, \ldots, 2N - 1 \tag{9.2.5}$$

It is important to note that when there is a change of scale, a function of x_r becomes a different function of r:

$$f(x_r) = g(r) \tag{9.2.6}$$

For example, in Illustrative Problem 9.2-1, suppose that $f(x_r) = 2x_r$. Then, substituting from (9.2.4),

$$f(x_r) = f\left(\frac{7\theta_r}{2\pi} - 2.7\right) = 2\left(\frac{7\theta_r}{2\pi} - 2.7\right) \tag{9.2.7}$$

Now, suppose there are $2N$ divisions. We use (9.2.5):

$$f(x_r) = f\left(\frac{7\theta_r}{2\pi} - 2.7\right) = f\left(\frac{7r\pi}{2\pi N} - 2.7\right) = f\left(\frac{7}{2N}r - 2.7\right) = g(r) \tag{9.2.8}$$

which, as you can see, is a function of r. From (9.2.7) and (9.2.5),

$$g(r) = 2\left(\frac{7r}{2N} - 2.7\right) \tag{9.2.9}$$

9.4.3 Orthogonal Functions: Even Finite Fourier Series

Suppose that the interval $[0, 2\pi]$ is divided into $2N$ parts:

$$\theta_r = \frac{r\pi}{N}, \qquad r = 0, 1, \ldots, 2N - 1 \tag{9.2.10}$$

Consider the product

$$S(r) = \sum_{r=0}^{2N-1} \cos m\theta_r \cos n\theta_r = \sum_{r=0}^{2N-1} \cos m\frac{r\pi}{N} \cos n\frac{r\pi}{N} \tag{9.2.11}$$

We convert the product into a sum:

$$S(r) = \frac{1}{2} \sum_{r=0}^{2N-1} \left[\cos \frac{r\pi}{N}(m+n) + \cos \frac{r\pi}{N}(m-n) \right] \tag{9.2.12}$$

Assuming $m \neq n$,

$$S_1(r) = \frac{1}{2} \sum_{r=0}^{2N-1} \cos\frac{r\pi}{N}(m+n)$$

$$= \frac{1}{2}\frac{\cos\frac{1}{2}\left[(\pi/N)(m+n)(2N-1)\right]\sin\left[(\pi/2N)(m+n)2N\right]}{\sin\left[(\pi/2N)(m+n)\right]}$$

(9.2.13)

$$S_2(r) = \frac{1}{2} \sum_{r=0}^{2N-1} \cos\frac{r\pi}{N}(m-n)$$

$$= \frac{1}{2}\frac{\cos\frac{1}{2}\left[(\pi/N)(m-n)(2N-1)\right]\sin\left[(\pi/2N)(m-n)2N\right]}{\sin\left[(\pi/2N)(m-n)\right]}$$

(9.2.14)

The two parts of $S(r)$, S_1 and S_2, both vanish because the sine of an integral multiple of π is 0. Therefore, in (9.2.11), $S=0$ whenever $m \neq n$.

In (9.2.11), there are two functions: $\cos(mr\pi/N)$ and $\cos(nr\pi/N)$. These are functions of r, but each has a *parameter*, in one case m and in the other n. We shall indicate this by writing, for example,

$$f(r,m) = \cos\frac{mr\pi}{N}$$ (9.2.15)

$$g(r,n) = \cos\frac{nr\pi}{N}$$ (9.2.16)

Since the sum of the products of these two functions is 0 on the interval $[0, 2\pi]$ for $m \neq n$, we say that they are orthogonal relative to the weighting function $w(r) = 1$ on the interval. More generally, we have the following definition.

ORTHOGONAL FUNCTIONS BY SUMS A set of functions $f(r,m)$ and $g(r,n)$ are orthogonal relative to a weighting function $w(r) \geq 0$ on the interval $r \in [a,b]$ iff

$$\sum_{r=a}^{b-1} w(r) f(r,m) g(r,n) = 0$$ (9.2.17)

for $m \neq n$.

By this definition, $\cos(mr\pi/N)$ and $\cos(nr\pi/N)$ are orthogonal on $[0, 2\pi]$ relative to the weighting function $w(r) = 1$.

We shall need to know the value of $S(r)$ in (9.2.11) when $m = n$. In that case, (9.2.11) can be written as

$$S(r) = \sum_{r=0}^{2N-1} \cos^2 \frac{mr\pi}{N} \tag{9.2.18}$$

$$= \frac{1}{2} \sum_{r=0}^{2N-1} \left(1 + \cos \frac{2mr\pi}{N}\right) \tag{9.2.19}$$

$$= \frac{1}{2} \sum_{r=0}^{2N-1} 1 + \frac{1}{2} \sum_{r=0}^{2N-1} \cos \frac{2mr\pi}{N} \tag{9.2.20}$$

There are now three cases:

    I. $m = n = 0$
    II. $m = n = N$
    III. $m \neq 0$ or N

In case I,

$$S(r) = N + N = 2N \qquad \text{for} \quad m = n = 0 \tag{9.2.21}$$

In case II,

$$S(r) = N + \frac{1}{2} \sum_{r=0}^{2N-1} \cos 2r\pi = N + N = 2N \tag{9.2.22}$$

In case III, we use formula (4) of Appendix Table A.6:

$$S(r) = N + \frac{1}{2} \frac{\cos\frac{1}{2}\left[2\pi m(2N-1)/N\right] \sin\left[\pi m 2N/N\right]}{\sin\frac{1}{2}\left[2\pi m/N\right]} \tag{9.2.23}$$

$$S(r) = N + 0 = N \qquad \text{for} \quad m = n \neq 0, N \tag{9.2.24}$$

We leave to the reader the very similar proofs for the orthogonal sets $\{\cos(mr\pi/N), \sin(nr\pi/N)\}$ and $\{\sin(mr\pi/N), \sin(nr\pi/N)\}$. The results are given in the following theorem.

HARMONIC-ANALYSIS THEOREM FOR $2N$ POINTS   If the interval $[0, 2\pi]$ is divided into $2N$ subintervals, with

$$\theta_r = \frac{r\pi}{N}, \qquad r = 0, 1, \ldots, (2N-1) \tag{9.2.25}$$

then the following table of products applies:

$$
\begin{array}{cccc}
 & m \neq n & m = n \neq 0, N & m = n = 0, N \\
\sum_{r=0}^{2N-1} \cos\frac{m\pi r}{N}\cos\frac{n\pi r}{N} = & 0 & N & 2N & (9.2.26) \\
\sum_{r=0}^{2N-1} \cos\frac{mr\pi}{N}\sin\frac{nr\pi}{N} = & 0 & 0 & 0 & (9.2.27) \\
\sum_{r=0}^{2N-1} \sin\frac{mr\pi}{N}\sin\frac{nr\pi}{N} = & 0 & N & 0 & (9.2.28)
\end{array}
$$

9.2.4 Fourier Coefficients: $2N$ Points

We shall start with the assumption that

$$
F(r) = \frac{1}{2}A_0 + \sum_{k=1}^{N-1}\left[A_k\cos\frac{k r\pi}{N} + B_k\sin\frac{k r\pi}{N}\right] + \frac{1}{2}A_n\cos r\pi \quad (9.2.29)
$$

We are assuming that $f(r)$, the given function, can be fitted precisely at $2N$ points with the function $F(r)$ so defined. An argument almost precisely like that in Section 9.1 (for the continuous case) can be made, using

$$
E = \sum_{r=0}^{2N-1}\left[f(r) - F(r)\right]^2 = \text{minimum} \quad (9.2.30)
$$

This leads to formulas for the Fourier coefficients $A_0, A_1, \ldots, A_N, B_1, \ldots, B_{N-1}$. (Note the use of capital letters for the *finite* case.) Instead, we shall use another derivation, not because it is easier, but because it shows the importance of the condition of orthogonality.

We multiply (9.2.29) by $\cos(mr\pi/N)$ and sum:

$$
\sum_{r=0}^{2N-1} F(r)\cos\frac{mr\pi}{N} = \frac{1}{2}A_0\sum_{r=0}^{2N-1}\cos\frac{mr\pi}{N}
$$

$$
+ \sum_{k=1}^{N-1}\sum_{r=0}^{2N-1}\left(A_k\cos\frac{k r\pi}{N}\cos\frac{mr\pi}{N} + B_k\sin\frac{k r\pi}{N}\cos\frac{mr\pi}{N}\right)
$$

$$
+ \frac{1}{2}A_N\sum_{r=0}^{2N-1}\cos r\pi\cos\frac{mr\pi}{N} \quad (9.2.31)
$$

Using the orthogonality relations, together with formulas from Table A.6, we have

Case I: $m = 0$.

$$
\sum_{r=0}^{2N-1} F(r) = \frac{1}{2}A_0 2N + 0 + 0 = NA_0 \quad (9.2.32)
$$

Case II: $m = N$.

$$\sum_{r=0}^{2N-1} F(r)\cos\pi r = 0 + 0 + \tfrac{1}{2}A_N 2N = NA_N \qquad (9.2.33)$$

Case III: $m \neq 0, N$. The only contribution in (9.2.31) is when $k = m$. Then

$$\sum_{r=0}^{2N-1} F(r)\cos\frac{kr\pi}{N} = 0 + \sum_{r=0}^{2N-1} A_k \cos^2\frac{kr\pi}{N} + 0 \qquad (9.2.34)$$

$$\sum_{r=0}^{2N-1} F(r)\cos\frac{kr\pi}{N} = NA_k \qquad (9.2.35)$$

This establishes the following theorem.

THEOREM: FINITE FOURIER SERIES (FOURIER COEFFICIENTS FOR $2N$ POINTS) If a finite Fourier series is written for $2N$ equally spaced points on $[0, 2\pi]$, then

$$\theta_r = \frac{r\pi}{N}, \qquad r = 0, 1, \ldots, (2N-1) \qquad (9.2.36)$$

$$F(r) = \tfrac{1}{2}A_0 + \sum_{k=1}^{N-1}\left(A_k\cos\frac{kr\pi}{N} + B_k\sin\frac{kr\pi}{N}\right) + \tfrac{1}{2}A_N\cos r\pi \qquad (9.2.37)$$

$$A_k = \frac{1}{N}\sum_{r=0}^{2N-1} F(r)\cos\frac{kr\pi}{N} \qquad k = 0, 1, \ldots, N \qquad (9.2.38)$$

$$B_k = \frac{1}{N}\sum_{r=0}^{2N-1} F(r)\sin\frac{kr\pi}{N} \qquad k = 1, 2, \ldots, N-1 \qquad (9.2.39)$$

Alternatively, in terms of θ_r,

$$G(\theta_r) = \tfrac{1}{2}A_0 + \sum_{k=1}^{N-1}(A_k\cos\theta_r + B_k\sin k\theta_r) + \tfrac{1}{2}A_N\cos N\theta_r \qquad (9.2.40)$$

$$A_k = \frac{1}{N}\sum_{r=0}^{2N-1} G(\theta_r)\cos r\theta_r \qquad (9.2.41)$$

$$B_k = \frac{1}{N}\sum_{r=0}^{2N-1} G(\theta_r)\sin r\theta_r \qquad (9.2.42)$$

ILLUSTRATIVE PROBLEM 9.2-2   We are given the following table of values, with θ in degrees. [It is customary to use the symbol F for both (9.2.37) and (9.2.40), although they are not the same function.]

r	0	1	2	3	4	5	6	7	8	9	10	11
θ	0	30	60	90	120	150	180	210	240	270	300	330
f	9.3	15.0	17.4	23.0	37.0	31.0	15.3	4.0	−8.0	−13.2	−14.2	−6.0

Derive a finite Fourier series which goes precisely through the given points.

Solution   Appendix Tables A.14, A.15, and A.16 provide the sines and cosines of the angles and their multiples. We note that there are 12 points and 12 intervals. Therefore $2N=12$ and $N=6$. Table 9.2-1 gives the appropriate values of the cosines. Note that we are using $F(r)=f(r)$ at the given points. This guarantees that there is no error except roundoff at the given points.

Table 9.2-1

Cosines for $N=6$*

r	θ_r	$F(r)$	$\cos k\theta_r$					
	(deg)		$k=1$	2	3	4	5	6
0	0	9.3	1	1	1	1	1	1
1	30	15.0	C	$\frac{1}{2}$	0	$-\frac{1}{2}$	$-C$	-1
2	60	17.4	$\frac{1}{2}$	$-\frac{1}{2}$	-1	$-\frac{1}{2}$	$\frac{1}{2}$	1
3	90	23.0	0	-1	0	1	0	-1
4	120	37.0	$-\frac{1}{2}$	$-\frac{1}{2}$	1	$-\frac{1}{2}$	$-\frac{1}{2}$	1
5	150	31.0	$-C$	$\frac{1}{2}$	0	$-\frac{1}{2}$	C	-1
6	180	15.3	-1	1	-1	1	-1	1
7	210	4.0	$-C$	$\frac{1}{2}$	0	$-\frac{1}{2}$	C	-1
8	240	−8.0	$-\frac{1}{2}$	$-\frac{1}{2}$	1	$-\frac{1}{2}$	$-\frac{1}{2}$	1
9	270	− 13.2	0	-1	0	1	0	-1
10	300	− 14.2	$\frac{1}{2}$	$-\frac{1}{2}$	-1	$-\frac{1}{2}$	$\frac{1}{2}$	1
11	330	−6.0	C	$\frac{1}{2}$	0	$-\frac{1}{2}$	$-C$	-1

*$C=\frac{1}{2}\sqrt{3}$ as in Table A.14.

We compute A_3 as an example of the brute-force method: $(9.3)(1)+(15.0)(0)+(17.4)(-1)+\cdots+(-14.2)(-1)+(-6.0)(0)$. The results are:

$$A_0=\tfrac{1}{6}(110.6)=18.4333\ldots$$

$$A_1=-6.90277\ 6750$$

$$A_2=3.45$$

$$A_3=3.30$$

$$A_4=-0.61666\ 6667$$

$$A_5=0.60277\ 67498$$

$$A_6=0.5$$

Table 9.2-2
Sines for $N = 6$*

r	θ_r	$F(r)$	$\sin k\theta_r$				
	(deg)		$k=1$	2	3	4	5
0	0	9.3	0	0	0	0	0
1	30	15.0	$\frac{1}{2}$	C	1	C	$\frac{1}{2}$
2	60	17.4	C	C	0	$-C$	$-C$
3	90	23.0	1	0	-1	0	1
4	120	37.0	C	$-C$	0	C	$-C$
5	150	31.0	$\frac{1}{2}$	$-C$	1	$-C$	$\frac{1}{2}$
6	180	15.3	0	0	0	0	0
7	210	4.0	$-\frac{1}{2}$	C	-1	C	$-\frac{1}{2}$
8	240	-8.0	$-C$	C	0	$-C$	C
9	270	-13.2	-1	0	1	0	-1
10	300	-14.2	$-C$	$-C$	0	C	C
11	330	-6.0	$-\frac{1}{2}$	$-C$	-1	$-C$	$-\frac{1}{2}$

*$C = \frac{1}{2}\sqrt{3}$ as in Table A.14.

Table 9.2-2 gives the appropriate values for the sines. From these values, by brute force,

$$B_1 = 21.08959099$$
$$B_2 = -2.800148807$$
$$B_3 = 1.966\ldots$$
$$B_4 = 1.068097998$$
$$B_5 = -1.022924324$$

Rounded to one more significant figure than the given data, the required series, in terms of θ, is as follows:

$$F(\theta) = 9.217 - 6.903\cos\theta + 3.450\cos 2\theta + 3.300\cos 3\theta - 0.6167\cos 4\theta$$
$$+ 0.6028\cos 5\theta + 0.2500\cos 6\theta + 21.09\sin\theta$$
$$- 2.800\sin 2\theta + 1.967\sin 3\theta$$
$$+ 1.068\sin 4\theta - 1.023\sin 5\theta$$

To check, we may test the given values of θ to see whether the Fourier series fits at the given points. It does.

9.2.5 Short-Cut Methods

Since 1903, many mathematicians, looking with jaundiced eyes at the brute-force procedure of Illustrative Problem 9.2-2, have devised schemes for reducing the number of steps. The method to be demonstrated can be used for any given number of points, but is easiest for 6, 12, 24 or 60 points. We shall display the algorithms for 6, 12 and 24 points.

9.2.6 The Six-Point Algorithm

It would probably not be very useful to write a Fourier series for six points, but the development shows, on a small scale, the general plan behind the formation of such an algorithm for any even number of points. In each case, you will see that the range of $F(r)$ is repeatedly divided into two equal parts until a satisfactory algorithm appears. This makes use of the symmetries of the sine and cosine curves.

For six points, $2N=6$ and $N=3$. We start with the following equation, which makes use of (9.2.38):

$$A_k = \frac{1}{3} \sum_{r=0}^{5} f(r) \cos \frac{k\pi r}{3} \qquad k=0,1,2,3 \qquad (9.2.43)$$

Multiplying by 3 and writing the summation as two summations, we have

$$3A_k = \sum_{r=0}^{2} F(r) \cos \frac{k\pi r}{3} + \sum_{r=3}^{5} F(r) \cos \frac{k\pi r}{3} \qquad (9.2.44)$$

In the second summation, we substitute

$$r' = 6 - r \qquad (9.2.45)$$

obtaining

$$\sum_{r=3}^{5} F(r) \cos \frac{k\pi r}{3} = \sum_{r'=1}^{3} F(6-r') \cos\left(\frac{k\pi 6}{3} - \frac{k\pi r'}{3}\right) \qquad (9.2.46)$$

But

$$\cos\left(\frac{k\pi 6}{3} - \frac{k\pi r'}{3}\right) = \cos 2\pi k \cos \frac{k\pi r'}{3} + \sin 2\pi k \sin \frac{k\pi r'}{3} \qquad (9.2.47)$$

$$= \cos \frac{k\pi r'}{3} \qquad (9.2.48)$$

We replace r' by r, since it is only a dummy, and substitute in (9.2.46) and (9.2.44):

$$3A_k = \sum_{r=0}^{2} F(r) \cos \frac{k\pi r}{3} + \sum_{r=1}^{3} F(6-r) \cos \frac{k\pi r}{3} \qquad (9.2.49)$$

$$3A_k = F(0) + \sum_{r=1}^{2} [F(r) + F(6-r)] \cos \frac{k\pi r}{3} + (-1)^k F(3) \qquad (9.2.50)$$

In (9.2.50), we used $\cos(3k\pi/3) = (-1)^k$. We leave as an exercise the similar derivation of

$$3B_k = \sum_{r=1}^{2} [F(r) - F(6-r)] \sin \frac{k\pi r}{3} \qquad (9.2.51)$$

We now set up a table to begin the algorithm:

$$
\begin{array}{ccccc}
 & F(0) & F(1) & F(2) & F(3) \\
 & & F(5) & F(4) & \\
\hline
\text{Sum} & s_0 & s_1 & s_2 & s_3 \\
\text{Diff} & & t_1 & t_2 &
\end{array}
\tag{9.2.52}
$$

Then (9.2.50) and (9.2.51) become

$$
3A_k = s_0 + \sum_{r=1}^{2} s_r \cos \frac{k\pi r}{3} + (-1)^k s_3
\tag{9.2.53}
$$

$$
3B_k = \sum_{r=1}^{2} t_r \sin \frac{k\pi r}{3}
\tag{9.2.54}
$$

Let us tabulate (9.2.53) for $k = 0, 1, 2, 3$:

$$
\begin{aligned}
3A_0 &= s_0 + s_1 \cos 0 + s_2 \cos 0 + (-1)^0 s_3 \\
3A_1 &= s_0 + s_1 \cos \frac{\pi}{3} + s_2 \cos \frac{2\pi}{3} + (-1)^1 s_3 \\
3A_2 &= s_0 + s_1 \cos \frac{2\pi}{3} + s_2 \cos \frac{4\pi}{3} + (-1)^2 s_3 \\
3A_3 &= s_0 + s_1 \cos \frac{3\pi}{3} + s_2 \cos \frac{6\pi}{3} + (-1)^3 s_3
\end{aligned}
\tag{9.2.55}
$$

Referring to Appendix Tables A.15 and A.16, this can be rewritten as

$$
\begin{aligned}
3A_0 &= s_0 + s_1 + s_2 + s_3 \\
3A_1 &= s_0 + \tfrac{1}{2}(s_1 - s_2) - s_3 \\
3A_2 &= s_0 - \tfrac{1}{2}(s_1 + s_2) + s_3 \\
3A_3 &= s_0 - (s_1 - s_2) - s_3
\end{aligned}
\tag{9.2.56}
$$

This suggests the following table:

$$
\begin{array}{ccc}
 & s_0 & s_1 \\
 & s_3 & s_2 \\
\hline
\text{Sum} & u_0 & u_1 \\
\text{Diff} & v_0 & v_1
\end{array}
\tag{9.2.57}
$$

Substituting in (9.2.56), we have

$$3A_0 = u_0 + u_1$$

$$3A_1 = v_0 + \tfrac{1}{2}v_1$$

$$3A_2 = u_0 - \tfrac{1}{2}u_1$$

$$3A_3 = v_0 - v_1 \tag{9.2.58}$$

We now turn our attention to (9.2.54):

$$3B_1 = t_1 \sin\frac{\pi}{3} + t_2 \sin\frac{2\pi}{3}$$

$$3B_2 = t_1 \sin\frac{2\pi}{3} + t_2 \sin\frac{4\pi}{3} \tag{9.2.59}$$

Using the symbol C to represent $\sqrt{3}/2$, we have

$$3B_1 = C(t_1 + t_2)$$

$$3B_2 = C(t_1 - t_2) \tag{9.2.60}$$

This suggests a continuation of the algorithm to

$$
\begin{array}{ll}
 & t_1 \\
 & t_2 \\
\hline
\text{Sum} & p \\
\text{Diff} & q
\end{array}
\tag{9.2.61}
$$

Substituting in (9.2.60), we have

$$3B_1 = Cp$$

$$3B_2 = Cq \tag{9.2.62}$$

The completed algorithm consists of the sequence: (9.2.52), (9.2.57), (9.2.61), and (9.2.62).

ILLUSTRATIVE PROBLEM 9.2-3   Write a finite Fourier series for the following data (θ in degrees):

θ	0	60	120	180	240	300
f	1.0	1.4	1.9	1.7	1.5	1.2

Then find $f(27°)$.

Solution

$$
\begin{array}{lcccc}
F(0\!\rightarrow\!3) & 1.0 & 1.4 & 1.9 & 1.7 \\
F(5,4) & & 1.2 & 1.5 & \\
\hline
s(0\!\rightarrow\!3) & 1.0 & 2.6 & 3.4 & 1.7 \\
t(1,2) & & 0.2 & 0.4 &
\end{array}
\tag{9.2.63}
$$

$$
\begin{array}{lcc|lc}
s(0,1) & 1.0 & 2.6 & t_1 & 0.2 \\
s(3,2) & 1.7 & 3.4 & t_2 & 0.4 \\
\hline
u(0,1) & 2.7 & 6.0 & p & 0.6 \\
v(0,1) & -0.7 & -0.8 & q & -0.2
\end{array}
\tag{9.2.64}
$$

$$
\begin{aligned}
3A_0 &= 2.7 + 6.0 = 8.7 \\
3A_1 &= -0.7 + \tfrac{1}{2}(-0.8) = -1.1 \\
3A_2 &= 2.7 - \tfrac{1}{2}(6.0) = -0.3 \\
3A_3 &= -0.7 - (-0.8) = 0.1 \\
3B_1 &= 0.6C = 0.5196 \\
3B_2 &= -0.2C = -0.1732
\end{aligned}
\tag{9.2.65}
$$

Using (9.2.40), the finite Fourier series, rounded, is

$$
\begin{aligned}
F(\theta) = 1.450 &+ (-0.3667\cos\theta + 0.5196\sin\theta) \\
&+ (-0.1000\cos 2\theta - 0.05773\sin 2\theta) + 0.01667\cos 3\theta
\end{aligned}
\tag{9.2.66}
$$

Substitute $\theta = 27°$; the result is approximately 1.256. This is $F(27°)$, our estimate of $f(27°)$, the true value.

9.2.7 Twelve-Point Algorithm

Here $2N = 12$. We start with

$$
6A_k = \sum_{r=0}^{11} F(r)\cos\frac{rk\pi}{6}
\tag{9.2.67}
$$

$$
6B_k = \sum_{r=0}^{11} F(r)\sin\frac{rk\pi}{6}
\tag{9.2.68}
$$

Proceeding as in Section 9.2.6 by separating each summation into two summations, we finally obtain

$$
6A_k = F(0) + \sum_{r=1}^{5} \left[F(r) + F(12-r) \right]\cos\frac{rk\pi}{6} + (-1)^k F(6)
\tag{9.2.69}
$$

$$
6B_k = \sum_{r=1}^{5} \left[F(r) - F(12-r) \right]\sin\frac{rk\pi}{6}
\tag{9.2.70}
$$

The beginning of the algorithm is, then,

	$F(0)$	$F(1)$	$F(2)$	$F(3)$	$F(4)$	$F(5)$	$F(6)$
		$F(11)$	$F(10)$	$F(9)$	$F(8)$	$F(7)$	
Sum	s_0	s_1	s_2	s_3	s_4	s_5	s_6
Diff		t_1	t_2	t_3	t_4	t_5	

$$(9.2.71)$$

We rewrite (9.2.69) and (9.2.70) as

$$6A_k = s_0 + \sum_{r=1}^{5} s_r \cos\frac{rk\pi}{6} + (-1)^k s_6 \tag{9.2.72}$$

$$6B_k = \sum_{r=1}^{5} t_r \sin\frac{rk\pi}{6} \tag{9.2.73}$$

The summations may be separated in a number of ways. Since each has an odd number of terms, it is not obvious whether to choose r as $\{1,2,3\}$ and $\{4,5\}$, or as $\{1,2\}$ and $\{3,4,5\}$. However, (9.2.72) can be written as

$$6A_k = s_0 + \sum_{r=1}^{6} s_r \cos\frac{rk\pi}{6} \tag{9.2.74}$$

and so we choose $\{1,2,3\}$ and $\{4,5,6\}$ for both. Then

$$\sum_{r=1}^{6} s_r \cos\frac{rk\pi}{6} = \sum_{r=1}^{3} s_r \cos\frac{rk\pi}{6} + \sum_{r=4}^{6} s_r \cos\frac{rk\pi}{6} \tag{9.2.75}$$

$$\sum_{r=1}^{5} t_r \sin\frac{rk\pi}{6} = \sum_{r=1}^{3} t_r \sin\frac{rk\pi}{6} + \sum_{r=4}^{5} t_r \sin\frac{rk\pi}{6} \tag{9.2.76}$$

In the last summations of these two equations, we now substitute $r' = 6 - r$. After simplifying and removing the primes from the dummies, we have

$$\sum_{r=4}^{6} s_r \cos\frac{rk\pi}{6} = \sum_{r=0}^{2} (-1)^k s_{6-r} \cos\frac{rk\pi}{6} \tag{9.2.77}$$

$$\sum_{r=4}^{5} t_r \sin\frac{rk\pi}{6} = -\sum_{r=1}^{2} (-1)^k t_{6-r} \sin\frac{rk\pi}{6} \tag{9.2.78}$$

We shall deal with A_k first. Substituting in (9.2.75),

$$\sum_{r=1}^{6} s_r \cos\frac{rk\pi}{6} = \sum_{r=1}^{3} s_r \cos\frac{rk\pi}{6} + \sum_{r=0}^{2}(-1)^k s_{6-r} \cos\frac{rk\pi}{6} \qquad (9.2.79)$$

$$= \sum_{r=1}^{2}\left[s_r + (-1)^k s_{6-r}\right]\cos\frac{rk\pi}{6} + s_3\cos\frac{k\pi}{2} + (-1)^k s_6 \quad (9.2.80)$$

Substituting into (9.2.74),

$$6A_k = s_0 + \sum_{r=1}^{2}\left[s_r + (-1)^k s_{6-r}\right]\cos\frac{rk\pi}{6} + s_3\cos\frac{k\pi}{2} + (-1)^k s_6 \quad (9.2.81)$$

which suggests a continuation of the algorithm as follows:

$$
\begin{array}{c|cccc}
 & s_0 & s_1 & s_2 & s_3 \\
 & s_6 & s_5 & s_4 & \\
\hline
\text{Sum} & u_0 & u_1 & u_2 & u_3 \\
\text{Diff} & v_0 & v_1 & v_2 &
\end{array}
\qquad (9.2.82)
$$

Turning our attention to B_k, we use (9.2.78) to deal with (9.2.76). Then

$$6B_k = \sum_{r=1}^{5} t_r \sin\frac{rk\pi}{6} = \sum_{r=1}^{3} t_r \sin\frac{rk\pi}{6} - \sum_{r=1}^{2}(-1)^k t_{6-r}\sin\frac{rk\pi}{6} \quad (9.2.83)$$

$$= \sum_{r=1}^{2}\left[t_r - (-1)^k t_{6-r}\right]\sin\frac{rk\pi}{6} + t_3\sin\frac{k\pi}{2} \qquad (9.2.84)$$

The continuation of the algorithm therefore includes

$$
\begin{array}{c|ccc}
 & t_1 & t_2 & t_3 \\
 & t_5 & t_4 & \\
\hline
\text{Sum} & p_1 & p_2 & p_3 \\
\text{Diff} & q_1 & q_2 &
\end{array}
\qquad (9.2.85)
$$

Now, let us consider the calculation of A_k using (9.2.81) and (9.2.82). In the following, the first few are displayed with $C = \sqrt{3}/2$.

$$6A_0 = s_0 + s_1 + s_2 + s_3 + s_4 + s_5 + s_6 = u_0 + u_1 + u_2 + u_3$$

$$6A_1 = s_0 + (s_1 - s_5)C + (s_2 - s_4)/2 - s_6 = v_0 + Cv_1 + \tfrac{1}{2}v_2$$

$$6A_2 = s_0 + (s_1 + s_5)/2 + (s_2 + s_4)\left(-\tfrac{1}{2}\right) - s_3 + s_6 \qquad (9.2.86)$$

$$= u_0 + \tfrac{1}{2}u_1 - \tfrac{1}{2}u_2 - u_3$$

$$6A_3 = s_0 + (s_2 - s_4)(-1) - s_6 = v_0 - v_2$$

When all these results are examined, it appears that further improvement can be made by using

$$
\begin{array}{cc}
u_0 & u_1 \\
u_3 & u_2 \\
\hline
\text{Sum} \quad g_0 & g_1 \\
\text{Diff} \quad h_0 & h_1
\end{array}
\tag{9.2.87}
$$

In the following, we make the substitutions and also use $m = v_0 + \frac{1}{2} v_2$ and $n = \frac{1}{2} p_1 + p_3$:

$$
\begin{aligned}
6A_0 &= g_0 + g_1, \\
6A_1 &= m + Cv, & 6B_1 &= n + Cp_2 \\
6A_2 &= h_0 + \tfrac{1}{2} h_1, & 6B_2 &= C(q_1 + q_2) \\
6A_3 &= v_0 - v_2, & 6B_3 &= p_1 - p_3 \\
6A_4 &= g_0 - \tfrac{1}{2} q_1, & 6B_4 &= C(q_1 - q_2) \\
6A_5 &= m - Cv_1, & 6B_5 &= n - Cp_2 \\
6A_6 &= h_0 - h_1,
\end{aligned}
\tag{9.2.88}
$$

The algorithm consists of the sequence (9.2.71), (9.2.82), (9.2.85), (9.2.87) and (9.2.88).

ILLUSTRATIVE PROBLEM 9.2-4   In Section 9.1.9, Illustrative Problem 9.1-7 displayed the following data:

$$
f(x) = \begin{cases} 0, & [-\pi, 0] \\ 1 - \cos 2x, & [0, \pi] \end{cases}
\tag{9.2.89}
$$

Choose 12 equally spaced points and write a finite Fourier series.

Solution   First, we shift the interval by the substitution

$$
y = x + \pi.
\tag{9.2.90}
$$

Then

$$
f(y) = \begin{cases} 0, & [0, \pi] \\ 1 - \cos 2y, & [\pi, 2\pi] \end{cases}
\tag{9.2.91}
$$

The twelve points are:

θ (deg)	0	30	60	90	120	150	180	210	240	270	300	330
$f(y)$	0	0	0	0	0	0	0	0.5	1.5	2	1.5	0.5

Using (9.2.71),

$$
\begin{array}{lccccccc}
F(0\rightarrow6) & 0 & 0 & 0 & 0 & 0 & 0 & 0 \\
F(11\rightarrow7) & & 0.5 & 1.5 & 2.0 & 1.5 & 0.5 \\
\hline
s(0\rightarrow6) & 0 & 0.5 & 1.5 & 2.0 & 1.5 & 0.5 & 0 \\
t(1\rightarrow5) & & -0.5 & -1.5 & -2.0 & -1.5 & -0.5
\end{array}
\tag{9.2.92}
$$

Using (9.2.82) and (9.2.85), we have

$s(0\to3)$	0	0.5	1.5	2		$t(1\to3)$	-0.5	-1.5	-2
$s(6\to4)$	0	0.5	1.5			$t(5\to4)$	-0.5	-1.5	
$u(0\to3)$	0	1	3	2		$p(1\to3)$	-1	-3	-2
$v(0\to2)$	0	0	0			$q(1\to2)$	0	0	

$$(9.2.93)$$

Using (9.2.87) and the definitions of m and n,

$u(0\to1)$	0	1	
$u(3\to2)$	2	3	$m=0,$
$g(0\to1)$	2	4	$n=-\frac{5}{2}$
$h(0\to1)$	-2	-2	

$$(9.2.94)$$

Substituting in (9.2.88),

$$6A_0=6$$
$$6A_1=0 \qquad 6B_1=-\tfrac{5}{2}-3C$$
$$6A_2=-3 \quad 6B_2=0$$
$$6A_3=0 \qquad 6B_3=1$$
$$6A_4=0 \qquad 6B_4=0$$
$$6A_5=0 \qquad 6B_5=-\tfrac{5}{2}+3C$$
$$6A_6=0$$

$$(9.2.95)$$

Using the theorem (9.2.37),

$$F(y)=0.5-0.5\cos 2y-0.8497\sin y+0.1667\sin 3y+0.01635\sin 5y \quad (9.2.96)$$

A computer program for the twelve-point algorithm is shown in Figure 9.2-1. The output for Illustrative Problem 9.2-4 is shown in Figure 9.2-2.

```
      DOUBLE PRECISION A(7),B(6),P(4),Q(3),S(7),T(6),U(4),V(3),PF(12),
     E(112),F(12),C,D,R,SUM,W,X,Z,ZC,ZN
      ZC=0.86602540378439D0
      ZN=6.0D0
      ZM=12.0D0
      C=0.5235987755982989D0
      D=0.5D0
C TO DO THE PROGRAM 7 TIMES.
      DO 12 M=1,7
C READ IN 3 ON A CARD, CC 1, 21, 41.
1     READ(1,100)(F(I),I=1,12)
100   FORMAT(3F20.0)
C COMPUTATION BY SHORTCUT METHOD.
      S(1)=F(1)
      DO 2 I=2,6
      J=14-I
      S(I)=F(I)+F(J)
2     T(I)=F(I)-F(J)
      S(7)=F(7)
      DO 3 I=1,3
      J=8-I
      U(I)=S(I)+S(J)
3     V(I)=S(I)-S(J)
      U(4)=S(4)
      DO 4 I=2,3
      J=8-I
      P(I)=T(I)+T(J)
```

```
4       Q(I)=T(I)-T(J)
        P(4)=T(4)
C CALCULATION OF COEFFICIENTS.
        A(1)=U(1)+U(2)+U(3)+U(4)
        A(1)=A(1)/ZM
        A(2)=V(1)+ZC*V(2)+V(3)*D
        A(3)=U(1)+(U(2)-U(3))*D-U(4)
        A(4)=V(1)-V(3)
        A(5)=U(1)-U(2)*D-U(3)*D+U(4)
        A(6)=V(1)-ZC*V(2)+V(3)*D
        DO 5 I=2,6
5       A(I)=A(I)/ZN
        A(7)=U(1)-U(2)+U(3)-U(4)
        A(7)=A(7)/ZM
C
        B(2)=P(2)*D+P(3)*ZC+P(4)
        B(3)=ZC*(Q(2)+Q(3))
        B(4)=P(2)-P(4)
        B(5)=ZC*(Q(2)-Q(3))
        B(6)=P(2)*D-ZC*P(3)+P(4)
        DO 6 I=2,6
6       B(I)=B(I)/ZN
C CALCULATION OF PREDICTED VALUES AND ERRORS.
        DO 9 N=1,12
        SUM=0.0D0
        R=N-1
        DO 7 I=2,7
        Z=I-1
        X=C*Z*R
7       SUM=SUM+A(I)*DCOS(X)
        DO 8 I=2,6
        Z=I-1
        X=C*Z*R
8       SUM=SUM+B(I)*DSIN(X)
        PF(N)=SUM+A(1)
9       E(N)=F(N)-PF(N)
C TO CALCULATE UNCORRECTED COEFFICIENTS.
        R=A(1)*2.0D0
        W=A(7)*2.0D0
C PRINT-OUT OF COEFFICIENTS.
        WRITE(3,101)
101     FORMAT(1H1)
        WRITE(3,102)
102     FORMAT(1H0,T25,'TABLE OF COEFFICIENTS')
        WRITE(3,103)R
103     FORMAT(1H ,T25,'A(0)',    F21.4)
        DO 10 I=2,6
        J=I-1
10      WRITE(3,104)J,A(I)
104     FORMAT(1H ,T25,'A(',I1,')',    F21.4)
        WRITE(3,105)W
105     FORMAT(1H ,T25,'A(6)',    F21.4)
        DO 11 I=2,6
        J=I-1
11      WRITE(3,106)J,B(I)
106     FORMAT(1H ,T25,'B(',I1,')',    F21.4)
C ERROR TABLE.
        WRITE(3,107)
107     FORMAT(1H0,T3,'R',T10,'ANGLE',T20,'F(R)',T32,'PF(R)',T41,'E = F -
       1PF',/)
        K=-30
        DO 12 I=1,12
        J=I-1
        K=K+30
12      WRITE(3,109)J,K,F(I),PF(I),E(I)
109     FORMAT(1H ,T2,I2,T10,I3,T17,F9.4    ,T29,F9.4    ,T41,1PE9.2)
        END
```

Fig. 9.2-1 Program for the 12-point algorithm.

```
                        TABLE OF COEFFICIENTS
                        A(0)              1.0000
                        A(1)              0.0
                        A(2)             -0.5000
                        A(3)              0.0
                        A(4)              0.0
                        A(5)              0.0
                        A(6)              0.0
                        B(1)             -0.8497
                        B(2)              0.0
                        B(3)              0.1667
                        B(4)              0.0
                        B(5)              0.0163
```

R	ANGLE	F(R)	PF(R)	E = F - PF
0	0	0.0	0.0	0.0
1	30	0.0	-0.0000	9.71D-17
2	60	0.0	-0.0000	2.78D-16
3	90	0.0	0.0000	-1.39D-17
4	120	0.0	-0.0000	1.39D-16
5	150	0.0	0.0000	-2.36D-16
6	180	0.0	0.0000	-6.94D-17
7	210	0.5000	0.5000	3.75D-16
8	240	1.5000	1.5000	2.22D-16
9	270	2.0000	2.0000	2.22D-16
10	3C0	1.5C00	1.5000	-1.11D-15
11	330	0.5000	0.5000	-5.13D-16

Fig. 9.2-2 Output for Illustrative Problem 9.2-4. PF is the predicted value of F.

9.2.8 Twenty-Four-Point Algorithm

We shall sketch the thinking and develop the algorithm briefly. Again, the general plan is to divide the region into halves repeatedly, then pick out sums and differences which occur more than once.

We start as usual:

$$12A_k = \sum_{r=0}^{23} F(r)\cos\frac{rk\pi}{12} \tag{9.2.97}$$

$$12B_k = \sum_{r=0}^{23} F(r)\sin\frac{rk\pi}{12} \tag{9.2.98}$$

Dividing the interval in the middle,

$$12A_k = \sum_{r=0}^{12} F(r)\cos\frac{rk\pi}{12} + \sum_{r=13}^{23} F(r)\cos\frac{rk\pi}{12} \tag{9.2.99}$$

$$12B_k = \sum_{r=0}^{12} F(r)\sin\frac{rk\pi}{12} + \sum_{r=13}^{23} F(r)\sin\frac{rk\pi}{12} \tag{9.2.100}$$

Letting $r' = 24 - r$ in the rightmost summations, and simplifying, we obtain

$$12A_k = F(0) + \sum_{r=1}^{11} \left[F(r) + F(24-r) \right]\cos\frac{rk\pi}{12} + F(12)\cdot(-1)^k \tag{9.2.101}$$

$$12B_k = \sum_{r=1}^{11} \left[F(r) - F(24-r) \right]\sin\frac{rk\pi}{12} \tag{9.2.102}$$

This suggests the beginning of the algorithm:

	$F(0)$	$F(1)$	$F(2)$	$\cdots$	$F(11)$	$F(12)$	
		$F(23)$	$F(22)$	$\cdots$	$F(13)$		
Sum	s_0	s_1	s_2	$\cdots$	s_{11}	s_{12}	(9.2.103)
Diff		t_1	t_2	$\cdots$	t_{11}		

Substituting in (9.2.101) and (9.2.102), we have

$$12A_k = s_0 + \sum_{r=1}^{11} s_r \cos\frac{rk\pi}{12} + (-1)^k s_{12} \qquad (9.2.104)$$

$$12B_k = \sum_{r=1}^{11} t_r \sin\frac{rk\pi}{12} \qquad (9.2.105)$$

Repeating the procedure by separating the summations, then letting $r' = 12 - r$ as we did previously, the result after simplifying is

$$12A_k = s_0 + \sum_{r=1}^{5} \left[s_r + (-1)^k s_{12-r} \right] \cos\frac{rk\pi}{12} + s_6 \cos\frac{k\pi}{2} + s_{12}(-1)^k \qquad (9.2.106)$$

$$12B_k = \sum_{r=1}^{5} \left[t_r - (-1)^k t_{12-r} \right] \sin\frac{rk\pi}{12} + t_6 \sin\frac{k\pi}{2} \qquad (9.2.107)$$

From this we can see that the s terms are being added and subtracted, suggesting the following:

	s_0	s_1	s_2	s_3	s_4	s_5	s_6			t_1	t_2	t_3	t_4	t_5	t_6
	s_{12}	s_{11}	s_{10}	s_9	s_8	s_7				t_{11}	t_{10}	t_9	t_8	t_7	
Sum	u_0	u_1	u_2	u_3	u_4	u_5	u_6		Sum	p_1	p_2	p_3	p_4	p_5	p_6
Diff	v_0	v_1	v_2	v_3	v_4	v_5			Diff	q_1	q_2	q_3	q_4	q_5	

$$(9.2.108)$$

At this point, we could write the 24 coefficients and search for repetitions, but our experience with the six-point and twelve-point algorithms suggests that we continue:

	u_0	u_1	u_2	u_3			v_0	v_1	v_2	v_3	
	u_6	u_5	u_4					v_5	v_4		
Sum	g_0	g_1	g_2	g_3		Sum	m_0	m_1	m_2	m_3	(9.2.109)
Diff	h_0	h_1	h_2			Diff		n_1	n_2	n_3	

One more round of sums and differences can be made. The result is the subroutine shown in Figure 9.2-3.

```
      SUBROUTINE F24
      DOUBLE PRECISION X(24),A(13),B(12),U(12),V(12),R(6),S(6),P(6),Q(6)
     1,ZL(3),ZM(3),G(3),H(3),E,F,C,D,CC,SS,AA,BB,V1,V7,EX(24),PX(24),Z
      EQUIVALENCE(PX(1),U(1),ZL(1)),(PX(4),ZM(1)),(PX(7),G(1)),(PX(10),H
     1(1)),(V(2),C),(V(3),D),(V(4),CC),(V(5),SS),(V(6),AA),(V(7),BB),(PX
     2(13),V(1)),(EX(1),R(1)),(EX(7),S(1)),(EX(13),P(1)),(EX(19),Q(1))
      COMMON X,A,B
      DO 1 I=2,12
      J=26-I
      U(I)=X(I)+X(J)
1     V(I)=X(I)-X(J)
      U(1)=X(1)+X(13)
      V(1)=X(1)-X(13)
C
      DO 2 I=2,6
      J=14-I
      R(I)=U(I)+U(J)
2     S(I)=U(I)-U(J)
      R(1)=U(1)+U(7)
      S(1)=U(1)-U(7)
C
      DO 3 I=2,6
      J=14-I
      P(I)=V(I)+V(J)
3     Q(I)=V(I)-V(J)
      V1=V(1)
      V7=V(7)
      ZL(1)=R(1)+R(4)
      ZL(2)=R(2)+R(6)
      ZL(3)=R(3)+R(5)
      ZM(1)=R(1)-R(4)
      ZM(2)=R(2)-R(6)
      ZM(3)=R(3)-R(5)
      G(2)=Q(2)+Q(6)
      H(2)=Q(2)-Q(6)
      G(3)=Q(3)+Q(5)
      H(3)=Q(3)-Q(5)
      E=ZL(2)+ZL(3)
      F=ZL(2)-ZL(3)
      C=H(2)+H(3)
      D=H(2)-H(3)
C DEFINITIONS
      SS=0.258819045102520D00
      CC=0.965925826289068D00
      BB=0.707106781186548D00
      AA=0.866025403784439D00
C
      A(1)=ZL(1)+E
      A(2)=V1+CC*S(2)+AA*S(3)+BB*S(4)+0.5*S(5)+SS*S(6)
      A(3)=S(1)+AA*ZM(2)+0.5*ZM(3)
      A(4)=V1+BB*(S(2)-S(4)-S(6))-S(5)
      A(5)=ZM(1)+0.5*F
      A(6)=V1+SS*S(2)-AA*S(3)-BB*S(4)+0.5*S(5)+CC*S(6)
      A(7)=S(1)-ZM(3)
      A(8)=V1-SS*S(2)-AA*S(3)+BB*S(4)+0.5*S(5)-CC*S(6)
      A(9)=ZL(1)-0.5*E
      A(10)=V1-BB*(S(2)-S(4)-S(6))-S(5)
      A(11)=S(1)-AA*ZM(2)+0.5*ZM(3)
      A(12)=V1-CC*S(2)+AA*S(3)-BB*S(4)+0.5*S(5)-SS*S(6)
      A(13)=ZM(1)-F
```

```
C
      B(2)=SS*P(2)+0.5*P(3)+BB*P(4)+AA*P(5)+CC*P(6)+V7
      B(3)=0.5*G(2)+AA*G(3)+Q(4)
      B(4)=P(3)-V7+BB*(P(2)+P(4)-P(6))
      B(5)=AA*C
      B(6)=CC*P(2)+0.5*P(3)-BB*P(4)-AA*P(5)+SS*P(6)+V7
      B(7)=G(2)-Q(4)
      B(8)=CC*P(2)-0.5*P(3)-BB*P(4)+AA*P(5)+SS*P(6)-V7
      B(9)=AA*D
      B(10)=V7-P(3)+BB*(P(2)+P(4)-P(6))
      B(11)=0.5*G(2)-AA*G(3)+Q(4)
      B(12)=SS*P(2)-0.5*P(3)+BB*P(4)-AA*P(5)+CC*P(6)-V7
C
      DO 4 I=1,12
      A(I)=A(I)/12.0
4     B(I)=B(I)/12.0
      A(13)=A(13)/12.0
C CALCULATION OF ERRORS AT GIVEN POINTS. THESE SHOULD BE CLOSE TO ZERO.
C UNIT ANGLE Z IS EQUIVALENT TO 15 DEGREES. E IS FLOATING FOR JR.
      PI=3.14159265358979D00
      Z=PI/12.0
      DO 41 JR=1,24
      E=JR-1
      PX(JR)=0.5*(A(1)+A(13)*DCOS(E*PI))
      DO 42 K=2,12
      F=K-1
      F=F*E
42    PX(JR)=PX(JR)+A(K)*DCOS(F*Z)+B(K)*DSIN(F*Z)
41    EX(JR)=X(JR)-PX(JR)
C PRINTOUT
      WRITE(3,100)
100   FORMAT(1H1,T25,'TABLE OF COEFFICIENTS',/)
      WRITE(3,101)A(1),A(1)
101   FORMAT(1H ,T25,'A( 0)',T34,F12.4,T70,1PE23.16)
      DO 5 I=2,13
      J=I-1
5     WRITE(3,102)J,A(I),A(I)
102   FORMAT(1H ,T25,'A(',I2,')',T34,F12.4,T70,1PE23.16)
      DO 6 I=2,12
      J=I-1
6     WRITE(3,103)J,B(I),B(I)
103   FORMAT(1H ,T25,'B(',I2,')',T34,F12.4,T70,1PE23.16)
C
      WRITE(3,104)
104   FORMAT(1H0,T25,'TABLE FOR CHECKING')
      WRITE(3,105)
105   FORMAT(1H0,T3,'R',T10,'ANGLE',T20,'F(R)',T32,'PF(R)',T42,'E=F-PF',
     1/)
      K=-15
      DO 7 I=1,24
      J=I-1
      K=K+15
7     WRITE(3,106)J,K,X(I),PX(I),EX(I)
106   FORMAT(1H ,T2,I2,T10,I3,T17,F9.4,T29,F9.4,T41,1PE9.2)
      RETURN
      END
```

Fig. 9.2-3 Subroutine for the 24-point algorithm.

TABLE OF COEFFICIENTS

A(0)	3.2842
A(1)	−1.0057
A(2)	−0.2558
A(3)	−0.1170
A(4)	−0.0685
A(5)	−0.0462
A(6)	−0.0343
A(7)	−0.0272
A(8)	−0.0228
A(9)	−0.0201
A(10)	−0.0184
A(11)	−0.0174
A(12)	−0.0171
B(1)	−0.0000
B(2)	−0.0000
B(3)	−0.0000
B(4)	−0.0000
B(5)	−0.0000
B(6)	0.0000
B(7)	−0.0000
B(8)	−0.0000
B(9)	−0.0000
B(10)	0.0000
B(11)	0.0000

TABLE FOR CHECKING

R	ANGLE	F(R)	PF(R)	E=F−PF
0	0	0.0	−0.0000	1.84D−16
1	15	0.3941	0.3941	1.07D−07
2	30	0.7539	0.7539	1.01D−07
3	45	1.0795	1.0795	2.22D−07
4	60	1.3708	1.3708	1.83D−07
5	75	1.6278	1.6278	2.74D−07
6	90	1.8506	1.8506	2.15D−07
7	105	2.0390	2.0390	2.68D−07
8	120	2.1932	2.1932	1.95D−07
9	135	2.3132	2.3132	2.05D−07
10	150	2.3989	2.3989	1.23D−07
11	165	2.4503	2.4503	8.31D−08
12	180	2.4674	2.4674	7.97D−14
13	195	2.4503	2.4503	−9.82D−08
14	210	2.3989	2.3989	−1.72D−07
15	225	2.3132	2.3132	−3.41D−07
16	240	2.1932	2.1932	−3.89D−07
17	255	2.0390	2.0390	−6.51D−07
18	270	1.8506	1.8506	−6.44D−07
19	285	1.6278	1.6278	−1.04D−06
20	3C0	1.3708	1.3708	−9.15D−07
21	315	1.0795	1.0795	−1.55D−06
22	330	0.7539	0.7539	−1.11D−06
23	345	0.3941	0.3941	−2.46D−06

Fig. 9.2-4 The 24-point algorithm applied to $f(x) = x(\pi - x)$ **for** $0 \leqslant x \leqslant \pi$.

To enter the subroutine, we define Y(24), A(13), B(12), PI, and Z $(=\pi/12)$ as double precision, read in the 24 given values at $0°, 15°, \ldots, 345°$, and then call the subroutine, which furnishes the answers. A sample output is given in Figure 9.2-4.

9.2.9 Comments about Programming

1. It must have occurred to you that a considerable amount of storage may be required for all the intermediate results. However, much of it can be *equivalenced*, as is shown in the fourth card of Figure 9.2-3.
2. In the usual FORTRAN, subscripts cannot start at 0. In writing programs for series, if a zero subscript is involved, it is then necessary to increase each subscript by 1. In the printout, the subscript must be reduced again. Other compilers do not have this problem.

9.2.10 Harmonic Analysis for $2N+1$ Points

Only rarely is it necessary to use an odd number of points. However, to give a full picture, we outline the theory. We start with $2N+1$ points dividing the interval at equal distances, and in the usual way prove the orthogonality and associated conditions.

Harmonic Analysis Theorem for $2N+1$ Points If the interval $[0, 2\pi]$ is divided into $2N+1$ points with

$$\theta_r = \frac{2r\pi}{2N+1}, \qquad r = 0, 1, \ldots, 2N \qquad (9.2.110)$$

then the following relationships hold:

	$m \neq n$	$m = n = 0$	$m = n = N$	$m = n \neq 0$
$\displaystyle\sum_{r=0}^{2N} \cos m\frac{2\pi r}{2N+1} \cos n\frac{2\pi r}{2N+1} =$	0	$2N+1$	$\dfrac{2N+1}{2}$	$\dfrac{2N+1}{2}$

$$(9.2.111)$$

| $\displaystyle\sum_{r=0}^{2N} \sin m\frac{2\pi r}{2N+1} \cos n\frac{2\pi r}{2N+1} =$ | 0 | 0 | 0 | 0 |

$$(9.2.112)$$

| $\displaystyle\sum_{r=0}^{2N} \sin m\frac{2\pi r}{2N+1} \sin n\frac{2\pi r}{2N+1} =$ | 0 | 0 | $\dfrac{2N+1}{2}$ | $\dfrac{2N+1}{2}$ |

$$(9.2.113)$$

Now we assume

$$F(r) = \tfrac{1}{2}A_0 + \sum_{k=1}^{N} \left[A_k \cos \frac{2\pi r}{2N+1} + B_k \sin \frac{2\pi r}{2N+1} \right] \qquad (9.2.114)$$

where there are $2N+1$ unknown coefficients to find. Multiplying and using the preceding theorem, we finally arrive at the theorem for the coefficients.

THEOREM: FINITE FOURIER SERIES (FOURIER COEFFICIENTS FOR $2N+1$ POINTS)   If a finite Fourier series is written for $2N+1$ points equally spaced on $[0, 2\pi]$, where (9.2.110) and (9.2.114) apply, then

$$A_k = \frac{2}{2N+1} \sum_{r=0}^{2N} F(r) \frac{2\pi rk}{2N-1}, \qquad k=0,1,\ldots,N \qquad (9.2.115)$$

$$B_k = \frac{2}{2N+1} \sum_{r=0}^{2N} F(r) \sin\frac{2\pi rk}{2N-1}, \qquad k=1,2,\ldots,N \qquad (9.2.116)$$

PROBLEMS SECTION 9.2

A. Using 6 points, find the finite Fourier series which fit:
   (1)

θ (deg)	0	30	60	90	120	150
y	0.8	0.6	0.4	0.7	0.9	1.1

(2)

θ (deg)	0	30	60	90	120	150
y	0.6	0.9	1.3	1.0	0.8	0.5

B. Using 12 points, find the finite Fourier series which fit the data in Table 9.2-3.

Table 9.2-3
Problem (3)–(8)

θ (deg)	(3)	(4)	(5)	(6)	(7)	(8)
0	9.3	45	1.21	9.55	149	38.4
30	15.0	142	1.32	18.99	128	11.8
60	17.4	138	1.46	18.60	128	4.3
90	23.0	−2	1.40	18.04	159	13.8
120	37.0	−25	1.34	17.00	189	3.9
150	31.0	−21	1.18	15.55	189	−18.1
180	15.3	−69	1.07	13.99	178	−22.9
210	4.0	−90	1.01	12.45	177	−27.2
240	−8.0	−92	1.05	11.19	181	−23.8
270	−13.2	−45	1.10	10.01	179	8.2
300	−14.2	68	1.14	9.02	182	31.7
330	−6.0	40	1.17	8.29	166	34.2

C. Using 24 points, find the finite Fourier series which fits the following:

(9) $y = \sqrt{2\pi x - x^2}$ from $x=0$ to $x=2\pi$

(10)

$$f(x) = \begin{cases} 0.1x, & 0 \leqslant x \leqslant 10 \\ 0.1(20-x), & 10 \leqslant x \leqslant 20 \end{cases}$$

(11) $f(x) = 4x - x^2,\ 0 \leqslant x \leqslant 4$

(12)

$$f(x) = \begin{cases} x, & -\pi \leqslant x < 0 \\ -x, & 0 \leqslant x < \pi \end{cases}$$

(13)

$$f(x) = \begin{cases} 1, & 0 < x < \pi/2 \\ 0, & \pi/2 < x < \pi \end{cases}$$

(14)

$$f(x) = \begin{cases} -1, & 0 < x < 2 \\ +1, & 2 < x < 4 \end{cases}$$

(15)

$$f(x) = \begin{cases} 0, & -\pi \leqslant x < -\pi/2 \\ \cos x, & -\pi/2 \leqslant x < \pi/2 \\ 0, & \pi/2 \leqslant x < \pi \end{cases}$$

D.
(16) Prove (a) (9.2.27), (b) (9.2.28), (c) (9.2.39), (d) (9.2.51).
(17) Prove (a) (9.2.120), (b) (9.2.121), (c) (9.2.122).
(18) Find a finite Fourier series for the following, assuming (a) $L=3$, (b) $L=4$:

x	0	1	2	3
y	0	1	1	0

Chapter 10

The Chebyshev Criterion

In an abbreviated course, just study Section 10.1 to furnish the background needed in Chapter 11.

This is the third of four chapters dealing with the representation of a function, $f(x)$, by an approximating expression.

In Chapter 7, we dealt with the Taylor series, probably the easiest to develop and program. If $f(x)$ is infinitely smooth and if reasonable care is observed in programming, the Taylor series is remarkably accurate in the neighborhood of the point about which the expansion is made. However, the Taylor-series error increases steadily as we depart from the vicinity of the point of expansion.

The irritating thing is not that the error is too great but rather it varies the way it does as we leave the point of expansion. It is always comforting to know the maximum error on an interval so that in a calculation this can be taken into account in evaluating the final answer.

In the previous chapter, we investigated the Fourier series. In this method, equally spaced points are chosen over an interval on which $f(x)$ is defined, and a least-squares formula on the interval is calculated. Apart from the practical fact that the Fourier series is extensively used in engineering, there are three important advantages:

1. $f(x)$ need not be continuous.
2. The error is spread over the interval instead of being small near one point and increasing as we leave that point.
3. The successive maximum errors oscillate in waves of approximately equal amplitude.

The main disadvantages are:

1. The maximum errors are not minimized.
2. The formula is in terms of sines and cosines.

When the formula is programmed for a computer, the machine functions are always used, and these have errors which may add to vitiate the final result. Each time a sine or cosine is used, a relative error of order 10^{-7} (single precision) or 10^{-16} (double precision) is contributed. Since a large number of sines and cosines are used, and since the arguments of these sines and cosines also have inherent errors, the total may build up into a sizable error.

In this chapter, we discuss the theory of approximations in the Chebyshev sense. An approximation in the Chebyshev sense has two important properties:

1. *The oscillation property*. Successive maximum errors alternate in sign and are approximately equal in absolute value; this is also called the *equal-ripple property*
2. *The minimax property*. The largest error on a given interval is as small as possible, i.e., the maximum error is minimum.

An approximation in the Chebyshev sense can be almost any kind of function. In the chapter after this one, we shall explain the practical methods for achieving such an approximation.

In this chapter, we shall make use of two theorems from the theory of equations:

> ROOT-DEGREE THEOREM   The number of distinct roots of a polynomial cannot exceed its degree.

For example, a polynomial with three zeros cannot be a quadratic. The proof can be found in algebra books; it is based upon the remainder theorem.

> IDENTITY THEOREM   If two polynomials, $P_n(x)$ and $Q_n(x)$, both of degree n or less, agree in value for more than n distinct values of x, then they are identical, i.e., they agree for all values of x.

The proof is easy. We consider $[P_n(x) - Q_n(x)]$, which must be a polynomial, and apply the root-degree theorem.

10.1 THE "BEST" POLYNOMIAL APPROXIMATION

10.1.1 Existence

It is customary to call a polynomial which satisfies the Chebyshev criterion (equal-ripple and minimax) a "best" polynomial approximation. We shall retain the quotation marks because, in fact, whether or not it is *best* depends upon the application to be made. For an interpolation problem (e.g., *given* $\sin 30°$ to 16 places, *find* $\sin 30.1°$), the Taylor series is actually *best*.

We shall prove that a "best" polynomial exists for any function $f(x)$ continuous on an interval. Note that there is no requirement of smoothness. The proof depends upon the fundamental theorem which states that if $f(x)$ is defined and continuous on a closed interval, it is bounded on the interval.

THEOREM   There exists an nth-degree polynomial approximation "best" in the Chebyshev sense to $f(x)$ if $f(x)$ is continuous for $x \in [a,b]$.

Proof   Let an approximating polynomial of degree n or less be defined:

$$P_n(x) = a_0 + a_1 x + \cdots + a_n x^n = \sum_{k=0}^{n} a_k x^k \qquad (10.1.1)$$

and let the error over an interval $[a,b]$ be defined by

$$E(x) = f(x) - P_n(x) \qquad (10.1.2)$$

The maximum *norm*, $D(f,P_n)$, is then defined as

$$D(f,P_n) = \|f(x) - P_n(x)\| = \max|E|, \ x \in [a,b] \qquad (10.1.3)$$

The symbol D comes from the word *deviation*. It is obvious that the maximum norm depends only on the set of coefficients, $A = \{a_0, a_1, \ldots, a_n\}$, in (10.1.1), so that for some function g,

$$D(f,P_n) = g(A) \qquad (10.1.4)$$

Note that $g(A)$ is defined and continuous, so that (by the theorem mentioned) it is bounded. We shall define m as the GLB of $g(A)$, noting [from (10.1.3)] that m is non-negative. We must prove that m is not only a GLB but also a minimum, i.e., we must show that for some set of coefficients, A, that $g(A) = m$.

The proof is by contradiction. We assume there is no such set A. Then for all A,

$$g(A) - m > 0 \qquad (10.1.5)$$

In fact, this inequality is a succinct expression of our assumption; we are going to show that it leads to a contradiction.

Since m is a GLB and $g(A)$ is continuous, it is certainly true that for every positive integer N, there exists at least one set of coefficients, A', for which

$$m < g(A') < m + \frac{1}{N} \qquad (10.1.6)$$

Because $g(A) - m$, according to (10.1.5), is always positive and $g(A)$ is continuous, it is also true that for any set A' we can find another set A'' so close that

$$g(A'') - g(A') < \frac{g(A'') - m}{N} \tag{10.1.7}$$

$$g(A') > \frac{g(A'') + m}{2} = m + \frac{g(A'') - m}{2} \tag{10.1.8}$$

Now, remembering that $g(A) - m$ cannot be 0 (by assumption), we choose N large enough in (10.1.6) so that

$$\frac{1}{N} < \frac{g(A'') - m}{2} \tag{10.1.9}$$

Combining (10.1.6) and (10.1.8),

$$m + \frac{1}{N} < g(A') < m + \frac{1}{N} \tag{10.1.10}$$

furnishing the contradiction and proving the theorem. $\square$

This shows that there really is a "best" polynomial in the Chebyshev sense for every continuous $f(x)$.

10.1.2 Reaching the Extrema

Before proceeding to the main Chebyshev theorem which defines the precise conditions for a "best" polynomial, we prove a little theorem, mostly to illustrate a common method of proof in this region of mathematics.

THEOREM If $P_n(x)$ is a "best" polynomial approximation for $f(x)$ with a maximum norm $D(f, P_n)$, then the error $E(x)$ takes on the values $-D$, 0 and D somewhere on the interval.

Proof We are told that the maximum norm is D. Therefore, the error $E(x)$ must actually reach either D or $-D$ somewhere. We must prove that $E(x)$ reaches both values.

The proof is by contradiction. Without loss of generality, suppose $E(x)$ reaches $-D$ as a minimum but falls short of D as a maximum, so that

$$-D \leqslant f(x) - P_n(x) \leqslant D - p^2, \qquad p^2 > 0 \tag{10.1.11}$$

We add $\frac{1}{2}p^2$ to each member and obtain

$$-\left(D - \tfrac{1}{2}p^2\right) \leqslant f(x) - \left[P_n(x) - \tfrac{1}{2}p^2\right] \leqslant D - \tfrac{1}{2}p^2 \tag{10.1.12}$$

But $[P_n(x) - \frac{1}{2}p^2]$ is a new polynomial of the same degree as $P_n(x)$ with a lower extreme error than $P_n(x)$. This contradicts the given statement that $P_n(x)$ is the "best" polynomial of degree n or less. Consequently, the "best" polynomial must have at least one minimum, $-D$, and one maximum, D, on the interval. Also, since $f(x)$ and $P(x)$ are continuous, $E(x)$ must vanish at least once between D and $-D$. This completes the proof. □

10.1.3   The Chebyshev Theorem

We are now ready to attack the main theorem.

CHEBYSHEV THEOREM   Let $P_n(x)$ be a polynomial approximation of degree n or less to $f(x)$, continuous for $x \in [a,b]$. Then $P_n(x)$ is "best" in the Chebyshev sense iff it has $n+2$ points of equal maximum or minimum error with alternating signs. The polynomial is unique.

To make sure this is understood, consider the problem of making a linear approximation (degree 1) to $f(x)$, as shown in Figure 10.1-1. We would need only two points to draw a line, but this might not be a "best" line.

Suppose we choose the extreme $n+2=3$ points at x_1, x_2 and x_3. (We haven't explained how to find these points as yet.) The line is placed so that successive maximum errors are $D, -D, D$. According to the theorem which we are about to prove, the resulting line would be a "best" polynomial approximation to the dotted curve. According to the theorem, there is no better *linear* approximation to $f(x)$, in the Chebyshev sense. If you were using $P_1(x)$ as a computer program to approximate $f(x)$, you could guarantee that the error of approximation is not worse than $\pm D$ everywhere on $[a,b]$.

We proceed to the three-part proof of the theorem.

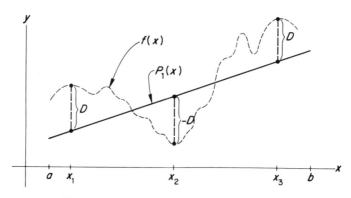

Fig. 10.1-1   A "best" linear approximation.

10.1.4 Sufficiency

If $P_n(x)$ has the oscillation property at least $n+2$ times—i.e.,

$$f(x_k) - P_n(x_k) = (-1)^k D(f, P_n) \text{ or } (-1)^{k+1} D(f, P_n) \quad (10.1.13)$$

where $k = 0, 1, \ldots, m$ and $m \geq n+2$—then $P_n(x)$ is a "best" approximation polynomial of degree n or less.

Proof Let $P_n(x) = \Sigma_{k=0}^n a_k x^k$, and suppose that there is another polynomial $Q_n(x) = \Sigma_{k=0}^n b_k x^k$ such that

$$D(f, Q_n) < D(f, P_n) \quad (10.1.14)$$

which would mean that Q_n is better than P_n, since it would have a smaller maximum error. Then

$$Q_n(x) - P_n(x) \equiv [f(x) - P_n(x)] - [f(x) - Q_n(x)] \quad (10.1.15)$$

We are going to investigate the behavior of $Q - P$ at $x_0, x_1, \ldots, x_m$. Let $D(f, P_n) = p^2 > 0$. By hypothesis, the value of $|f(x) - Q_n(x)|$ everywhere, including the selected points, is certainly less than p^2. For example, we may say that $E_Q(x_k) = p^2 - c_k$ where $0 \leq c_k < p^2$. There are four possible combinations at the selected points. Using (10.1.15), we have

$E_P = f(x) - P_n(x)$	$E_Q = f(x) - Q_n(x)$	$E_P - E_Q = Q_n(x) - P_n(x)$
$+p^2 > 0$	$+(p^2 - c_1)$	$p^2 - (p^2 - c_1) = c_1 \geq 0$
$+p^2 > 0$	$-(p^2 - c_2)$	$p^2 + (p^2 - c_2) = 2p^2 - c_2 > 0$
$-p^2 < 0$	$+(p^2 - c_3)$	$-p^2 - (p^2 - c_3) = -2p^2 + c_3 < 0$
$-p^2 < 0$	$-(p^2 - c_4)$	$-p^2 + (p^2 - c_4) = -c_4 \leq 0$

From the above, $Q_n(x_k) - P_n(x_k)$ either has the same sign as $f(x_k) - P_n(x_k)$, or else it vanishes. But we are given that $f(x) - P_n(x)$ oscillates at least $n+2$ times. This means it has at least $n+1$ zeros. Then $Q_n(x)$, because of this demonstrated agreement, also has at least $n+1$ zeros.

Is this possible? No, because

$$Q_n(x) - P_n(x) = \sum_{k=0}^n (b_k - a_k) x^k \quad (10.1.16)$$

which is a polynomial of degree no higher than n. By the root-degree theorem, it is impossible for P_n and Q_n to be distinct. The contradiction shows that there cannot be a polynomial of degree n or less better than $P_n(x)$, as constructed. $\square$

10.1.5 Necessity

Let $P_n(x)$ be a "best" approximation polynomial of degree n or less for $f(x)$—i.e., for any other polynomial $Q_n(x)$,

$$D(f,P_n) < D(f,Q_n) \tag{10.1.17}$$

Then we shall prove that $P_n(x)$ must have the oscillation property at least $n+2$ times, i.e.,

$$f(x) - P_n(x_k) = (-1)^k D(f,P_n) \text{ or } (-1)^{k+1} D(f,P_n) \tag{10.1.18}$$

for $k=0,1,\dots,m$, where $m \geqslant n+2$.

Proof We already know that the error of a "best" polynomial must actually reach D and $-D$ at least once each, and that $E(x)$ must vanish somewhere between them. Let us assume that for one or more points, the error $E_P(x) = f(x) - P_n(x)$ does not quite reach $\pm D(f,P_n)$.

We shall lead up to a general proof slowly in order to show how it evolved. We note from (10.1.18) that the error curve may start (at $k=0$) with a (positive) maximum or a (negative) minimum. Without loss of generality, we shall assume it starts with a (positive) maximum. In Figure 10.1-2, we show a special case with four selected points such so that

$$a < x_1 < x_2 < x_3 < x_4 < b \tag{10.1.19}$$

at which there are successive maxima and minima.

Note that the extrema fall short of $\pm D$ at x_1 and x_2 and that they reach $\pm D$ at x_3 and x_4. We shall show that this cannot be the error curve for a "best" polynomial.

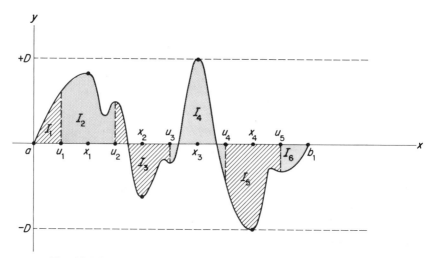

Fig. 10.1-2 An error curve with successive maxima and minima.

We now divide $[a, b]$ into six intervals by defining

$$u_1 = \tfrac{1}{2}(a + x_1)$$
$$u_2 = \tfrac{1}{2}(x_1 + x_2)$$
$$\vdots$$
$$u_5 = \tfrac{1}{2}(x_4 + b) \tag{10.1.20}$$

This defines six subintervals:

$$I_1 = [a, u_1)$$
$$I_2 = [u_1, u_2)$$
$$\vdots$$
$$I_6 = [u_5, b] \tag{10.1.21}$$

By hypothesis, the first extremum is a (positive) maximum at x_1, so that I_1 does not have an extremum at all. From our previous work on errors, we are familiar with the product

$$\pi(x) = (x - u_1)(x - u_2)(x - u_3)(x - u_4)(x - u_5)(x - b) \tag{10.1.22}$$

It will turn out to be more convenient if we define

$$g(x) = (-1)^5 (x - u_1)(x - u_2)(x - u_3)(x - u_4)(x - u_5)(x - u_5)(x - b) \tag{10.1.23}$$

Like $\pi(x)$, $g(x)$ is of the sixth degree and can be expressed as

$$g(x) = \sum_{k=0}^{6} c_k x^k \tag{10.1.24}$$

The sign of $g(x)$ evidently depends upon the position of x. This is shown in the following:

x IN	SIGN OF $g(x)$
I_1	$(-1)^5 (+)^0 (-)^6 = -$
I_2	$(-1)^5 (+)^1 (-)^5 = +$
I_3	$(-1)^5 (+)^2 (-)^4 = -$
I_4	$(-1)^5 (+)^3 (-)^3 = +$
I_5	$(-1)^5 (+)^4 (-)^2 = -$
I_6	$(-1)^5 (+)^5 (-)^1 = +$

In short, $g(x)$ is negative when the subscript for I is odd, and positive otherwise. Now form a new polynomial

$$Q_n(x) = P_n(x) + \epsilon g(x), \qquad \epsilon > 0 \tag{10.1.25}$$

We choose ϵ sufficiently small that the degree of $Q_n(x)$ is the same as that of $P_n(x)$. This is obviously always possible. We merely make sure that in

$$\epsilon g(x) = \epsilon \sum_{k=0}^{6} c_k x^k = \epsilon c_0 + \epsilon c_1 x + \cdots \epsilon c_6 x^6 \qquad (10.1.26)$$

the coefficient, ϵc_6, of the highest degree term is small enough to avoid canceling the corresponding coefficient of $P_n(x)$. From (10.1.25),

$$[f(x) - Q_n(x)] = [f(x) - P_n(x)] - \epsilon g(x) \qquad (10.1.27)$$

$$E_Q = E_P - \epsilon g(x) \qquad (10.1.28)$$

By construction, I_1 and I_6 have no extrema. In I_2 and I_4, $g(x)$ is positive, so that

$$E_Q < E_P \qquad (10.1.29)$$

In I_3 and I_5, $g(x)$ is negative, so that

$$E_Q > E_P \qquad (10.1.30)$$

This shows that for the case shown, $Q(x)$ must have positive errors lower than $D(f, P_n)$ and negative errors higher than $-D(f, P_n)$. But if this were so, $P_n(x)$ would not be the "best" polynomial, thus contradicting the assumption that the errors fall short of $\pm D$.

Let's see (Figure 10.1-3) what happens if there is another extremum. Then

$$\pi(x) = (x - u_1)(x - u_2) \cdots (x - u_6)(x - b) \qquad (10.1.31)$$

$$g(x) = (-1)^6 \pi(x) \qquad (10.1.32)$$

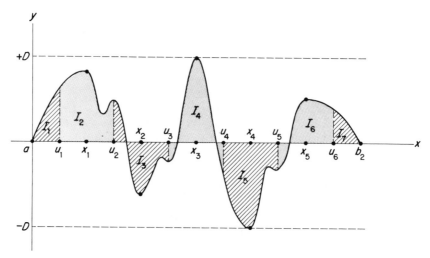

Fig. 10.1-3 The situation for 7 subintervals.

The sign of $g(x)$ varies as follows:

x IN	SIGN OF $g(x)$
I_1	$(-1)^6(+)^0(-)^7 = -$
I_2	$(-1)^6(+)^1(-)^6 = +$
I_3	$(-1)^6(+)^2(-)^5 = -$
$\vdots$	
I_6	$(-1)^6(+)^5(-)^2 = +$
I_7	$(-1)^6(+)^6(-)^1 = -$

$$(10.1.33)$$

The same equations (10.1.17)–(10.1.30) evidently hold, so that the proof is good for either case.

The general proof proceeds in the same fashion except that $g(x)$ is defined as follows:

$$g(x) = (-1)^r(x - u_1)(x - u_2) \cdots (x - u_r)(x - b),$$

where the first (positive) maximum is found at x_1. If the first extremum, at x_1, is a (negative) minimum, the proof is quite similar. This completes the proof of necessity. $\square$

10.1.6 Uniqueness

Assume that $P_n(x)$ and $Q_n(x)$ are different "best" polynomials approximating $f(x)$ with degree n. Then

$$-D \leqslant f(x) - P_n(x) \leqslant D \qquad (10.1.34)$$

$$-D \leqslant f(x) - Q_n(x) \leqslant D \qquad (10.1.35)$$

It is permissible to add or multiply inequalities in the same sense (but not to subtract or divide them). Adding these two, we have

$$-2D \leqslant 2f(x) - \left[P_n(x) + Q_n(x) \right] \leqslant 2D \qquad (10.1.36)$$

Dividing by 2,

$$-D \leqslant f(x) - \tfrac{1}{2}\left[P_n(x) + Q_n(x) \right] \leqslant D \qquad (10.1.37)$$

showing that if $P_n(x)$ and $Q_n(x)$ are both "best", then half their sum (which is obviously of degree no greater than n) is also "best". We shall now prove this is impossible.

If $\tfrac{1}{2}[P_n(x) + Q_n(x)]$ is "best", it must have an error of $\pm D$ at least $n + 2$ times. Let its value at x_1 be D. Then

$$f(x_1) - \tfrac{1}{2}\left[P_n(x_1) + Q_n(x_1) \right] = D \qquad (10.1.38)$$

Multiplying by 2,

$$[f(x_1) - P_n(x_1)] + [f(x_1) - Q_n(x_1)] = 2D \qquad (10.1.39)$$

Since $P_n(x)$ and $Q_n(x)$ are each assumed "best", the maximum value for each binomial is D, and this is the only value which will satisfy (10.1.39). Therefore

$$f(x_1) - P_n(x_1) = D$$
$$f(x_1) - Q_n(x_1) = D \qquad (10.1.40)$$

from which

$$P_n(x_1) = Q_n(x_1) \qquad (10.1.41)$$

The argument can be repeated $n+2$ times, showing that $P_n(x)$ and $Q_n(x)$ agree at $n+2$ values of x. Since $P_n(x)$ and $Q_n(x)$ are of degree no greater than n it then follows from the identity theorem that $P_n(x)$ and $Q_n(x)$ are identical, contrary to the first assumption. Therefore there is a *unique* "best" polynomial approximation of degree n to $f(x)$. $\square$

PROBLEMS SECTION 10.1

Show by a graph based upon a table of values that in each of the following cases, $C(x)$ is a "best" approximation to $f(x)$ on the interval I.

(1) $f(x) = |x|$, $C(x) = x^2 + \frac{1}{6}$, $I = (-1, +1)$

(2) $f(x) = e^x$, $C(x) = 0.55404x^2 + 1.13018x + 0.98904$, $I = (-1, +1)$

(3) $f(x) = \arcsin x$, $C(x) = \pi/2 - (1.5707288 - 0.2121144x + 0.0742610x^2 - 0.0187293x^3)\sqrt{1-x}$, $I = (0, 1)$

(4) $f(x) = \arctan x$, $C(x) = 0.9998660x - 0.3302995x^3 + 0.1801410x^5 - 0.0851330x^7 + 0.0208351x^9$, $I = (-1, +1)$

(5) $f(x) = 10^x$, $C(x) = (1 + 1.1499196x + 0.6774323x^2 + 0.2080030x^3 + 0.1268089x^4)^2$, $I = (0.1)$

(6) $f(x) = \log_{10} x$, $C(x) = 0.86305(x-a)/(x+1) + 0.36415(x-1)^3/(x+1)^3$, $I = (10^{-1/2}, 10^{1/2})$

10.2.   THE CHEBYSHEV SERIES BY INTEGRATION

10.2.1   Derivation

From the preceding discussion, it should be clear that we are looking for a polynomial which is "best" in the Chebyshev sense. We know in advance that it is not the polynomial derived by truncating the Taylor series. (Remember that the Taylor series is based upon the variables $x^0, x^1, x^2, x^3, \ldots$) We also know that it is not the Fourier series. That series is trigonometric, based upon the variables $\sin kx$ and $\cos kx$.

But there are other possible sets of variables to use as a basis. We shall call them $T_0(x), T_1(x), \ldots$, where every $T_k(x)$ is itself a polynomial of some kind. Then, perhaps we can find a series of the form

$$f(x) = a_0 T_0(x) + a_1 T_1(x) + \cdots = \sum_{k=0}^{\infty} a_k T_k(x) \qquad (10.2.1)$$

which can be truncated to form a useful polynomial. This possibility is what we shall investigate. The letter T comes from one of the many spellings, Tschebyshcheff, of the Russian mathematician's name.

It is obvious that if we knew all the values of $T_k(x)$—in general, an infinity of values—it would be possible to set up a set of equations to find the values of a_k. (We assume that we do know all the values of $f(x_k)$ on the interval.) However, not only do we not know the *values* of $T_k(x)$; at this point we don't even know the *form* of $T_k(x)$.

Because the Fourier series has approximately equal ripples, which we want, we shall employ a device which was successful in finding the Fourier coefficients, and then make adjustments as necessary. We shall assume that the functions $T_k(x)$ form an orthogonal set under integration relative to some weighting factor $w(x) \geqslant 0$. Then, for example, if we wish to solve for a_2, we need merely multiply both members of (10.2.1) by $T_2(x)$ and integrate over the interval. If the $T_k(x)$ are orthogonal, then

$$\int_a^b w(x) f(x) T_2(x) \, dx = a_0 \cdot 0 + a_1 \cdot 0 + a_2 \int_a^b w(x) T_2^2(x) \, dx + a_3 \cdot 0 + \cdots$$

$$(10.2.2)$$

so that

$$a_2 = \frac{\displaystyle\int_a^b w(x) f(x) T_2(x) \, dx}{\displaystyle\int_a^b w(x) T_2^2(x) \, dx} \qquad (10.2.3)$$

This removes the problem of solving an infinite set of equations, but, of course, we can still do nothing unless we know $w(x)$ and $T(x)$. We shall have to make some more guesses which we hope will be lucky.

Looking back at the Taylor series, we note that the errors increase as we approach the ends of an interval with the point of expansion at the center. Perhaps we can reduce the errors by supplying a suitable weighting function. What we wish to do is to give the values of $f(x)$ at the ends of the interval more "importance," so to speak, than those at the center. We accomplish this by guessing that the weighting function is of the form

$$w(x) = \frac{1}{\sqrt{x-a}\,\sqrt{b-x}}, \qquad x \in [a, b] \qquad (10.2.4)$$

By arbitrary agreement among mathematicians, it is customary to "normalize" the interval, in working with Chebyshev approximations, to either

$[-1, +1]$ or else $[0, 1]$. We shall choose the first of these. Then the weighting function becomes

$$w(x) = \frac{1}{\sqrt{1-x^2}}, \qquad x \in [-1, +1] \tag{10.2.5}$$

To show how this weights the ends of the interval, examine Table 10.2-1. Note that the value of the integrand in (10.2.3) is multiplied by 1 at $x = 0$, and that the integrands are multiplied by larger and larger numbers as the ends of the interval are approached.

Table 10.2-1

x	$w(x)$
± 0.9999	70.7
± 0.999	22.4
± 0.99	7.1
± 0.95	3.2
± 0.90	2.3
± 0.80	1.7
± 0.50	1.2
± 0.20	1.02
0	1

Now, in order to obtain the useful results of (10.2.2) and (10.2.3), consider

$$J = \int_{-1}^{+1} \frac{T_m(x) T_n(x)}{\sqrt{1-x^2}} dx = 0, \qquad m \neq n \tag{10.2.6}$$

which we hope will solve our problem. In trying to solve this kind of problem, it is natural to make the substitution $x = \cos\theta$, with $\theta \in [0, \pi]$. (Of course, $x = \sin\theta$, with $\theta \in [-\pi/2, +\pi/2]$ would serve just as well.) Then

$$J = \int_0^\pi \frac{T_m(\cos\theta) T_n(\cos\theta)}{\sin\theta} \sin\theta \, d\theta, \qquad m \neq n \tag{10.2.7}$$

$$J = \int_0^\pi T_m(\cos\theta) T_n(\cos\theta) \, d\theta = 0, \qquad m \neq n \tag{10.2.8}$$

There are many orthogonal functions that will satisfy (10.2.8). Using our experience with Fourier series, we try an obvious one:

$$T_k(\cos\theta) = \cos k\theta, \qquad \theta \in [0, \pi] \tag{10.2.9}$$

which we know, from previous experience, will satisfy (10.2.8). Now, we use $x = \cos\theta$ to obtain

$$T_k(x) = \cos(k \arccos x), \qquad x \in [-1, +1] \tag{10.2.10}$$

which we shall proceed to investigate. For example, we wish to know whether (10.2.10) is really a polynomial.

10.2.2 Properties of $T_n(x)$

I.

THEOREM $T_n(x)$ is a polynomial of nth degree in x.

Proof We start with De Moivre's theorem:

$$\cos n\theta + i \sin n\theta = (\cos\theta + i \sin\theta)^n \qquad (10.2.11)$$

We expand the right member and use

$$\sin^2\theta = 1 - \cos^2\theta \qquad (10.2.12)$$

Then, equating the real parts of the two members of (10.2.11), it is obvious that $\cos n\theta$ is a polynomial in $\cos\theta$, beginning with $\cos^n\theta$ [see formulas (6) to (10) of Appendix Table A.8]. Substituting $x = \cos\theta$ in (10.2.10), we have

$$T_n(x) = \cos n\theta = \cos^n\theta + \cdots = x^n + \cdots \qquad (10.2.13)$$

completing the proof that $T_n(x)$ is a polynomial of nth degree in x. $\square$

CHEBYSHEV POLYNOMIAL $T_n(x)$ The nth Chebyshev polynomial, $T_n(x)$, is defined by

$$T_n(x) = \cos(n \arccos x), \qquad x \in [-1, +1]$$

II. *Recursion Formula for $T_n(x)$*

From formula (15) of Table A.8,

$$\cos(n+1)\theta + \cos(n-1)\theta = 2\cos\theta\cos n\theta \qquad (10.2.14)$$

we have immediately

$$T_{n+1}(x) + T_{n-1}(x) = 2xT_n(x) \qquad (10.2.15)$$

and the recursion relation

$$T_{n+1}(x) = 2xT_n(x) - T_{n-1}(x) \qquad (10.2.16)$$

Starting with $T_0 = 1$ and $T_1 = x$, we quickly find

$$T_2(x) = 2x(x) - 1 = 2x^2 - 1 \qquad (10.2.17)$$

$$T_3(x) = 2x(2x^2 - 1) - x = 4x^3 - 3x \qquad (10.2.18)$$

and so on. The same result can be obtained from Table A.8 by using $T_n(x) = \cos nA$ and $x = \cos A$. For example,

$$\cos 5A = 16\cos^5 A - 20\cos^3 A + 5\cos A \tag{10.2.19}$$

becomes

$$T_5(x) = 16x^5 - 20x^3 + 5x \tag{10.2.20}$$

The first few Chebyshev polynomials are listed in Table A.17.

III

THEOREM   The coefficient of x^n in $T_n(x)$ is 2^{n-1}.

Proof We prove this by induction. Note that $T_1(x) = 2^0 x$ and $T_2(x) = 2^1 x^2 - 1$. Suppose that for any $k \geqslant 1$

$$T_k(x) = 2^{k-1} x^k + O(x^{k-1}) \tag{10.2.21}$$

We wish to show that this implies

$$T_{k+1}(x) = 2^k x^{k+1} + O(x^k) \tag{10.2.22}$$

Using (10.2.21) and the recursion relation (10.2.16), we have

$$T_{k+1}(x) = 2xT_k(x) - T_{k-1}(x) \tag{10.2.23}$$
$$= 2x\left[2^{k-1}x^k + O(x^{k-1})\right] - \left[2^{k-2}x^{k-1} + O(x^{k-2})\right] \tag{10.2.24}$$
$$= 2^k x^{k+1} + O(x^k) - 2^{k-2} x^{k-1} + O(x^{k-2}) \tag{10.2.25}$$
$$= 2^k x^{k+1} + O(x^k) \tag{10.2.26}$$

completing the induction. $\square$

IV. *Orthogonality*

We leave as an easy exercise the proof of the following:

$$\int_{-1}^{+1} \frac{T_m(x) T_n(x)}{\sqrt{1-x^2}} dx = \begin{cases} 0, & m \neq n \\ \pi/2, & m = n \neq 0 \\ \pi, & m = n = 0 \end{cases} \tag{10.2.27}$$

V. *Zeros of $T_n(x)$*

We leave as another easy exercise the proof of the following theorem:

ZEROS THEOREM   For $x \in [-1, +1]$, $T_n = 0$ when

$$x_k = \cos\left(\frac{2k+1}{n}\right)\frac{\pi}{2}, \qquad k = 0, 1, 2, \ldots, n-1 \tag{10.2.28}$$

As an example, suppose $n = 3$. Then, let

$$T_3(x) = \cos(3 \arccos x) = 0 \tag{10.2.29}$$

$$\therefore \quad 3 \arccos x = \frac{\pi}{2}, \frac{3\pi}{2}, \frac{5\pi}{2}, \frac{7\pi}{2}, \dots \tag{10.2.30}$$

$$\arccos x = \frac{\pi}{6}, \frac{3\pi}{6}, \frac{5\pi}{6}, \frac{7\pi}{6}, \dots \tag{10.2.31}$$

$$x = \cos\frac{\pi}{6}, \cos\frac{3\pi}{6}, \cos\frac{5\pi}{6}, \cos\frac{7\pi}{6}, \dots \tag{10.2.32}$$

But $T_3(x)$ is a third-degree polynomial in x. Therefore there are only three roots. All the others are repetitions. For example,

$$\cos\frac{7\pi}{6} = \cos\frac{5\pi}{6} \tag{10.2.33}$$

VI. *Extrema of $T_n(x)$*

Because $T_n(x)$ is a cosine function, there are no complications, such as double roots, in finding its zeros. The zeros are distinct and (by using Rolle's theorem), the extrema alternate and are between the successive zeros. For example (Figure 10.2-1), T_2 has two zeros and three extrema. (There are two at the endpoints.) T_3 has three zeros and four extrema, including the endpoints. In general, $T_n(x)$ has n zeros, $n-1$ extrema between them and two extrema at the endpoints. Because $T_n(x)$ is a cosine, the extreme values are always ± 1.

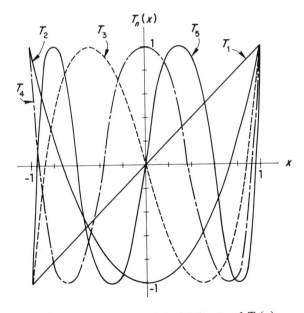

Fig. 10.2-1 The zeros and extrema of $T_n(x)$.

> EXTREMA THEOREM For $x \in [-1, +1]$, $T_n(x)$ has $n+1$ alternating extrema equal to ± 1.

10.2.3 The Chebyshev Series

Using the orthogonal properties of $T_n(x)$ and precisely the same techniques as those we employed in the Fourier series, we obtain the following Chebyshev coefficients:

$$a_0 = \frac{1}{\pi} \int_{-1}^{1} \frac{f(x)}{\sqrt{1-x^2}} dx = \frac{1}{\pi} \int_0^{\pi} f(\cos\theta) d\theta \qquad (10.2.34)$$

$$a_k = \frac{2}{\pi} \int_{-1}^{+1} \frac{f(x) T_k(x)}{\sqrt{1-x^2}} dx = \frac{2}{\pi} \int_{-1}^{+1} f(\cos\theta) \cos k\theta \, d\theta \qquad k = 1, 2, \cdots$$

$$(10.2.35)$$

These are the coefficients for the Chebyshev series

$$f(x) = a_0 T_0(x) + a_1 T_1(x) + \cdots = \sum_{k=0}^{\infty} a_k T_k(x) \qquad (10.2.36)$$

We haven't proved that the Chebyshev series is of any use at all. It will be helpful in the discussion if we have actual problems to look at. For that purpose, we provide three illustrative problems. We are going to do the first one by direct application of (10.2.34) and (10.2.35). As it happens, there is a very much easier method, which we shall display in Section 10.3.

We are not going to worry about errors of truncation or convergence. These are the reasons:

1. If you compare (10.2.34) and (10.2.35) with the corresponding definitions for the Fourier series, you will see that we have actually formed a least-square series for the function

$$g(x) = f(x) / \sqrt{1-x^2} \qquad (10.2.37)$$

   so that the theorems which we proved for the Fourier series should hold except at $x = \pm 1$. Of course (10.2.36) is *not* a least-squares series for $f(x)$.

2. In a practical sense, we are more interested in the actual errors of (10.2.36) than in any estimates or proofs of convergence. We can always compute the actual errors by calculating $f(x) - P(x)$.

3. The third reason is devastating. In fact, the *truncated* Chebyshev series usually does *not* satisfy the Chebyshev criterion. However, in some practical cases, it almost does. Then, it may either be used "as is" or else used as a first step in an empirical procedure to be displayed in the next chapter.

10.2.4 The Chebyshev Series for a Polynomial

ILLUSTRATIVE PROBLEM 10.2-1   Find the Chebyshev series for

$$f(x) = 5x^3 + 3x^2 - 2x + 8 \qquad \text{on } [-1, +1] \tag{10.2.38}$$

Solution   Using (10.2.34) and (10.2.35),

$$a_0 = \frac{1}{\pi} \int_{-1}^{+1} \frac{5x^3 + 3x^2 - 2x + 8}{\sqrt{1 - x^2}} \, dx \tag{10.2.39}$$

$$a_k = \frac{2}{\pi} \int_{-1}^{+1} \frac{(5x^3 + 3x^2 - 2x + 8) T_k(x)}{\sqrt{1 - x^2}} \, dx \tag{10.2.40}$$

Let $x = \cos\theta$, $\theta \in [0, \pi]$. Then $dx = -\sin\theta$, and $\sqrt{1 - x^2} = \sin\theta$, which is non-negative on $[0, \pi]$. Then

$$a_0 = \frac{1}{\pi} \int_{\pi}^{0} \frac{5\cos^3\theta + 3\cos^2\theta - 2\cos\theta + 8}{\sin\theta} (-\sin\theta) \, d\theta \tag{10.2.41}$$

$$a_0 = \frac{1}{\pi} \int_{0}^{\pi} (5\cos^3\theta + 3\cos^2\theta - 2\cos\theta + 8) \, d\theta. \tag{10.2.42}$$

Of course, these could have been written directly from the alternative forms of (10.2.34) and (10.2.35). From a handbook of integrals, we use:

THEOREM   Let $J(n) = \int_0^\pi (\cos\theta)^n \, d\theta$, $n = 2, 3, \ldots$. Then

$$J(n) = [(n-1)/n] J(n-2).$$

Using this, we have, since $\int_0^\pi \cos\theta \, d\theta = 0$, that for all the odd values of n, $J(n) = 0$. Also, since $J(2) = \frac{1}{2}\pi$, then $J(4) = \frac{3}{8}\pi$, $J(6) = \frac{5}{16}\pi$, and so on. Substituting in (10.2.42),

$$a_0 = \frac{1}{\pi} \left[\frac{3\pi}{2} + 8\pi \right] = \frac{19}{2} \tag{10.2.43}$$

$$a_1 = \frac{2}{\pi} \int_{-1}^{+1} \frac{f(x) T_1(x)}{\sqrt{1 - x^2}} \, dx \tag{10.2.44}$$

Using $T_1(x) = x$,

$$a_1 = \frac{2}{\pi} \int_{-1}^{+1} \frac{5x^4 + 3x^3 - 2x^2 + 8x}{\sqrt{1 - x^2}} \, dx \tag{10.2.45}$$

Using the same substitutions,

$$a_1 = \frac{2}{\pi}\left(5 \cdot \tfrac{3}{8}\pi - 2 \cdot \tfrac{1}{2}\pi\right) = \frac{7}{4} \tag{10.2.46}$$

To find a_2, we use $T_2(x) = 2x^2 - 1$:

$$a_2 = \frac{2}{\pi}\int_{-1}^{+1} \frac{f(x)(2x^2 - 1)}{\sqrt{1 - x^2}}\,dx$$

$$= \frac{4}{\pi}\int_{-1}^{+1} \frac{f(x)x^2}{\sqrt{1 - x^2}}\,dx - \frac{2}{\pi}\int_{-1}^{+1} \frac{f(x)}{\sqrt{1 - x^2}}\,dx \tag{10.2.47}$$

Note that the last integral is $2a_0$. For the next to last we have

$$\int_{-1}^{+1} \frac{f(x)x^2}{\sqrt{1 - x^2}}\,dx = \int_{-1}^{+1} \frac{5x^5 + 3x^4 - 2x^3 + 8x^2}{\sqrt{1 - x^2}}\,dx \tag{10.2.48}$$

which, after the usual substitutions, comes to $\frac{41}{8}\pi$. Then

$$a_2 = \frac{4}{\pi}\left(\frac{41\pi}{8}\right) - 2\left(\frac{19}{2}\right) = \frac{3}{2} \tag{10.2.49}$$

In the same way, we find

$$a_3 = \tfrac{5}{4} \tag{10.2.50}$$

Now consider a_4, using $T_4(x) = 8x^4 - 8x^2 + 1$:

$$a_4 = \frac{2}{\pi}\int_{-1}^{+1} \frac{f(x)(8x^4 - 8x^2 + 1)}{\sqrt{1 - x^2}}\,dx$$

$$= \frac{16}{\pi}\int_{-1}^{+1} \frac{f(x)x^4}{\sqrt{1 - x^2}}\,dx - \frac{16}{\pi}\int_{-1}^{+1} \frac{f(x)x^2}{\sqrt{1 - x^2}}\,dx + \frac{2}{\pi}\int_{-1}^{+1} \frac{f(x)}{\sqrt{1 - x^2}}\,dx$$

$$\tag{10.2.51}$$

A close look shows that, using (10.2.48) and (10.2.49),

$$a_4 = \frac{16}{\pi}\int_{-1}^{+1} \frac{5x^7 + 3x^6 - 2x^5 + 8x^4}{\sqrt{1 - x^2}}\,dx - \frac{16}{\pi}\frac{41\pi}{8} + 2 \cdot \frac{19}{2} \tag{10.2.52}$$

The value of the integral is 63, so that

$$a_4 = 63 - 82 + 19 = 0 \tag{10.2.53}$$

This could have been expected, of course, since if $a_4 \neq 0$, there would be a non-vanishing term $a_4 T_4(x)$ in the Chebyshev series, meaning that the series representing a third-degree polynomial would have a term in x^4. The Chebyshev series for the given polynomial is, therefore, exactly

$$\frac{19}{2}T_0(x) + \frac{7}{4}T_1(x) + \frac{3}{2}T_2(x) + \frac{5}{4}T_3(x) \qquad (10.2.54)$$

We repeat that this is a dreadful way to do the problem. We are going to use it to illustrate the concept of a "best" polynomial.

Suppose now we wish to represent the function in this problem by the "best" polynomial of degree 3. Then (10.2.54) is undoubtedly the answer because (1) it is of third degree owing to the term containing $T_3(x)$, which is actually $4x^3 - 3x$, and (2) the error is 0.

Let us ask for the "best" polynomial of degree 2. Then we chop off the $T_3(x)$ term and try the polynomial

$$P_2(x) = \frac{19}{2} + \frac{7}{4}T_1(x) + \frac{3}{2}T_2(x) \qquad (10.2.55)$$

The error is

$$f(x) - P_2(x) = \frac{5}{4}T_3(x) \qquad (10.2.56)$$

We already know that $T_3(x)$ has 4 equal alternating extrema equal to ± 1. This is two more than the degree of (10.2.55). Therefore, by the Chebyshev theorem, it must be the "best" polynomial.

Suppose we want the "best" polynomial of degree 1. This would require *three* alternating equal extrema. It is natural to guess

$$P_1(x) = \frac{19}{2} + \frac{7}{4}T_1(x) \qquad (10.2.57)$$

which has an error equal to

$$f(x) - P_1(x) = \frac{3}{2}T_2(x) + \frac{5}{4}T_3(x) \qquad (10.2.58)$$

Unfortunately, there is absolutely no guarantee that this error has the required extrema. Consequently, we cannot say that $P_1(x)$ is the "best" polynomial.

10.2.5 The Chebyshev Series for Arcsin x

ILLUSTRATIVE PROBLEM 10.2-2 Find the Chebyshev series for arcsin x on $[-1, +1]$.

Solution Using (10.2.34) and 10.2.35),

$$a_0 = \frac{1}{\pi}\int_0^\pi \arcsin(\cos\theta)\,d\theta = 0 \qquad (10.2.59)$$

$$a_k = \frac{2}{\pi}\int_0^\pi \arcsin(\cos\theta)\cos k\theta\,d\theta, \qquad k \geqslant 1 \qquad (10.2.60)$$

Integrating by parts,

$$a_k = \frac{4}{\pi}\left(\frac{1}{k^2}\right), \qquad k = 1,3,5,\ldots \tag{10.2.61}$$

from which

$$\arcsin x = \frac{4}{\pi}\left[T_1(x) + \frac{1}{9}T_3(x) + \frac{1}{25}T_5(x) + \cdots\right] \tag{10.2.62}$$

If we replace the Ts by xs, using Appendix Table A.17, this becomes

$$\arcsin x = \frac{4}{\pi}\left[\frac{789}{445}x - \frac{656}{945}x^3 + \frac{5808}{1575}x^5 - \frac{2560}{441}x^7 + \cdots\right]$$

These are, of course, infinite series. For practical use on a computer, they have to be truncated somewhere. A rule of thumb quoted in various textbooks states that the error of truncation is approximately the first term neglected. Since $|T_k(x)| \leqslant 1$, this implies that

$$|E_n| = |f(x) - P_n(x)| \sim |a_{n+1}| \tag{10.2.63}$$

We can test this by evaluating (10.2.62) for various numbers of terms. This is done in Figure 10.2-2, where the errors for two to five terms are displayed.

Now let us calculate the error using the rule of thumb. After two terms, the error should be about $(4/\pi) \times \frac{1}{25} \sim 0.05$; after three, about $(4/\pi) \times \frac{1}{49} \sim 0.03$; after four, about $(4/\pi) \times \frac{1}{81} \sim 0.02$; and after five, about 0.01. This doesn't even resemble the actual errors. In fact the best way to test a truncated Chebyshev series is to print a table of actual errors.

Furthermore, if we were to believe the rule of thumb and wanted a maximum error $\leqslant 10^{-16}$, we would have, from (10.2.63),

$$|a_{n+1}| \leqslant 10^{-16} \tag{10.2.64}$$

$$\frac{4}{\pi}\frac{1}{(2n-1)^2} \leqslant 10^{-16} \tag{10.2.65}$$

from which

$$n > 56,000,000 \tag{10.2.66}$$

which is unreasonable.

As a matter of curiosity, we may rewrite (10.2.62) as a power series and examine the truncated series to see whether the rule of thumb, or something like it, works for the power series. The result is shown in Figure 10.2-3. As you can see, the rule of thumb does not work for that, either.

ERROR TABLE

ARCSIN X BY CHEBYSHEV SERIES

X	TWO TERMS	THREE TERMS	FOUR TERMS	FIVE TERMS
-1.00	-1.56D-01	-1.05D-01	-7.92D-02	-6.35D-02
-0.95	3.83D-02	3.75D-02	2.17D-02	6.60D-03
-0.90	5.67D-02	2.45D-02	-1.47D-03	-1.10D-02
-0.85	5.30D-02	5.51D-03	-1.36D-02	-9.28D-03
-0.80	4.15D-02	-9.28D-03	-1.46D-02	-7.90D-04
-0.75	2.73D-02	-1.81D-02	-9.24D-03	6.10D-03
-0.70	1.29D-02	-2.13D-02	-1.67D-03	8.40D-03
-0.65	-4.41D-04	-2.01D-02	5.13D-03	6.47D-03
-0.60	-1.20D-02	-1.58D-02	9.59D-03	2.17D-03
-0.55	-2.14D-02	-9.76D-03	1.11D-02	-2.43D-03
-0.50	-2.85D-02	-2.99D-03	1.00D-02	-5.71D-03
-0.45	-3.32D-02	3.58D-03	6.84D-03	-6.87D-03
-0.40	-3.58D-02	9.24D-03	2.54D-03	-5.84D-03
-0.35	-3.62D-02	1.35D-02	-1.97D-03	-3.18D-03
-0.30	-3.48D-02	1.61D-02	-5.88D-03	2.32D-04
-0.25	-3.16D-02	1.69D-02	-8.57D-03	3.42D-03
-0.20	-2.71D-02	1.60D-02	-9.67D-03	5.59D-03
-0.15	-2.13D-02	1.35D-02	-9.10D-03	6.25D-03
-0.10	-1.47D-02	9.74D-03	-7.03D-03	5.30D-03
-0.05	-7.51D-03	5.10D-03	-3.82D-03	3.02D-03
0.00	3.25D-17	-2.23D-17	1.69D-17	-1.36D-17
0.05	7.51D-03	-5.10D-03	3.82D-03	-3.02D-03
0.10	1.47D-02	-9.74D-03	7.03D-03	-5.30D-03
0.15	2.13D-02	-1.35D-02	9.10D-03	-6.25D-03
0.20	2.71D-02	-1.60D-02	9.67D-03	-5.59D-03
0.25	3.16D-02	-1.69D-02	8.57D-03	-3.42D-03
0.30	3.48D-02	-1.61D-02	5.88D-03	-2.32D-04
0.35	3.62D-02	-1.35D-02	1.97D-03	3.18D-03
0.40	3.58D-02	-9.24D-03	-2.54D-03	5.84D-03
0.45	3.32D-02	-3.58D-03	-6.84D-03	6.87D-03
0.50	2.85D-02	2.99D-03	-1.00D-02	5.71D-03
0.55	2.14D-02	9.76D-03	-1.11D-02	2.43D-03
0.60	1.20D-02	1.58D-02	-9.59D-03	-2.17D-03
0.65	4.41D-04	2.01D-02	-5.13D-03	-6.47D-03
0.70	-1.29D-02	2.13D-02	1.67D-03	-8.40D-03
0.75	-2.73D-02	1.81D-02	9.24D-03	-6.10D-03
0.80	-4.15D-02	9.28D-03	1.46D-02	7.90D-04
0.85	-5.30D-02	-5.51D-03	1.36D-02	9.28D-03
0.90	-5.67D-02	-2.45D-02	1.47D-03	1.10D-02
0.95	-3.83D-02	-3.75D-02	-2.17D-02	-6.60D-03
1.00	1.56D-01	1.05D-01	7.92D-02	6.35D-02

Fig. 10.2-2 Errors for a truncated Chebyshev series.

10.2.6 The Chebyshev Series for e^x

ILLUSTRATIVE PROBLEM 10.2-3 Find the Chebyshev Series for e^x on $[-1, +1]$.

Discussion Using (10.2.34) and (10.2.35),

$$a_0 = \frac{1}{\pi} \int_{-1}^{1} \frac{e^x}{\sqrt{1-x^2}} \, dx \qquad (10.2.67)$$

$$a_k = \frac{2}{\pi} \int_{-1}^{1} \frac{e^x T_k(x)}{\sqrt{1-x^2}} \, dx \qquad (10.2.68)$$

ARCSIN X BY CHEBYSHEV SERIES

X	TWO TERMS	THREE TERMS	FOUR TERMS	FIVE TERMS
-1.00	-1.39D 00	3.30D 00	-4.09D 00	-6.35D-02
-0.95	-1.00D 00	2.63D 00	-2.53D 00	6.60D-03
-0.90	-8.07D-01	1.97D 00	-1.57D 00	-1.10D-02
-0.85	-6.55D-01	1.43D 00	-9.41D-01	-9.28D-03
-0.80	-5.29D-01	1.01D 00	-5.41D-01	-7.90D-04
-0.75	-4.24D-01	6.91D-01	-2.96D-01	6.10D-03
-0.70	-3.34D-01	4.55D-01	-1.54D-01	8.40D-03
-0.65	-2.59D-01	2.85D-01	-7.69D-02	6.47D-03
-0.60	-1.97D-01	1.69D-01	-3.84D-02	2.17D-03
-0.55	-1.45D-01	9.16D-02	-2.10D-02	-2.43D-03
-0.50	-1.03D-01	4.42D-02	-1.36D-02	-5.71D-03
-0.45	-6.89D-02	1.77D-02	-9.91D-03	-6.87D-03
-0.40	-4.29D-02	5.22D-03	-6.89D-03	-5.84D-03
-0.35	-2.34D-02	1.26D-03	-3.49D-03	-3.18D-03
-0.30	-9.64D-03	1.77D-03	1.52D-04	2.32D-04
-0.25	-7.27D-04	3.86D-03	3.41D-03	3.42D-03
-0.20	4.18D-03	5.68D-03	5.59D-03	5.59D-03
-0.15	5.91D-03	6.26D-03	6.25D-03	6.25D-03
-0.10	5.25D-03	5.30D-03	5.30D-03	5.30D-03
-0.05	3.02D-03	3.02D-03	3.02D-03	3.02D-03
0.00	-1.36D-17	-1.36D-17	-1.36D-17	-1.36D-17
0.05	-3.02D-03	-3.02D-03	-3.02D-03	-3.02D-03
0.10	-5.25D-03	-5.30D-03	-5.30D-03	-5.30D-03
0.15	-5.91D-03	-6.26D-03	-6.25D-03	-6.25D-03
0.20	-4.18D-03	-5.68D-03	-5.59D-03	-5.59D-03
0.25	7.27D-04	-3.86D-03	-3.41D-03	-3.42D-03
0.30	9.64D-03	-1.77D-03	-1.52D-04	-2.32D-04
0.35	2.34D-02	-1.26D-03	3.49D-03	3.18D-03
0.40	4.29D-02	-5.22D-03	6.89D-03	5.84D-03
0.45	6.89D-02	-1.77D-02	9.91D-03	6.87D-03
0.50	1.03D-01	-4.42D-02	1.36D-02	5.71D-03
0.55	1.45D-01	-9.16D-02	2.10D-02	2.43D-03
0.60	1.97D-01	-1.69D-01	3.84D-02	-2.17D-03
0.65	2.59D-01	-2.85D-01	7.69D-02	-6.47D-03
0.70	3.34D-01	-4.55D-01	1.54D-01	-8.40D-03
0.75	4.24D-01	-6.91D-01	2.96D-01	-6.10D-03
0.80	5.29D-01	-1.01D 00	5.41D-01	7.90D-04
0.85	6.55D-01	-1.43D 00	9.41D-01	9.28D-03
0.90	8.07D-01	-1.97D 00	1.57D 00	1.10D-02
0.95	1.00D 00	-2.63D 00	2.53D 00	-6.60D-03
1.00	1.39D 00	-3.30D 00	4.09D 00	6.35D-02

Fig. 10.2-3 Errors for a truncated Chebyshev series which has been rewritten as a power series.

These are very difficult integrals. The substitutions $x = \cos\theta$, $T_k(x) = \cos k\theta$ lead to

$$a_0 = \frac{1}{\pi} \int_0^\pi e^{\cos\theta}\, d\theta \qquad (10.2.69)$$

$$a_k = \frac{2}{\pi} \int_0^\pi e^{\cos\theta} \cos k\theta\, d\theta \qquad (10.2.70)$$

which also occasion no great joy. In the next chapter, we shall describe a practical method for satisfying the Chebyshev criterion.

10.2.7 Shifted Chebyshev Polynomials

We mentioned (in Section 10.2.1) that for interval $[a,b]$ the weight function for a Chebyshev approximation is $[(x-a)(b-x)]^{-1/2}$. For $a = -1$

and $b=1$, the interval is $[-1, +1]$, $w(x)=(1-x^2)^{-1/2}$, and the Chebyshev polynomials are $T_k(x)=\cos(k\arccos x)$.

If the interval is $[0,1]$, the weight function is $[(x-0)(1-x)]^{-1/2}$ or $(x-x^2)^{-1/2}$, and the polynomials are said to be *shifted*.

SHIFTED CHEBYSHEV POLYNOMIALS $T_k^*(x)$

$$T_k^*(x)=\cos[k\arccos(2x-1)] \qquad (10.2.71)$$

The following little theorem is immediate.

THEOREM

$$T_k^*(x)=T_k(2x-1) \qquad (10.2.72)$$

The recursion relation is also immediate, using (10.2.15) and (10.2.72):

THEOREM

$$T_k^*(x)=(4x-2)T_k^*(x)-T_{k-1}^*(x) \qquad (10.2.73)$$

The first few shifted polynomials are as follows:

$$
\begin{aligned}
T_0^*(x) &= 1\\
T_1^*(x) &= 2x-1\\
T_2^*(x) &= 8x^2-8x+1\\
T_3^*(x) &= 32x^3-48x^2+18x-1\\
T_4^*(x) &= 128x^4-256x^3+160x^2-32x+1 \qquad (10.2.74)
\end{aligned}
$$

The first few reverse transformations are

$$
\begin{aligned}
1 &= T_0^*(x)\\
x &= \tfrac{1}{2}(T_0^*+T_1^*)\\
x^2 &= \tfrac{1}{8}(3T_0^*+4T_1^*+T_2^*)\\
x^3 &= \tfrac{1}{32}(10T_0^*+15T_1^*+6T_2^*+T_3^*) \qquad (10.2.75)
\end{aligned}
$$

We have used the interval $[-1, +1]$, but it may be more convenient, from time to time, to use the shifted interval $[0,1]$. Then the substitution

$$x=\frac{1+t}{2} \qquad (10.2.76)$$

together with (10.2.71) serves to make the equivalent computation.

10.2.8 Change of Scale

More generally, suppose the function $f(x)$ is to be investigated on the interval $[a, b]$, but we wish to use the (ordinary) Chebyshev polynomials. Then our problem is to find m and c such that

$$x = mt + c \tag{10.2.77}$$

where $t = -1 \; (+1)$ when $x = a \; (b)$. In the usual fashion, we set up a pair of equations

$$\begin{cases} a = -m + c \\ b = m + c \end{cases} \tag{10.2.78}$$

and obtain

$$x = \left(\frac{b-a}{2}\right)t + \left(\frac{b+a}{2}\right) \tag{10.2.79}$$

$$t = \frac{2x - (b+a)}{b-a} \tag{10.2.80}$$

In the following problem, we demonstrate the application of these formulas.

ILLUSTRATIVE PROBLEM 10.2-4 Write a Chebyshev series for $f(x) = (1+x)^{-1}$, where $x \in [0, 1]$.

Solution We shall change the interval so that the (ordinary) $T_k(x)$ can be used. Using $t = 2x - 1$, or $x = \frac{1}{2}(t+1)$, the corresponding function is

$$g(t) = \frac{2}{t+3}, \qquad t \in [-1, +1] \tag{10.2.81}$$

Then, in terms of t,

$$a_0 = \frac{1}{\pi} \int_0^\pi g(\cos\theta)\, d\theta = \frac{1}{\pi} \int_0^\pi \frac{2}{3+\cos\theta}\, d\theta \tag{10.2.82}$$

$$a_k = \frac{2}{\pi} \int_0^\pi g(\cos\theta) \cos k\theta\, d\theta = \frac{2}{\pi} \int_0^\pi \frac{2\cos k\theta}{3+\cos\theta}\, d\theta \tag{10.2.83}$$

From a handbook of integrals,

$$\int_0^\pi \frac{\cos k\theta}{a + b\cos\theta}\, d\theta = \frac{\pi(c-a)^k}{cb^k}, \qquad c = \sqrt{a^2 - b^2}, \quad k \geqslant 0 \tag{10.2.84}$$

Here $a = 3$, $b = 1$, $c = 2\sqrt{2}$. Therefore

$$a_0 = \tfrac{1}{2}\sqrt{2} \tag{10.2.85}$$

$$a_k = \sqrt{2}\,(2\sqrt{2} - 3)^k, \qquad k \geqslant 1 \tag{10.2.86}$$

from which

$$g(t) = a_0 T_0(t) + a_1 T_1(t) + \cdots + a_k T_k(t) + \cdots \qquad (10.2.87)$$

$$g(t) = \sqrt{2} \left[\tfrac{1}{2} + pt + p^2 T_1(t) + \cdots + p^k T_k(t) + \cdots \right] \qquad (10.2.88)$$

where $p = 2\sqrt{2} - 3 \sim -0.1715728753$ and $t = 2x - 1$.

In this specific case, for $x = 0.9$, a Taylor series would require over 100 terms for an error $< 5 \times 10^{-6}$, but the Chebyshev series to only 8 terms is equally efficient. Of course, it would be even easier to calculate $1/(1+x)$ directly, but this example shows that from time to time a truncated Chebyshev series may be surprisingly good.

PROBLEM SECTION 10.2

(1) Prove property IV, Section 10.2.2.
(2) Prove property V, Section 10.2.2.
(3) Prove that $2T_m(x)T_n(x) = T_{n+m}(x) - T_{n-m}(x)$, $n > m$.
   [*Hint*: Let $T_k(x) = \cos kx$.]
(4) (a) Prove that

$$T_k'(x) = \frac{k \sin k\theta}{\sin \theta}$$

   (b) Show that $T_1'(x) = T_0(x)$ and $T_2'(x) = 4T_1(x)$.
(5) Using the previous problem, prove that

$$(1 - x^2)T_k'(x) = k[T_{k-1}(x) - xT_k(x)]$$

(6) Use De Moivre's theorem to show that $\cos 3\theta = 4\cos^3\theta - 3\cos\theta$ and therefore

$$T_3(x) = 4x^3 - 3x$$

(7) Use De Moivre's theorem to show that

$$T_4(x) = 8x^4 - 8x^2 + 1$$

(8) Show that $x^2 = \tfrac{1}{2}[T_0(x) + T_2(x)]$.
(9) (a) Find the Chebyshev series on $[-1, +1]$ for

$$f(x) = -4x^3 + 2x^2 - 5x + 2$$

   (b) What is the "best" polynomial approximation of second degree?
   (c) What is the maximum error of this approximation?
(10) (a) Find the Chebyshev series on $[-1, +1]$ to

$$f(x) = 12x^3 - 8x^2 - 8x + 6$$

   (b) What is the "best" polynomial approximation of second degree?
   (c) What is the maximum error of this approximation?
(11) Prove that $T_n^*(x^2) = T_{2n}(x)$. [*Hint*: Let $x = \cos\theta$.]
(12) Prove that

$$\int_0^1 \frac{T_m^*(x) T_n^*(x)}{\sqrt{x(1-x)}} dx = \begin{cases} 0, & m \neq n \\ \pi, & m = n = 0 \\ \pi/2, & m = n \neq 0 \end{cases}$$

(13) Find the coefficient of x^n in $T_n^*(x)$.

(14) Find the Chebyshev series for $\arccos x$ on $[-1, +1]$.

(15) Find the Chebyshev series for $f(x) = \sqrt{1 - x^2}$ on $[-1, +1]$.

10.3 THE CHEBYSHEV SERIES FOR A POLYNOMIAL

10.3.1 Introduction

We now return to the problem in Section 10.2.4 to introduce an easier method and to emphasize an important concept.

ILLUSTRATIVE PROBLEM 10.3-1   (a) Find the Chebyshev series for

$$f(x) = 5x^3 + 3x^2 - 2x + 8 \tag{10.3.1}$$

on $[1, +1]$. (b) Find the "best" polynomial approximation of second degree.

Solution   (a) We should remind you that if we really wanted to compute (10.3.1), the best way would be to nest the original polynomial as follows:

$$f(x) = 8 + x(-2 + x(3 + 5x)) \tag{10.3.2}$$

so that this problem is merely illustrative. We now use Table A.17 and find that

$$f(x) = 5\left[\frac{1}{4}(3T_1 + T_3)\right] + 3\left[\frac{1}{2}(T_0 + T_2)\right] - 2T_1 + 8T_0 \tag{10.3.3}$$

from which, immediately,

$$f(x) = \frac{5}{4}T_3(x) + \frac{3}{2}T_2(x) + \frac{7}{4}T_1(x) + \frac{19}{2}T_0(x) \tag{10.3.4}$$

the same result as that obtained by the laborious procedure of the previous section. Note that (10.3.4) is a *finite* series. (b) The "best" polynomial of second degree is obtained by dropping the $T_3(x)$ term, as explained in Section 10.2. The result is

$$P_2(x) = \frac{3}{2}T_2(x) + \frac{7}{4}T_1(x) + \frac{19}{2}T_0(x) \tag{10.3.5}$$

with an error

$$E_2(x) \leqslant \frac{5}{4}T_3(x) \tag{10.3.6}$$

which clearly satisfies the Chebyshev criterion (Section 10.1.3). It is better computing practice to express (10.3.5) as a polynomial in nested form. Using Table A.17 again,

$$P_2(x) = \frac{3}{2}(2x^2 - 1) + \frac{7}{4}x + \frac{19}{2} = 3x^2 + \frac{7}{4}x + 8 \tag{10.3.7}$$

$$P_2(x) = 8 + x\left(\frac{7}{4} + 3x\right) \tag{10.3.8}$$

10.3.2 Economization

We shall now describe and illustrate a method which is supposed to allow us to approximate the Chebyshev series without doing the integration. Contrary to some claims in the literature, it does not always do so. However, the method has other values, in particular, it speeds up a computation. It can be used whenever there is a known polynomial expression (such as a Maclaurin expansion) for $f(x)$. The process is called *economization* because it economizes on the number of terms needed for computation. The steps are:

Step I. Change scale, if necessary, so that the argument is in $[-1, +1]$.

Step II. Write the Maclaurin expansion, truncating when the actual error of truncation is satisfactorily small. This gives us the polynomial. (Any method of obtaining the polynomial is satisfactory.)

Step III. Using Table A.17, rewrite the *last* term of the polynomial in terms of T_k.

Step IV. Drop the highest T_k. This provides the "best" polynomial approximation to the truncated polynomial.

Step V. Substitute and simplify.

Step VI. Convert to nested polynomial form for computing.

ILLUSTRATIVE PROBLEM 10.3-2 Economize four terms of the Maclaurin expansion for $\arcsin x$ with the interval $[-1, +1]$.

Solution *Step I.* No adjustment is needed in this case.

Step II. The Maclaurin expansion is

$$\arcsin x = x + \frac{x^3}{2 \times 3} + \frac{1 \times 3}{2 \times 4 \times 5} x^5 + \frac{1 \times 3 \times 5}{2 \times 4 \times 6 \times 7} x^7 + R_4, \qquad x^2 < 1 \quad (10.3.9)$$

$$\arcsin x = x + \frac{1}{6} x^3 + \frac{3}{40} x^5 + \frac{15}{336} x^7 + R_4 \qquad (10.3.10)$$

Step III. The last term kept in (10.3.10) is

$$\frac{15}{336} x^7 = \frac{15}{336} \left[\frac{1}{64} (35 T_1 + 21 T_3 + 7 T_5 + T_7) \right] \qquad (10.3.11)$$

Step IV. We drop the T_7 term. This *adds* an error no greater than $\frac{15}{336} \times \frac{1}{64} \sim 7 \times 10^{-4}$. Now we replace T_1, T_3 and T_5 in (10.3.10), using Table A.17:

$$\frac{15}{336} x^7 \sim \frac{15}{336} \cdot \frac{1}{64} \left[35x + 21(4x^3 - 3x) + 7(16x^5 - 20x^3 + 5x) \right] \quad (10.3.12)$$

Step V. Returning to (10.3.10), we drop R_4 and substitute from (10.3.12). The new coefficients are as follows:

$$x:\ 1 + \frac{15}{(336)(64)}[35 + 21(-3) + 7(5)] = \frac{1029}{1024}$$

$$x^3:\ \frac{1}{6} + \frac{15}{(336)(64)}[21(4) + 7(-20)] = \frac{343}{2688}$$

$$x^5:\ \frac{3}{40} + \frac{15}{(336)(64)}[7(16)] = \frac{343}{2240}$$

thus

$$P(x) = \frac{1029}{1024}x + \frac{343}{2688}x^3 + \frac{343}{2240}x^5 \qquad (10.3.13)$$

Step VI. For computation,

$$P(x) = x\left(\frac{1029}{1024} + 343x^2\left(\frac{1}{2688} + \frac{1}{2240}x^2\right)\right) \qquad (10.3.14)$$

ERROR TABLE (ARCSIN X)

X	MACLAURIN	ECONOMIZED
-1.00	-2.84D-01	-2.85D-01
-0.95	-7.11D-02	-7.07D-02
-0.90	-3.26D-02	-3.19D-02
-0.85	-1.60D-02	-1.55D-02
-0.80	-8.02D-03	-7.88D-03
-0.75	-3.99D-03	-4.23D-03
-0.70	-1.95D-03	-2.48D-03
-0.65	-9.23D-04	-1.60D-03
-0.60	-4.19D-04	-1.10D-03
-0.55	-1.81D-04	-7.42D-04
-0.50	-7.29D-05	-4.22D-04
-0.45	-2.71D-05	-1.15D-04
-0.40	-9.04D-06	1.71D-04
-0.35	-2.63D-06	4.13D-04
-0.30	-6.41D-07	5.90D-04
-0.25	-1.22D-07	6.84D-04
-0.20	-1.60D-08	6.88D-04
-0.15	-1.19D-09	6.06D-04
-0.10	-3.07D-11	4.50D-04
-0.05	-6.32D-14	2.39D-04
0.00	0.0	-1.05D-18
0.05	6.32D-14	-2.39D-04
0.10	3.07D-11	-4.50D-04
0.15	1.19D-09	-6.06D-04
0.20	1.60D-08	-6.88D-04
0.25	1.22D-07	-6.84D-04
0.30	6.41D-07	-5.90D-04
0.35	2.63D-06	-4.13D-04
0.40	9.04D-06	-1.71D-04
0.45	2.71D-05	1.15D-04
0.50	7.29D-05	4.22D-04
0.55	1.81D-04	7.42D-04
0.60	4.19D-04	1.10D-03
0.65	9.23D-04	1.60D-03
0.70	1.95D-03	2.48D-03
0.75	3.99D-03	4.23D-03
0.80	8.02D-03	7.88D-03
0.85	1.60D-02	1.55D-02
0.90	3.26D-02	3.19D-02
0.95	7.11D-02	7.07D-02
1.00	2.84D-01	2.85D-01

Fig. 10.3-1 A comparison of the errors when $\arcsin x$ is computed from a Maclaurin series and when it is computed from an economized Maclaurin series.

Figure 10.3-1 compares the errors of the original Maclaurin series with those of the economized series, and Figure 10.3-2 shows the simple program. Note that the effect of the economization was to spread the errors more evenly over the interval. Note the alternation of positive and negative errors in Figure 10.3-1.

Is the economized series, which uses three terms rather than the original four of the Maclaurin series, better than the original? Actually, it is not, in this case, but the additional error (about 0.0007) was offset by the reduced roundoff error. In some cases, there is an even greater improvement because of the reduction in roundoff error. The worse the roundoff error, the better the effect of economization.

The following table compares the coefficients of the Maclaurin series with those for the economized series and those for the Chebyshev series:

	MACLAURIN	ECONOMIZED	CHEBYSHEV
x	1.00000	1.00488	0.83492
x^3	0.66667	0.12760	-0.69418
x^5	0.75000	0.15313	3.68762
x^7	0.04464	0.00000	-5.80500
x^9	0.03038		3.16049

It should be clear that the economized Maclaurin series makes use of the Chebyshev polynomials, but it is not a Chebyshev series at all. If we

```
      DOUBLE PRECISION ASM(41),ASC(41),EM(41),EC(41),X,Y,A1,A2,A3,Z,B1,B2
     1,B3
C COEFFICIENTS FOR ECONOMIZED SERIES
      A1=1029.D00/1024.D00
      A2=343.D00/2688.D00
      A3=343.D00/2240.D00
      B1=1.0D00/6.0D00
      B2=3.0D00/40.0D00
      B3=15.0D00/336.0D00
C CALCULATION OF VALUES AND ERRORS
      X=-1.05D00
      DO 1 I=1,41
      X=X+0.05D00
      Y=X**2
      ASM(I)=X*(1.0D00+Y*(B1+Y*(B2+B3*Y)))
      ASC(I)=X*(A1+Y*(A2+Y*A3))
      Z=DARSIN(X)
      EM(I)=Z-ASM(I)
    1 EC(I)=Z-ASC(I)
C HEADING
      WRITE(3,100)
  100 FORMAT(1H1,T24,'ERROR TABLE (ARCSIN X)')
      WRITE(3,101)
  101 FORMAT(1H0,T16,'X',T28,'MACLAURIN',T46,'ECONOMIZED')
C
      X=-1.05D00
      DO 2 I=1,41
      X=X+0.05D00
    2 WRITE(3,102)X,EM(I),EC(I)
  102 FORMAT(1H ,T14,F5.2,T28,1PD9.2,T47,D9.2)
      STOP
      END
```

Fig. 10.3-2 Program for Illustrative Problem 10.3-2.

compare the results of a computation by all three methods, the worst one is the truncated infinite Chebyshev series.

10.3.3 Change of Scale

ILLUSTRATIVE PROBLEM 10.3-3   It is required to economize a Maclaurin series on $[0,2]$. Describe the change of scale.

Solution   We want to use a new argument, t, which is -1 $(+1)$ when x is 0 (2). We start with the substitution formula

$$x = mt + b \tag{10.3.15}$$

Then

$$\begin{cases} 2 = m + b & (10.3.16) \\ 0 = -m + b & (10.3.17) \end{cases}$$

from which $b = 1$ and $m = 1$. Therefore

$$x = t + 1 \tag{10.3.18}$$

$$t = x - 1 \tag{10.3.19}$$

If we wished to economize the arcsine series on $[0,2]$, we could not do so, of course, because arcsine exists for real numbers only when $x^2 \leqslant 1$, and the series converges only for $x^2 < 1$. However, if we wished to economize the series for e^x on $[0,2]$, we would start with

$$e^x = 1 + x + \frac{1}{2}x^2 + \frac{1}{6}x^3 + \cdots, \qquad 0 \leqslant x \leqslant 2 \tag{10.3.20}$$

and substitute for x.

The following illustrative problem demonstrates how this is done. It also shows that economization does not transform an inefficient series into an efficient one. All it does is spread the errors over the interval.

ILLUSTRATIVE PROBLEM 10.3-4   Economize six terms of the e^x series on the interval $[0,2]$.

Solution   Using $x = t + 1$, with $t \in [-1, +1]$,

$$e^x = e^{t+1} = 1 + (t+1) + \frac{1}{2}(t+1)^2 + \frac{1}{6}(t+1)^3 + \frac{1}{24}(t+1)^4 + \frac{1}{120}(t+1)^5 + R_6 \tag{10.3.21}$$

$$= \frac{163}{60} + \frac{65}{24}t + \frac{4}{3}t^2 + \frac{5}{12}t^3 + \frac{1}{12}t^4 + \frac{1}{120}t^5 + R_6 \tag{10.3.22}$$

We prepare to spread the t^5 term:

$$\frac{1}{120}t^5 = \frac{1}{120}\left[\frac{1}{16}(10T_1 + 5T_3 + T_5)\right] \tag{10.3.23}$$

Dropping the T_5 contribution will add an error no greater than $\frac{1}{20} \times \frac{1}{16} \sim$ 5.2×10^{-4}. We therefore use

$$\frac{1}{120} t^5 \sim \frac{1}{1920} \left[10t + 5(4t^3 - 3t) \right] = \frac{1}{384} [-t + 4t^3] \qquad (10.3.24)$$

and replace this term in (10.3.22). Then

$$e^x \sim \frac{163}{60} + \left(\frac{65}{24} - \frac{1}{384} \right) t + \frac{4}{3} t^2 + \left(\frac{5}{12} + \frac{4}{384} \right) t^3 + \frac{1}{12} t^4 \qquad (10.3.25)$$

$$e^x \sim \frac{163}{60} + \frac{1039}{384} t + \frac{4}{3} t^2 + \frac{41}{96} t^3 + \frac{1}{12} t^4 \qquad (10.3.26)$$

For programming purposes, we might very well work with (10.3.26), using $t = x - 1$, but we shall rewrite this equation in order to see what happened to the coefficients:

$$e^x \sim \frac{163}{60} + \frac{1039}{384} (x-1) + \frac{4}{3} (x-1)^2 + \frac{41}{96} (x-1)^3 + \frac{1}{12} (x-1)^4 \qquad (10.3.27)$$

$$e^x \sim \frac{1921}{1920} + \frac{379}{384} x + \frac{53}{96} x^2 + \frac{3}{32} x^3 + \frac{1}{12} x^4 \qquad (10.3.28)$$

Figure 10.3-3 compares the errors for the Maclaurin series with those for the economized Maclaurin series. In this case, economization increased the error.

For comparison, the coefficients were altered as follows:

	Maclaurin	Economized
1	1.00000	1.00005
x	1.00000	0.98698
x^2	0.50000	0.55208
x^3	0.16667	0.09375
x^4	0.04167	0.08333
x	0.00833	—

Trigonometric functions are often economized over regions identified with principal values.

ILLUSTRATIVE PROBLEM 10.3-5   Prepare the Maclaurin series for $\cos x$ to be economized, where $x \in [0, \pi]$.

Solution

$$x = mt + b \qquad (10.3.29)$$

so

$$\begin{cases} \pi = m + b & (10.3.30) \\ 0 = -m + b & (10.3.31) \end{cases}$$

ERROR TABLE FOR EXP(X) ECONOMIZED

X	MACLAURIN	ECONOMIZED
0.0	0.0	-5.21D-04
0.05	2.19D-11	8.86D-06
0.10	1.41D-09	3.29D-04
0.15	1.62D-08	4.86D-04
0.20	9.15D-08	5.19D-04
0.25	3.52D-07	4.64D-04
0.30	1.06D-06	3.50D-04
0.35	2.69D-06	2.04D-04
0.40	6.03D-06	4.55D-05
0.45	1.23D-05	-1.06D-04
0.50	2.34D-05	-2.37D-04
0.55	4.17D-05	-3.35D-04
0.60	7.08D-05	-3.90D-04
0.65	1.15D-04	-3.93D-04
0.70	1.81D-04	-3.39D-04
0.75	2.76D-04	-2.20D-04
0.80	4.10D-04	-2.99D-05
0.85	5.95D-04	2.39D-04
0.90	8.45D-04	5.95D-04
0.95	1.18D-03	1.05D-03
1.00	1.62D-03	1.62D-03
1.05	2.18D-03	2.31D-03
1.10	2.91D-03	3.16D-03
1.15	3.83D-03	4.18D-03
1.20	4.98D-03	5.42D-03
1.25	6.42D-03	6.91D-03
1.30	8.18D-03	8.71D-03
1.35	1.03D-02	1.09D-02
1.40	1.30D-02	1.34D-02
1.45	1.62D-02	1.65D-02
1.50	2.00D-02	2.02D-02
1.55	2.45D-02	2.46D-02
1.60	2.99D-02	2.99D-02
1.65	3.63D-02	3.61D-02
1.70	4.38D-02	4.34D-02
1.75	5.26D-02	5.21D-02
1.80	6.28D-02	6.23D-02
1.85	7.47D-02	7.42D-02
1.90	8.84D-02	8.81D-02
1.95	1.04D-01	1.04D-01
2.00	1.22D-01	1.23D-01

Fig. 10.3-3 A comparison of the errors when e^x is computed from a Maclaurin series and when it is computed from an economized Maclaurin series.

from which $b = \pi/2$ and $m = \pi/2$, so that

$$x = \frac{\pi}{2}(t+1) \tag{10.3.32}$$

$$\cos x = 1 - \frac{x^2}{2} + \frac{x^4}{24} - \frac{x^6}{720} + \cdots \tag{10.3.33}$$

$$\cos x = 1 - \frac{\pi^2}{4}(t+1)^2 + \frac{\pi^4}{384}(t+1)^4 - \frac{\pi^6}{46080}(t+1)^6 + \ldots \tag{13.3.34}$$

and so on.

Not every series will "economize". This is shown in the following.

ILLUSTRATIVE PROBLEM 10.3-6   Economize four terms of the series for $\ln(1+x)$ on the interval $[-1, +1]$.

Discussion

$$\ln(1+x) = x - \frac{1}{2}x^2 + \frac{1}{3}x^3 - \frac{1}{4}x^4 + \cdots$$

$$-\frac{1}{4}x^4 = -\frac{1}{4}\left[\frac{1}{8}(3T_0 + 4T_2 + T_4)\right]$$

Dropping the T_4 term adds an error of $\frac{1}{4} \times \frac{1}{8}$ or about 0.03:

$$-\frac{1}{4}x^4 \sim -\frac{1}{32}[3T_0 + 4T_2] = \frac{1}{32} - \frac{x^2}{4}$$

Then

$$\ln(1+x) \sim \left(x - \frac{1}{2}x^2 + \frac{1}{3}x^3\right) + \left(\frac{1}{32} - \frac{x^2}{4}\right)$$

$$\ln(1+x) \sim \frac{1}{32} + x - \frac{3}{4}x^2 + \frac{1}{3}x^3$$

which has as many terms as the original.

10.3.4 Other Polynomials

Chebyshev polynomials are by no means the only ones which can be used for approximations. The Legendre polynomials, Laguerre polynomials, and Hermite polynomials and their properties are discussed in Hildebrand, pp. 272–279.

PROBLEM SECTION 10.3

Economize each of the following by one term and estimate the maximum additional error due to the economization. Sketch an error curve over the interval for the original Maclaurin series and for the economized series.

A. For the following, the interval is $[-1, +1]$:

(1) $\sin x = x - \dfrac{x^3}{6} + \dfrac{x^5}{120} - \dfrac{x^7}{5040} + R_4$

(2) $\cosh x = 1 + \dfrac{x^2}{2} + \dfrac{x^4}{24} + \dfrac{x^6}{720} + R_4$

(3) $\tan x = x + \dfrac{x^3}{3} + \dfrac{2x^5}{15} + \dfrac{17x^7}{315} + R_4$

(4) $\tan^{-1}x = x - \dfrac{1}{3}x^3 + \dfrac{1}{5}x^5 - \dfrac{1}{7}x^7 + R_4$

(5) $e^{\sin x} = 1 + x + \dfrac{x^2}{2} - \dfrac{x^4}{8} + R_4$

(6) $e^{\cos x} = e[1 - \dfrac{x^2}{2} + \dfrac{x^4}{6} - \dfrac{31x^6}{720}] + R_4$

B. In the following, the interval is $[-\pi/2, +\pi/2]$:
(7) $\sin x$ (8) $\tan x$

Chapter 11

The Chebyshev Summation

This is a very important chapter. Don't omit anything.

On the whole, the preceding chapter, while it has perhaps illuminated the nature of the Chebyshev criterion, did not display practical methods for satisfying it. One great hurdle is the integral with the denominator $\sqrt{1-x^2}$. Another, equally uncomfortable, is that the *truncated* infinite series does not, except accidentally, satisfy the Chebyshev criterion.

We shall reduce one difficulty in the first section, which deals with a *finite* Chebyshev series, and the second in the second section, which deals with a remarkably efficient empirical method, the *Hastings method*. The third section deals with an empirical method for producing a rational fraction to represent a given function.

11.1 THE FINITE SERIES

11.1.1 Introduction

Having defined and investigated $T_k(x)$ for the infinite series, we shall turn our attention to the finite series, which we shall call the *Chebyshev summation*. As in the similar situation with the Fourier series, we assume that we can choose N values of x: $x_0, x_1, \ldots, x_{N-1}$, to find the coefficients of the summation

$$P(x) = \sum_{k=0}^{N-1} A_k T_k(x) = A_0 T_0(x) + \cdots + A_{N-1} T_{N-1}(x) \quad (11.1.1)$$

Clearly, $P(x)$ is of degree $N-1$ or less, since $T_{N-1}(x)$ is a polynomial of degree $N-1$. Our intention is to approximate $f(x)$ by $P(x)$ using the N

chosen values of x:

$$f(x_0) = P(x_0)$$
$$f(x_1) = P(x_1)$$
$$\vdots$$
$$f(x_{N-1}) = P(x_{N-1})$$

(11.1.2)

where P is of degree $N - 1$.

11.1.2 The Basic Assumption

We shall assume, *for the purpose of choosing suitable values of x*, that $f(x)$ is a polynomial $Q(x)$ of Nth degree. In general, this is not true, but it gives us a basis for beginning to think about the values of x. If $Q(x)$ is written as a Chebyshev sum of Nth degree, i.e., with $N + 1$ points, we have

$$Q(x) = A_0 T_0(x) + A_1 T_1(x) + \cdots + A_N T_N(x)$$

(11.1.3)

If the last term is dropped, the result is (11.1.1), and $P(x)$ is the "best" polynomial approximation to $Q(x)$. By definition, the error $A_N T_N(x)$ satisfies the Chebyshev criterion. If $f(x)$ and $Q(x)$ were really close (sometimes a fiction), then $P(x)$ would be close to a Chebyshev expression for $f(x)$.

We now go one step farther, using the same line of thought. We know neither $P(x)$ nor $Q(x)$, but we have at hand all values of $f(x)$ for $x \in [-1, +1]$. By assumption, $Q(x) \sim f(x)$. Also, the difference between $Q(x)$ and $P(x)$ is $A_N T_N(x)$. But, by property V of Chebyshev polynomials, $T_N(x)$ vanishes when

$$x_r = \cos\left(\frac{2r+1}{N}\right)\frac{\pi}{2}, \qquad r = 0, 1, 2, \ldots, N-1$$

(11.1.4)

so that if we choose these N special values of x_r, we are as justified in approximating $f(x)$ by $P(x)$ *at those points* as we are in approximating $f(x)$ by $Q(x)$.

This may seem like an odd way to proceed, but, as in any empirical method, the end justifies the means. In any case, we shall discuss the error further in Section 11.1.5.

11.1.3 Orthogonality under Summation

Solving N equations of form (11.1.1) is easy by computer. However, as we found in the similar situation with Fourier series, it is easier to write a formula for the A_k coefficients if the $T_k(x)$ are orthogonal. We shall prove that with a weighting function, $w(x) = 1$,

$$\sum_{r=0}^{N-1} T_m(x_r) T_n(x_r) = 0, \qquad \begin{cases} x_r \in [-1, +1] \\ m \neq n \\ m, n = 0, 1, \ldots, N-1 \end{cases} \tag{11.1.5}$$

We use

$$\theta_r = \frac{(2r+1)\pi}{2N}, \qquad r = 0, 1, \ldots, N-1 \tag{11.1.6}$$

$$x_r = \cos\theta_r \tag{11.1.7}$$

$$T_k(x_r) = \cos k\theta_r = \cos k\frac{(2r+1)\pi}{2N}, \qquad k = 0, 1, \ldots, N-1 \tag{11.1.8}$$

The left member of (11.1.5) is, then, equivalent to

$$\sum_{r=0}^{N-1} \cos m\theta_r \cos n\theta_r \tag{11.1.9}$$

which, from Appendix Table A.8, is equivalent to

$$\frac{1}{2} \sum_{r=0}^{N-1} \cos(m+n)\theta_r + \frac{1}{2} \sum_{r=0}^{N-1} \cos(m-n)\theta_r \tag{11.1.10}$$

We leave as exercises the proofs of the following two lemmas.

Lemma I. *For $\theta_r = (2r+1)\pi/(2N)$,*

$$\sum_{r=0}^{N-1} \cos t\theta_r = 0 \tag{11.1.11}$$

provided that $t \neq 0$ or any other multiple of $2N$.

Lemma II. *If $t = 0$,*

$$\sum_{r=0}^{N-1} \cos t\theta_r = N \tag{11.1.12}$$

Using the lemmas, it follows immediately that

(1) if $m \neq n$, (11.1.10) becomes $0 + 0 = 0$
(2) if $m = n \neq 0$, (11.1.10) becomes $0 + \frac{1}{2}N$ (11.1.13)
(3) if $m = n = 0$, (11.1.10) becomes $\frac{1}{2}N + \frac{1}{2}N$

This completes the proof of the following theorem.

CHEBYSHEV-SUMMATION ORTHOGONALITY THEOREM   For $x_r \in [-1, +1]$
such that $x_r = \cos\theta_r$ where $\theta_r = (2r+1)\pi/(2N)$,

$$\sum_{r=0}^{N-1} T_m(x_r)T_n(x_r) = \begin{cases} 0, & m \neq n \\ N/2, & m = n \neq 0 \\ N, & m = n = 0 \end{cases} \qquad (11.1.14)$$

$$r = 0, 1, \ldots, N-1, \qquad m, n = 0, 1, \ldots, N-1$$

11.1.4 Chebyshev-Summation Coefficients

We now return to (11.1.1.1), multiply by $T_0(x_r)$ and form the sum from $r = 0$ to $r = N-1$. We assume that $P(x_k)$, which we don't know, is approximately equal to $f(x_k)$, which we do know. Then

$$\sum_{r=0}^{N-1} f(x_r)T_0(x_r) = A_0 \sum_{r=0}^{N-1} T_0(x_r)T_0(x_r)$$

$$+ A_1 \sum_{r=0}^{N-1} T_0(x_r)T_1(x_r) + \cdots$$

$$+ A_{N-1} \sum_{r=0}^{N-1} T_0(x_r)T_{N-1}(x_r) \qquad (11.1.15)$$

Using $T_0(x_r) = 1$ for the left member and (11.1.14) for the right member, we have

$$A_0 = \frac{1}{N} \sum_{r=0}^{N-1} f(x_r) \qquad (11.1.16)$$

For $k \neq 0$, we multiply by $T_k(x_r)$ and form the sum:

$$\sum_{r=0}^{N-1} f(x_r)T_k(x_r) = A_0 \sum_{r=0}^{N-1} T_0(x_r)T_k(x_r) + \cdots$$

$$+ A_k \sum_{r=0}^{N-1} T_k(x_r)T_k(x_r) + \cdots$$

$$+ A_{N-1} \sum_{r=0}^{N-1} T_k(x_r)T_{N-1}(x_r) \qquad (11.1.17)$$

from which

$$A_k = \frac{2}{N} \sum_{r=0}^{N-1} f(x_r)T_k(x_r), \qquad k = 1, 2, \ldots, N-1 \qquad (11.1.18)$$

This completes the proof of the following theorem.

CHEBYSHEV-SUMMATION COEFFICIENT THEOREM   For the Chebyshev summation using N points,

$$f(x) \sim \sum_{k=0}^{N-1} A_k T_k(x), \qquad k = 0, 1, \ldots, N-1 \qquad (11.1.19)$$

the coefficients are given by

$$A_0 = \frac{1}{N} \sum_{r=0}^{N-1} f(x_r) = \frac{1}{N} \sum_{r=0}^{N-1} f(\cos\theta_r) \qquad (11.1.20)$$

$$A_k = \frac{2}{N} \sum_{r=0}^{N-1} f(x_r) T_k(x_r) = \frac{2}{N} \sum_{r=0}^{N-1} f(\cos\theta_r) \cos k\theta_r,$$

$$k = 1, 2, \ldots, N-1 \qquad (11.1.21)$$

where

$$\theta_r = \frac{(2r+1)\pi}{2N} \qquad (11.1.22)$$

11.1.5 Theoretical Error of the Chebyshev Summation

In using N points and substituting $f(x)$ for $P(x)$, we have introduced an error of truncation as follows:

$$E_{N-1}(x) = f(x) - P(x) = A_N T_N(x) + A_{N+1} T_{N+1}(x) + \cdots \qquad (11.1.23)$$

The theoretical error of truncation can be estimated if $f(x)$ has N derivatives. We use the interpolating-polynomial error theorem:

$$E_{N-1}(x) = f(x) - P(x) = \frac{f^{(N)}(c)}{N!} \pi_{N-1}(x), \qquad c \in [-1, +1] \qquad (11.1.24)$$

where

$$\pi_{N-1}(x) = (x - x_0)(x - x_1) \cdots (x - x_{N-1}) = x^N + O(x^{N-1}) \qquad (11.1.25)$$

We would like to find out whether the error of truncation satisfies the Chebyshev criterion, i.e., whether it has $N+1$ ripples of approximately equal size. Using properties III and IV for Chebyshev polynomials,

$$\pi_{N-1}(x) = \frac{T_N(x)}{2^{N-1}} \qquad (11.1.26)$$

so that (11.1.24) becomes

$$E_{N-1}(x) = \frac{T_N(x)}{2^{N-1}N!} f^{(N)}(c), \qquad c \in [-1, +1] \qquad (11.1.27)$$

If $f^{(N)}$ is a constant, E has alternating maximum and minimum ripples equal to $1/(2^{N-1}N!)$ and therefore satisfies the Chebyshev criterion. Then the Chebyshev summation is, indeed, the "best" polynomial. In general, $f^{(N)}$ is not constant and the summation may or may not be the "best" polynomial approximation.

In a practical situation, it is best to find the *actual* error by a computer run comparing true values and summation values.

ILLUSTRATIVE PROBLEM 11.1-1 For $f(x) = e^x$, $f^{(N)}(c) = e^c$. The maximum value of $f^{(N)}$ is therefore e^1. Using (11.1.27),

$$|E_{N-1}(x)| \leqslant \frac{e}{2^{N-1}N!} \leqslant 10^{-6} \qquad (11.1.28)$$

when

$$2^{N-1}N! \leqslant e \times 10^6 < 3 \times 10^6 \qquad (11.1.29)$$

Then N is about 7. An actual computer run for $N = 6$ shows that the maximum total error (including roundoff error) is about 4×10^{-5}, and, for $N = 12$, about 10^{-12}. We shall demonstrate the programs.

There are many methods for finding the coefficients and then evaluating a summation. For $N = 6$, 12 and 24, short methods have been devised using the same techniques as those used for Fourier series. For the more general case, a clever algorithm due to Clenshaw can be employed. We shall explain these methods.

11.1.6 Special Case: Six Points

For $N = 6$,

$$A_k = \frac{2}{6} \sum_{r=0}^{5} F(r) \cos k\theta_r, \qquad k = 1, 2, 3, 4, 5 \qquad (11.1.30)$$

where

$$\theta_r = \frac{(2r+1)\pi}{12}, \qquad F(r) = f(\cos \theta_r) \qquad (11.1.31)$$

Then

$$3A_k = \sum_{r=0}^{2} F(r) \cos k\theta_r + \sum_{r=3}^{5} F(r) \cos k\theta_r \qquad (11.1.32)$$

We substitute $r' = 5 - r$ in the second summation:

$$\sum_{r=3}^{5} F(r)\cos k\theta_r = \sum_{r'=0}^{2} F(5-r')\cos\frac{k(10-2r'+1)}{12} \qquad (11.1.33)$$

But $10 - 2r' + 1 = 12 - (2r' + 1)$. Replacing r' by r,

$$\sum_{r=3}^{k} F(r)\cos k\theta_r = \sum_{r=0}^{2} F(5-r)\cos\left[k\pi - \frac{k(2r+1)\pi}{12}\right] \qquad (11.1.34)$$

$$= (-1)^k \sum_{r=0}^{2} F(5-r)\cos\frac{k(2r+1)\pi}{12} \qquad (11.1.35)$$

Substituting in (11.1.32),

$$3A_k = \sum_{r=0}^{2} \left[F(r)+(-1)^k F(5-r)\right]\cos\frac{k(2r+1)\pi}{12} \qquad (11.1.36)$$

The arguments of the cosine are $k\pi/12$, $3k\pi/12$ and $5k\pi/12$. Using the symbols from Appendix Table A.14, the cosines are as follows:

k	$k\pi/12$	$3k\pi/12$	$5k\pi/12$	
1	D	B	A	
2	C	0	$-C$	
3	B	$-B$	$-B$	$(11.1.37)$
4	$\frac{1}{2}$	-1	$\frac{1}{2}$	
5	A	$-B$	D	

We now define u and v by the following table:

	$F(0)$	$F(1)$	$F(2)$	
	$F(5)$	$F(4)$	$F(3)$	
Sum	$u(0)$	$u(1)$	$u(2)$	$(11.1.38)$
Diff	$v(0)$	$v(1)$	$v(2)$	

with which we have

$$6A_0 = u(0) + u(1) + u(2)$$
$$3A_1 = Dv(0) + Bv(1) + Av(2)$$
$$3A_2 = \left[u(0) - u(2)\right]C$$
$$3A_3 = \left[v(0) - v(1) - v(2)\right]B \qquad (11.1.39)$$
$$3A_4 = \tfrac{1}{2}u(0) - u(1) + \tfrac{1}{2}u(1)$$
$$3A_5 = Av(0) - Bv(1) + Dv(2)$$

ILLUSTRATIVE PROBLEM 11.1-2 Find the Chebyshev coefficients for a six-point summation approximating $f(x) = \arcsin x$.

Solution The calculation, even by calculator, is quite simple:

$$
\begin{aligned}
F(0) &= \arcsin(\cos \pi/12) &&= &&5\pi/12 \\
F(1) &= \arcsin(\cos 3\pi/12) &&= &&3\pi/12 \\
F(2) &= \arcsin(\cos 5\pi/12) &&= &&\pi/12 \\
F(3) &= \arcsin(\cos 7\pi/12) &&= &&-\pi/12 \qquad (11.1.40)\\
F(4) &= \arcsin(\cos 9\pi/12) &&= &&-3\pi/12 \\
F(5) &= \arcsin(\cos 11\pi/12) &&= &&-5\pi/12
\end{aligned}
$$

From these,

$$
\begin{aligned}
u(0) &= 0, & u(1) &= 0, & u(2) &= 0 \\
v(0) &= 10\pi/12, & v(1) &= 6\pi/12, & v(2) &= 2\pi/12
\end{aligned}
$$

Then, using (11.1.39),

$$
\begin{aligned}
6A_0 &= 0, & A_0 &= 0 \\
3A_1 &= \frac{\pi}{6}(A + 3B + 5D), & A_1 &\sim 1.25834\,1990 \\
3A_2 &= 0, & A_2 &= 0 \\
3A_3 &= B\pi/6, & A_3 &\sim 0.12341\,34149 \\
3A_4 &= 0, & A_4 &= 0 \\
3A_5 &= \frac{\pi}{6}(5A - 3B + D), & A_5 &\sim 0.02420\,78405
\end{aligned}
$$

and the approximation is

$$
\arcsin x \sim 1.25834\,1990\,T_1 + 0.12341\,34149\,T_3 + 0.02420\,78405\,T_5 \quad (11.1.41)
$$

We prefer to rewrite the Chebyshev summation as a power series, so that

$$
\sum_{k=0}^{N-1} A_k T_k(x) = \sum_{k=0}^{N-1} B_k x^k \quad (11.1.42)
$$

Substituting the equivalents of $T_k(x)$ from Table A.17, we find, after some arithmetic, that

$$
\begin{aligned}
B_0 &= A_0 - A_2 + A_4 \\
B_1 &= A_1 - 3A_3 + 5A_5 \\
B_2 &= 2A_2 - 8A_4 \\
B_3 &= 4A_3 - 20A_5 \qquad (11.1.43)\\
B_4 &= 8A_4 \\
B_5 &= 16A_5
\end{aligned}
$$

In using the power series, the polynomial should, of course, be programmed in nested form:

$$f(x) = B_0 + x\left(B_1 + x\left(B_2 + x\left(B_3 + x\left(B_4 + xB_5\right)\right)\right)\right) \qquad (11.1.44)$$

The coefficients, derived by computer, are as follows.

ARCSIN X

CHEBYSHEV SUMMATION COEFFICIENTS, 6 TERMS

K	FOR T(K)	FOR POWER SERIES
0	0.0	0.0
1	1.258341989825150D 00	1.009140946662193D 00
2	0.0	0.0
3	1.234134149488433D-01	9.496853061077642D-03
4	0.0	0.0
5	2.420784033671478D-02	3.873254453874365D-01

When the computed values and true values are compared, the results are as shown in Figure 11.1-1.

We call special attention to the changes of sign in the error table. These are characteristic of both Fourier and Chebyshev approximations. They show the ripple effect. The *extremals* (extreme errors) are:

x	Error	
-1.000	-1650	$\times 10^{-4}$
-0.900	$+241$	$\times 10^{-4}$
-0.550	-62.6	$\times 10^{-4}$
-0.150	$+8.64$	$\times 10^{-4}$
$+0.154$	-8.64	$\times 10^{-4}$
$+0.550$	$+62.6$	$\times 10^{-4}$
$+0.900$	-241	$\times 10^{-4}$
$+1.000$	$+1650$	$\times 10^{-4}$

Obviously the extremals do *not* show an equal-ripple quality and the Chebyshev criterion is *not* satisfied.

The computer program used for 6 and 12 points is shown in Figures 11.1-2 through 11.1-6. For convenience, four subroutines shown in Figures 11.1-3 through 11.1-6, were catalogued in the computer to speed up the compilation process.

ERROR TABLE

X	ERROR	X	ERROR
-1.000	-1.65D-01	0.0	0.0
-0.975	-1.27D-02	0.025	-2.26D-04
-0.950	1.33D-02	0.050	-4.37D-04
-0.925	2.22D-02	0.075	-6.20D-04
-0.900	2.41D-02	0.100	-7.60D-04
-0.875	2.26D-02	0.125	-8.45D-04
-0.850	1.95D-02	0.150	-8.64D-04
-0.825	1.57D-02	0.175	-8.08D-04
-0.800	1.18D-02	0.200	-6.70D-04
-0.775	8.08D-03	0.225	-4.45D-04
-0.750	4.71D-03	0.250	-1.32D-04
-0.725	1.79D-03	0.275	2.69D-04
-0.700	-6.44D-04	0.300	7.53D-04
-0.675	-2.60D-03	0.325	1.31D-03
-0.650	-4.09D-03	0.350	1.93D-03
-0.625	-5.16D-03	0.375	2.60D-03
-0.600	-5.85D-03	0.400	3.29D-03
-0.575	-6.20D-03	0.425	3.98D-03
-0.550	-6.26D-03	0.450	4.64D-03
-0.525	-6.09D-03	0.475	5.24D-03
-0.500	-5.74D-03	0.500	5.74D-03
-0.475	-5.24D-03	0.525	6.09D-03
-0.450	-4.64D-03	0.550	6.26D-03
-0.425	-3.98D-03	0.575	6.20D-03
-0.400	-3.29D-03	0.600	5.85D-03
-0.375	-2.60D-03	0.625	5.16D-03
-0.350	-1.93D-03	0.650	4.09D-03
-0.325	-1.31D-03	0.675	2.60D-03
-0.300	-7.53D-04	0.700	6.44D-04
-0.275	-2.69D-04	0.725	-1.79D-03
-0.250	1.32D-04	0.750	-4.71D-03
-0.225	4.45D-04	0.775	-8.08D-03
-0.200	6.70D-04	0.800	-1.18D-02
-0.175	8.08D-04	0.825	-1.57D-02
-0.150	8.64D-04	0.850	-1.95D-02
-0.125	8.45D-04	0.875	-2.26D-02
-0.100	7.60D-04	0.900	-2.41D-02
-0.075	6.20D-04	0.925	-2.22D-02
-0.050	4.37D-04	0.950	-1.33D-02
-0.025	2.26D-04	0.975	1.27D-02
0.0	0.0	1.000	1.65D-01

Fig. 11.1-1   Error table for Illustrative Problem 11.1-2.

```
C THE CALLING PROGRAM MUST START WITH F(12),FCH(81),C6(5),C12(11),H(105).
C THE H(105) IS A FILLER TO MAKE THE COMMON AREAS EQUAL IN LENGTH. FOR
C USE WITH DIFFERENT FUNCTIONS, ONLY 4 CARDS NEED BE CHANGED - 140,150,
C 160 AND 190.
      DOUBLE PRECISION F(12),FCH(81),C6(5),C12(11),H(105),G,X
      COMMON F,FCH,C6,C12,H
C STATEMENT FUNCTION. THE FOLLOWING THREE CARDS MUST BE CHANGED. CARD 140
C DEFINES THE FUNCTION. ITYPE=0 FOR THE GENERAL CASE, 1 IF NO ERROR
C TABLES ARE WANTED, 2 IF NEITHER ERROR TABLES NOR CHEBYSHEV COEF-
C FICIENTS ARE WANTED.  JSTART=0 IF THE ERROR TABLE BEGINS WITH -1
C AND 1 IF, AS IN LOG, IT STARTS WITH 0.025
      G(X)=DARSIN(X)
      ITYPE=0
      JSTART=0
C HEADING
      WRITE(3,100)
100   FORMAT(1H1,T36,'ARCSIN X'/)
C COSINES, SUPPLIED BY A SUBROUTINE. C6(K) IS COS(K*PI/12) AND C12(K) IS
C COS(K*PI/24).
      CALL SBRCA
C CALCULATION OF F(1) TO F(6) FOR N=6
      DO 1 I=1,3
      J=2*I-1
      X=C6(J)
1     F(I)=G(X)
C
      DO 2 I=4,6
      J=13-2*I
      X=-C6(J)
2     F(I)=G(X)
C TRUE VALUES. TV FROM A TABLE MAY BE USED.
      IX=-1025
      IT=81
      IF(JSTART.EQ.1)IX=0
      IF(JSTART.EQ.1)IT=40
C
      DO 3 I=1,IT
      IX=IX+25
      X=IX
      X=X*1.D-3
      X=G(X)
3     FCH(I)=X
C FOR N=6
      CALL CSUM6(ITYPE,JSTART)
C CALCULATION OF F(1) TO F(12) FOR N=12
      DO 4 I=1,6
      J=2*I-1
      X=C12(J)
4     F(I)=G(X)
C
      DO 5 I=7,12
      J=25-2*I
      X=-C12(J)
5     F(I)=G(X)
C FOR N=12
      WRITE(3,100)
      CALL CSUM12(ITYPE,JSTART)
      END
```

Fig. 11.1-2 Main program for Chebyshev six- and twelve-point summations.

```
      SUBROUTINE SBRCA
      DOUBLE PRECISION F(12),FCH(81),C6(5),C12(11),H(105),A,X
      COMMON F,FCH,C6,C12,H
C
      A=3.141592653589793D0/1.2D1
C COSINES USED FOR N=6. A IS 15 DEGREES.
      C6(1)=DCOS(A)
```

```
        X=2.D0*A
        C6(2)=DCOS(X)
        X=3.D0*A
        C6(3)=DCOS(X)
        C6(4)=5.D-1
        X=5.D0*A
        C6(5)=DCOS(X)
C COSINES USED FOR N=12. A IS 7.5 DEGREES.
        A=A/2.D0
        C12(1)=DCOS(A)
        C12(2)=C6(1)
        X=3.D0*A
        C12(3)=DCOS(X)
        C12(4)=C6(2)
        X=5.D0*A
        C12(5)=DCOS(X)
        C12(6)=C6(3)
        X=7.D0*A
        C12(7)=DCOS(X)
        C12(8)=5.D-1
        X=9.D0*A
        C12(9)=DCOS(X)
        C12(10)=C6(5)
        X=1.1D1*A
        C12(11)=DCOS(X)
C
        RETURN
        END
```

Fig. 11.1-3   Subroutine SBRCA, which supplied the cosines.

```
      SUBROUTINE CSUM6(ITYPE,JSTART)
      DOUBLE PRECISION F(12),FCH(81),C6(5),C12(11),G(81),A(12),B(12),U(3
     1),V(3),Y
      COMMON F,FCH,C6,C12,G,A,B
C
      DO 1 I=1,3
      J=7-I
      U(I)=F(I)+F(J)
1     V(I)=F(I)-F(J)
C COEFFICIENTS FOR T(K)
      A(1)=(U(1)+U(2)+U(3))/6.D0
      IF(DABS(A(1)).LE.5.D-15)A(1)=0.D0
C
      A(2)=V(1)*C6(1)+V(2)*C6(3)+V(3)*C6(5)
      A(3)=(U(1)-U(3))*C6(2)
      A(4)=(V(1)-V(2)-V(3))*C6(3)
      A(5)=(U(1)+U(3))/2.D0-U(2)
      A(6)=V(1)*C6(5)-V(2)*C6(3) +V(3)*C6(1)
C
      DO 2 I=2,6
      Y=A(I)/3.D0
      IF(DABS(Y).LE.5.D-15)Y=0.D0
2     A(I)=Y
C COEFFICIENTS FOR THE POWER SERIES
      B(1)=A(1)+A(5)-A(3)
      B(2)=A(2)+5.D0*A(6)-3.D0*A(4)
      B(3)=2.D0*A(3)-8.D0*A(5)
      B(4)=4.D0*A(4)-2.D1*A(6)
      B(5)=8.D0*A(5)
```

Fig. 11.1-4   (Continued on next page)

```
       B(6)=1.6D1*A(6)
C
       WRITE(3,100)
100    FORMAT(1H0,T20,'CHEBYSHEV SUMMATION COEFFICIENTS, 6 TERMS'/)
       N=6
       CALL DODPR(ITYPE,JSTART,N)
       RETURN
       END
```

Fig. 11.1-4   Subroutine CSUM6, which calculated the coefficients for 6 points.

```
       SUBROUTINE CSUM12(ITYPE,JSTART)
       DOUBLE PRECISION F(12),FCH(81),C6(5),C12(11),G(81),A(12),B(12),U(6
      1),V(6),P(3),Q(3),R(3),            S(3),Y
       COMMON F,FCH,C6,        C12,G,A,B
C
       DO 1 I=1,6
       J=13-I
       U(I)=F(I)+F(J)
1      V(I)=F(I)-F(J)
C
       DO 2 I=1,3
       J=7-I
       P(I)=U(I)+U(J)
       Q(I)=U(I)-U(J)
       R(I)=V(I)+V(J)
2      S(I)=V(I)-V(J)
C COEFFICIENTS FOR T(K)
       A(1)=(P(1)+P(2)+P(3))/1.2D1
         IF(DABS(A(1)).LE.5.D-16)A(1)=0.D0
C
       A(2)=V(1)*C12(1)+V(2)*C12(3)+V(3)*C12(5)+V(4)*C12(7)+V(5)*C12(9)+V
      1(6)*C12(11)
       A(3)=Q(1)*C12(2)+Q(2)*C12(6)+Q(3)*C12(10)
       A(4)=(V(1)-V(4)-V(5))*C12(3)+(V(2)-V(3)-V(6))*C12(9)
       A(5)=(P(1)-P(3))*C12(4)
       A(6)=V(1)*C12(5)-V(2)*C12(9)-V(3)*C12(1)-V(4)*C12(11)+V(5)*C12(3)+
      1V(6)*C12(7)
       A(7)=(Q(1)-Q(2)-Q(3))*C12(6)
       A(8)=V(1)*C12(7)-V(2)*C12(3)-V(3)*C12(11)+V(4)*C12(1)-V(5)*C12(9)-
      1V(6)*C12(5)
       A(9)=(P(1)+P(3))/2.D0-P(2)
       A(10)=(V(1)-V(4)-V(5))*C12(9)-(V(2)-V(3)-V(6))*C12(3)
       A(11)=Q(1)*C12(10)-Q(2)*C12(6)+Q(3)*C12(2)
       A(12)=V(1)*C12(11)-V(2)*C12(9)+V(3)*C12(7)-V(4)*C12(5)+V(5)*C12(3)
      1-V(6)*C12(1)
C
       DO 3 I=2,12
       Y=A(I)/6.D0
       IF(DABS(Y).LE.5.D-16)Y=0.D0
3      A(I)=Y
C COEFFICIENTS FOR THE POWER SERIES
       B(1)=A(1)-A(3)+A(5)-A(7)+A(9)-A(11)
       B(2)=A(2)-3.D0*A(4)+5.D0*A(6)-7.D0*A(8)+9.D0*A(10)-1.1D1*A(12)
       B(3)=2.D0*A(3)-8.D0*A(5)+1.8D1*A(7)-3.2D1*A(9)+5.D1*A(11)
       B(4)=4.D0*A(4)-2.D1*A(6)+5.6D1*A(8)-1.2D2*A(10)+2.2D2*A(12)
       B(5)=8.D0*A(5)-4.8D1*A(7)+1.6D2*A(9)-4.D2*A(11)
       B(6)=1.6D1*A(6)-1.12D2*A(8)+4.32D2*A(10)-1.232D3*A(12)
       B(7)=3.2D1*A(7)-2.56D2*A(9)+1.12D3*A(11)
       B(8)=6.4D1*A(8)-5.76D2*A(10)+2.816D3*A(12)
       B(9)=1.28D2*A(9)-1.28D3*A(11)
       B(10)=2.56D2*A(10)-2.816D3*A(12)
       B(11)=5.12D2*A(11)
```

```
       B(12)=1.024D3*A(12)
C
       WRITE(3,100)
100    FORMAT(1H0,T20,'CHEBYSHEV SUMMATION COEFFICIENTS, 12 TERMS'/)
       N=12
       CALL DODPR(ITYPE,JSTART,N)
       RETURN
       END
```

Fig. 11.1-5   Subroutine CSUM12, which calculated the coefficients for 12 points.

```
       SUBROUTINE DODPR(ITYPE,JSTART,N)
C KT IS THE NUMBER OF COPIES WANTED.  IF JSTART=0, THE RANGE IS (-1,+1).
C IF JSTART=1, THE RANGE IS (0.025,1).  N IS THE NUMBER OF TERMS.
       DOUBLE PRECISION F(12),FCH(81),C6(5),C12(11),G(81),A(12),B(12),Y,X
       COMMON F,FCH,C6,        C12,G,A,B
C
       KOUNT=0
10     WRITE(3,101)
101    FORMAT(1H0,T11,'K',T30,'FOR T(K)',T57,'FOR POWER SERIES',/)
C COEFFICIENTS
       DO 1 I=1,N
       K=I-1
1      WRITE(3,102)K,A(I),B(I)
102    FORMAT(1H ,T10,I2,T16,1PD22.15,T57,D22.15)
C ERROR TABLE. IN THE GENERAL CASE, X STARTS AT -1 AND IS INCREMENTED BY
C CALCULATION TO 1 BY INTERVALS OF 0.025. IF JSTART =1, THE INTERVAL
C IS (0.025,1.000).
       IX=-1025
       IT=81
C
       IF(JSTART.EQ.1)IX=0
       IF(JSTART.EQ.1)IT=40
C
       DO 3 I=1,IT
       IX=IX+25
       X=DFLOAT(IX)
       X=X*1.D-3
C
       Y=B(N)
       M=N-1
       DO 4 L=1,M
       J=N-L
4      Y=B(J)+X*Y
       G(I)=Y
3      CONTINUE
C ERRORS
       DO 5 I=1,IT
5      G(I)=FCH(I)-G(I)
C
       IF(JSTART.EQ.1)GO TO 7
C NORMAL PRINT-OUT
       WRITE(3,103)
103    FORMAT(1H1,T30,'ERROR TABLE'/)
       WRITE(3,104)
104    FORMAT(1H0,T16,'X',T26,'ERROR',T46,'X',T56,'ERROR'/)
C
       IX=-1025
       IY=-25
       DO 6 I=1,41
       IX=IX+25
       X=DFLOAT(IX)
```

Fig. 11.1-6   (Continued on next page)

```
        X=X*1.D-3
        IY=IY+25
        Y=DFLOAT(IY)
        Y=Y*1.D-3
C
        J=I+40
C
6       WRITE(3,105)X,G(I),Y,G(J)
105     FORMAT(1H ,T11,F6.3,T22,1PD9.2,T41,0PF6.3,T52,1PD9.2)
        GO TO 9
C SHORT TABLE
7       WRITE(3,106)
106     FORMAT(1H1,T16,'ERROR TABLE'/)
        WRITE(3,107)
107     FORMAT(1H0,T12,'X',T27,'ERROR'/)
C
        IX=0
        DO 8 I=1,40
        IX=IX+25
        X=DFLOAT(IX)
        X=X*1.D-3
8       WRITE(3,108)X,G(I)
108     FORMAT(1H ,T10,F6.3,T25,1PD9.2)
C
9       CONTINUE
11      RETURN
        END
```

Fig. 11.1-6   Subroutine DODPR, which controlled the output.

11.1.7   Special Case: Twelve Points

We start with

$$A_k = \frac{2}{12} \sum_{r=0}^{11} F(r) \cos k\theta_r, \qquad k = 1, 2, \ldots, 11 \qquad (11.1.45)$$

where

$$\theta_r = \frac{(2r+1)\pi}{24}, \qquad F(r) = f(\cos\theta_r) \qquad (11.1.46)$$

Then

$$6A_k = \sum_{r=0}^{5} F(r) \cos k\theta_r + \sum_{r=6}^{11} F(r) \cos k\theta_r \qquad (11.1.47)$$

Substitute $r' = 11 - r$ in the second summation:

$$\sum_{r=6}^{11} F(r) \cos k\theta_r = \sum_{r'=0}^{5} F(11-r') \cos \frac{k(22-2r'+1)\pi}{24} \qquad (11.1.48)$$

But $22 - 2r' + 1 = 24 - 2r' - 1$. Replacing r' by r,

$$\sum_{r=6}^{11} F(r)\cos k\theta_r = \sum_{r=0}^{5} F(11-r)\cos\left[k\pi - \frac{k(2r+1)\pi}{24}\right] \quad (11.1.49)$$

$$= (-1)^k \sum_{r=0}^{5} F(11-r)\cos\frac{k(2r+1)\pi}{24} \quad (11.1.50)$$

Substituting in (11.1.47),

$$6A_k = \sum_{r=0}^{5}\left[F(r) + (-1)^k F(11-r)\right]\cos\frac{k(2r+1)\pi}{24} \quad (11.1.51)$$

The arguments of the cosine are $\pi/24$, $3\pi/24$, $5\pi/24$, $7\pi/24$, $9\pi/24$ and $11\pi/24$. For brevity in the explanation, we shall use the symbols

$$P = \cos\pi/24$$
$$Q = \cos 3\pi/24$$
$$R = \cos 5\pi/24$$
$$S = \cos 7\pi/24 \quad\quad (11.1.52)$$
$$T = \cos 9\pi/24$$
$$U = \cos 11\pi/24$$

Then the cosines are given in Table 11.1-1.

Table 11.1-1
Cosines in (11.1.51)

k	Cosine[a]					
	$\dfrac{k\pi}{24}$	$\dfrac{3k\pi}{24}$	$\dfrac{5k\pi}{24}$	$\dfrac{7k\pi}{24}$	$\dfrac{9k\pi}{24}$	$\dfrac{11k\pi}{24}$
1	P	Q	R	S	T	U
2	D	B	A	$-A$	$-B$	$-D$
3	Q	T	$-T$	$-Q$	$-Q$	$-T$
4	C	0	$-C$	$-C$	0	C
5	R	$-T$	$-P$	$-U$	Q	S
6	B	$-B$	$-B$	B	B	$-B$
7	S	$-Q$	$-U$	P	$-T$	$-R$
8	$\frac{1}{2}$	-1	$\frac{1}{2}$	$\frac{1}{2}$	-1	$\frac{1}{2}$
9	T	$-Q$	Q	$-T$	$-T$	Q
10	A	$-B$	D	$-D$	B	$-A$
11	U	$-T$	S	$-R$	Q	$-P$

[a] Symbols are defined in (11.1.52)

We define u and v as sums and differences:

$$
\begin{array}{ccccccc}
 & F(0) & F(1) & F(2) & F(3) & F(4) & F(5) \\
 & F(11) & F(10) & F(9) & F(8) & F(7) & F(6) \\
\hline
\text{Sum} & u(0) & u(1) & u(2) & u(3) & u(4) & u(5) \\
\text{Diff} & v(0) & v(1) & v(2) & v(3) & v(4) & v(5)
\end{array}
\qquad (11.1.53)
$$

and continue the algorithm as we did with Fourier series:

$$
\begin{array}{cccc}
 & u(0) & u(1) & u(2) \\
 & u(5) & u(4) & u(3) \\
\hline
\text{Sum} & p(0) & p(1) & p(2) \\
\text{Diff} & q(0) & q(1) & q(2)
\end{array}
\qquad (11.1.54)
$$

$$
\begin{array}{cccc}
 & v(0) & v(1) & v(2) \\
 & v(5) & v(4) & v(3) \\
\hline
\text{Sum} & r(0) & r(1) & r(2) \\
\text{Diff} & s(0) & s(1) & s(2)
\end{array}
\qquad (11.1.55)
$$

Then we have the following relationships:

$$
\left\{
\begin{aligned}
12A_0 &= p_0 + p_1 + p_2 \\
6A_1 &= Pv_0 + Qv_1 + Rv_2 + Sv_3 + Tv_4 + Uv_5 \\
6A_2 &= Dq_0 + Bq_1 + Aq_2 \\
6A_3 &= Q(v_0 - v_3 - v_4) + T(v_1 - v_2 - v_5) \\
6A_4 &= C(p_0 - p_2) \\
6A_5 &= Rv_0 - Tv_1 - Pv_2 - Uv_3 + Qv_4 + Sv_5 \\
6A_6 &= B(q_0 - q_1 - q_2) \\
6A_7 &= Sv_0 - Qv_1 - Uv_2 + Pv_3 - Tv_4 - Rv_5 \\
6A_8 &= \tfrac{1}{2}(p_0 + p_2) - p_1 \\
6A_9 &= T(v_0 - v_3 - v_4) - Q(v_1 - v_2 - v_5) \\
6A_{10} &= Aq_0 - Bq_1 + Dq_2 \\
6A_{11} &= Uv_0 - Tv_1 + Sv_2 - Rv_3 + Qv_4 - Pv_5
\end{aligned}
\right.
\qquad (11.1.56)
$$

ILLUSTRATIVE PROBLEM 11.1-3   Find the Chebyshev coefficients for a twelve-point summation of $f(x) = \arcsin x$, $x \in [-1, +1]$.

Solution

$$F(0) = \arcsin(\cos \pi/24) = 11\pi/24$$
$$F(1) = \arcsin(\cos 3\pi/24) = 9\pi/24$$
$$F(2) = \arcsin(\cos 5\pi/24) = 7\pi/24$$
$$F(3) = \arcsin(\cos 7\pi/24) = 5\pi/24$$
$$F(4) = \arcsin(\cos 9\pi/24) = 3\pi/24$$
$$F(5) = \arcsin(\cos 11\pi/24) = \pi/24$$
$$F(6) = \arcsin(\cos 13\pi/24) = -\pi/24$$
$$F(7) = \arcsin(\cos 15\pi/24) = -3\pi/24$$
$$F(8) = \arcsin(\cos 17\pi/24) = -5\pi/24$$
$$F(9) = \arcsin(\cos 19\pi/24) = -7\pi/24$$
$$F(10) = \arcsin(\cos 21\pi/24) = -9\pi/24$$
$$F(11) = \arcsin(\cos 23\pi/24) = -11\pi/24$$

In the following, $\pi/24$ has been factored on each line:

$F(0) \rightarrow F(5)$	11	9	7	5	3	1	$\times \pi/24$
$F(11) \rightarrow F(6)$	-11	-9	-7	-5	-3	-1	$\times \pi/24$
$u(0) \rightarrow u(5)$	0	0	0	0	0	0	
$v(0) \rightarrow v(5)$	22	18	14	10	6	2	$\times \pi/24$

$u(0) \rightarrow u(2)$	0	0	0	$v(0) \rightarrow v(2)$	22	18	14	$\times \pi/24$
$u(5) \rightarrow u(3)$	0	0	0	$v(5) \rightarrow v(3)$	2	6	10	$\times \pi/24$
$p(0) \rightarrow p(2)$	0	0	0	$r(0) \rightarrow r(2)$	24	24	24	$\times \pi/24$
$q(0) \rightarrow q(2)$	0	0	0	$s(0) \rightarrow s(2)$	20	12	4	$\times \pi/24$

$12A_0 = 0$, $A_0 = 0$

$6A_1 = \dfrac{\pi}{24}[22P + 18Q + 14R + 10S + 6T + 2U]$, $A_1 = 1.26958\,1570$

$6A_2 = 0$, $A_2 = 0$

$6A_3 = [Q(22 - 10 - 6) + T(18 - 14 - 2)]\dfrac{\pi}{24}$, $A_3 = 0.13763\,32628$

$6A_4 = 0$, $A_4 = 0$

$6A_5 = [22R - 18T - 14P - 10U + 6Q + 2S]\dfrac{\pi}{24}$, $A_5 = 0.04670\,46133$

$6A_6 = 0$, $A_6 = 0$

$6A_7 = [22S - 18Q - 14U + 10P - 6T - 2R]\dfrac{\pi}{24}$, $A_7 = 0.02110\,09108$

$6A_8 = 0$, $A_8 = 0$

$6A_9 = [T(22 - 10 - 6) - Q(18 - 14 - 2)]\dfrac{\pi}{24}$, $A_9 = 0.00978\,12948$

$6A_{10} = 0$, $A_{10} = 0$

$6A_{11} = [22U - 18T + 14S - 10R + 6Q - 2P]\dfrac{\pi}{24}$, $A_{11} = 0.00289\,69961$

This results in a finite series in $T_k(x)$. To do the actual calculations on the computer, we prefer a power series. Upon making the substitutions for $T_k(x)$, we find that

$$B_0 = A_0 - A_2 + A_4 - A_6 + A_8 - A_{10}$$
$$B_1 = A_1 - 3A_3 + 5A_5 - 7A_7 + 9A_9 - 11A_{11}$$
$$B_2 = 2A_2 - 8A_4 + 18A_6 - 32A_8 + 50A_{10}$$
$$B_3 = 4A_3 - 20A_5 + 56A_7 - 120A_9 + 220A_{11}$$
$$B_4 = 8A_4 - 48A_6 + 160A_8 - 400A_{10}$$
$$B_5 = 16A_5 - 112A_7 + 432A_9 - 1232A_{11}$$
$$B_6 = 32A_6 - 256A_9 + 2816A_{11}$$
$$B_7 = 64A_7 - 576A_9 + 2816A_{11}$$
$$B_8 = 128A_8 - 1280A_{10}$$
$$B_9 = 256A_9 - 2816A_{11}$$
$$B_{10} = 512A_{10}$$
$$B_{11} = 1024A_{11}$$

(11.1.57)

This leads to a finite series of the form $\sum B_k x^k$, which is, of course, computed in nested form. The result of the computation is given in Figure 11.1-7. The error table for the 12-point Chebyshev summation is given in Figure 11.1-8.

ARCSIN X

CHEBYSHEV SUMMATION COEFFICIENTS, 12 TERMS

K	FOR T(K)	FOR POWER SERIES
0	0.0	0.0
1	1.269581570029405D 00	9.986631688207135D-01
2	0.0	0.0
3	1.376332627248190D-01	2.616755583041848D-01
4	0.0	0.0
5	4.670461318616952D-02	-9.596080706126411D-01
6	0.0	0.0
7	2.110091068427239D-02	3.874373604637979D 00
8	0.0	0.0
9	9.781294821759128D-03	-5.653929663807468D 00
10	0.0	0.0
11	2.896996142818822D-03	2.966524050246475D 00

Fig. 11.1-7   Summation coefficients for Illustrative Problem 11.1-3.

11.1.8   Special Case: Twenty-Four Points

We shall merely start the discussion for the 24-point case because the empirical method to be explained in Section 11.2 is far better.

According to the Chebyshev coefficient theorem,

$$f(x) \sim \sum_{k=0}^{23} A_k T_k(x)$$

(11.1.58)

ERROR TABLE

X	ERROR	X	ERROR
-1.000	-8.31D-02	0.0	0.0
-0.975	1.27D-02	0.025	3.19D-05
-0.950	6.91D-03	0.050	5.53D-05
-0.925	2.27D-04	0.075	6.26D-05
-0.900	-3.27D-03	0.100	4.86D-05
-0.875	-4.07D-03	0.125	1.13D-05
-0.850	-3.29D-03	0.150	-4.79D-05
-0.825	-1.84D-03	0.175	-1.24D-04
-0.800	-3.51D-04	0.200	-2.08D-04
-0.775	8.22D-04	0.225	-2.88D-04
-0.750	1.54D-03	0.250	-3.53D-04
-0.725	1.82D-03	0.275	-3.87D-04
-0.700	1.71D-03	0.300	-3.81D-04
-0.675	1.35D-03	0.325	-3.25D-04
-0.650	8.53D-04	0.350	-2.16D-04
-0.625	3.22D-04	0.375	-5.74D-05
-0.600	-1.59D-04	0.400	1.41D-04
-0.575	-5.35D-04	0.425	3.60D-04
-0.550	-7.80D-04	0.450	5.76D-04
-0.525	-8.87D-04	0.475	7.57D-04
-0.500	-8.71D-04	0.500	8.71D-04
-0.475	-7.57D-04	0.525	8.87D-04
-0.450	-5.76D-04	0.550	7.80D-04
-0.425	-3.60D-04	0.575	5.35D-04
-0.400	-1.41D-04	0.600	1.59D-04
-0.375	5.74D-05	0.625	-3.22D-04
-0.350	2.16D-04	0.650	-8.53D-04
-0.325	3.25D-04	0.675	-1.35D-03
-0.300	3.81D-04	0.700	-1.71D-03
-0.275	3.87D-04	0.725	-1.82D-03
-0.250	3.53D-04	0.750	-1.54D-03
-0.225	2.88D-04	0.775	-8.22D-04
-0.200	2.08D-04	0.800	3.51D-04
-0.175	1.24D-04	0.825	1.84D-03
-0.150	4.79D-05	0.850	3.29D-03
-0.125	-1.13D-05	0.875	4.07D-03
-0.100	-4.86D-05	0.900	3.27D-03
-0.075	-6.26D-05	0.925	-2.27D-04
-0.050	-5.53D-05	0.950	-6.91D-03
-0.025	-3.19D-05	0.975	-1.27D-02
0.0	0.0	1.000	8.31D-02

Fig. 11.1-8   Error table for Illustrative Problem 11.1-3.

for $N=24$, with coefficients as follows:

$$A_0 = \frac{1}{24} \sum_{r=0}^{23} f(\cos\theta_r) \tag{11.1.59}$$

$$A_k = \frac{1}{12} \sum_{r=0}^{23} f(\cos\theta_r) \cos k\theta_r, \qquad k=1,2,\ldots,23 \tag{11.1.60}$$

where

$$\theta_r = \frac{(2r+1)\pi}{48}, \qquad r=0,1,\ldots,23 \tag{11.1.61}$$

The values of θ_r are

$$\frac{\pi}{48}, \ \frac{3\pi}{48}, \ \frac{5\pi}{48}, \ \ldots, \ \frac{47\pi}{48} \tag{11.1.62}$$

Then

$$12A_k = \sum_{r=0}^{11} f(r)\cos k\theta_r + \sum_{r=0}^{23} f(r)\cos k\theta_r \qquad (11.1.63)$$

In the second summation, let $r' = 23 - r$:

$$\sum_{r=12}^{23} F(r)\cos k\theta_r = \sum_{r'=0}^{11} F(23-r')\cos\left[\frac{k(46-2r'+1)\pi}{48}\right] \qquad (11.1.64)$$

But

$$46 - 2r' + 1 = 48 - (2r' + 1) \qquad (11.1.65)$$

Replacing r' by r,

$$\sum_{r=12}^{23} F(r)\cos k\theta_r = \sum_{r=0}^{11} F(23-r)\cos\left[k\pi - \frac{k(2r+1)}{48}\right] \qquad (11.1.66)$$

$$= (-1)^k \sum_{r=0}^{11} F(23-r)\cos\frac{(2r+1)k\pi}{48} \qquad (11.1.67)$$

Substituting in (11.1.63),

$$12A_k = \sum_{r=0}^{11} \left[F(r) + (-1)^k F(23-r) \right]\cos k\theta_r \qquad (11.1.68)$$

For brevity, we define

$$C_k = \cos\frac{k\pi}{48} \qquad (11.1.69)$$

We also define

$$
\begin{aligned}
u_k &= F_k + F_{23-k}, & k &= 0, 1, \ldots, 11 \\
v_k &= F_k - F_{23-k}, & k &= 0, 1, \ldots, 11 \\
p_k &= u_k + u_{11-k}, & k &= 0, 1, \ldots, 5 \\
q_k &= u_k - u_{11-k}, & k &= 0, 1, \ldots, 5 \\
t_k &= p_k + p_{5-k}, & k &= 0, 1, 2 \\
w_k &= p_k - p_{5-k}, & k &= 0, 1, 2 \\
r_0 &= v_0 - v_7 - v_8 \\
r_1 &= v_1 - v_6 - v_9 \\
r_2 &= v_2 - v_5 - v_{10} \\
r_3 &= v_3 - v_4 - v_{11}
\end{aligned}
\qquad (11.1.70)
$$

After some tedious manipulations, the following expressions for the Chebyshev coefficients are obtained:

$$24A_0 = \sum_{k=0}^{2} t_k \tag{11.1.71}$$

$$12A_1 = \sum_{k=0}^{11} v_k C_{2k+1} \tag{11.1.72}$$

$$12A_2 = \sum_{k=0}^{5} q_k C_{4k+2} \tag{11.1.73}$$

$$12A_3 = \sum_{k=0}^{3} r_k C_{6k+3} \tag{11.1.74}$$

$$12A_4 = \sum_{k=0}^{2} w_k C_{8k+4} \tag{11.1.75}$$

$$12A_5 = v_0 C_5 + v_1 C_{15} - v_2 C_{23} - v_3 C_{13} - v_4 C_3 - v_5 C_7 - v_6 C_{17} + v_7 C_{21}$$
$$+ v_8 C_{11} + v_9 C_1 + v_{10} C_9 + v_{11} C_{19} \tag{11.1.76}$$

$$12A_6 = (t_0 - t_2) C_8 \tag{11.1.77}$$

$$12A_7 = v_0 C_7 + v_1 C_{21} - v_2 C_{13} - v_3 C_1 - v_4 C_{15} + v_5 C_{19}$$
$$+ v_6 C_5 + v_7 C_9 + v_8 C_{23} - v_9 C_{11} - v_{10} C_3 - v_{11} C_{17} \tag{11.1.78}$$

$$A_8 = A_6 \tag{11.1.79}$$

$$12A_9 = r_0 C_9 - r_1 C_{21} - r_2 C_3 - r_3 C_{15} \tag{11.1.80}$$

$$12A_{10} = q_0 C_{10} - q_1 C_{18} - q_2 C_2 - q_3 C_{22} + q_4 C_6 + q_5 C_{14} \tag{11.1.81}$$

$$12A_{11} = v_0 C_{11} - v_1 C_{15} - v_2 C_7 + v_3 C_{19} + v_4 C_3 - v_5 C_{23} - v_6 C_1 - v_7 C_{21}$$
$$+ v_8 C_5 + v_9 C_{17} - v_{10} C_9 - v_{11} C_{13} \tag{11.1.82}$$

$$A_{12} = A_4 \tag{11.1.83}$$

$$12A_{13} = v_0 C_{13} - v_1 C_9 - v_2 C_{17} + v_3 C_5 + v_4 C_{21} - v_5 C_1 + v_6 C_{23} + v_7 C_3$$
$$- v_8 C_{19} - v_9 C_7 + v_{10} C_{15} + v_{11} C_{11} \tag{11.1.84}$$

$$12A_{14} = q_0 C_{14} - q_1 C_6 - q_2 C_{22} + q_3 C_2 - q_4 C_{18} - q_5 C_{10} \tag{11.1.85}$$

$$12A_{15} = r_0 C_{15} - r_1 C_3 + r_2 C_{21} + r_3 C_9 \tag{11.1.86}$$

$$12A_{16} = \tfrac{1}{2}(t_0 + t_2) - t_1 \tag{11.1.87}$$

$$12A_{17} = v_0 C_{17} - v_1 C_3 + v_2 C_{11} + v_3 C_{23} - v_4 C_9 + v_5 C_5 - v_6 C_{19} - v_7 C_{15}$$
$$+ v_8 C_1 - v_9 C_{13} - v_{10} C_{21} + v_{11} C_7 \tag{11.1.88}$$

$$12A_{18} = (q_0 - q_3 - q_4) C_{18} - (q_1 - q_2 - q_5) C_6 \tag{11.1.89}$$

$$12A_{19} = v_0 C_{19} - v_1 C_9 + v_2 C_1 - v_3 C_{11} + v_4 C_{21} + v_5 C_{17} - v_6 C_7$$
$$+ v_7 C_3 - v_8 C_{13} + v_9 C_{23} + v_{10} C_{15} - v_{11} C_5 \tag{11.1.90}$$

$$12A_{20} = w_0 C_{20} - w_1 C_{12} + w_2 C_4 \tag{11.1.91}$$

$$12A_{21} = r_0 C_{21} - r_1 C_{15} + r_2 C_9 - r_3 C_3 \tag{11.1.92}$$

$$12A_{22} = \sum_{k=0}^{5} (-1)^k q_k C_{22-4k} \tag{11.1.93}$$

$$12A_{23} = \sum_{k=0}^{11} (-1)^k v_k C_{23-2k} \tag{11.1.94}$$

11.1.9   Clenshaw's Method

Suppose $f(x)$ is represented by the Chebyshev series

$$f(x) = \sum_{k=0}^{\infty} A_k T_k(x) \tag{11.1.95}$$

We wish to calculate $f(x)$ directly without the intermediate substitution for or calculation of the $T_k(x)$. To do this, we seek functions $C_k(x)$ such that

$$f(x) = C_0 + C_1 x + C_2 x^2 + \cdots \tag{11.1.96}$$

We shall call the C_k *Clenshaw coefficients* after the one who thought of this algorithm.[1]

Note that we are *not* requiring that the C_k be constants. We start with the recursion relation (property II) for the Chebyshev polynomials:

$$T_{k-2}(x) - 2x T_{k-1}(x) + T_k(x) = 0 \tag{11.1.97}$$

Multiplying by $C_k(x)$ and summing,

$$\sum_{k=0}^{\infty} C_k T_{k-2} - 2x \sum_{k=0}^{\infty} C_k T_{k-1} + \sum_{k=0}^{\infty} C_k T_k = 0 \tag{11.1.98}$$

In order to use (11.1.95) we must try to obtain from (11.1.98) a single summation which resembles (11.1.95). We do this by adjusting the dummies. We shall need to use the little theorem

$$T_{-k}(x) = T_k(x) \tag{11.1.99}$$

which follows immediately from the fact that $T_k(x)$ is a cosine. Then

$$\sum_{k=0}^{\infty} C_k T_{k-2} = C_0 T_{-2} + C_1 T_{-1} + \sum_{k=0}^{\infty} C_{k+2} T_k \tag{11.1.100}$$

$$= C_0 T_2 + C_1 T_1 + \sum_{k=0}^{\infty} C_{k+2} T_k \tag{11.1.101}$$

$$= C_0 (2x^2 - 1) + C_1 x + \sum_{k=0}^{\infty} C_{k+2} T_k \tag{11.1.102}$$

---

[1] *Math Tables and Other Aids to Computations* **9**, 118 (1955).

For (11.1.98), we also need a relationship for $\sum\limits_{k=0}^{\infty} C_k T_{k-1}$:

$$\sum_{k=0}^{\infty} C_k T_{k-1} = C_0 T_{-1} + \sum_{k=0}^{\infty} C_{k+1} T_k \qquad (11.1.103)$$

$$= C_0 x + \sum_{k=0}^{\infty} C_{k+1} T_k \qquad (11.1.104)$$

Substituting from (11.1.102) and (11.1.104) into (11.1.98) and simplifying,

$$\sum_{k=0}^{\infty} (C_{k+2} - 2x C_{k+1} + C_k) T_k = C_0 - C_1 x \qquad (11.1.105)$$

Comparing (11.1.106) with (11.1.95), we see that if

$$A_k = C_{k+2} - 2x C_{k+1} + C_k \qquad (11.1.106)$$

then

$$f(x) = C_0 - C_1 x \qquad (11.1.107)$$

We use (11.1.106) in a backwards recursion, to be demonstrated.

THEOREM If $f(x = \sum\limits_{k=0}^{\infty} A_k T_k(x)$, and if C_k is computed recursively by

$$C_k = A_k + 2x C_{k+1} - C_{k+2} \qquad (11.1.108)$$

then the numerical value of $f(x)$ at $x = x_0$ is

$$f(x_0) = C_0 - C_1 x_0 \qquad (11.1.109)$$

In Clenshaw's method for a Chebyshev summation to N terms, we assume that all the Chebyshev coefficients, A_k, are known, the last non-vanishing one being A_N. Then

$$A_{N+1} = A_{N+2} = \cdots = 0 \qquad (11.1.110)$$

and, of course,

$$C_{N+1} = C_{N+2} = \cdots = 0 \qquad (11.1.111)$$

as well. Otherwise, the degree of the polynomial would be $> N$. Then

$$C_N = A_N + 2x C_{N+1} - C_{N+2} = A_N \qquad (11.1.112)$$

Now, since C_N is known,

$$C_{N-1}=A_{N-1}+2xC_N-C_{N+1}=A_{N-1}+2xC_N \qquad (11.1.113)$$

After the first two Clenshaw coefficients have been calculated, (11.1.108) serves to find the others.

ILLUSTRATIVE PROBLEM 11.1-4   Calculate

$$f(x)=T_0+2T_1-3T_2+4T_3$$

by Clenshaw's method, for $x=0.5$.

Solution   In this simple problem, it is easy to substitute for the T_k and find $f(x)=16x^3-6x^2-10x+4$ directly. Then, for $x=0.5$, $f(x)=-0.5$.
   By Clenshaw's method,

$$C_3=A_3=4$$

from (11.1.112). From (11.1.113),

$$C_2=A_2+2xC_3=-3+2(0.5)4=1$$

and from (11.1.108),

$$C_1=A_1+2xC_2-C_3=2+2(0.5)1-4=-1$$
$$C_0=A_0+2xC_1-C_2=1+2(0.5)(-1)-1=-1$$

Then, from (11.1.107), $f(x)=-1-(-1)(0.5)=0.5$.

11.1.10   Number of Terms

   There are some theoretical devices for estimating the number of terms to be used to attain a desired accuracy. On the whole, the best practical method is to program the computer to test various numbers of points, printing an error table for each. Table 11.1-2 is excerpted from a printout involving the computation of the coefficients and displaying 27 error tables, each with 81 points. The computation required less than 10 minutes on a small IBM 360/30. The table shows the extremals (extreme errors) over the interval $[-1,+1]$ except for the endpoints.

Table 11.1-2

FUNCTION	6 POINTS	12 POINTS	24 POINTS
$\sin x$	5×10^{-6}	6×10^{-14}	7×10^{-16}
e^x	4×10^{-5}	1×10^{-12}	4×10^{-15}
$\cosh x$	4×10^{-5}	1×10^{-12}	3×10^{-15}
$e^{\sin x}$	4×10^{-4}	4×10^{-8}	3×10^{-15}
$\arctan x$	1×10^{-3}	3×10^{-6}	3×10^{-11}
$e^{\cos x}$	3×10^{-3}	3×10^{-7}	5×10^{-15}
$\tan x$	4×10^{-3}	9×10^{-6}	4×10^{-11}
$\arcsin x$	2×10^{-2}	3×10^{-3}	4×10^{-4}
$\ln(1+x)$	3×10^{-1}	7×10^{-2}	2×10^{-2}

PROBLEM SECTION 11.1

A. For each of the following, write a Chebyshev summation and an error table for $N = 6$ and $N = 12$ with $x \in [-1, +1]$.

The table in Section 1.10 can be used as a rough check.

(1) $\sin x$ (6) $e^{\sin x}$

(2) e^x (7) $e^{\cos x}$

(3) $\sinh x$ (8) e^{x^2}

(4) $\cosh x$ (9) $e^{\arcsin x}$

(5) $\arctan x$ (10) $\ln(1 + x)$

B.

(11) Prove Lemma I, (11.1.11).

(12) Prove Lemma II, (11.1.12).

(13) For $N = 2m$, show that

$$a_k = \sum_{r=0}^{m-1} \left[F(r) + (-1)^k F(N-1-r) \cos \frac{k\pi(2r+1)}{2N} \right]$$

11.2 THE HASTINGS METHOD

11.2.1 Acknowledgment

The present section describes a practical, empirical, computer-oriented, iterative method for obtaining a useful Chebyshev approximation for a function. The function may or may not be differentiable or continuous, but has to be defined over the chosen interval. The method to be described is adapted from an astonishingly clear and beautifully written monograph by Hastings.

The description in this section differs in some respects from that in the monograph: (1) it is not as complete; (2) a few definitions are slightly different—e.g., *error*; (3) we use the standard interval $[-1, +1]$ instead of the shifted interval $[0, 1]$.

In the following, the procedure is illustrated by a step-by-step display of the Hastings method for the Chebyshev approximation, with $N = 6$, to $f(x) = e^x$.

11.2.2 Step I. The Basic Assumption

We start by assuming, as explained in the preceding section, that a polynomial of Nth degree fits well and that we are making a Chebyshev approximation of degree $(N-1)$ using N points.

11.2.3 Step II. The No-Error Points

We have already explained that the no-error points for the interval $[-1, +1]$ are at

$$x_r = \cos \frac{(2r+1)\pi}{2N}, \qquad r = 0, 1, \ldots, N-1 \qquad (11.2.1)$$

The angles corresponding to these values are

$$\theta_r = \frac{\pi}{12}, \frac{3\pi}{12}, \frac{5\pi}{12}, \frac{7\pi}{12}, \frac{9\pi}{12}, \frac{11\pi}{12} \qquad (11.2.2)$$

Using Appendix Table A.14, the corresponding cosines are

$$D, B, A, -A, -B, -D \qquad (11.2.3)$$

11.2.4 Step III. No-Error Equations, Trial 1

Assuming there is no error at each of the x_r in (11.2.1), i.e., that $f(x_r) = P(x_r)$ at each of these values, we have a set of simultaneous linear equations of the form

$$h_0 + h_1 x_r + h_2 x_r^2 + \cdots + h_{n-1} x_r^{n-1} = P(x_r) = f(x_r) \qquad (11.2.4)$$

For the illustrative problem, $f(x) = e^x$ and $N = 6$, so that, using (11.2.2) and (11.2.3), we have

$$\begin{cases} h_0 + h_1 D + h_2 D^2 + h_3 D^3 + h_4 D^4 + h_5 D^5 = e^D \\ h_0 + h_1 B + h_2 B^2 + h_3 B^3 + h_4 B^4 + h_5 D^5 = e^B \\ \qquad\qquad\qquad\qquad\qquad \vdots \\ h_0 - h_1 D + h_2 D^2 - h_3 D^3 + h_4 D^4 - h_5 D^5 = e^{-D} \end{cases} \qquad (11.2.5)$$

We solve this by the Gauss-Jordan process. For the illustrative problem, the coefficients are as follows:

$$H(0) = 1.0000445799970960\,00$$
$$H(1) = 1.C000063309015320\,00$$
$$H(2) = 4.9919834813006600\,-01$$
$$H(3) = 1.6655279773853660\,-01$$
$$H(4) = 4.3792329836905280\,-02$$
$$H(5) = 8.6356460072253100\,-03$$

Rounding the coefficients for clarity, the first trial leads to the following approximation:

$$e^x \sim 1.0004 + 1.00001 x + 0.49920 x^2 + 0.16655 x^3 + 0.043792 x^4 + 0.0086356 x^5 \qquad (11.2.7)$$

Note that in this method it is unnecessary to use $T_k(x)$ explicitly, thus saving a step.

11.2.5 Step IV. Error Table, Trial 1

The error table for trial 1 is given in Figure 11.2-1. The error curve corresponding to this error table is shown in Figure 11.2-2.

ERROR TABLE FOR EXP X

X	ERROR	X	ERROR
-1.000	3.895785D-05	0.0	-4.458000D-05
-0.975	8.770851D-06	0.025	-4.423629D-05
-0.950	-1.287450D-05	0.050	-4.289154D-05
-0.925	-2.731938D-05	0.075	-4.056521D-05
-0.900	-3.577987D-05	0.100	-3.729688D-05
-0.875	-3.935327D-05	0.125	-3.314594D-05
-0.850	-3.902426D-05	0.150	-2.819107D-05
-0.825	-3.567105D-05	0.175	-2.252940D-05
-0.800	-3.007134D-05	0.200	-1.627546D-05
-0.775	-2.290821D-05	0.225	-9.559794D-06
-0.750	-1.477594D-05	0.250	-2.527294D-06
-0.725	-6.185659D-06	0.275	4.664757D-06
-0.700	2.429097D-06	0.300	1.184895D-05
-0.675	1.070691D-05	0.325	1.885031D-05
-0.650	1.835318D-05	0.350	2.548925D-05
-0.625	2.513488D-05	0.375	3.158473D-05
-0.600	3.087551D-05	0.400	3.695787D-05
-0.575	3.545012D-05	0.425	4.143586D-05
-0.550	3.878048D-05	0.450	4.485622D-05
-0.525	4.083040D-05	0.475	4.707143D-05
-0.500	4.160117D-05	0.500	4.795395D-05
-0.475	4.112714D-05	0.525	4.740161D-05
-0.450	3.947146D-05	0.550	4.534339D-05
-0.425	3.672194D-05	0.575	4.174558D-05
-0.400	3.298712D-05	0.600	3.661843D-05
-0.375	2.839242D-05	0.625	3.002313D-05
-0.350	2.307648D-05	0.650	2.207928D-05
-0.325	1.718770D-05	0.675	1.297278D-05
-0.300	1.088091D-05	0.700	2.964216D-06
-0.275	4.314202D-06	0.725	-7.602354D-06
-0.250	-2.354034D-06	0.750	-1.829007D-05
-0.225	-8.967900D-06	0.775	-2.855948D-05
-0.200	-1.537660D-05	0.800	-3.775826D-05
-0.175	-2.143687D-05	0.825	-4.511035D-05
-0.150	-2.701517D-05	0.850	-4.970466D-05
-0.125	-3.198975D-05	0.875	-5.048316D-05
-0.100	-3.625243D-05	0.900	-4.622841D-05
-0.075	-3.971022D-05	0.925	-3.555057D-05
-0.050	-4.228673D-05	0.950	-1.687380D-05
-0.025	-4.392330D-05	0.975	1.157797D-05
0.0	-4.458000D-05	1.000	5.179585D-05

Fig. 11.2-1   Error table for trial 1.

It is clear that the error curve is of the Chebyshev type, but the Chebyshev criterion is not satisfied sufficiently. The ripples are not equal.

11.2.6   Step V. Estimating the Minimax Deviation

From the first-cycle error table, the extremals are approximately as follows:

x	$|\text{EXTREMAL}|$
-0.875	394×10^{-7}
-0.500	416×10^{-7}
$0.$	446×10^{-7}
$+0.500$	480×10^{-7}
$+0.875$	505×10^{-7}

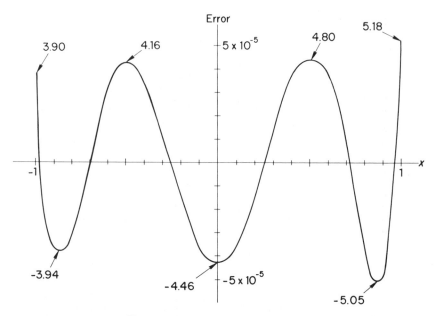

Fig. 11.2-2   Error curve for trial 1.

To continue, we shall need some improved estimates of the extremals. To arrive at these, we must digress to prove the *extremal-point theorem*.

EXTREMAL-POINT THEOREM   Given three points: $(x_0 - h, y_{-1})$, (x_0, y_0), $(x_0 + h, y_1)$, then the extreme values, $\hat{x}$ and $\hat{y}$, can be estimated as

$$\hat{x} = x_0 - \frac{ah}{2b} \qquad (11.2.8)$$

$$\hat{y} = y_0 - \frac{a^2}{8b} \qquad (11.2.9)$$

where

$$a = y_1 - y_{-1}$$

$$b = y_1 - 2y_0 + y_{-1} \qquad (11.2.10)$$

Proof   We shall assume that the curve is actually a second-degree function. This is called *quadratic interpolation*. Then

$$y = Ax^2 + Bx + C \qquad (11.2.11)$$

For the three given points,

$$\begin{cases} A(x_0 - h)^2 + B(x_0 + h) + C = y_{-1} \\ Ax_0^2 \qquad\quad + Bx_0 \qquad\; + C = y_0 \\ A(x_0 + h)^2 + B(x_0 + h) + C = y_1 \end{cases} \qquad (11.2.12)$$

Upon solving these three equations, we find that

$$A = \frac{b}{2h^2} \tag{11.2.13}$$

$$B = \frac{ah - 2x_0 b}{2h^2} \tag{11.2.14}$$

$$C = \frac{x_0^2 b - hx_0 a + 2h^2 y_0}{2h^2} \tag{11.2.15}$$

where a and b are given by (11.2.10). Now we use (11.2.11) and differentiate:

$$\frac{dy}{dx} = 0 \quad \Rightarrow \quad 2Ax + B = 0 \tag{11.2.16}$$

from which (11.2.8) follows directly. Substituting this value into (11.2.11), we obtain (11.2.9), completing the proof. $\square$

Applying the extremal-point theorem to the error table for e^x, we obtain the following improved estimates:

x	$|\text{EXTREMAL}|$
-0.86461	396.904×10^{-7}
-0.49702	416.100×10^{-7}
$0.$	$445.8 \quad \times 10^{-7}$
$+0.50288$	479.634×10^{-7}
$+0.86637$	507.833×10^{-7}

We must now estimate the minimal deviation, d. Hastings suggests that the first and last extremals be weighted by 1 and the others by 2, because the distances between successive values of $\hat{x}$ are approximately $1:2:2:\cdots:2:1$. When this is done, we have

$$d = \left[(396.904 + 507.833) + 2(416.100 + 445.8 + 479.634) \right] \times 10^{-7} \tag{11.2.17}$$

$$d = 448.476 \times 10^{-7} \tag{11.2.18}$$

11.5.7 Step VI. Forming the Extremal Equations

According to (11.2.18), the actual minimax errors at the points of extreme error should be 4.48476×10^{-5}. If we accept this as a fact, then the Chebyshev polynomial, at these points of extreme error, should yield values according to

$$h_0 + h_1 x + \cdots + h_{n-1} x^{n-1} = f(x) \pm d \qquad (11.2.19)$$

For the illustrative problem, we have, then,

$$h_0 + h_1(-0.86461) + h_2(-0.86461)^2 + \cdots + h_5(-0.86461)^5 = e^{-0.86461} + d$$
$$h_0 + h_1(-0.49702) + h_2(-0.49702)^2 + \cdots + h_5(-0.49702)^5 = e^{-0.49702} - d$$
$$h_0 + h_1(0) \qquad + h_2(0)^2 \qquad + \cdots + h_5(0)^5 \qquad = e^0 \qquad + d$$
$$h_0 + h_1(+0.50288) + h_2(+0.50288)^2 + \cdots + h_5(+0.50288)^5 = e^{0.50288} - d$$
$$h_0 + h_1(+0.86637) + h_2(+0.86637)^2 + \cdots + h_5(+0.86637)^5 = e^{0.86637} + d$$
$$\qquad\qquad\qquad\qquad\qquad\qquad\qquad\qquad\qquad\qquad (11.2.20)$$

Note that there are six unknowns $(h_0, h_1, \ldots, h_5)$ and only five equations. We are free to choose another equation at will. We choose, arbitrarily, the value at $x = 1$. Then

$$h_0 + h_1(1) + h_2(1)^2 + \cdots + h_5(1)^5 = e^1 - d \qquad (11.2.21)$$

We now solve the six equations, using the Gauss-Jordan method:

$$
\begin{aligned}
H(0) &= 1.000044847600000D\ 00 \\
H(1) &= 1.0000392768416110D\ 00 \\
H(2) &= 4.991973850789573D\text{-}01 \\
H(3) &= 1.664194894506621D\text{-}01 \\
H(4) &= 4.379249352930872D\text{-}02 \\
H(5) &= 8.743488358505729D\text{-}03
\end{aligned}
$$

Rounding the coefficients for clarity, the resulting equation for e^x is as follows:

$$e^x \sim 1.00004 + 1.00004x + 0.499197x^2 + 0.166419x^3$$
$$+ 0.0437925x^4 + 0.00874349x^5 \qquad (11.2.22)$$

The error table is given in Figure 11.2-3. The extremals are approximately given by

```
EXP X                     ERROR TABLE

      X              ERROR                    X              ERROR

   -1.000        4.696961D-05             0.0          -4.484760D-05
   -0.975        1.285451D-05             0.025        -4.532486D-05
   -0.950       -1.195648D-05             0.050        -4.478740D-05
   -0.925       -2.888563D-05             0.075        -4.324236D-05
   -0.900       -3.922537D-05             0.100        -4.071723D-05
   -0.875       -4.414462D-05             0.125        -3.725970D-05
   -0.850       -4.469519D-05             0.150        -3.293725D-05
   -0.825       -4.181793D-05             0.175        -2.783645D-05
   -0.800       -3.634895D-05             0.200        -2.206203D-05
   -0.775       -2.902552D-05             0.225        -1.573562D-05
   -0.750       -2.049211D-05             0.250        -8.994203D-06
   -0.725       -1.130607D-05             0.275        -1.988300D-06
   -0.700       -1.943435D-06             0.300         5.120177D-06
   -0.675        7.195601D-06             0.325         1.216037D-05
   -0.650        1.578116D-05             0.350         1.895527D-05
   -0.625        2.354838D-05             0.375         2.532478D-05
   -0.600        3.029219D-05             0.400         3.108921D-05
   -0.575        3.586221D-05             0.425         3.607302D-05
   -0.550        4.015786D-05             0.450         4.010897D-05
   -0.525        4.312346D-05             0.475         4.304266D-05
   -0.500        4.474361D-05             0.500         4.473737D-05
   -0.475        4.503862D-05             0.525         4.507936D-05
   -0.450        4.406011D-05             0.550         4.398349D-05
   -0.425        4.188680D-05             0.575         4.139931D-05
   -0.400        3.862037D-05             0.600         3.731752D-05
   -0.375        3.438154D-05             0.625         3.177685D-05
   -0.350        2.930629D-05             0.650         2.487143D-05
   -0.325        2.354223D-05             0.675         1.675850D-05
   -0.300        1.724517D-05             0.700         7.666726D-06
   -0.275        1.057584D-05             0.725        -2.095187D-06
   -0.250        3.696772D-06             0.750        -1.212927D-05
   -0.225       -3.230609D-06             0.775        -2.193862D-05
   -0.200       -1.004871D-05             0.800        -3.091725D-05
   -0.175       -1.660634D-05             0.825        -3.833938D-05
   -0.150       -2.276103D-05             0.850        -4.334823D-05
   -0.125       -2.838118D-05             0.875        -4.494424D-05
   -0.100       -3.334806D-05             0.900        -4.197277D-05
   -0.075       -3.755745D-05             0.925        -3.311180D-05
   -0.050       -4.092125D-05             0.950        -1.685539D-05
   -0.025       -4.336873D-05             0.975         8.494251D-06
    0.0         -4.484760D-05             1.000         4.484760D-05
```

Fig. 11.2-3   Error table for step VI.

x	$|\text{EXTREMAL}|$
-0.850	447×10^{-7}
-0.500	447×10^{-7}
$0.$	448×10^{-7}
$+0.525$	451×10^{-7}
$+0.875$	449×10^{-7}

which is almost equi-ripple. The process can be repeated several times to iron out the ripples until they are as even as you wish.

For convenience, the programs for the first cycle and for subsequent cycles are shown in Figures 11.2-4 and 11.2-5.

```
      DOUBLE PRECISION A(10,11),B(10),F,G,D,BB,AA,ONE,TWO,X1,X2,Y1,Y2,Y,   040
     1RP01,RP02,RP11,RP12,RP21,RP22,SA,SB,S,XMAX1,XMAX2,YMAX1,YMAX2,ZERO    045
      COMMON A,B                                                           050
C STATEMENT FUNCTION. CARD 070 MUST BE CHANGEC FOR EACH PROGRAM.           060
      G(Y)=DEXP(Y)                                                         070
      ITYPE=0                                                              080
105   FORMAT(1H1,T2,'COEFFICIENTS',T20,'EXP X'//)                          085
100   FORMAT(1H1,T12,'ERROR TABLE FOR EXP X',T88,'EXTREMALS'//)            086
C CONSTANTS.                                                               090
      ONE=1.D0                                                             100
      TWO=2.D0                                                             110
      AA=DSQRT(3.D0)                                                       120
      D=(DSQRT(TWO+AA))/TWO                                                130
      BB=(DSQRT(TWO))/TWO                                                  140
      IT=41                                                                150
      AA=(DSQRT(TWO-AA))/TWO                                               151
C INITIAL VALUES.                                                          152
      ZERO=0.D0                                                            153
      RP11=ZERO                                                            153
      RP12=ZERO                                                            154
      RP21=ZERO                                                            155
      RP22=ZERO                                                            156
      DO 3 KOUNT=1,3                                                       157
C MATRIX ELEMENTS.                                                         160
      DO 1 I=1,6                                                           170
1     A(I,1)=ONE                                                           180
C                                                                          190
      DO 2 J=2,6                                                           200
      K=J-1                                                                205
      A(1,J)=D**K                                                          210
      A(2,J)=BB**K                                                         220
      A(3,J)=AA**K                                                         230
      A(4,J)=(-AA)**K                                                      240
      A(5,J)=(-BB)**K                                                      250
      A(6,J)=(-D)**K                                                       260
C                                                                          270
      A(1,7)=G(D)                                                          280
      A(2,7)=G(BB)                                                         290
      A(3,7)=G(AA)                                                         300
```

```
      A(4,7)=G(-AA)                                                    310
      A(5,7)=G(-BB)                                                    320
      A(6,7)=G(-D)                                                     330
C ECHO MATRIX                                                          351
      WRITE(3,103)                                                     352
103   FORMAT(1H1,T35,'ECHO MATRIX'/)                                   353
      DO 5 K=1,6                                                       354
      WRITE(3,104)(A(K,J),J=1,7)                                       355
5                                                                      356
104   FORMAT(1H0,10X,7(1PD15.5))                                       340
C SOLUTION.                                                            350
      CALL FGJDOD(7)                                                   357
C ANSWERS                                                              358
      WRITE(3,105)                                                     360
      DO 7 I=1,6                                                       360
      L=I-1                                                            361
      WRITE(3,106)L,B(I)                                               362
7                                                                      363
106   FORMAT(1H0,'H(',I1,')= ',1PD22.15)                              370
C ERROR TABLE AND EXTREMALS                                            390
      WRITE(3,100)                                                     400
      WRITE(3,101)                                                     405
101   FORMAT(1H0,T13,'X',T26,'ERROR',T43,'X',T56, 'ERROR',T74,'X',T86,'Y   410
     1',T98,'X',T110,'Y'/)                                             420
C                                                                      430
      IX=-1025                                                         421
      IY=-25                                                           422
      IF(ITYPE.EQ.1)IX=-1000                                           423
      IF(ITYPE.EQ.1)IY=0                                               440
      IF(ITYPE.EQ.1)IT=40                                              450
      DO 3 I=1,IT                                                      460
      IX=IX+25                                                         470
      IY=IY+25                                                         480
      X1=IX                                                            490
      X2=IY                                                            500
      X1=X1*1.D-3                                                      510
      X2=X2*1.D-3                                                      520
C                                                                      521
      Y=X1
      Y1 =B(1)+Y*(B(2)+Y*(B(3)+Y*(B(4)+Y*(B(5)+B(6)*Y))))
C
```

Fig. 11.2-4   (Continued on next page)

```
        Y=X2                                                              522
        Y2 =B(1)+Y*(B(2)+Y*(B(3)+Y*(B(4)+Y*(B(5)+B(6)*Y))))              523
        Y1=G(X1)-Y1                                                       530
        Y2=G(X2)-Y2                                                       531
C TWO PREVIOUS VALUES OF R(X) ARE RP2 AND RP1. THE PRESENT VALUE IS RP0   532
        RP01=Y1                                                          A532
        RP02=Y2                                                          A532
        SA=RP21-RP01                                                      533
        SB=(RP01-TWO*RP11+RP21)*TWO                                       534
        S=SA/SB                                                           535
        XMAX1=X1+(2.5D-2)*(S-ONE)                                         536
        SA=SA**2                                                          537
        SB=4.D0*SB                                                        538
        YMAX1=RP11-SA/SB                                                  539
C
        SA=RP22-RP02                                                      541
        SB=(RP02-TWO*RP12+RP22)*TWO                                       542
        S=SA/SB                                                           543
        XMAX2=X2+(2.5D-2)*(S-ONE)                                         544
        SA=SA**2                                                          545
        SB=4.D0*SB                                                        546
        YMAX2=RP12-SA/SB                                                  547
C
        RP21=RP11                                                         548
        RP11=RP01                                                         549
        RP22=RP12
        RP12=RP02
C
3       WRITE(3,102)X1,Y1,X2,Y2,XMAX1,YMAX1,XMAX2,YMAX2                   540
102     FORMAT(1H ,T11,F6.3,T22,1PD13.6,T41,0PF6.3,T52,1PD13.6,T70,0PF8.5, 550
       1T80,1PD12.5,T94,0PF8.5,T104,1PD12.5)                             555
        END                                                              560
```

Fig. 11.2-4   Program to find the first estimate of the Hastings coefficients.

```
      DOUBLE PRECISION A(10,11),H(10),G,Y,D,R,S,T,U,X1,X2,Y1,Y2,Z,W
      COMMON A,H
C STATEMENT FUNCTION. THIS CARD MUST BE CHANGED FOR EACH PROBLEM.
      G(Y)=DEXP(Y)
      ITYPE=0
C CONSTANTS. THESE MUST BE CHANGED FOR EACH PROBLEM.
      D=4.48476D-5
      R=-8.6461D-1
      S=-4.9702D-1
      Z=0.D0
      T=5.02880-1
      U=8.6637D-1
      W=1.D0
C
106   FORMAT(1H1,T15,'EXP X'/)
      IT=41
C MATRIX ELEMENTS.
      DO 1 I=1,6
1     A(I,1)=1.D0
C
      DO 2 J=2,6
      K=J-1
      A(1,J)=R**K
      A(2,J)=S**K
      IF(Z.EQ.0.D0)A(3,J)=Z
      IF(Z.EQ.0.D0)GO TO 21
      A(3,J)=Z**K
21    A(4,J)=T**K
      A(5,J)=U**K
2     A(6,J)=W**K
C
      A(1,7)=G(R)+D
      A(2,7)=G(S)-D
      A(3,7)=G(Z)+D
      A(4,7)=G(T)-D
      A(5,7)=G(U)+D
      A(6,7)=G(W)-D
C ECHO MATRIX. THE REST CAN BE MADE INTO A SUBROUTINE.
      WRITE(3,106)
      WRITE(3,103)
103   FORMAT(1H0,T50,'ECHO MATRIX'/)
      DO 5 K=1,6
5     WRITE(3,104)(A(K,J),J=1,7)
104   FORMAT(1H0,10X,7(1PD15.5))
C SOLUTION
      CALL FGJDOD(7)
C ANSWERS
      WRITE(3,106)
      WRITE(3,105)
105   FORMAT(1H0,T15,'COEFFICIENTS'/)
      DO 7 I=1,6
      L=I-1
7     WRITE(3,108)L,H(I)
108   FORMAT(1H0,'H(',I1,')= ',1PD22.15)
C ERROR TABLE
      WRITE(3,106)
      WRITE(3,100)
100   FORMAT(1H0,T30,'ERROR TABLE'/)
      WRITE(3,101)
101   FORMAT(1H0,T16,'X',T26,'ERROR',T46,'X',T56,'ERROR'/)
C
      IX=-1025
      IY=-25
      IF(ITYPE.EQ.1)IX=-1000
      IF(ITYPE.EQ.1)IY=0
      IF(ITYPE.EQ.1)IT=40
```

Fig. 11.2-5 (Continued on next page)

```
C
        DO 3 I=1,IT
        IX=IX+25
        IY=IY+25
        X1=IX
        X2=IY
        X1=X1*1.D-3
        X2=X2*1.D-3
C
        Y=X1
        Y1=H(1)+Y*(H(2)+Y*(H(3)+Y*(H(4)+Y*(H(5)+Y*H(6)))))
        Y=X2
        Y2=H(1)+Y*(H(2)+Y*(H(3)+Y*(H(4)+Y*(H(5)+Y*H(6)))))
        Y1=G(X1)-Y1
        Y2=G(X2)-Y2
3       WRITE(3,102)X1,Y1,X2,Y2
102     FORMAT(1H ,T11,F6.3,T22,1PD13.6,T41,0PF6.3,T52,1PD13.6)
        END
```

Fig. 11.2-5   Program to find subsequent estimates of the Hastings coefficients.

PROBLEM SECTION 11.2

Using the Hastings method with $N=6$, find a Chebyshev polynomial approximation for each of the following functions. The process is self-checking.

A.   Use $x \in [-1, +1]$:
(1) $\sin x$ (6) $e^{\cos x}$
(2) $\sinh x$ (7) e^{x^2}
(3) $\cosh x$ (8) $e^{\arcsin x}$
(4) $\arctan x$ (9) $\ln(1+x)$
(5) $e^{\sin x}$ (10) $(\sin x)/x$

B.   Change the variable; then find a Hastings approximation:
(11) $\sin x, \ x \in [0, \pi]$
(12) $\cos x, \ x \in [-\pi/2, +\pi/2]$
(13) $\tan x, \ x \in [-\pi/2, +\pi/2]$
(14) $\sec x, \ x \in [-\pi/2, +\pi/2]$

11.3   THE METHOD OF RATIONAL FRACTIONS

11.3.1   Introduction

It can be shown[2] that

$$e^x = 1 + \cfrac{x}{1 + \cfrac{x}{-2 + \cfrac{x}{-3 + \cfrac{x}{2 + \cfrac{x}{5 + \cfrac{x}{-2 + \cdots}}}}}} \qquad (11.3.1)$$

---

[2]L. M. Milne-Thomson, p. 121.

This is called a *continued fraction*. Equation (11.3.1) is obtained by using the theory of reciprocal differences.

If this fraction is truncated, e.g.,

$$e^x \sim 1 + \cfrac{x}{1 + \cfrac{x}{-2}} \qquad (11.3.2)$$

then the truncated continued fraction can be simplified to

$$e^x \sim \frac{-x - 2}{x - 2} \qquad (11.3.3)$$

Every truncated continued fraction can obviously be reduced to a rational fraction of the form

$$\frac{\sum_{k=0}^{n} p_k x^k}{\sum_{k=0}^{m} q_k x^k} \qquad (11.3.4)$$

If $q_0 \neq 0$, we can divide numerator and denominator by q_0 and obtain a denominator with a leading coefficient equal to 1. If $q_0 = 0$ and $q_1 \neq 0$, we can divide by q_1. In general, we can always obtain a denominator with its leading coefficient equal to 1.

Experimentation shows that fractions are quite sensitive to small changes in the coefficients. Therefore, it is often possible to obtain a better approximation with a fraction than with a polynomial using the same number of coefficients. Furthermore, if $f(x)$ has singularities, a fraction can approximate it, whereas a polynomial cannot. For example, if $f(x)$ becomes infinite at $x = -2$ and $x = +5$, a function of the form

$$\frac{N(x)}{(x+2)(x-5)Q(x)} \qquad (11.3.5)$$

will have a similar characteristic provided that the numerator does not have the factors $(x+2)$ or $(x-5)$.

It is possible to go through a theoretical discussion via continued fractions and the theory of reciprocal differences to develop a fraction to approximate $f(x)$. However, we shall omit this development because in practice an empirical method using the Chebyshev no-error points will serve as well or better.

In our demonstration, we limit ourselves to *six* independent coefficients in order to permit an easy comparison with the other methods in this chapter. In our illustrative problem, to avoid unnecessary work, we have used IBM subprogram values instead of true values. In a practical problem, it would be necessary to use true values and to test different numbers of independent coefficients.

11.3.2 Eleven Types

We shall define eleven *types* of rational functions with six independent coefficients, $h_0, h_1, \ldots, h_5$:

Type 1:
$$R(x) = \frac{h_0}{1 + h_1 x + h_2 x^2 + h_3 x^3 + h_4 x^4 + h_5 x^5} \qquad (11.3.6)$$

Type 2:
$$R(x) = \frac{h_0}{x + h_1 x^2 + h_2 x^3 + h_3 x^4 + h_4 x^5 + h_5 x^6} \qquad (11.3.7)$$

Type 3:
$$R(x) = \frac{h_0 + h_1 x}{1 + h_2 x + h_3 x^2 + h_4 x^3 + h_5 x^4} \qquad (11.3.8)$$

Type 4:
$$R(x) = \frac{h_0 + h_1 x}{x + h_2 x^2 + h_3 x^3 + h_4 x^4 + h_5 x^5} \qquad (11.3.9)$$

Type 5:
$$R(x) = \frac{h_0 + h_1 x + h_2 x^2}{1 + h_3 x + h_4 x^2 + h_5 x^3} \qquad (11.3.10)$$

Type 6:
$$R(x) = \frac{h_0 + h_1 x + h_2 x^2}{x + h_3 x^2 + h_4 x^3 + h_5 x^4} \qquad (11.3.11)$$

Type 7:
$$R(x) = \frac{h_0 + h_1 x + h_2 x^2 + h_3 x^3}{1 + h_4 x + h_5 x^2} \qquad (11.3.12)$$

Type 8:
$$R(x) = \frac{h_0 + h_1 x + h_2 x^2 + h_3 x^3}{x + h_4 x^2 + h_5 x^3} \qquad (11.3.13)$$

Type 9:
$$R(x) = \frac{h_0 + h_1 x + h_2 x^2 + h_3 x^3 + h_4 x^4}{1 + h_5 x} \qquad (11.3.14)$$

Type 10:
$$R(x) = \frac{h_0 + h_1 x + h_2 x^2 + h_3 x^3 + h_4 x^4}{x + h_5 x^2} \qquad (11.3.15)$$

Type 11:
$$R(x) = \frac{h_0 + h_1 x + h_2 x^2 + h_3 x^3 + h_4 x^4 + h_5 x^5}{x} \qquad (11.3.16)$$

As in the preceding discussions, we assume that the no-error points are at

$$\theta_r = \pi/12, 3\pi/12, 5\pi/12, 7\pi/12, 9\pi/12, 11\pi/12 \qquad (11.3.17)$$

i.e., that at these points

$$R(x) = f(x) \qquad (11.3.18)$$

The corresponding values of $x_r = \cos\theta_r$ are then $D, B, A, -A, -B, -D$, where the symbols refer to Appendix Table A.14.

For each type, a set of six simultaneous equations can be written using the Chebyshev points (11.3.17) and the relationship (11.3.18). The equations are as follows:

Type 1:	$h_0 - xRh_1 - x^2Rh_2 - x^3Rh_3 - x^4Rh_4 - x^5Rh_5 = R$	(11.3.19)
Type 2:	$h_0 - x^2Rh_1 - x^3Rh_2 - x^4Rh_3 - x^5Rh_4 - x^6Rh_5 = xR$	(11.3.20)
Type 3:	$h_0 + xh_1 - xRh_2 - x^2Rh_3 - x^3Rh_4 - x^4Rh_5 = R$	(11.3.21)
Type 4:	$h_0 + xh_1 - x^2Rh_2 - x^3Rh_3 - x^4Rh_4 - x^5Rh_5 = xR$	(11.3.22)
Type 5:	$h_0 + xh_1 + x^2h_2 - xRh_3 - x^2Rh_4 - x^3Rh_5 = R$	(11.3.23)
Type 6:	$h_0 + xh_1 + x^2h_2 - x^2Rh_3 - x^3Rh_4 - x^4Rh_5 = xR$	(11.3.24)
Type 7:	$h_0 + xh_1 + x^2h_2 + x^3h_3 - xRh_4 - x^2Rh_5 = R$	(11.3.25)
Type 8:	$h_0 + xh_1 + x^2h_2 + x^3h_3 - x^2Rh_4 - x^3Rh_5 = xR$	(11.3.26)
Type 9:	$h_0 + xh_1 + x^2h_2 + x^3h_3 + x^4h_4 - x^2Rh_5 = R$	(11.3.27)
Type 10:	$h_0 + xh_1 + x^2h_2 + x^3h_3 + x^4h_4 - x^2Rh_5 = xR$	(11.3.28)
Type 11:	$h_0 + xh_1 + x^2h_2 + x^3h_3 + x^4h_4 + x^5h_5 = xR$	(11.3.29)

11.3.3 Cycle 1

ILLUSTRATIVE PROBLEM 11.3-1 Find a six-point rational fraction to approximate $f(x) = e^x$ for $x \in [-1, +1]$.

Solution Cycle 1 We program the computer to calculate the coefficients h_k and to print an error table for each type. Strictly speaking, since e^x has no discontinuity at $x = 0$, it is unnecessary to test types 2, 4, 6, 8, 10 and 11, but it was easier to run the complete program (Figure 11.3-1) than to alter it.

```
      DOUBLE PRECISION A(10,11),H(10),R(6),C(6),R1,R2,R3,R4,R5,R6,R7,R8,
     1R9,R10,R11,H1,H2,H3,H4,H5,H6,X,X1,X2,Y,Y1,Y2,P,Q,G,CK,TWO,RP01,
     1RP02,RP11,RP21,     RP22,SA,SB,S,XMAX1,XMAX2,YMAX1,YMAX2,ZERO
      COMMON A,H
C STATEMENT FUNCTIONS.
      R1(X)=H1/(1.D0+X*(H2+X*(H3+X*(H4+X*(H5+H6*X)))))
      R2(X)=H1/(X*(1.D0+X*(H2+X*(H3+X*(H4+X*(H5+H6*X))))))
      R3(X)=(H1+H2*X)/(1.D0+X*(H3+X*(H4+X*(H5+H6*X))))
      R4(X)=(H1+H2*X)/(X*(1.D0+X*(H3+X*(H4+X*(H5+H6*X)))))
      R5(X)=(H1+X*(H2+H3*X))/(1.D0+X*(H4+X*(H5+H6*X)))
      R6(X)=(H1+X*(H2+H3*X))/(X*(1.D0+X*(H4+X*(H5+H6*X))))
      R7(X)=(H1+X*(H2+X*(H3+H4*X)))/(1.D0+X*(H5+H6*X))
      R8(X)=(H1+X*(H2+X*(H3+H4*X)))/(X*(1.D0+X*(H5+H6*X)))
      R9(X)=(H1+X*(H2+X*(H3+X*(H4+H5*X))))/(1.D0+H6*X)
      R10(X)=(H1+X*(H2+X*(H3+X*(H4+H5*X))))/(X*(1.D0+H6*X))
      R11(X)=(H1+X*(H2+X*(H3+X*(H4+X*(H5+H6*X)))))/X
C THE FOLLOWING THREE CARDS ARE CHANGED FOR EACH RUN.
      G(X)=DEXP(X)
100   FORMAT(1H1,T7,'RATIONAL FUNCTION FOR EXP X'//)
      JSTART=0
C CHEBYSHEV POINTS.
```

Fig. 11.3-1 (Continued on next page)

```
                TWO=2.DO
                CK=DSQRT(3.DO)
                C(1)=DSQRT(TWO+CK)/TWO
                C(2)=DSQRT(TWO)/TWO
                C(3)=DSQRT(TWO-CK)/TWO
                C(4)=-C(3)
                C(5)=-C(2)
                C(6)=-C(1)
        C TRUE VALUES.
                DO 1 I=1,6
                Y=C(I)
        1       R(I)=G(Y)
        C COMPUTATION OF COEFFICIENTS ELEVEN WAYS.
                DO 7 N=1,11
        C INITIAL VALUES.
                ZERO=0.DO
                RP11=ZERO
                RP12=ZERO
                RP21=ZERO
                RP22=ZERO
                DO 2 I=1,6
                A(I,1)=1.DO
                X=C(I)
                A(I,2)=X
                A(I,3)=X**2
                A(I,4)=A(I,2)*A(I,3)
                A(I,5)=A(I,3)**2
                A(I,6)=A(I,3)*A(I,4)
                A(I,7)=X*R(I)
        C
                GO TO (21,22,23,24,25,26,27,28,29,30,2),N
        21      A(I,2)=-X*R(I)
                DO 211 J=3,6
                L=J-1
        211     A(I,J)=A(I,L)*X
                A(I,7)=R(I)
                GO TO 2
        C
        22      A(I,2)=-X*A(I,7)
                DO 221 J=3,6
                L=J-1
        221     A(I,J)=A(I,L)*X
                GO TO 2
        C
        23      A(I,3)=-X*R(I)
                DO 231 J=4,6
                L=J-1
        231     A(I,J)=A(I,L)*X
                A(I,7)=R(I)
                GO TO 2
        C
        24      A(I,3)=-X**2*R(I)
                DO 241 J=4,6
                L=J-1
        241     A(I,J)=A(I,L)*X
                GO TO 2
        C
        25      A(I,4)=-X*R(I)
                A(I,5)=X*A(I,4)
                A(I,6)=X*A(I,5)
                A(I,7)=R(I)
                GO TO 2
        C
        26      A(I,4)=-A(I,3)*R(I)
                A(I,5)=X*A(I,4)
```

```
        A(I,6)=X*A(I,5)
        GO TO 2
C
27      A(I,5)=-A(I,7)
        A(I,6)=X*A(I,5)
        A(I,7)=R(I)
        GO TO 2
C
28      A(I,5)=-X*A(I,7)
        A(I,6)=X*A(I,5)
        GO TO 2
C
29      A(I,6)=-A(I,7)
        A(I,7)=R(I)
        GO TO 2
C
30      A(I,6)=-X*A(I,7)
C
2       CONTINUE
C ECHO MATRIX.
        WRITE(3,200)N
200     FORMAT(1H1,T45,'ECHO MATRIX, TYPE ',I2/)
        DO 3 I=1,6
3       WRITE(3,201)(A(I,J),J=1,7)
201     FORMAT(1H0,10X,7(1PD15.5))
C SOLUTION.
        CALL FGJDOD(7)
        H1=H(1)
        H2=H(2)
        H3=H(3)
        H4=H(4)
        H5=H(5)
        H6=H(6)
C PRINT-OUT OF COEFFICIENTS.
        WRITE(3,100)
        WRITE(3,202)N
202     FORMAT(1H0,T17,'COEFFICIENTS, TYPE ',I2/)
        DO 4 I=1,6
        L=I-1
4       WRITE(3,203)L,H(I)
203     FORMAT(1H0,10X,'H(',I1,')=',1PD22.15)
C ERROR TABLES. EXTREMALS BY QUADRATIC INTERPOLATION.
        WRITE(3,100)
        WRITE(3,204)N
204     FORMAT(1H0,T15,'ERROR TABLE, TYPE ',I2,T88,'EXTREMALS'/)
        WRITE(3,205)
205     FORMAT(1H0,T16,'X',T26,'ERROR',T46,'X',T56,'ERROR',T74,'X',T86,
       1'Y',T98,'X',T110,'Y',/)
C
        IX=-1025
        IY=-25
        IT=41
        IF(JSTART.EQ.1)IX=-1000
        IF(JSTART.EQ.1)IT=40
        IF(N.EQ.2.OR.N.EQ.4.OR.N.EQ.6.OR.N.EQ.8.OR.N.EQ.10.OR.N.EQ.11)IY=0
        IF(N.EQ.2.OR.N.EQ.4.OR.N.EQ.6.OR.N.EQ.8.OR.N.EQ.10.OR.N.EQ.11)IT=4
       10
C
        DO 7 I=1,IT
        IX=IX+25
        IY=IY+25
        X1=IX
        X2=IY
        X1=X1*1.D-3
        X2=X2*1.D-3
```

Fig. 11.3-1   (Continued on next page)

```
C
      P=G(X1)
      Q=G(X2)
C CALCULATION ELEVEN WAYS.
      GO TO (301,302,303,304,305,306,307,308,309,310,311),N
301   Y1=P-R1(X1)
      Y2=Q-R1(X2)
      GO TO 6
302   Y1=P-R2(X1)
      Y2=Q-R2(X2)
      GO TO 6
303   Y1=P-R3(X1)
      Y2=Q-R3(X2)
      GO TO 6
304   Y1=P-R4(X1)
      Y2=Q-R4(X2)
      GO TO 6
305   Y1=P-R5(X1)
      Y2=Q-R5(X2)
      GO TO 6
306   Y1=P-R6(X1)
      Y2=Q-R6(X2)
      GO TO 6
307   Y1=P-R7(X1)
      Y2=Q-R7(X2)
      GO TO 6
308   Y1=P-R8(X1)
      Y2=Q-R8(X2)
      GO TO 6
309   Y1=P-R9(X1)
      Y2=Q-R9(X2)
      GO TO 6
310   Y1=P-R10(X1)
      Y2=Q-R10(X2)
      GO TO 6
311   Y1=P-R11(X1)
      Y2=Q-R11(X2)
C TWO PREVIOUS VALUES OF R(X) ARE RP2 AND RP1. THE PRESENT VALUE IS RPO.
6     RP01=Y1
      SA=RP21-RP01
      SB=(RP01-2.D0*RP11+RP21)*2.D0
      S=SA/SB
      XMAX1=X1+(2.5D-2)*(S-1.D0)
      SA=SA**2
      SB=4.D0*SB
      YMAX1=RP11-SA/SB
C
      RP02=Y2
      SA=RP22-RP02
      SB=(RP02-2.D0*RP12+RP22)*2.D0
      S=SA/SB
      XMAX2=X2+(2.5D-2)*(S-1.D0)
      SA=SA**2
      SB=4.D0*SB
      YMAX2=RP12-SA/SB
C
      RP21=RP11
      RP11=RP01
      RP22=RP12
      RP12=RP02
C
7     WRITE(3,206)X1,Y1,X2,Y2,XMAX1,YMAX1,XMAX2,YMAX2
206   FORMAT(1H ,T11,F6.3,T22,1PD13.6,T41,0PF6.3,T52,1PD13.6,T70,0PF8.5,
     1T80,1PD12.5,T94,0PF8.5,T104,1PD12.5)
      END
```

Fig. 11.3-1   Program to compute the Chebyshev-type coefficients for a rational fraction, and to produce error tables for eleven types of rational fractions.

A glance at the error tables produced by the computer showed the maximum extremals for the various types to be as in Table 11.3-1.

Table 11.3-1

TYPE	|EXTREMAL|
1	117×10^{-6}
2	$946{,}849 \times 10^{-6}$
3	331×10^{-6}
4	1117×10^{-6}
5	12.2×10^{-6}
6	262.3×10^{-6}
7	9.0×10^{-6}
8	169.8×10^{-6}
9	13.4×10^{-6}
10	256.8×10^{-6}
11	$107{,}176 \times 10^{-6}$

The best one is type 7. The error table for type 7 is given in Figure 11.3-2.

The computer was programmed to calculate the extremals at the same time, using the extremal-point theorem. The extremals are given by

x	EXTREMAL
-0.86066	-2.15272×10^{-6}
-0.48401	2.86034×10^{-6}
0.02270	-4.28890×10^{-6}
0.51752	6.55005×10^{-6}
0.87089	-9.03480×10^{-6}

from which $d = 4.82326 \times 10^{-6}$. We can set up six equations: the first five of which use these values of x along with $L \pm d$, and the last of which uses some other value of x, say $x = 1$. The procedure for subsequent cycles is precisely as demonstrated already and will be omitted here.

The computer values of the coefficients, for type 7, were as follows:

```
H(0)= 1.0000042483520720 00

H(1)= 6.0148224998197100-01

H(2)= 1.5051943799487310-01

H(3)= 1.6489242256608820-02

H(4)=-3.9851956929182350-01

H(5)= 4.9115022986765510-02
```

RATIONAL FUNCTION FOR EXP X

ERROR TABLE, TYPE 7

X	ERROR
-1.000	1.941595D-06
-0.975	4.438880D-07
-0.950	-6.616807D-07
-0.925	-1.425914D-06
-0.900	-1.896635D-06
-0.875	-2.118687D-06
-0.850	-2.133929D-06
-0.825	-1.981248D-06
-0.800	-1.696571D-06
-0.775	-1.312881D-06
-0.750	-8.602461D-07
-0.725	-3.658522D-07
-0.700	1.459605D-07
-0.675	6.536491D-07
-0.650	1.138415D-06

X	ERROR
0.0	-4.248352D-06
0.025	-4.288486D-06
0.050	-4.230214D-06
0.075	-4.070340D-06
0.100	-3.807645D-06
0.125	-3.443041D-06
0.150	-2.979696D-06
0.175	-2.423147D-06
0.200	-1.781370D-06
0.225	-1.064832D-06
0.250	-2.864975D-07
0.275	5.382053D-07
0.300	1.391469D-06
0.325	2.253245D-06
0.350	3.101446D-06

EXTREMALS

X	Y	X	Y
-1.03750	-2.42699D-07	-0.03750	5.31044D-07
-0.99839	1.94876D-06	0.01274	-4.79464D-06
-0.89202	-1.71640D-06	0.02270	-4.28890D-06
-0.88153	-1.94200D-06	0.02316	-4.28876D-06
-0.87241	-2.07543D-06	0.02363	-4.28742D-06
-0.86518	-2.13789D-06	0.02306	-4.29031D-06
-0.86066	-2.15272D-06	0.02019	-4.31084D-06
-0.86023	-2.14799D-06	0.01322	-4.37473D-06
-0.86642	-2.16239D-06	-0.00075	-4.52924D-06
-0.88438	-2.26053D-06	-0.02711	-4.86629D-06
-0.92663	-2.58100D-06	-0.07738	-5.58508D-06
-1.03348	-3.54486D-06	-0.18214	-7.21382D-06
-1.44707	-7.63137D-06	-0.45938	-1.17844D-05
2.39005	3.16490D-05	-2.21825	-4.17976D-05
-0.13381	6.02468D-06	1.89949	2.91767D-05

x					
0.375	3.912215D-06	-0.35207	3.91065D-06	0.90401	1.22924D-05
0.400	4.660281D-06	-0.42542	3.25843D-06	0.68576	8.75650D-06
0.425	5.319398D-06	-0.45773	3.00577D-06	0.59775	7.44299D-06
0.450	5.862879D-06	-0.47307	2.90567D-06	0.55500	6.88278D-06
0.475	6.264251D-06	-0.48011	2.87015D-06	0.53311	6.64814D-06
0.500	6.498019D-06	-0.48284	2.86085D-06	0.52237	6.56511D-06
0.525	6.540573D-06	-0.48346	2.85986D-06	0.51806	6.54793D-06
0.550	6.371241D-06	-0.48345	2.85983D-06	0.51752	6.55005D-06
0.575	5.973506D-06	-0.48401	2.86034D-06	0.51897	6.54723D-06
0.600	5.336403D-06	-0.48637	2.86701D-06	0.52096	6.53273D-06
0.625	4.456113D-06	-0.49209	2.89247D-06	0.52200	6.51990D-06
0.650	3.337780D-06	-0.50343	2.95981D-06	0.52005	6.55369D-06
0.675	1.997560D-06	-0.52424	3.11091D-06	0.51150	6.74295D-06
0.700	4.649345D-07	-0.56191	3.42776D-06	0.48836	7.35944D-06
0.725	-1.214693D-06	-0.63322	4.09550D-06	0.42685	9.23911D-06
0.750	-2.977086D-06	-0.78668	5.64616D-06	0.20517	1.66781D-05
0.775	-4.735929D-06	-1.24690	1.05325D-05	13.14813	-4.39544D-04
0.800	-6.379207D-06	********	1.19030D-04	1.14299	-1.72555D-05
0.825	-7.765189D-06	1.00460	-1.42928D-05	0.94717	-1.08373D-05
0.850	-8.717964D-06	0.37725	-7.58758D-06	0.89248	-9.34347D-06
0.875	-9.022514D-06	0.19316	-5.70818D-06	0.87425	-9.02281D-06
0.900	-8.419262D-06	0.11121	-4.93113D-06	0.87089	-9.03480D-06
0.925	-6.598066D-06	0.06867	-4.56909D-06	0.87512	-9.02253D-06
0.950	-3.191602D-06	0.04530	-4.39825D-06	0.88378	-8.75294D-06
0.975	2.231899D-06	0.03251	-4.32294D-06	0.89528	-8.02346D-06
1.000	1.017663D-05	0.02598	-4.29520D-06	0.90872	-6.62833D-06

x	
-0.625	1.584141D-06
-0.600	1.977324D-06
-0.575	2.306999D-06
-0.550	2.564650D-06
-0.525	2.744119D-06
-0.500	2.841506D-06
-0.475	2.855056D-06
-0.450	2.785043D-06
-0.425	2.633641D-06
-0.400	2.404790D-06
-0.375	2.104052D-06
-0.350	1.738461D-06
-0.325	1.316358D-06
-0.300	8.472323D-07
-0.275	3.415376D-07
-0.250	-1.894836D-07
-0.225	-7.339906D-07
-0.200	-1.279738D-06
-0.175	-1.814276D-06
-0.150	-2.325151D-06
-0.125	-2.800116D-06
-0.100	-3.227337D-06
-0.075	-3.595606D-06
-0.050	-3.894546D-06
-0.025	-4.114824D-06
0.0	-4.248352D-06

Fig. 11.3-2 Error table, type 7.

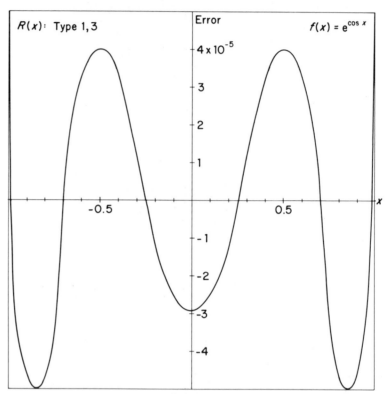

Fig. 11.3-3 Error curve for $e^{\cos x}$ **using rational fraction types 1 and 3.** $R(x)\sim$
$2.7183/(1+0.50014x^2+0.083513x^4)$, **Chebyshev error** $\sim 4.05\times10^{-5}$.

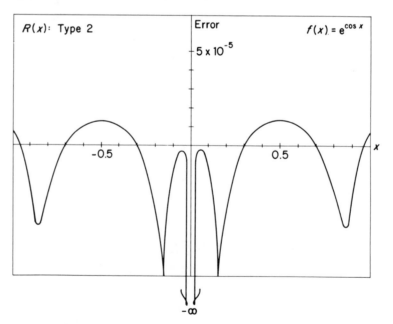

Fig. 11.3-4 Error curve for $e^{\cos x}$ **using rational fraction type 2.** $R(x)\sim 0.20903/(x$
$+1.42260x^2-3.68461x^4+2.46069x^6)$.

11.3.5 Another Example

A function like e^x is handled efficiently in many ways because it is monotonic and continuous, whereas $e^{\cos x}$ is less tractable. For comparison, the first-cycle curves, the first-cycle fraction and the probable minimax errors for various types are shown in Figures 11.3-3 to 11.3-6.

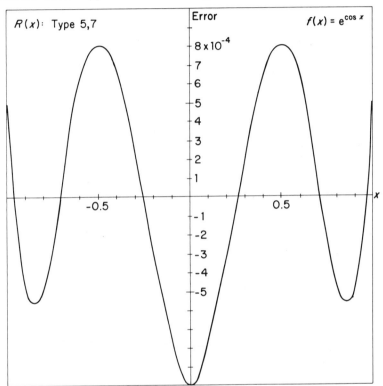

Fig. 11.3-5 Error curve for $e^{\cos x}$ **using rational fraction type 5 or 7.** $R(x) \sim$ $(2.71926 - 0.36311 x^2)/(1 + 0.37015 x^2)$. **Chebyshev error** $\sim 8.22 \times 10^{-4}$.

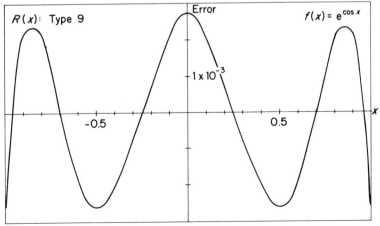

Fig. 11.3-6 Error curve for $e^{\cos x}$ **using rational fraction type 9.** $R(x) \sim (2.71560 -$ $1.31031 x^2 + 0.31341 x^4)/1$, **Chebyshev error** $\sim 2.59 \times 10^{-3}$.

PROBLEM SECTION 11.3

Using six equations, find a rational fraction to approximate the following on the interval $[-1, +1]$. Our guess about the correct minimax error for the best rational fraction is given in brackets. These are all for the first cycle.

(1) $e^{\sin x}$, $[4 \times 10^{-5}]$

(2) $\ln(1 + x)$, $[7 \times 10^{-3}]$

(3) $\sin x$, $[3 \times 10^{-6}]$

(4) $\arccos x$, $[8 \times 10^{-3}]$

(5) $\arcsin x$, $[2 \times 10^{-3}]$

(6) $\tan x$, $[2 \times 10^{-5}]$

(7) $\cos x$, $[4 \times 10^{-5}]$

(8) $\exp x^2$, $[4 \times 10^{-3}]$

(9) $\arctan x$, $[9 \times 10^{-5}]$

(10) $\sinh x$, $[3 \times 10^{-6}]$

(11) $\cosh x$, $[5 \times 10^{-5}]$

Chapter 12

Integration by Computer

There is a considerable body of literature concerning integration by computer. We shall discuss three fairly common methods used by computer experts:

1. Simpson's rule, with and without the Richardson extrapolation
2. The trapezoidal rule with Romberg extrapolation to the limit
3. Approximation of the integrand by a Chebyshev summation

Of these, the first is probably the most popular and least reliable.

One of the best general methods, in terms of reliability, is a difficult one: the approximation of the integrand by the Hastings method. To show the power of this method, consider

$$J(x) = \int_0^x e^t \, dt, \qquad x \in (0,1) \tag{12.1}$$

We developed a six-term Hastings-type approximation of the form

$$P_1(x) = h_0 + h_1(x) + \cdots + h_5 x^5 \tag{12.2}$$

in a previous chapter. The actual minimax error was 4.5×10^{-5}. Integrating (12.2), we have

$$P_2(x) = h_0 x + \tfrac{1}{2} h_1 x^2 + \cdots + \tfrac{1}{6} h_5 x^6 \tag{12.3}$$

with a theoretical maximum error of

$$E = \int_0^x (4.5 \times 10^{-5}) \, dt = 4.5 \times 10^{-5} x \tag{12.4}$$

The actual maximum error is not nearly this large. The Hastings approximation is such that the errors are alternately positive and negative over the interval, and approximately equal in magnitude. Under integration over the full range, or over an even number of ripples, the errors should tend to cancel. In the case of (12.1), we know the exact answer, namely $e^x - 1$, and it is easy to compute the actual error. Figure 12-1 compares the error of P_1 with that of P_2. Note that the error was reduced by a factor of more than 5. It is clear that if the approximating function is of the equal-ripple type, the error of the integral can be considerably less than that of the approximating function.

431

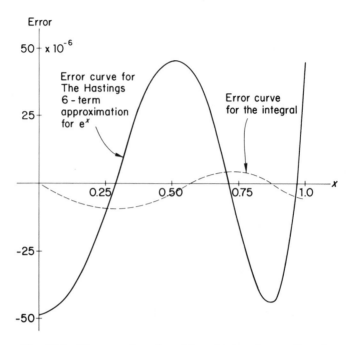

Fig. 12-1   The error is reduced by a factor of more than 5.

If the integration is very important and the correct answer is not known, it may be worth the time and effort to develop a sufficiently precise Hastings approximation with a large number of small, almost equal, ripples.

We shall not pursue this further, since the method has already been explored in great detail. However, we shall close this chapter with a discussion of the use of the Chebyshev summation, which, although not nearly as good, is very easy to program and affords a ballpark estimate of the error.

In this chapter, we shall pretend that the job is to solve the following thirteen problems:

I. Compute a table of values for the *complete elliptic integral of the first kind*:

$$K(p) = \int_0^{\pi/2} (1 - p\sin^2 t)^{-1/2} \, dt$$

from $p=0$ to $p=0.9$ at intervals of 0.1. This is written symbolically as

$$p = 0(0.1)0.9.$$

II. Compute a table of values for the *error function*

$$\operatorname{erf} x = \frac{2}{\sqrt{\pi}} \int_0^x e^{-t^2} \, dt, \qquad x = 0.1(0.1)1.0$$

III. Compute a table of values for the *gamma function*

$$\Gamma(x) = \int_0^\infty t^{p-1} e^{-t} \, dt, \qquad p = 1.25(0.05)1.70$$

These three will be used for illustrative problems. The problem sections will have the following integrals to be computed by three different methods:

1. $\displaystyle\int_0^1 e^{-x^2} dx$

2. $\displaystyle\int_0^{2\pi} \ln(1+t) \sin 10t \, dt$

3. *Clausen's integral:*

$$\psi(x) = \int_0^x -\ln\left(2\sin\frac{t}{2}\right) dt, \qquad x = 16°(2°)34°$$

4. *Complete elliptic integral of the second kind:*

$$E(p) = \int_0^{\pi/2} (1 - p\sin^2 t)^{1/2} \, dt, \qquad p = 0.05(0.05)0.50$$

5. *Sievert integral at* 60°:

$$S(p) = \int_0^{\pi/3} e^{-p\sec t} \, dt, \qquad p = 0(0.2)1.8$$

6. *Bessel function of order* 0:

$$J_0(p) = \frac{1}{\pi} \int_0^\pi \cos(p\cos t) \, dt, \qquad p = 0.5(0.5)5.0$$

7. *Fresnel integral:*

$$C(x) = \int_0^x \cos\left(\frac{\pi}{2} t^2\right) dt, \qquad x = 0.5(0.5)5.0$$

8. *Sine integral:*

$$\mathrm{Si}(x) = \int_0^x \frac{\sin t}{t} \, dt, \qquad x = 1.1(0.1)2.0$$

9. *Debye integral:*

$$D(x) = \frac{4}{x^4} \int_0^x \frac{t^4}{e^t - 1} \, dt, \qquad x = 0.1(0.1)1.0$$

10. *Dilogarithm:*

$$\mathrm{Di}(x) = \int_1^x \frac{\ln t}{1 - t} \, dt, \qquad x = 0.05(0.05)0.50$$

Table 12-1
Errors for Four Methods of Integration

FUNCTION	VARIABLE	SIMPSON'S RULE	SIMPSON-RICHARDSON	TRAPEZOIDAL-ROMBERG	CHEBYSHEV SUMMATION
COMPLETE	0.0	0.0	2.1 D-8	0.0	1.3 D-15
ELLIPTIC	0.1	1.6 D-14	1.9 D-8	0.0	0.0
INTEGRAL	0.2	1.7 D-14	3.0 D-8	0.0	1.1 D-15
OF THE	0.3	1.7 D-14	2.2 D-8	0.0	1.3 D-15
FIRST KIND.	0.4	1.7 D-14	9.7 D-9	1.8 D-15	1.3 D-15
D-15	0.5	1.8 D-14	2.1 D-7	2.7 D-15	1.8 D-15
	0.6	1.9 D-14	9.7 D-7	2.4 D-15	4.0 D-15
	0.7	1.9 D-14	9.9 D-7	1.6 D-15	4.5 D-14
	0.8	2.0 D-14	3.4 D-7	2.7 D-15	1.5 D-12
	0.9	1.9 5-14	5.3 D-7	2.7 D-15	9.7 D-11
BESSEL	0.5	0.0	2.6 D-9	0.0	0.0
FUNCTION	1.0	0.0	3.5 D-9	0.0	0.0
OF ORDER 0.	1.5	0.0	2.2 D-9	0.0	1.8 D-14
D-15	2.0	0.0	3.8 D-9	0.0	2.5 D-13
	2.5	0.0	2.5 D-11	0.0	2.1 D-12
	3.0	0.0	3.3 D-9	0.0	1.3 D-11
	3.5	0.0	8.0 D-10	0.0	6.3 D-11
	4.0	0.0	5.7 D-10	0.0	2.6 D-10
	4.5	0.0	3.6 D-9	0.0	9.0 D-10
	5.0	0.0	1.6 D-9	0.0	2.8 D-9
ERROR	0.1	0.0	2.5 D-9	0.0	0.0
FUNCTION.	0.2	0.0	1.8 D-9	0.0	0.0
D-10	0.3	0.0	0.0	0.0	0.0
	0.4	0.0	0.0	0.0	0.0
	0.5	0.0	3.5 D-9	0.0	0.0
	0.6	0.0	0.0	0.0	0.0
	0.7	0.0	0.0	0.0	0.0
	0.8	0.0	7.2 D-10	0.0	0.0
	0.9	0.0	8.7 D-10	0.0	0.0
	1.0	0.0	9.5 D-10	0.0	0.0
SINE	1.1	0.0	4.0 D-8	0.0	0.0
INTEGRAL.	1.2	0.0	4.5 D-8	0.0	0.0
D-10	1.3	0.0	6.1 D-8	0.0	0.0
	1.4	0.0	1.3 D-8	0.0	0.0
	1.5	0.0	2.3 D-8	0.0	0.0
	1.6	0.0	2.0 D-8	0.0	0.0
	1.7	0.0	4.4 D-8	0.0	0.0
	1.8	0.0	2.1 D-8	0.0	0.0
	1.9	0.0	5.1 D-8	0.0	0.0
	2.0	0.0	3.3 D-8	0.0	0.0
GAMMA	1.25	3.0 D-5	2.8 D-5	2.6 D-5	9.6 D-6
FUNCTION.	1.30	1.8 D-5	1.7 D-5	1.6 D-5	8.6 D-6
D-10	1.35	1.1 D-5	1.0 D-5	9.8 D-6	7.3 D-6
	1.40	6.8 D-5	6.2 D-6	5.9 D-6	6.0 D-6
	1.45	4.1 D-6	3.7 D-6	3.5 D-6	4.8 D-6
	1.50	2.5 D-6	2.2 D-6	2.1 D-6	3.8 D-6
	1.55	1.5 D-6	1.3 D-6	3.6 D-6	2.9 D-6
	1.60	8.6 D-7	7.6 D-7	2.2 D-6	2.3 D-6
	1.65	5.0 D-7	4.4 D-7	1.3 D-6	1.7 D-6
	1.70	2.8 D-7	2.4 D-7	7.4 D-7	1.4 D-6

Function	Variable	Simpson's Rule	Simpson-Richardson	Trapezoidal-Romberg	Chebyshev Summation
COMPLETE	0.05	0.0	2.7 D-8	0.0	0.0
ELLIPTIC	0.10	0.0	4.6 D-8	0.0	0.0
INTEGRAL	0.15	1.1 D-9	3.4 D-8	1.1 D-9	1.1 D-9
OF THE	0.20	0.0	2.7 D-8	0.0	0.0
SECOND KIND.	0.25	0.0	4.2 D-8	0.0	0.0
D-9	0.30	0.0	1.7 D-8	0.0	0.0
	0.35	0.0	5.0 D-8	0.0	0.0
	0.40	0.0	0.0	0.0	0.0
	0.45	0.0	3.2 D-8	0.0	0.0
	0.50	0.0	4.8 D-8	0.0	0.0
DILOGARITHM	0.05	0.0	1.5 D-9	0.0	0.0
D-9	0.10	0.0	4.2 D-8	0.0	0.0
	0.15	0.0	1.9 D-9	0.0	0.0
	0.20	0.0	5.2 D-8	0.0	0.0
	0.25	0.0	1.3 D-9	0.0	0.0
	0.30	0.0	2.3 D-9	0.0	0.0
	0.35	0.0	2.1 D-9	0.0	0.0
	0.40	0.0	2.3 D-9	0.0	0.0
	0.45	0.0	3.2 D-9	0.0	0.0
	0.50	0.0	7.7 D-10	0.0	0.0
LN(1+X)SIN(10X). FROM 0 TO 2 PI D-8	NONE	8.0 D-7	8.0 D-7	5.8 D-7	4.4 D-5
EXP(-X**2). 0 TO 1,D-7	NONE	0.0	0.0	0.0	0.0
FRESNEL	0.5	0.0	0.0	0.0	0.0
INTEGRAL.	1.0	0.0	0.0	0.0	0.0
D-7	1.5	0.0	0.0	0.0	0.0
	2.0	0.0	0.0	0.0	0.0
	2.5	0.0	0.0	0.0	0.0
	3.0	0.0	0.0	0.0	0.0
	3.5	0.0	0.0	0.0	0.0
	4.0	0.0	0.0	0.0	1.2 D-5
	4.5	0.0	0.0	0.0	6.3 D-5
	5.0	0.0	0.0	0.0	2.2 D-3
DEBYE	0.1	0.0	0.0	0.0	0.0
INTEGRAL.	0.2	0.0	0.0	0.0	0.0
D-6	0.3	0.0	0.0	0.0	0.0
	0.4	0.0	0.0	0.0	0.0
	0.5	0.0	0.0	0.0	0.0
	0.6	0.0	5.1 D-7	0.0	0.0
	0.7	0.0	0.0	0.0	0.0
	0.8	0.0	6.1 D-7	0.0	0.0
	0.9	0.0	0.0	0.0	0.0
	1.0	0.0	0.0	0.0	0.0
CLAUSEN'S INTEGRAL. D-6	FROM 16 DEG. TO 34	0.0	0.0	0.0	0.0
SIEVERT INTEGRAL. D-5	FROM 0.0 TO 1.8	0.0	0.0	0.0	0.0

The "true values" to check the programs were obtained mainly from Abramowitz and Stegun. These true values are listed in the problems for Section 12.3 to enable checking programs.

Table 12-1 shows the errors obtained by four methods using the IBM 360, double precision. In each case, a zero error means that the computed value and the "true value" agreed to all places. The number under the name of the integral describes the precision of the table of true values. For example, in the first entry of the table the source table is supposed to be correct in the form $\boxed{1}$. $\boxed{15} \times 10^n$, i.e., in scientific notation with an integer followed by 15 decimal places.

12.1 PRELIMINARY STEPS FOR COMPUTER INTEGRATION

12.1.1 The Problem

Having decided to integrate by computer, it is often necessary to make certain adjustments. The most common are:

1. The definition of endpoint values
2. A change of variable to make the limits of integration finite
3. A change of integrand to avoid computing either logarithms or terms of the form x^y (which become $e^{y \ln x}$).

12.1.2 Definition of an Endpoint Value

ILLUSTRATIVE PROBLEM 12.1-1   Discuss the problem of integrating the sine integral

$$\text{Si}(x) = \int_0^x f(t) \, dt = \int_0^x \frac{\sin t}{t} \, dt \qquad (12.1.1)$$

Discussion The actual computation, by any of the methods to be discussed, is fairly straightforward. However, at the lower limit, the form of the integral is $0/0$. Using L'Hôpital's rule, $f(0) = 1$. This must be defined (*not calculated*) in the computer program.

12.1.3 Change of Variable to Eliminate Removable Singularities

ILLUSTRATIVE PROBLEM 12.1-2   Discuss the problem of integrating

$$\Gamma(p) = \int f(t,p) \, dt = \int_0^\infty t^{p-1} e^{-t} \, dt \qquad (12.1.2)$$

Discussion There are many problems associated with this computation, and as a matter of fact it is not a good formulation for a computer. There is no way to get rid of the factor t^{p-1}, which, as we mentioned, causes a very large error. There are alternative formulations in the literature—some, not involving integration, which are more useful. However, we shall discuss this integral for illustrative purposes.

To remove the infinite upper limit, let

$$x = \frac{1}{t+1}, \qquad t = \frac{1-x}{x} \qquad (12.1.3)$$

Then

$$dt = -\frac{1}{x^2}dx \qquad (12.1.4)$$

The integral becomes

$$\Gamma(p) = \int_0^1 \left(\frac{1-x}{x}\right)^{p-1} \exp\left(-\frac{1-x}{x}\right)\frac{1}{x^2}dx \qquad (12.1.5)$$

Note that $f(x,p)$ is 0 at both endpoints.

In other cases, substitutions of the form $x = (t-1)/(t+1)$, $x = e^t$, $x = t^{-n}$, $x = $ (the denominator of the integrand)^n are useful, n being chosen for convenience.

12.1.4 Eliminating Logarithms

ILLUSTRATIVE PROBLEM 12.1-3   Discuss the problem of Clausen's integral

$$\psi(x) = \int_0^x -\ln\left(2\sin\frac{t}{2}\right)dt \qquad (12.1.6)$$

Discussion   The results are quite poor if the job is done in the obvious fashion. Instead, let

$$u = -\ln\left(2\sin\frac{t}{2}\right), \qquad dv = dt \qquad (12.1.7)$$

then

$$du = -\frac{1}{2}\cot\frac{t}{2}dt, \qquad v = t \qquad (12.1.8)$$

$$\psi(x) = -t\ln\left(2\sin\frac{t}{2}\right)\Big|_0^x + \int_0^x \frac{1}{2}t\cot\frac{t}{2}dt \qquad (12.1.9)$$

At $t = 0$, the first contribution in the right member vanishes. Then

$$\psi(x) = -x\ln\left(2\sin\frac{x}{2}\right) + \int_0^x \frac{1}{2}t\cot\frac{t}{2}dt \qquad (12.1.10)$$

For the integral in the right member, the value of the integrand at $t = 0$ is 1.

ILLUSTRATIVE PROBLEM 12.1-4   Discuss the problem of the dilogarithm

$$\text{Di}(x) = \int_1^x \frac{\ln t}{1-t}dt \qquad (12.1.11)$$

Discussion   Let

$$\ln t = u, \qquad dt = e^u\,du \qquad (12.1.12)$$

Then the integral becomes

$$\text{Di}(x) = \int_0^{\ln x} \frac{u\,du}{e^{-u}-1} \tag{12.1.13}$$

The integrand is -1 at the lower limit.

PROBLEM SECTION 12.1

A. (1) Discuss the limits for the Debye integral.

(2) $\int_0^x \frac{f(x)(1-\cos x)}{\sin^2 x}\,dx$ has a singularity at $x=0$. Assuming that $f(x)$ has no singularity, remove the singularity if $x < \pi$.

B. Rewrite the integral to remove the logarithm:

(3) $\int_0^{2\pi} \ln(1+t)\sin 10t\,dt$

(4) $\int_0^a \frac{\ln(1+x)}{\sqrt{1-x^2}}\,dx$

C. Devise a substitution to improve the computability. Discuss the endpoints.

(5) $\int_0^a \frac{e^{-x}}{\sqrt{x}}\,dx,\ a>0$

(6) $\int_0^a \frac{e^{-x}}{\sqrt{b^2-x^2}}\,dx,\ a>0$

(7) $\ln\Gamma(p) = \int_0^\infty \left[(p-1)e^{-t} - \frac{e^{-t}+e^{-pt}}{1+e^{-t}} \right]\frac{dt}{t}$

D. In the following, the two substitutions change the integral to a known integral times a constant. Discuss.

(8) $\int_0^\infty \frac{e^{-ax}}{\sqrt{1+x}}\,dx,\ a>0$. Substitute $u^2=1+x$, then $t^2=au^2$.

(9) $\int_0^\infty \frac{e^{-bt}}{1+t}\,dt,\ b>0$. Substitute $t=s-1$, then $bs=x$.

12.2 NEWTON-COTES FORMULAS, CLOSED TYPE

12.2.1 Introduction

The general assumption in this section is that the integrand can be approximated satisfactorily by a polynomial. Then the polynomial can easily be integrated. If the integrand cannot be approximated by a polynomial (e.g. if it has discontinuities or non-removable singularities), then a Fourier series or, sometimes, a Hastings approximation can be used.

We shall show how to write such a polynomial and then display a general theorem for the theoretical error of one class of formulas called *Newton-Cotes formulas, closed type.* In this class of formulas, the interval of integration is divided into n equal subintervals beginning at one endpoint and ending at the other.

12.2.2 The General Procedure

It might seem logical to decide upon a suitable approximating polynomial by means of a forward-difference table. In a previous chapter, we tested ten functions and found that on the intervals tested, e^x, $\sinh x$ and $\cosh x$ might possibly be fitted by an 11th-degree polynomial, $\sin x$ and $\cos x$ by a 13th-degree polynomial, and e^{x^2} by a 17th-degree polynomial; and that neither $\ln(1+x)$ nor $\ln[(1+x)/(1-x)]$ was satisfactorily approximated even by a 19th-degree polynomial (which was as far as we went).

The alternative scheme actually used is to approximate small adjacent intervals by *different* polynomials of the same (low) degree. In Figure 12.2-1, $f(x)$ is approximated by a set of line segments, forming trapezoids. In Figure 12.2-2, $f(x)$ is approximated by successive parabolic segments.

In general, $n+1$ conditions are needed to determine a polynomial of degree n. For example, two points suffice to fix a straight line segment; one point and the slope make two conditions, and this would be equally good. Three points suffice to fix a parabolic segment. Note that a polynomial of

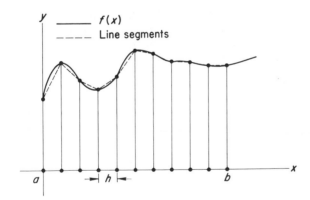

Fig. 12.2-1 The curve is approximated by successive line segments.

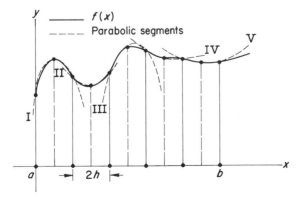

Fig. 12.2-2 The curve is approximated by successive parabolic segments.

degree n,

$$P(x) = a_0 + a_1 x + a_2 x^2 + \cdots + a_n x^n \qquad (12.2.1)$$

has $n+1$ unknowns, namely the coefficients $\{a_k\}$.

12.2.3 Fitting a Polynomial of First Degree

We shall concentrate upon one interval, say $[a, a+h]$, and fit a polynomial of first degree. We shall use two points, namely the endpoints of the interval, in order to obtain a Newton-Cotes formula of the closed type. We assume that there exist *weights*, w_0 and w_1, such that

$$\int_a^{a+h} f(x) = w_0 f(a) + w_1 f(a+h) \qquad (12.2.2)$$

We are assuming, in effect, that

$$f(x) = c_0 + c_1 x \qquad (12.2.3)$$

where the cs are constants, and that there will be *no error* for any *f(x) up to a linear function*. Because we are finding the coefficients w_0 and w_1, this method is called the *method of undetermined coefficients*. In computer language, we say that (12.2.2) is *exact* for $f(x) \equiv 1$ and $f(x) \equiv x$.

1. Suppose $f(x) \equiv 1$. Then $f(a) = 1$ and $f(a+h) = 1$. Substituting in (12.2.2), we have

    $$\int_a^{a+h} 1 \, dx = w_0 \cdot 1 + w_2 \cdot 1 \qquad (12.2.4)$$

    from which

    $$w_0 + w_1 = h \qquad (12.2.5)$$

2. Suppose $f(x) \equiv x$. Then $f(a) = a$, and $f(a+h) = a+h$. Substituting in (12.2.2),

    $$\int_a^{a+h} x \, dx = w_0 a + w_1 (a+h) \qquad (12.2.6)$$

    from which

    $$a w_0 + (a+h) w_1 = ah + \tfrac{1}{2} h^2 \qquad (12.2.7)$$

Solving (12.2.5) and (12.2.7) simultaneously, we obtain $w_0 = w_1 = \tfrac{1}{2} h$. Therefore, for one interval,

$$\int_a^{a+h} f(x) \, dx = \tfrac{1}{2} h \left[f(a) + f(a+h) \right] \qquad (12.2.8)$$

provided $f(x)$ is of degreee 0 or 1. Otherwise, there may be an error:

$$\int_a^{a+h} f(x)\,dx = \tfrac{1}{2}h\big[f(a)+f(a+h)\big] + \text{error} \qquad (12.2.9)$$

If there are two intervals,

$$\int_a^{a+2h} f(x)\,dx = \int_a^{a+h} + \int_{a+h}^{a+2h}$$

$$= \tfrac{1}{2}h\big[f(a)+2f(a+h)+f(a+2h)\big] + \text{error} \qquad (12.2.10)$$

In general, we have the following theorem.

TRAPEZOIDAL RULE   If the interval of integration is divided into n parts, then

$$\int_a^b f(x)\,dx = h\left[\frac{1}{2}(f_0+f_n) + \sum_{k=1}^{n-1} f_k\right] - \frac{nh^3}{12} f''(\theta), \quad \theta \in (a,b) \quad (12.2.11)$$

where $f_k = f(a+kh)$ and $h = (b-a)/n$.

12.2.4 Fitting a Polynomial of Degree 2

In Figure 12.2-2, we illustrated the fitting of five parabolic segments to adjacent parts of the curve. Each second-degree curve requires three conditions. We shall use the two endpoints of the interval and the midpoint. To find the relationship for a single interval, we assume that there exist weights w_0, w_1 and w_2 such that

$$\int_a^{a+2h} f(x)\,dx = w_0 f(a) + w_1 f(a+h) + w_2 f(a+2h) \qquad (12.2.12)$$

which is equivalent to the assumption that

$$f(x) = c_0 + c_1 x + c_2 x^2 \qquad (12.2.13)$$

where the cs are constants. We are assuming, then, that (12.2.12) is *exact* for

$$f(x) \equiv 1 \qquad (12.2.14)$$
$$f(x) \equiv x \qquad (12.2.15)$$
$$f(x) \equiv x^2 \qquad (12.2.16)$$

The three equations and the solution are left as an exercise. The result, for a single segment, is

$$\int_a^{a+2h} f(x)\,dx = \frac{h}{3}\big[f(a)+4f(a+h)+f(a+2h)\big] + \text{error} \qquad (12.2.17)$$

For two intervals, we have

$$\int_a^{a+4h} f(x)dx = \int_a^{a+2h} + \int_{a+2h}^{a+4h}$$

$$= \frac{h}{3}\big[f(a)+4f(a+h)+2f(a+2h)$$

$$+f(a+3h)+f(a+4h)\big]+\text{error} \qquad (12.2.18)$$

In general, we have the following theorem.

SIMPSON'S RULE   If the interval of integration is divided into m intervals, and if $n=2m$, then

$$\int_a^b f(x)dx = \frac{h}{3}\left[(f_0+f_n)+4\sum_{\text{odd}} f_k+2\sum_{\text{even}} f_k\right] - \frac{nh^5}{180} f^{iv}(\theta), \quad \theta \in (a,b)$$

(12.2.19)

where $h=(b-a)/n$, $f_k=f(a+kh)$, $\Sigma_{\text{odd}} = [f(a+h)+f(a+3h)+f(a+5h)+ \cdots +f(a+[n-1]h)]$, and $\Sigma_{\text{even}} = [f(a+2h)+f(a+4h)+ \cdots +f(a+[n-2]h)]$.

Because this error term contains a fourth derivative, it follows that Simpson's rule will be exact for a polynomial of *third* degree (even though it was derived for *second* degree), since the fourth derivative of third-degree polynomial vanishes. We mention this because every textbook does, but it is, in fact, trivial, because it would be senseless to use Simpson's rule to integrate a third-degree polynomial.

12.2.5   The Error Term

The original proof of the general theorems for the error term was given by J. F. Steffensen (*Interpolation*, Williams and Wilkins, Baltimore, 1927, pp. 154–160). Unfortunately, it is quite lengthy, and we regretfully omit it. The theorem for an odd number of points is as follows.

GENERAL REMAINDER THEOREM FOR NEWTON-COTES INTEGRATION (ODD)   If $f(x)$ is approximated by a polynomial through $2r+1$ points equally-spaced over $[a,b]$ including the endpoints, the error for $\int_a^b f(x)dx$ is given by

$$R = \frac{h^{2r+3}}{(2r+2)!} f^{(2r+2)}(\theta) \int_{-r}^r s^2(s^2-1)(s^2-4)\cdots(s^2-r^2)ds \qquad (12.2.20)$$

where $\theta \in (a,b)$.

Because of the symmetry of the integrand, this may be written as

$$R = \frac{2h^{2r+3}}{(2r+2)!} f^{(2r+2)}(\theta) \int_0^r s^2(s^2-1)(s^2-4)\cdots(s^2-r^2)\,ds \quad (12.2.21)$$

ILLUSTRATIVE PROBLEM 12.2-1 If three equally spaced points are used (Simpson's rule), find the remainder for the interval.

Solution Here $2r+1=3$, $r=1$

$$R = \frac{2}{4!} h^5 f^{iv}(\theta) \int_0^1 s^2(s^2-1)\,ds \qquad\qquad (12.2.22)$$

$$= \frac{2}{24} h^5 f^{iv}(\theta) \left(\frac{s^5}{5} - \frac{s^3}{3}\right)\Big|_0^1 \qquad\qquad (12.2.23)$$

$$R = -\frac{1}{90} h^5 f^{iv}(\theta) \qquad\qquad (12.2.24)$$

For an even number of interpolating points, Steffensen's theorem is as follows:

GENERAL REMAINDER THEOREM FOR NEWTON-COTES INTEGRATION (EVEN) If $f(x)$ is a approximated by a polynomial through $2r$ points equally spaced over $[a,b]$, including the endpoints, the error for $\int_a^b f(x)\,dx$ is given by

$$R = \frac{2h^{2r+1}}{(2r)!} f^{(2r)}(\theta) \int_0^{r-1/2} \left(s^2 - \frac{1}{4}\right)\left(s^2 - \frac{9}{4}\right)\cdots\left(s^2 - \left[r-\frac{1}{2}\right]^2\right)ds$$
$$(12.2.25)$$

where $\theta \in (a,b)$.

ILLUSTRATIVE PROBLEM 12.2-2 For four equally spaced points on $[a,b]$, find the remainder.

Solution $2r=4$, $r=2$

$$R = \frac{2}{4!} h^5 f^{iv}(\theta) \int_0^{3/2} \left(s^2 - \frac{1}{4}\right)\left(s^2 - \frac{9}{4}\right)ds$$

$$= \frac{2}{24} h^5 f^{iv}(\theta) \left(\frac{s^5}{5} - \frac{10s^3}{4\times3} + \frac{9s}{16}\right)\Big|_0^{3/2}$$

$$R = -\frac{3}{80} h^5 f^{iv}(\theta)$$

12.2.6 Composite Trapezoidal Formula

In Figure 12.2-1, the baseline has been divided into 10 intervals, so that $h = (b-a)/10$. In Problem (4) at the end of this section, the reader is asked to prove that the error for each interval is

$$E_k = -\frac{h^3 f''(\theta)}{12} \tag{12.2.26}$$

For n intervals, there are n such errors. Assuming that f'' is bounded on $[a,b]$, we can state that there exists some value of θ on (a,b) such that the total error is given by

$$E_{\text{trap}} = -\frac{n h^3 f''(\theta)}{12} \tag{12.2.27}$$

where θ is some value inside the interval (a,b).
    Replacing h by $(b-a)/n$,

$$E_{\text{trap}} = -\frac{(b-a)^3 f''(\theta)}{12n^2} \tag{12.2.28}$$

This shows that the theoretical error of the trapezoidal rule varies as $1/n^2$. If it were not for conversion and roundoff errors, the error would be divided by 4 every time the number of intervals was doubled.
    Substituting $h=(b-a)/n$ in (12.2.28), this can also be expressed as

$$E_{\text{trap}} = -\frac{b-a}{12} h^2 f''(\theta) \tag{12.2.29}$$

12.2.7   Alternative Form for the Trapezoidal Error

In Section 12.4 we shall need another form for the trapezoidal error. The error in computing the error under a curve, $f(x)$, from a to b is unchanged if the entire curve is shifted right or left. This is geometrically obvious. For convenience, let us take another look at the single interval of length h placed symmetrically. Using the trapezoidal rule,

$$\int_{-h/2}^{+h/2} f(x)\,dx = \frac{1}{2} h\left[f\left(-\frac{h}{2}\right) + f\left(+\frac{h}{2}\right) \right] + E \tag{12.2.30}$$

where E is the error for which we are seeking an alternative formulation.
    Assuming that $f(x)$ has a Maclaurin series, we have

$$f(x) = f(0) + xf'(0) + \frac{x^2}{2!} f''(0) + \cdots \tag{12.2.31}$$

$$= \sum_{k=0}^{\infty} f^{(k)}(0)\frac{x^k}{k!} \tag{12.2.32}$$

Integrating (12.2.31) between $-h/2$ and $+h/2$, we have

$$\int_{-h/2}^{+h/2} f(x)\,dx = xf(0) + \frac{x^2}{2!}\,f'(0) + \frac{x^3}{3!}\,f''(0) + \cdots \Big|_{-h/2}^{+h/2} \tag{12.2.33}$$

$$\int_{-h/2}^{+h/2} f(x)\,dx = hf(0) + \frac{h^3}{2^2 3!}\,f''(0) + \frac{h^5}{2^4 5!}\,f^{\text{iv}}(0) + \cdots \tag{12.2.34}$$

$$= \sum_{k=0}^{\infty} \frac{h^{2k+1}}{2^{2k}(2k+1)!}\,f^{(2k)}(0) \tag{12.2.35}$$

In the right member of (12.2.30), we can expand $f(-h/2)$ and $f(+h/2)$ for comparison. Then

$$f\left(+\frac{h}{2}\right) = f(0) + \frac{h}{2}\,f'(0) + \frac{h^2}{2^2 2!}\,f''(0) + \frac{h^3}{2^3 3!}\,f'''(0) + \cdots \tag{12.2.36}$$

$$f\left(-\frac{h}{2}\right) = f(0) - \frac{h}{2}\,f'(0) + \frac{h^2}{2^2 2!}\,f''(0) - \frac{h^3}{2^3 3!}\,f'''(0) + \cdots \tag{12.2.37}$$

Adding, multiplying by h and dividing by 2, we have the trapezoidal approximation:

$$\mathrm{TR} = hf(0) + \frac{h^3}{2^2 2!}\,f''(0) + \frac{h^5}{2^4 4!}\,f^{\text{iv}}(0) + \cdots \tag{12.2.38}$$

$$\mathrm{TR} = \sum_{k=0}^{\infty} \frac{h^{2k+1}}{2^{2k}(2k)!}\,f^{(2k)}(0) \tag{12.2.39}$$

The error is (12.2.35) minus (12.2.39):

$$E = -\sum_{k=0}^{\infty} \frac{h^{2k+1}}{2^{2k}} \left[\frac{1}{(2k+1)!} - \frac{1}{(2k)!} \right] f^{(2k)}(0) \tag{12.2.40}$$

$$E = -\sum_{k=0}^{\infty} \frac{h^{2k+1}}{2^{2k-1}} \frac{k}{(2k+1)!}\,f^{(2k)}(0) \tag{12.2.41}$$

When $k=0$, (12.2.41) gives the first estimate of the error as 0. When $k=1$, the contribution is

$$-\frac{h^3}{2}\cdot\frac{1}{3!}\,f''(0) = -\frac{h^3 f''(0)}{12} \tag{12.2.42}$$

which resembles (12.2.26). In general, we see that the error for an interval can be written as

$$E = a_1 h^3 + a_2 h^5 + a_3 h^7 + \cdots = \sum_{k=1}^{\infty} a_k h^{2k+1} \tag{12.2.43}$$

For *n* intervals, there are *n* such errors. Multiplying (12.2.43) by

$$n = \frac{b-a}{h} \tag{12.2.44}$$

we have

$$E_{\text{trap}} = Ah^2 + Bh^4 + Ch^6 + \cdots \tag{12.2.45}$$

which is the form we will need.

12.2.8 Error Term in the Composite Simpson's Rule

In figure 12.2-2, the baseline is again divided into ten parts, but there are only *five* intervals, since each interval requires two parts. The error for each interval (from Illustrative Problem 12.2-1) is

$$E_k = -\frac{1}{90} h^5 f^{\text{iv}}(\theta) \tag{12.2.46}$$

For *n* divisions (or *n*/2 intervals), using the same assumption as that in Section 12.2.6,

$$E_{\text{SR}} = -\frac{n/2}{90} h^5 f^{\text{iv}}(\theta), \qquad \theta \in (a,b) \tag{12.2.47}$$

which explains the error term in (12.2.19).

Replacing *h* by $(b-a)/n$,

$$E_{\text{SR}} = -\frac{(b-a)^5}{180n^4} f^{\text{iv}}(\theta) \tag{12.2.48}$$

which shows that the theoretical error of Simpson's rule varies as $1/n^4$. Substituting $h = (b-a)/n$, this can be written as

$$E_{\text{SR}} = -\frac{b-a}{180} h^4 f^{\text{iv}}(\theta) \tag{12.2.49}$$

PROBLEM SECTION 12.2

(1) Solve (12.2.5) and (12.2.7) to obtain (12.2.8).

(2) Derive Simpson's rule for one segment:

$$\int_a^{a+2h} f(x)\,dx = \frac{h}{3}[f(a) + 4f(a+h) + f(a+2h)] + E$$

(3) Derive the three-eighths rule for one segment:

$$\int_a^{a+3h} f(x)\,dx = \frac{3h}{8}[f(a) + 3f(a+h) + 3f(a+2h) + f(a+3h)] + E$$

(4) Show that the error term for a single subinterval, using the trapezoidal rule, is

$$-\frac{h^3}{12} f''(\theta)$$

(5) The Newton-Cotes 7-point formula leads to the following. For exercise, derive the formula and error term for a single interval:

$$\int_a^{a+7h} f(x)\,dx = \frac{h}{140}(41f_0 + 216f_1 + 27f_2$$

$$+272f_3 + 27f_4 + 216f_5 + 41f_6) - \frac{9h^9}{1400} f^{\text{viii}}(\theta)$$

(6) In Simpson's rule, the sum of the weights is $(h/3)(1+4+1)=2h$, which is the length of one interval. (a) Show that the sum of the weights is the length of one interval for the trapezoidal rule, the three-eighths rule and the 7-point rule. (b) Using the fact that all the formulas are exact when $f(x)\equiv 1$, prove the general theorem.

12.3 SIMPSON'S RULE AND THE RICHARDSON EXTRAPOLATION

12.3.1 Introduction

Simpson's rule is the easiest to program and is usually satisfactory if f^{iv} in the error term

$$E_{\text{SR}} = -\frac{nh^5 f^{\text{iv}}(\theta)}{180} \tag{12.3.1}$$

is reasonably small. It is obvious that if f^{iv} is bounded, the Simpson's-rule integration will theoretically converge as $h \to 0$, except that conversion and roundoff errors inevitably drown the reduced truncation error.

In practice, the computer program is arranged so that the first calculation is made with n equal to a small even number, say $n=8$, and then doubled until either a critical value (the difference between two successive values) is reached or else the patience and time of the programmer is exhausted. (We shall explain.) In the illustrative problem, we shall do a hand calculation for the complete elliptic integral of the first kind,

$$K(p) = \int_0^{\pi/2} (1 - p\sin^2 t)^{-1/2}\,dt \tag{12.3.2}$$

using $p=0.8$. The true value is $2.2572053268\,20854$. The purpose of this demonstration is to clarify the computer procedure and to assist in debugging a program by furnishing intermediate values.

12.3.2 Step I. Calculation for $n=8$

Using a calculator, we have

$$h = \frac{\pi/2 - 0}{8} \sim 0.19634\,95408 \tag{12.3.3}$$

Simpson's rule is conveniently expressed as

$$\int_a^b f(t)\,dt = \frac{h}{3}(W + 4S_1 + 2S_2) + \text{error} \tag{12.3.4}$$

where

$$W = f_0 + f_n \tag{12.3.5}$$
$$S_1 = \Sigma(\text{odd}) \tag{12.3.6}$$
$$S_2 = \Sigma(\text{even}) \tag{12.3.7}$$

For $n = 8$,

$$E(0.8) \sim \frac{0.19634\,95408}{3}\left[W + 4(f_1 + f_3 + f_5 + f_7) + 2(f_2 + f_4 + f_6)\right] \tag{12.3.8}$$

$$W = f_0 + f_n = (1 - 0.8\sin^2 0)^{-1/2} + \left(1 - 0.8\sin^2\frac{\pi}{2}\right)^{-1/2} = 3.23606\,7977 \tag{12.3.9}$$

$$f_1 = \left(1 - 0.8\sin^2\frac{\pi}{16}\right)^{-1/2}, \qquad f_2 = \left(1 - 0.8\sin^2\frac{2\pi}{16}\right)^{-1/2}$$

$$f_3 = \left(1 - 0.8\sin^2\frac{3\pi}{16}\right)^{-1/2}, \qquad f_4 = \left(1 - 0.8\sin^2\frac{4\pi}{16}\right)^{-1/2}$$

$$f_5 = \left(1 - 0.8\sin^2\frac{5\pi}{16}\right)^{-1/2}, \qquad f_6 = \left(1 - 0.8\sin^2\frac{6\pi}{16}\right)^{-1/2}$$

$$f_7 = \left(1 - 0.8\sin^2\frac{7\pi}{16}\right)^{-1/2} \tag{12.3.10}$$

from which

$$S_1 = 5.74686\,7005 \tag{12.3.11}$$

$$S_2 = 4.13095\,2020 \tag{12.3.12}$$

Substituting in (12.3.4),

$$K(0.8) \sim \frac{0.19634\,95408}{3}\left[3.236\,06\,7977\right.$$

$$\left. + 4(5.746\,86\,7005) + 2(4.13095\,2\,020)\right] \tag{12.3.13}$$

$$K(0.8) \sim 2.25706\,6772 \tag{12.3.14}$$

which is the first estimate.

12.3.3 Step II. Calculation for $n = 16$

For $n = 16$, we have

$$h = \frac{\pi/2 - 0}{16} \sim 0.09817\,47704\,2 \qquad (12.3.15)$$

$$K(0.8) \sim \frac{0.09817\,47704\,2}{3} \big[W + 4(f_1 + f_3 + \cdots + f_{15})$$
$$+ 2(f_2 + f_4 + \cdots + f_{14}) \big] \qquad (12.3.16)$$

W does not have to be recalculated, because it is the sum of the ordinates at the endpoints, which are fixed. Let us examine the other coordinates:

$$f_1 = \left(1 - 0.8\sin^2\frac{\pi}{32}\right)^{-1/2}, \qquad f_2 = \left(1 - 0.8\sin^2\frac{\pi}{16}\right)^{-1/2}$$

$$f_3 = \left(1 - 0.8\sin^2\frac{3\pi}{32}\right)^{-1/2}, \qquad f_4 = \left(1 - 0.8\sin^2\frac{2\pi}{16}\right)^{-1/2}$$

$$f_5 = \left(1 - 0.8\sin^2\frac{5\pi}{32}\right)^{-1/2}, \qquad f_6 = \left(1 - 0.8\sin^2\frac{3\pi}{16}\right)^{-1/2}$$

$$f_7 = \left(1 - 0.8\sin^2\frac{7\pi}{32}\right)^{-1/2}, \qquad f_8 = \left(1 - 0.8\sin^2\frac{4\pi}{16}\right)^{-1/2}$$

$$f_9 = \left(1 - 0.8\sin^2\frac{9\pi}{32}\right)^{-1/2}, \qquad f_{10} = \left(1 - 0.8\sin^2\frac{5\pi}{16}\right)^{-1/2}$$

$$f_{11} = \left(1 - 0.8\sin^2\frac{11\pi}{32}\right)^{-1/2}, \qquad f_{12} = \left(1 - 0.8\sin^2\frac{6\pi}{16}\right)^{-1/2}$$

$$f_{13} = \left(1 - 0.8\sin^2\frac{13\pi}{32}\right)^{-1/2}, \qquad f_{14} = \left(1 - 0.8\sin^2\frac{7\pi}{16}\right)^{-1/2}$$

$$f_{15} = \left(1 - 0.8\sin^2\frac{15\pi}{32}\right)^{-1/2} \qquad (12.3.17)$$

Comparing (12.3.17) with (12.3.10), we see that the new S_2 is the sum of the old S_1 and the old S_2. The explanation for this should be clear from Figure 12.3-1, where the points have been labeled S_1 and S_2 to identify them. In each iteration, $W = f_0 + f_n$ is constant, and in each case the new S_2 is the sum of the previous S_1 and the previous S_2. Only S_1 need be calculated anew, using part of (12.3.17).

Using this fact, we have

$$W = 3.23606\,7977 \qquad (12.3.18)$$

$$S_1 = 11.49585\,164 \qquad (12.3.19)$$

$$S_2 = 5.74686\,7005 + 4.13095\,2020 = 9.87781\,9025 \qquad (12.3.20)$$

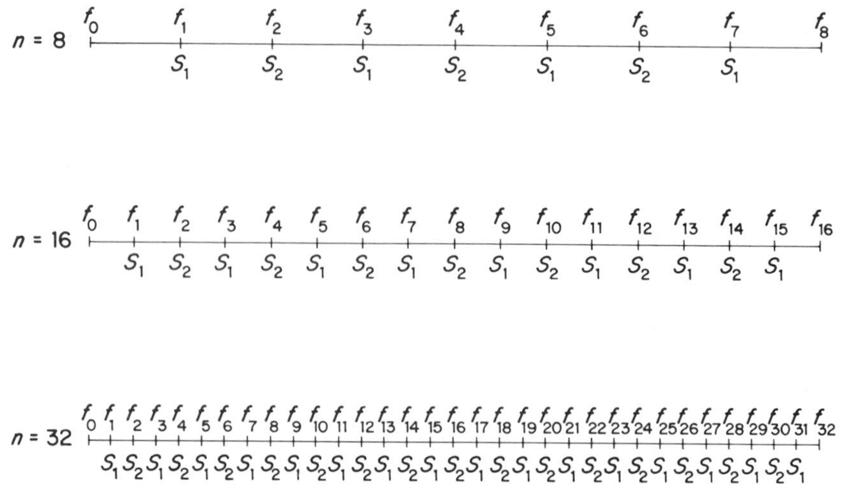

$n = 8$: f_0 f_1 f_2 f_3 f_4 f_5 f_6 f_7 f_8 with S_1 S_2 S_1 S_2 S_1 S_2 S_1

$n = 16$: f_0 through f_{16} with S_1 S_2 S_1 S_2 S_1 S_2 S_1 S_2 S_1 S_2 S_1 S_2 S_1 S_2 S_1

$n = 32$: f_0 through f_{32} with $S_1 S_2 S_1 S_2 S_1 S_2 S_1 S_2 S_1 S_2 S_1 S_2 S_1 S_2 S_1 S_2 S_1 S_2 S_1 S_2 S_1 S_2 S_1 S_2 S_1 S_2 S_1 S_2 S_1 S_2 S_1$

Fig. 12.3-1   New S_2 is the sum of old S_1 and old S_2.

Using (12.3.4),

$$K(0.8) \sim \frac{0.09817\,47704\,2}{3} \big[3.23606\,7977$$

$$+ 4(11.49585\,164) + 2(9.87781\,9025) \big] \quad (12.3.21)$$

$$K(0.8) \sim 2.25720\,5280 \quad (12.3.22)$$

which is the second estimate.

12.3.4   The Richardson Extrapolation

Suppose that for Simpson's rule, the computed value of an integral is I_1 when n divisions are used, and I_2 when $2n$ divisions are used. Using the error value in (12.2.48), we have

$$I = I_1 - \frac{(b-a)^5}{180 n^4} f^{iv}(\theta_1) \quad (12.3.23)$$

$$I = I_2 - \frac{(b-a)^5}{180(2n)^4} f^{iv}(\theta_2) \quad (12.3.24)$$

where I is the true value of the integral. Assuming that $f^{iv}(\theta_1)$ is approximately equal to $f^{iv}(\theta_2)$, and substituting

$$C = -\frac{(b-a)^5}{180} f^{iv}(\theta) \quad (12.3.25)$$

we have

$$I = I_1 + \frac{C}{n^4} \tag{12.3.26}$$

$$I = I_2 + \frac{C}{16n^4} \tag{12.3.27}$$

Eliminating C between these two equations, we obtain

THEOREM: RICHARDSON EXTRAPOLATION If f^{iv} is reasonably constant when the number of divisions is doubled in Simpson's rule, then

$$I = \frac{16I_2 - I_1}{15} \tag{12.3.28}$$

where I is the corrected value, I_1 the computed value with n divisions, and I_2 the computed value with $2n$ divisions.

Applying this to the results in our illustrative problem, we have

$$I = \frac{16(2.257205280) - 2.257066772}{15} \tag{12.3.29}$$

$$= 2.257214513 \tag{12.3.30}$$

The actual errors, to the number of places calculated, are as follows:

Simpson's rule, $n = 8$:	1.4×10^{-4}
Simpson's rule, $n = 16$:	4.7×10^{-8}
Richardson:	9.2×10^{-6}

It is evident from this, and from Table 12-1, that the Richardson extrapolation procedure does not necessarily improve matters. In fact, the assumption on which it is based is not very likely and we do not recommend this procedure.

12.3.5 Computer Procedure

The computer procedure is quite straightforward. Typically, $W = f_0 + f_n$ is set, either by computation or, if there is a singularity, by definition. Then S_1 and S_2 are calculated for a small even number, e.g., $n = 8$. Equation (12.3.4) is used to obtain the first estimate.

The second estimate is done after doubling n. W is constant, and S_2 is simply the sum of the previous S_1 and S_2. Only the new S_1 has to be calculated fully, leading to a second estimate.

Figure 12.3-2 shows the program used for Simpson's rule, and Figure 12.3-3 shows the successive approximations. The program was set to

```
C THE PURPOSE OF THIS PROGRAM IS TO INTEGRATE BY SIMPSON'S RULE.      040
C INPUT CARDS ARE AS FOLLOW.                                          050
C CARD 1. NAME OF INTEGRAL, STARTING AT CC1 UP TO CC 48.              060
C         ITYPE IN CC 51, NDEC IN CC 52-53.  IF ITYPE =0, VARIABLE    061
C         PARAMETER.  IF ITYPE=1, VARIABLE UPPER LIMIT.  IF ITYPE=2,  062
C         NEITHER. NDEC IS THE NUMBER OF DECIMAL PLACES REQUIRED.     063
C CARD 2. LOWER LIMIT IN CC1.  FIRST UPPER LIMIT IN CC 21. INCREMENT  070
C         IN CC 41. PARAMETERS, IF ANY, IN CC 61                      075
C CARDS 3 TO 6, IF NECESSARY. UP TO 10 TRUE VALUES IN CC 1, 21,41.    100
C IF F(A) OR F(B) CANNOT BE COMPUTED, CARDS MUST BE INSERTED TO       110
C BYPASS CARD 530.  THESE INSERTED CARDS ARE OF THE FORM -            111
C         W = ........                                                112
C         GO TO 52                                                    113
C THESE CARDS ARE INSERTED AFTER CARD 520                             114
C IF THE PROBLEM INVOLVES INTEGRATION BY PARTS, A STATEMENT FUNCTION  115
C FOR THE PART IS INSERTED AFTER CARD 160.                            116
C CCCCCCCCCCCCCCCCCCCCCCCCCCCCCCCCCCCCCCCCCCCCCCCCCCCCCCCCCCCCCCCCCCCCC117
      DOUBLE PRECISION CV(10),TV(10),E(10),F,X,W,CR,PI,A,B,P,D,Y,S,S1,S2 120
     1,S3,TN,H,T,V(10),Z,TE(10),PART,Q,R                              130
      DIMENSION HEAD(12)                                              140
C VARIABLE STATEMENT FUNCTION.                                        150
      F(X)=1.D0/DSQRT(1.D0-P*(DSIN(X))**2)                            160
C STATEMENT FUNCTION FOR PART(X), IF NECESSARY.                       160
      PART(X)=0.D0                                                    165
C INPUT 1. LIMITS, PARAMETER(IF ANY), INCREMENT, AND CODES.          170
      READ(1,100)HEAD,ITYPE,NDEC                                     180
  100 FORMAT(12A4,2X,I1,I2)                                          190
      READ(1,101)A,B,D,P                                             200
  101 FORMAT(4D20.0)                                                 210
C DEFINITIONS                                                        220
      CR=0.5D0*1.D1**(-NDEC)                                         230
      PI=3.141592653589793D0                                         240
      NDEC=NDEC-4                                                    241
      LINE=1                                                         245
C DECISION AND INPUT 2.                                              250
      IF(ITYPE-2)91,92,91                                           260
   91 READ(1,102)TV                                                 270
  102 FORMAT(3D20.0/3D20.0/3D20.0/D20.0)                            280
      GO TO 1                                                       290
   92 READ(1,103)TV(1)                                              300
  103 FORMAT(D20.0)                                                 310
C FIRST HEADING.                                                    320
    1 WRITE(3,104)HEAD                                              330
```

```
104     FORMAT(1H1,T20,'SIMPSONS RULE',10X,                                    340
       112A4//)                                                                341
C   DECISION AND SECOND HEADING.                                               350
        IF(ITYPE)94,93,94                                                      360
93      WRITE(3,105)                                                           370
105     FORMAT(1H0,T7,'N',T14,'PARAMETER',T35,'COMPUTED VALUE',T67,'TRUE V      380
       1ALUE',T85,'CHANGE',T98,'TRUE ERROR'/)                                  390
        GO TO 2                                                                400
94      WRITE(3,106)                                                           410
106     FORMAT(1H0,T7,'N',T12,'UPPER LIMIT',T35,'COMPUTED VALUE',T67,'TRUE      420
       1 VALUE',T85,'CHANGE',T98,'TRUE ERROR'/)                                430
C   SET INITIAL CONDITIONS.                                                    440
2       I=1                                                                    450
        NT=1024                                                                460
C   SETTING VARIABLE AND ERROR CODE                                           470
        IF(ITYPE.EQ.0)V(1)=P                                                   480
        IF(ITYPE.NE.0)V(1)=B                                                   490
C   PRELIMINARY CALCULATION                                                    500
50      N=8                                                                    510
        TN=8.D0                                                                520
51      W=F(A)+F(B)                                                            530
52      Z=B-A                                                                  540
        H=Z/TN                                                                 550
C                                                                              560
        S1=0.D0                                                                570
        NP=N-1                                                                 580
        DO 21 K=1,NP,2                                                         590
        X=DFLOAT(K)                                                            600
        T=A+X*H                                                                610
21      S1=S1+F(T)                                                             620
C                                                                              630
        S2=0.D0                                                                640
        NP=N-2                                                                 650
        DO 22 K=2,NP,2                                                         660
        X=DFLOAT(K)                                                            670
        T=A+X*H                                                                680
22      S2=S2+F(T)                                                             690
        S=H*(W+4.D0*S1+2.D0*S2)/3.D0                                           700
        CV(I)=S                                                                710
33      N=N*2                                                                  720
        PV=CV(I)                                                               730
C   MAIN CALCULATION.                                                          740
3       TN=DFLOAT(N)                                                           750
        H=Z/TN                                                                 760
```

```
C
      S2=S1+S2                                                          770
      S1=0.D0                                                           780
      NP=N-1                                                            790
      DO 31 K=1,NP,2                                                    800
      X=DFLOAT(K)                                                       810
      T=A+X*H                                                           820
      S1=S1+F(T)                                                        830
   31                                                                   840
C                                                                       850
      S=H*(W+4.D0*S1+2.D0*S2)/3.D0                                      860
C                                                                       870
      CV(I)=S                                                           880
      E(I)=PV-CV(I)                                                     890
      Q=CV(I)+PART(B)                                                   891
      TE(I)=TV(I)-Q                                                     895
      IF(DABS(TE(I)).LT.CR)TE(I)=0.D0                                   896
      GO TO(905,906,907,908,909,910,911,912,913,914,915),NDEC          901
  905 WRITE(3,925)N,V(I),Q     ,TV(I),E(I),TE(I)                        902
      GO TO 6                                                           903
  906 WRITE(3,926)N,V(I),Q     ,TV(I),E(I),TE(I)                        904
      GO TO 6                                                           905
  907 WRITE(3,927)N,V(I),Q     ,TV(I),E(I),TE(I)                        906
      GO TO 6                                                           907
  908 WRITE(3,928)N,V(I),Q     ,TV(I),E(I),TE(I)                        908
      GO TO 6                                                           909
  909 WRITE(3,929)N,V(I),Q     ,TV(I),E(I),TE(I)                        910
      GO TO 6                                                           911
  910 WRITE(3,930)N,V(I),Q     ,TV(I),E(I),TE(I)                        912
      GO TO 6                                                           913
  911 WRITE(3,931)N,V(I),Q     ,TV(I),E(I),TE(I)                        914
      GO TO 6                                                           915
  912 WRITE(3,932)N,V(I),Q     ,TV(I),E(I),TE(I)                        916
      GO TO 6                                                           917
  913 WRITE(3,933)N,V(I),Q     ,TV(I),E(I),TE(I)                        918
      GO TO 6                                                           919
  914 WRITE(3,934)N,V(I),Q     ,TV(I),E(I),TE(I)                        920
      GO TO 6                                                           921
  915 WRITE(3,935)N,V(I),Q     ,TV(I),E(I),TE(I)                        922
  925 FORMAT(1H ,T3,I6,T15,F7.3,T27,1PD22.5 ,T55,D22.5 ,T84,D9.2,T99,D9. 923
     12)                                                                924
  926 FORMAT(1H ,T3,I6,T15,F7.3,T27,1PD22.6 ,T55,D22.6 ,T84,D9.2,T99,D9. 925
     12)                                                                926
  927 FORMAT(1H ,T3,I6,T15,F7.3,T27,1PD22.7 ,T55,D22.7 ,T84,D9.2,T99,D9. 927
     12)                                                                928
```

```
928   FORMAT(1H ,T3,I6,T15,F7.3,T27,1PD22.8 ,T55,D22.8 ,T84,D9.2,T99,D9.    929
     12)                                                                     930
929   FORMAT(1H ,T3,I6,T15,F7.3,T27,1PD22.9 ,T55,D22.9 ,T84,D9.2,T99,D9.    931
     12)                                                                     932
930   FORMAT(1H ,T3,I6,T15,F7.3,T27,1PD22.10,T55,D22.10,T84,D9.2,T99,D9.    933
     12)                                                                     934
931   FORMAT(1H ,T3,I6,T15,F7.3,T27,1PD22.11,T55,D22.11,T84,D9.2,T99,D9.    935
     12)                                                                     936
932   FORMAT(1H ,T3,I6,T15,F7.3,T27,1PD22.12,T55,D22.12,T84,D9.2,T99,D9.    937
     12)                                                                     938
933   FORMAT(1H ,T3,I6,T15,F7.3,T27,1PD22.13,T55,D22.13,T84,D9.2,T99,D9.    939
     12)                                                                     940
934   FORMAT(1H ,T3,I6,T15,F7.3,T27,1PD22.14,T55,D22.14,T84,D9.2,T99,D9.    941
     12)                                                                     942
935   FORMAT(1H ,T3,I6,T15,F7.3,T27,1PD22.15,T55,D22.15,T84,D9.2,T99,D9.    943
     12)                                                                     944
6     LINE=LINE+1                                                            945
      IF(LINE.NE.40)GO TO 7                                                  950
      WRITE(3,104)HEAD                                                       955
      IF(ITYPE.EQ.0)WRITE(3,105)                                            960
      IF(ITYPE.NE.0)WRITE(3,106)                                            965
      LINE=1                                                                 970
C DECISIONS REGARDING ERROR AND NUMBER OF DIVISIONS.                        980
7     IF(DABS(E(I)).LE.CR)GO TO 4                                           990
      IF(N.NE.NT)GO TO 33                                                   1000
4     WRITE(3,109)                                                          1010
109   FORMAT(1H0)                                                           1020
41    IF(ITYPE.EQ.2)GO TO 999                                              1030
      IF(I.EQ.10)GO TO 999                                                  1040
C RETURN TO STEP BOX.                                                       1070
5     I=I+1                                                                 1080
      IF(ITYPE.EQ.0)P=P+D                                                  1100
      IF(ITYPE.EQ.0)V(I)=P                                                 1110
      IF(ITYPE.EQ.1)B=B+D                                                  1120
      IF(ITYPE.EQ.1)V(I)=B                                                 1130
      GO TO 50                                                             1140
999   STOP                                                                 1430
      END                                                                  1440
```

Fig. 12.3-2 Computer program for Simpson's rule. This program allows for different formats, different definitions for the endpoints if there are singularities, and integrating by parts.

SIMPSONS RULE WITH RICHARDSON EXTRAPOLATION   COMPLETE ELLIPTIC INTEGRAL K(M)

N	PARAMETER	COMPUTED VALUE	TRUE VALUE	CHANGE	TRUE ERROR
16	0.200	1.6596236287807540 D 00	1.6596235986105280 D 00	-4.83D-07	-3.02D-08
32	0.200	1.6596236287807540 00	1.6596235986105280 00	-4.83D-07	-3.02D-08
64	0.200	1.6596236287807540 00	1.6596235986105280 00	-4.83D-07	-3.02D-08
128	0.200	1.6596236287807520 00	1.6596235986105280 00	-4.83D-07	-3.02D-08
256	0.200	1.6596236287807520 00	1.6596235986105280 00	-4.83D-07	-3.02D-08
512	0.200	1.6596236287807500 00	1.6596235986105280 00	-4.83D-07	-3.02D-08
1024	0.200	1.6596236287807360 00	1.6596235986105280 00	-4.83D-07	-3.02D-08
16	0.300	1.7138894699923425 00	1.7138894481787910 00	-3.48D-07	-2.17D-08
32	0.300	1.7138894699923425 00	1.7138894481787910 00	-3.48D-07	-2.17D-08
64	0.300	1.7138894699923424 00	1.7138894481787910 00	-3.48D-07	-2.17D-08
128	0.300	1.7138894699923423 00	1.7138894481787910 00	-3.48D-07	-2.17D-08
256	0.300	1.7138894699923423 00	1.7138894481787910 00	-3.48D-07	-2.17D-08
512	0.300	1.7138894699923421 00	1.7138894481787910 00	-3.48D-07	-2.17D-08
1024	0.300	1.7138894699923407 00	1.7138894481787910 00	-3.48D-07	-2.17D-08
16	0.400	1.7775193811857210 00	1.7775193714912530 00	-1.55D-07	-9.69D-09
32	0.400	1.7775193811857220 00	1.7775193714912530 00	-1.55D-07	-9.69D-09
64	0.400	1.7775193811857210 00	1.7775193714912530 00	-1.55D-07	-9.69D-09
128	0.400	1.7775193811857200 00	1.7775193714912530 00	-1.55D-07	-9.69D-09
256	0.400	1.7775193811857200 00	1.7775193714912530 00	-1.55D-07	-9.69D-09
512	0.400	1.7775193811857160 00	1.7775193714912530 00	-1.55D-07	-9.69D-09
1024	0.400	1.7775193811857030 00	1.7775193714912530 00	-1.55D-07	-9.69D-09
16	0.500	1.8540746905780640 00	1.8540746773013720 00	-2.12D-07	-1.33D-08
32	0.500	1.8540746905781680 00	1.8540746773013720 00	-2.12D-07	-1.33D-08
64	0.500	1.8540746905781680 00	1.8540746773013720 00	-2.12D-07	-1.33D-08
128	0.500	1.8540746905781670 00	1.8540746773013720 00	-2.12D-07	-1.33D-08

n	x				
256	0.500	1.8540746905781660D 00	1.8540746773013720D 00	-2.12D-07	-1.33D-08
512	0.500	1.8540746905781620D 00	1.8540746773013720D 00	-2.12D-07	-1.33D-08
1024	0.500	1.8540746905781500D 00	1.8540746773013720D 00	-2.12D-07	-1.33D-08
16	0.600	1.9495678739544520D 00	1.9495677498006026D 00	-1.99D-06	-1.24D-07
32	0.600	1.9495677468064400D 00	1.9495677498006026D 00	4.80D-08	3.00D-09
64	0.600	1.9495678103884727D 00	1.9495677498006026D 00	-9.69D-07	-6.06D-08
128	0.600	1.9495677468064438D 00	1.9495677498006026D 00	4.80D-08	3.00D-09
256	0.600	1.9495677468047260D 00	1.9495677498006026D 00	-9.69D-07	-6.06D-08
512	0.600	1.9495677468064330D 00	1.9495677498006026D 00	4.80D-08	3.00D-09
1024	0.600	1.9495678103847080D 00	1.9495677498006026D 00	-9.69D-07	-6.06D-08
16	0.700	2.0753641503561810D 00	2.0753631352924690D 00	-1.62D-05	-1.02D-06
32	0.700	2.0753630701216990D 00	2.0753631352924690D 00	1.04D-06	6.52D-08
64	0.700	2.0753631972782740D 00	2.0753631352924690D 00	-9.92D-07	-6.20D-08
128	0.700	2.0753631336999850D 00	2.0753631352924690D 00	2.55D-08	1.59D-09
256	0.700	2.0753631972782730D 00	2.0753631352924690D 00	-9.92D-07	-6.20D-08
512	0.700	2.0753631369997900D 00	2.0753631352924690D 00	2.55D-08	1.59D-09
1024	0.700	2.0753631972782550D 00	2.0753631352924690D 00	-9.92D-07	-6.20D-08
16	0.800	2.2572145189573430D 00	2.2572053268208540D 00	-1.48D-04	-9.19D-06
32	0.800	2.2572047757736160D 00	2.2572053268208540D 00	8.82D-06	5.51D-07
64	0.800	2.2572054115565000D 00	2.2572053268208540D 00	-1.36D-06	-8.47D-08
128	0.800	2.2572053479782110D 00	2.2572053268208540D 00	-3.39D-07	-2.12D-08
256	0.800	2.2572053479782110D 00	2.2572053268208540D 00	-3.39D-07	-2.12D-08
512	0.800	2.2572053479782050D 00	2.2572053268208540D 00	-3.39D-07	-2.12D-08
1024	0.800	2.2572053479781930D 00	2.2572053268208540D 00	-3.39D-07	-2.12D-08
16	0.900	2.5782109505772650D 00	2.5780921133481730D 00	-2.02D-03	-1.19D-04
32	0.900	2.5780841987030140D 00	2.5780921133481730D 00	1.27D-04	7.91D-06
64	0.900	2.5780926547710900D 00	2.5780921133481730D 00	-8.66D-06	-5.41D-07
128	0.900	2.5780920825664990D 00	2.5780921133481730D 00	4.93D-07	3.08D-08
256	0.900	2.5780921461447870D 00	2.5780921133481730D 00	-5.25D-07	-3.28D-08
512	0.900	2.5780921461447800D 00	2.5780921133481730D 00	-5.25D-07	-3.28D-08
1024	0.900	2.5780921461447690D 00	2.5780921133481730D 00	-5.25D-07	-3.28D-08

Fig. 12.3-3   A portion of the output for Simpson's rule.

terminate when the number of divisions was 1024 or when the difference between successive estimates was less than one-half the last place of the true value. Experience with other iterative procedures seems to cast considerable doubt upon the reliability of adjacent differences as an estimate of error, but there appears to be nothing better at this time for Simpson's rule. However, note that in the results (Figure 12.3-3), the adjacent differences were actually much less than the true errors. While this proves nothing, it is heartening for those who insist upon using Simpson's rule.

When the same program was run for Clausen's integral, for which the true values are given to only six significant figures, the error was 0 when n became 16. For the error function, with true values given to 10 significant figures, the program terminated after 1024 divisions with all figures correct.

PROBLEM SECTION 12.3

Figure 12.3-4 is a tabulation of "true values" called from various sources for your convenience. (It is, of course, possible that more precise "true values" will be available by the time this book is published.) Choose one of the problems defined at the beginning of this chapter, and integrate using Simpson's rule. You may wish to try the Richardson extrapolation procedure as well. The errors in Table 12-1 are probably not very different from those on other computers of the binary fixed-word-length type.

COMPLETE ELLIPTIC INTEGRAL OF THE FIRST KIND

PARAMETER	TRUE VALUE	PARAMETER	TRUE VALUE
0.0	1.570796326794897D 00	5.000	1.854074677301372D 00
0.100	1.612441348720219D 00	6.000	1.949567749806026D 00
0.200	1.659623598610528D 00	7.000	2.075363135292469D 00
0.300	1.713889448178791D 00	8.000	2.257205326820854D 00
0.400	1.777519371491253D 00	9.000	2.578092113348173D 00

ERROR FUNCTION

UPPER LIMIT	TRUE VALUE	UPPER LIMIT	TRUE VALUE
0.100	1.124629160D-01	6.000	6.038560908D-01
0.200	2.227025892D-01	7.000	6.778011938D-01
0.300	3.286267595D-01	8.000	7.421009647D-01
0.400	4.283923550D-01	9.000	7.969082124D-01
0.500	5.204998778D-01	10.000	8.427007929D-01

GAMMA FUNCTION

PARAMETER	TRUE VALUE	PARAMETER	TRUE VALUE
1.250	9.064024771D-01	15.000	8.862269255D-01
1.300	8.974706963D-01	15.500	8.886834780D-01
1.350	8.911514420D-01	16.000	8.935153493D-01
1.400	8.872638175D-01	16.500	9.001168163D-01
1.450	8.856613803D-01	17.000	9.086387329D-01

EXP(-X**2) FROM 0 TO 1

7.46824D-01

LOG (1+X) SIN 10X FROM 0 TO 36J DEGREES

-1.976276D-01

COMPLETE ELLIPTIC INTEGRAL OF THE SECOND KIND

PARAMETER	TRUE VALUE	PARAMETER	TRUE VALUE
0.050	1.550973352D 00	3.000	1.445363064D 00
0.100	1.530757637D 00	3.500	1.422691133D 00
0.150	1.510121831D 00	4.000	1.399392139D 00
0.200	1.489035058D 00	4.500	1.375401972D 00
0.250	1.467462209D 00	5.000	1.350643881D 00

SIEVERT INTEGRAL AT 60 DEGREES

PARAMETER	TRUE VALUE	PARAMETER	TRUE VALUE
0.0	1.04720D 00	10.000	3.07694D-01
0.200	8.15477D-01	12.000	2.42523D-01
0.400	6.36769D-01	14.000	1.91533D-01
0.600	4.98504D-01	16.000	1.51541D-01
0.800	3.91204D-01	18.000	1.20105D-01

BESSEL FUNCTION ORDER O

PARAMETER	TRUE VALUE	PARAMETER	TRUE VALUE
0.500	9.38469807240813D-01	30.000	-2.600519549019330-01
1.000	7.65197686557967D-01	35.000	-3.80127739987263D-01
1.500	5.11827671735918D-01	40.000	-3.97149809863847D-01
2.000	2.238907791412360D-01	45.000	-3.205425089851210-01
2.500	-4.838377646819800-02	50.000	-1.775967713143380-01

CLAUSENS INTEGRAL

UPPER LIMIT	TRUE VALUE	UPPER LIMIT	TRUE VALUE
0.0	6.35781D-01	100.000	8.13635D-01
2.000	6.78341D-01	120.000	8.40230D-01
4.000	7.17047D-01	140.000	8.64379D-01
6.000	7.52292D-01	160.000	8.86253D-01
8.000	7.84398D-01	180.000	9.06001D-01

Fig. 12.3-4 (Continued on next page)

FRESNEL COSINE INTEGRAL

UPPER LIMIT	TRUE VALUE	UPPER LIMIT	TRUE VALUE
0.500	4.923442D-01	30.000	6.057208D-01
1.000	7.798934D-01	35.000	5.325724D-01
1.500	4.452612D-01	40.000	4.984260D-01
2.000	4.882534D-01	45.000	5.260259D-01
2.500	4.574130D-01	50.000	5.636312D-01

SINE INTEGRAL

UPPER LIMIT	TRUE VALUE	UPPER LIMIT	TRUE VALUE
1.100	1.0286852187D 00	16.000	1.3891804859D 00
1.200	1.1080471990D 00	17.000	1.4495922897D 00
1.300	1.1839580091D 00	18.000	1.5058167803D 00
1.400	1.2562267328D 00	19.000	1.5577753137D 00
1.500	.32468353120D 00	20.000	1.6054129768D 00

DEBYE FUNCTION, N=4

UPPER LIMIT	TRUE VALUE	UPPER LIMIT	TRUE VALUE
0.100	9.60555D-01	6.000	7.79911D-01
0.200	9.22221D-01	7.000	7.47057D-01
0.300	8.84994D-01	8.000	7.15275D-01
0.400	8.48871D-01	9.000	6.84551D-01
0.500	8.13846D-01	10.000	6.54874D-01

DILOGARITHM

UPPER LIMIT	TRUE VALUE	UPPER LIMIT	TRUE VALUE
0.050	1.440633797D 00	3.000	8.893776240D-01
0.100	1.299714723D 00	3.500	8.060826890D-01
0.150	1.180581124D 00	4.000	7.275863080D-01
0.200	1.074794600D 00	4.500	6.531576310D-01
0.250	9.784693930D-01	5.000	5.822405260D-01

Fig. 12.3-4 "True values" for the problem sections.

12.4 THE TRAPEZOIDAL RULE AND THE ROMBERG EXTRAPOLATION PROCEDURE

12.4.1 Introduction

According to Section 12.2.6, the trapezoidal rule involves the following error term:

$$E_{\text{trap}} = -\frac{nh^3 f''(\theta)}{12} = -\frac{(b-a)^3 f''(\theta)}{12n^2}, \qquad \theta \in (a,b) \quad (12.4.1)$$

If $f''(\theta)$ is reasonably small, the theoretical error should be small. If f'' is bounded, the theoretical error should decrease as $1/n^2$ and, except for conversion and roundoff errors, should converge to the true value.

In practice, the computer program is arranged so that the first calculation is made with n equal to a small even number (we started with 4). Then n is doubled again and again and the *Romberg extrapolation to the limit*, which is extremely good, is applied. This Romberg procedure, which we shall explain and illustrate by a hand calculation, is continued until either two successive values differ by less than a critical value, or else the limit for size or time is reached. This is a very fast program, highly recommended.

The illustrative calculation uses the *error function*:

$$\operatorname{erf} x = \frac{2}{\sqrt{\pi}} \int_0^x e^{-t^2}\, dt \qquad (12.4.2)$$

at $x = 0.1$, the true value being $0.11246\,29160$.

12.4.2 Step I. Calculation for $n = 4$

For $n = 4$, using a calculator throughout,

$$h = \frac{0.1 - 0}{4} = 0.025 \qquad (12.4.3)$$

The trapezoidal rule, from (12.2.11), is

$$\int_a^b f(t)\, dt \sim h \left[W + \sum_{k=1}^{n-1} f_k \right] \qquad (12.4.4)$$

where

$$W = \tfrac{1}{2}(f_0 + f_n) \qquad (12.4.5)$$
$$f_k = f(a + kh) \qquad (12.4.6)$$

For $n = 4$,

$$\operatorname{erf} 0.1 \sim h \left[W + \sum_{k=1}^{3} f_k \right] \qquad (12.4.7)$$

$$W = \frac{1}{2} \frac{2}{\sqrt{\pi}} (e^{-0} + e^{-0.01}) \sim 1.12276\,5387 \qquad (12.4.8)$$

$$\sum_{k=1}^{3} f_k = \frac{2}{\sqrt{\pi}} (e^{-0.025^2} + e^{-0.050^2} + e^{-0.075^2}) \qquad (12.4.9)$$

$$= 3.37528\,5747 \qquad (12.4.10)$$

Substituting in (12.4.4), the first estimate is

$$\text{erf}\, 0.1 \sim 0.11245\,12783 \tag{12.4.11}$$

12.4.3 Step II. Calculation for $n = 8$

For $n = 8$, W is unchanged:

$$h = \frac{0.1 - 0}{8} = 0.0125 \tag{12.4.12}$$

$$\text{erf}\, 0.1 \sim h \left[W + \sum_{k=1}^{7} f_k \right] \tag{12.4.13}$$

$$\sum_{k=1}^{7} f_k = \frac{2}{\sqrt{\pi}} \left[e^{-0.0125^2} + e^{-0.025^2} + e^{-0.0375^2} + \cdots + e^{-0.0875^2} \right] \tag{12.4.14}$$

$$= 7.87403\,5155 \tag{12.4.15}$$

As in the case of Simpson's rule, there are repeated values which need not be recomputed. Substituting in (12.4.13), the second estimate is

$$\text{erf}\, 0.1 \sim 0.11246\,00067 \tag{12.4.16}$$

12.4.4 Romberg Extrapolation to the Limit

Suppose that for the trapezoidal rule, the computed value of an integral is I_1 when n divisions are used and I_2 when $2n$ divisions are used. Using the error value in (12.4.1), the true value, I, is related to I_1 and I_2 by

$$I = I_1 - \frac{(b-a)^3}{12 n^2} f''(\theta_1) \tag{12.4.17}$$

$$I = I_2 - \frac{(b-a)^3}{12(2n)^3} f''(\theta_2) \tag{12.4.18}$$

Assuming that $f''(\theta_1)$ and $f''(\theta_2)$ are approximately equal, and substituting

$$C = -\frac{(b-a)^3}{12} f''(\theta) \tag{12.4.19}$$

we have

$$I = I_1 + \frac{C}{n^2} \tag{12.4.20}$$

$$I = I_2 + \frac{C}{4 n^2} \tag{12.4.21}$$

Eliminating C between these, we have

$$I = \frac{4I_2 - I_1}{3} \tag{12.4.22}$$

which is clearly a Richardson extrapolation. We have seen that this was not a successful method of correction for Simpson's Rule, and it is no better for the trapezoidal rule.

Our problem, in short, is that the trapezoidal rule would converge to a satisfactory value if we made n large enough, but we always lose in the race between smaller truncation errors and larger roundoff errors. What we need to do is to find a way of *simulating* larger values of n, decreasing the error without actually increasing n. We turn our attention to (12.2.45), which states that the error of the trapezoidal rule is

$$E_{\text{trap}} = Ah^2 + Bh^4 + Ch^6 + \cdots \tag{12.4.23}$$

We write the true value of the integral, with intervals of length h_1, as

$$I = \text{TR}(h_1) + Ah_1^2 + Bh_1^4 + Ch_1^6 + \cdots \tag{12.4.24}$$

For intervals of length h_2,

$$I = \text{TR}(h_2) + Ah_2^2 + Bh_2^4 + Ch_2^6 + \cdots \tag{12.4.25}$$

We now have two estimates: $\text{TR}(h_1)$ and $\text{TR}(h_2)$. We can eliminate the h^2 term between (12.4.24) and (12.4.25) by multiplying (12.4.24) by h_2^2 and (12.4.25) by h_1^2, then subtracting:

$$\left[h_2^2 - h_1^2 \right] I = \left[h_2^2 \text{TR}(h_1) - h_1^2 \text{TR}(h_2) \right] + B \left[h_2^2 h_1^4 - h_1^2 h_2^4 \right] + \cdots \tag{12.4.26}$$

Although we do not know $B, C, \ldots$, this serves to give us a new estimate if we neglect those unknown quantities. Replacing I by $\text{TR}(h_1, h_2)$, we solve to obtain

$$\text{TR}(h_1, h_2) = \frac{h_2^2 \text{TR}(h_1) - h_1^2 \text{TR}(h_2)}{h_2^2 - h_1^2} \tag{12.4.27}$$

If $h_2 = \frac{1}{2} h_1$, this reduces to (12.4.22), which we rewrite in the following form:

$$\text{TR}\left(1, \tfrac{1}{2}\right) = \frac{2^2 \text{TR}\left(\tfrac{1}{2}\right) - \text{TR}(1)}{2^2 - 1} \tag{12.4.28}$$

This is the estimate based upon intervals h and $\frac{1}{2} h$.

We now cut the interval in half again, eliminate in the same way, and obtain another estimate:

$$\text{TR}\left(\tfrac{1}{2}, \tfrac{1}{4}\right) = \frac{2^2 \text{TR}\left(\tfrac{1}{4}\right) - \text{TR}\left(\tfrac{1}{2}\right)}{2^2 - 1} \tag{12.4.29}$$

Equations (12.4.28) and (12.4.29) are approximations which [from (12.4.26)] do not involve the Ah^2 error term. Therefore, we say that the true value is given by equations of the form

$$I = TR\left(1, \tfrac{1}{2}\right) + Bh^4 + Ch^6 + \cdots \tag{12.4.30}$$

$$I = TR\left(\tfrac{1}{2}, \tfrac{1}{4}\right) + B\left(\frac{h}{2}\right)^4 + C\left(\frac{h}{2}\right)^6 + \cdots \tag{12.4.31}$$

We now eliminate the h^4 term between these equations by eliminating B. The result is

$$(2^4 - 1)I = \left[2^4 TR\left(\tfrac{1}{2}, \tfrac{1}{4}\right) - TR\left(1, \tfrac{1}{2}\right)\right] + Ch^6\left[\frac{1}{2^2} - 1\right] + \cdots \tag{12.4.32}$$

Although we do not know $C, D, \ldots$, this serves to furnish a new estimate. Replacing I by $TR(1, \tfrac{1}{2}, \tfrac{1}{4})$, we have

$$TR\left(1, \tfrac{1}{2}, \tfrac{1}{4}\right) = \frac{2^4 TR\left(\tfrac{1}{2}, \tfrac{1}{4}\right) - TR\left(1, \tfrac{1}{2}\right)}{2^4 - 1} \tag{12.4.33}$$

Looking ahead, the next step would eliminate the Ch^6 term between $TR(1, \tfrac{1}{2}, \tfrac{1}{4})$ and $TR(\tfrac{1}{2}, \tfrac{1}{4}, \tfrac{1}{8})$ and lead to

$$TR\left(1, \tfrac{1}{2}, \tfrac{1}{4}, \tfrac{1}{8}\right) = \frac{2^6 TR\left(\tfrac{1}{2}, \tfrac{1}{4}, \tfrac{1}{8}\right) - TR\left(1, \tfrac{1}{2}, \tfrac{1}{4}\right)}{2^6 - 1} \tag{12.4.34}$$

This process, called *extrapolation to the limit*, can be continued. Note that the successive coefficients 2^2 [in (12.4.28)], 2^4 [in (12.4.33)] and 2^6 [in (12.4.34)] can be written as 4^1, 4^2, and 4^3. With this in mind, we display the following theorem, which can be proved by an easy induction.

THEOREM: ROMBERG EXTRAPOLATION TO THE LIMIT If TR_1 and TR_2 are successive trapezoidal estimates of an integral obtained by doubling the number of divisions, a first corrected estimate can be obtained by the formula

$$TR_{12} = \frac{4 TR_2 - TR_1}{4^1 - 1} \tag{12.4.35}$$

A second corrected estimate can be obtained by the formula

$$TR_{123} = \frac{4^2 TR_{23} - TR_{12}}{4^2 - 1} \tag{12.4.36}$$

and the nth corrected estimate becomes

$$TR_{123\ldots(n+1)} = \frac{4^n TR_{23\ldots n} - TR_{123\ldots(n-1)}}{4^n - 1} \tag{12.4.37}$$

It turns out that (12.4.36) is the same as the Richardson extrapolation for Simpson's rule, but the later extrapolations are not Richardson extrapolations.

12.4.5 The Romberg Tableau

If we arrange successive calculations in matrix form, the result can be written as follows:

$$
\begin{array}{cccccc}
A_{11} & A_{12} & A_{13} & A_{14} & A_{15} & \cdots \\
 & A_{22} & A_{23} & A_{24} & A_{25} & \cdots \\
 & & A_{33} & A_{34} & A_{35} & \cdots \\
 & & & A_{44} & A_{45} & \cdots \\
 & & & & A_{55} & \cdots
\end{array}
$$

The first row represents the trapezoidal estimates for, say, $n=4$, 8, 16, 32 and 64. The second row represents the first extrapolation: For example, $A_{22}, A_{23}, A_{24}, \ldots$ are obtained by applying (12.4.35) to $\{A_{11}, A_{12}\}$, $\{A_{12}, A_{13}\}$, $\{A_{13}, A_{14}\}$, etc. The third row represents the second extrapolation, obtained by applying (12.4.36) to $\{A_{22}, A_{23}\}$, $\{A_{23}, A_{24}\}$, $\{A_{24}, A_{25}\}$, etc. Successive rows are calculated column by column until either the difference between two row values in a column satisfies the criterion or else the new subscript equals the column subscript, in which case the next column must be started. (We shall explain and illustrate.)

We already know, for the illustrative problem, the trapezoidal values for $n=4$ and $n=8$. These are A_{11} and A_{12}. Using (12.4.35), or else (12.4.37) with $n=1$, we calculate A_{22}:

$$
A_{22} = \frac{4^1 A_{12} - A_{11}}{4^1 - 1} \tag{12.4.38}
$$

$$
A_{22} = \frac{4(0.1124600067) - 0.112572783}{4^1 - 1} \tag{12.4.39}
$$

$$
A_{22} = 0.1124629161 \tag{12.4.40}
$$

The tableau, to this point, is

	Column 1: $n=4$	Column 2: $n=8$
Trap.	0.1124512783	0.1124600067
1st approx.		0.1124629161

The last two values calculated were A_{12} and A_{22}. Their difference is 2.9×10^{-6}, too large. We must continue. We cannot continue in the second column. We therefore start a new column for $n=16$. W is unchanged,

$\dot{h} = (0.1 - 0)/16 = 0.00625$, and

$$\text{erf } 0.1 \sim h \left[W + \sum_{k=1}^{15} f_k \right] \qquad (12.4.41)$$

We find that $A_{13} = 0.112621887$. Using A_{12} and A_{13} with (12.4.35) or with (12.4.37) using $n = 1$, we have

$$A_{23} = \frac{4^1(0.112621887) - 0.112600067}{4^1 - 1} \qquad (12.4.42)$$

$$A_{23} = 0.112629160 \qquad (12.4.43)$$

The tableau now extends as follows:

Trap.	0.112512783	0.112600067	0.112621887
1st Approx.		0.112629161	0.112629160

The difference between the last two values calculated, A_{13} and A_{23}, is 7×10^{-7}, still not satisfactory. We calculate A_{33}, using (12.4.36) or else (12.4.37) with $n = 2$ [note that n in (12.4.37) is one less than the row number]:

$$A_{33} = \frac{4^2(0.112629160) - 0.112629161}{4^2 - 1} \qquad (12.4.44)$$

$$A_{33} = 0.112629160 \qquad (12.4.45)$$

The tableau now is

0.112512783	0.112600067	0.112621887
	0.112629161	0.112629160
		0.112629160

The difference between the last two values calculated is 0. The answer is correct to the number of places shown. As can be seen in Table 12-1, the Romberg procedure was very successful in almost every case.

Figure 12.4-1 shows an abbreviated flowchart for the computer program shown in Figure 12.4-2. Figure 12.4-3 displays the output of the program for the illustrative problem. The program was set to terminate when the number of rows was 9 or when the difference between two successive values was less than $\frac{1}{2}$ the last place of the true value. Fixed-point arithmetic was used, wherever possible, to reduce the error of conversion and roundoff.

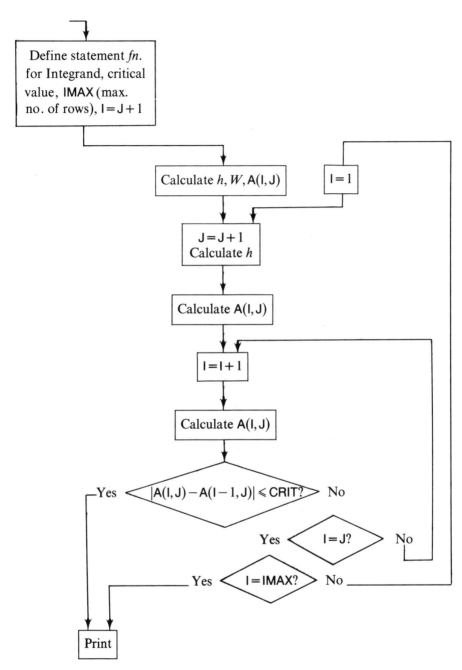

Fig. 12.4-1 Flowchart for the Romberg integration.

```
      DOUBLE PRECISION R(9,9),TV(10),A,B,CRIT,D,E,F,H,P,S1,S2,TE,TI,V,X,   040
     1Y,Z,PI,TN,PART,X1,RR                                                050
      DIMENSION HEAD(12)                                                 060
      COMMON NR,NC,V,X1,TV,Z,TE,NDEC                                     065
C  THE PURPOSE OF THIS PROGRAM IS TO PERFORM AN INTEGRATION STARTING WITH 070
C  THE TRAPEZOIDAL RULE, THEN EMPLOYING THE ROMBERG ALGORITHM. THE NEXT  080
C  CARD DEFINES THE INTEGRAND.                                          090
C  IF F(A) OR F(B) CANNOT BE COMPUTED, CARDS MUST BE INSERTED TO        091
C  SKIP OVER CARD 450. THIS IS DONE BY INSERTING CARDS LIKE             092
C           S1= (INSERT PROPER VALUES)                                  092
C           GO TO 31                                                    093
      F (X)=2.D0*DEXP(-X*X)/DSQRT(PI)                                    VAR
C  CCCCCCCCCCCCCCCCCCCCCCCCCCCCCCCCCCCCCCCCCCCCCCCCCCCCCCCCCCCCCCCCCC094
C  CCCCCCCCCCCCCCCCCCCCCCCCCCCCCCCCCCCCCCCCCCCCCCCCCCCCCCCCCCCCCCCCCC095
C  IF THE PROBLEM USES INTEGRATION BY PARTS, A STATEMENT FUNCTION       100
C  MUST REPLACE CARD 105.  OTHERWISE, PART(X) = 0                       101
C  CCCCCCCCCCCCCCCCCCCCCCCCCCCCCCCCCCCCCCCCCCCCCCCCCCCCCCCCCCCCCCCCCC102
      PART(X) = 0.D0                                                    105
C  DEFINITIONS.                                                         110
      PI=3.141592653589793D0                                           130
      MAX=9                                                            140
      J=1                                                              145
C  FIRST DATA CARD. THE NAME OF THE INTEGRAL CENTERED IN CC 1-48, AND THE 150
C  TYPE IN CC 51. IF ITYPE=0, THE PARAMETER IS VARIABLE. IF ITYPE=1, THE 160
C  UPPER LIMIT IS VARIABLE. IF ITYPE=2, NEITHER.NDEC IS THE NUMBER OF   170
C  DECIMAL PLACES DESIRED, SHOWN IN CC 52-53.                          175
      READ(1,100)HEAD,ITYPE,NDEC                                       180
100   FORMAT(12A4,2X,I1,I2)                                            190
      CRIT=0.5D0*1.D1**(-NDEC)                                         191
      NDEC=NDEC-4                                                      192
C  SECOND DATA CARD. LOWER LIMIT (A) IN CC 1. FIRST UPPER LIMIT (B) IN  200
C  CC 21. INCREMENT (D) IN CC 41. PARAMETER (P), IF ANY, IN CC 61.     210
      READ(1,101)A,B,D,P                                               220
```

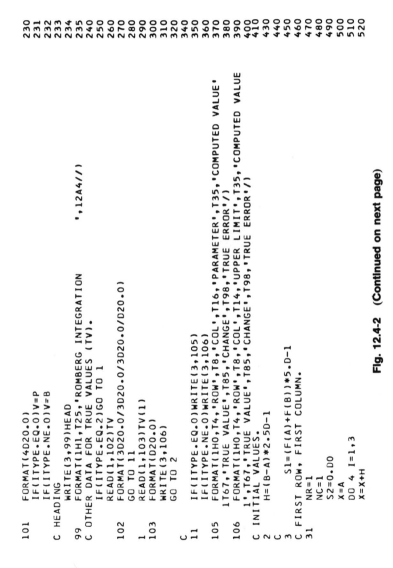

```
101    FORMAT(4D20.0)                                                     230
       IF(ITYPE.EQ.0)V=P                                                  231
       IF(ITYPE.NE.0)V=B                                                  232
C HEADING                                                                 233
       WRITE(3,99)HEAD                                                    234
99     FORMAT(1H1,T25,'ROMBERG INTEGRATION        ',12A4//)              235
C OTHER DATA FOR TRUE VALUES (TV).                                        240
       IF(ITYPE.EQ.2)GO TO 1                                             250
       READ(1,102)TV                                                      260
102    FORMAT(3D20.0/3D20.0/3D20.0/D20.0)                                 270
       GO TO 11                                                           280
1      READ(1,103)TV(1)                                                   290
103    FORMAT(D20.0)                                                      300
       WRITE(3,106)                                                       310
       GO TO 2                                                            320
C                                                                         340
11     IF(ITYPE.EQ.0)WRITE(3,105)                                         350
       IF(ITYPE.NE.0)WRITE(3,106)                                         360
105    FORMAT(1H0,T4,'ROW',T8,'COL',T16,'PARAMETER',T35,'COMPUTED VALUE'  370
      1T67,'TRUE VALUE',T85,'CHANGE',T98,'TRUE ERROR'/)                  380
106    FORMAT(1H0,T4,'ROW',T8,'COL',T14,'UPPER LIMIT',T35,'COMPUTED VALUE 390
      1',T67,'TRUE VALUE',T85,'CHANGE',T98,'TRUE ERROR'/)                400
C INITIAL VALUES.                                                         410
2      H=(B-A)*2.5D-1                                                     430
C                                                                         440
3      S1=(F(A)+F(B))*5.D-1                                               450
C FIRST ROW, FIRST COLUMN.                                                460
31     NR=1                                                               470
       NC=1                                                               480
       S2=0.D0                                                            490
       X=A                                                                500
       DO 4 I=1,3                                                         510
       X=X+H                                                              520
```

Fig. 12.4-2 (Continued on next page)

```
4     S2=S2+F(X)                                                      530
      R(1,1)=H*(S1+S2)                                                540
C                                                                     550
C THE NEXT CARD IS A DUMMY TO FORCE ASTERISKS.                        555
      Z=1234567890123 4.D0                                            560
      X=R(NR,NC)                                                      561
      X1=X+PART(B)                                                    562
      TE=TV(J)-X1                                                     565
      IF(DABS(TE).LT.CRIT)TE=0.D0                                     566
      IF(ITYPE.EQ.2)CALL NDECSB(J)                                    570
C NEXT COLUMN.                                                        580
5     NC=NC+1                                                         590
      N=2**(NC+1)                                                     600
      TN=DFLOAT(N)                                                    610
      H=(B-A)/TN                                                      620
C                                                                     630
      NP=N-1                                                          635
      DO 6 I=1,NP,2                                                   640
      TI=DFLOAT(I)                                                    650
      X=A+TI*H                                                        660
      S2=S2+F(X)                                                      670
6     R(1,NC)=H*(S1+S2)                                               680
C ERROR CALCULATIONS.                                                 690
      X=R(NR,NC)                                                      692
      Z=X-R(NR-1,NC)                                                  693
      X1=X+PART(B)                                                    694
      TE=TV(J)-X1                                                     700
      IF(DABS(TE).LT.CRIT)TE=0.D0                                     705
      IF(ITYPE.EQ.2)CALL NDECSB(J)                                    710
C NEXT ROW.                                                           720
7     E=DFLOAT(4**NR)                                                 721
      NR=NR+1                                                         725
      R(NR,NC)=(E*R(NR-1,NC)-R(NR-1,NC-1))/(E-1.D0)                   730
```

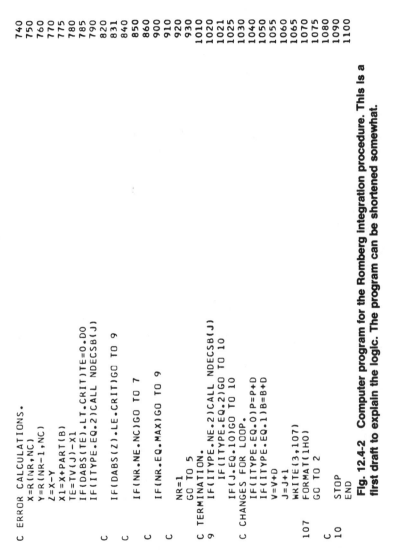

```
                                                              740
C ERROR CALCULATIONS.                                         750
      X=R(NR,NC)                                              760
      Y=R(NR-1,NC)                                            770
      Z=X-Y                                                   775
      X1=X+PART(B)                                            780
      TE=TV(J)-X1                                             785
      IF(DABS(TE).LT.CRIT)TE=0.D0                             790
      IF(ITYPE.EQ.2)CALL NDECSB(J)                            820
C                                                             831
C                                                             840
      IF(DABS(Z).LE.CRIT)GO TO 9                              850
C                                                             860
      IF(NR.NE.NC)GO TO 7                                     900
C                                                             910
      IF(NR.EQ.MAX)GO TO 9                                    920
C                                                             930
      NR=1                                                   1010
      GO TO 5                                                1020
C TERMINATION.                                               1021
9     IF(ITYPE.NE.2)CALL NDECSB(J)                           1025
      IF(ITYPE.EQ.2)GO TO 10                                 1030
      IF(J.EQ.10)GO TO 10                                    1040
C CHANGES FOR LOOP.                                          1050
      IF(ITYPE.EQ.0)P=P+D                                    1055
      IF(ITYPE.EQ.1)B=B+D                                    1060
      V=V+D                                                  1065
      J=J+1                                                  1070
107   WRITE(3,107)                                           1075
      FORMAT(1H0)                                            1080
      GO TO 2                                                1090
C                                                            1100
10    STOP
      END
```

Fig. 12.4-2 Computer program for the Romberg integration procedure. This is a first draft to explain the logic. The program can be shortened somewhat.

ROMBERG INTEGRATION   ERROR FUNCTION

ROW	COL	UPPER LIMIT	COMPUTED VALUE	TRUE VALUE	CHANGE	TRUE ERROR
3	3	0.100	1.1246291600-01	1.1246291600-01	-1.130-11	0.0
3	3	0.200	2.2270258920-01	2.2270258920-01	-3.440-10	0.0
3	4	0.300	3.2862675950-01	3.2862675950-01	-1.500-10	0.0
4	4	0.400	4.2839235500-01	4.2839235500-01	2.970-13	0.0
4	4	0.500	5.2049987780-01	5.2049987780-01	1.120-12	0.0
4	4	0.600	6.0385609080-01	6.0385609080-01	2.910-12	0.0
4	4	0.700	6.7780119380-01	6.7780119380-01	5.570-12	0.0
4	4	0.800	7.4210096470-01	7.4210096470-01	7.570-12	0.0
4	4	0.900	7.9690821240-01	7.9690821240-01	5.200-12	0.0
4	4	1.000	8.427C079290-01	8.4270079290-01	-7.370-12	0.0

Fig. 12.4-3 Output of the Romberg computer program.

PROBLEM SECTION 12.4

Using the data displayed in the problems for Section 12.3, select the same integral that was done by the Simpson's-rule procedure and do it with the Romberg procedure. The errors should be compared with those in Table 12-1.

12.5 INTEGRATION BY CHEBYSHEV SUMMATION

12.5.1 Introduction

From an examination of Table 12-1, it may be hard to see why one should not be perfectly satisfied with a rapid, easy, accurate application of the trapezoidal rule using Romberg extrapolation to the limit. The difficulty is, of course, that if we do not know the correct answer in advance, we have no practical way to estimate the accuracy and precision of the results. The theoretical formulas are virtually useless, and the rule of thumb concerning successive differences untrue and unreliable.

However, if the Chebyshev summation is used, we can derive a formula which appears to give us a reasonable estimate of the error.

In this section, we shall tackle our worst result, that for the *gamma function*, to see whether (if we did not know the correct answers) we could have known that the results were not very accurate.

The gamma function is a very interesting one. It is important in practical work and has been studied intensively by the greatest mathematicians. It is also interesting because it is a solution of the finite difference equation

$$f(x+1) - xf(x) = 0$$

if $f(1) = 1$. In spite of all the investigation, no really good method exists for computing $\Gamma(x)$. For very large x, De Moivre (1667–1754) and Stirling (1692–1770) used an asymptotic series. For small and moderate-sized x, Legendre (1752–1833) used different series for different intervals. For example, he used one for x in $(1.001, 1.250)$, another in $(1.250, 1.375)$, another in $(1.375, 1.625)$, and so on. From our discussion of error analysis in the opening chapter, it is easy to see why the integrand, involving a factor of the form x^y, is error-prone.

The first step in the numerical integration is the handling of the integrand.

12.5.2 Step I. Approximating the Integrand

We have already discussed the Chebyshev summation in great detail. For the purpose of illustration, we take

$$\Gamma(p) = \int_0^\infty x^{p-1} e^{-x} dx \tag{12.5.1}$$

with the parameter p equal to 1.25. As explained in Section 12.1.3, this can be rewritten as

$$\Gamma(p) = \int_0^1 \left(\frac{1-t}{t}\right)^{p-1} \exp\left(-\frac{1-t}{t}\right) \frac{dt}{t^2} \qquad (12.5.2)$$

For the standard Chebyshev interval $[-1, +1]$, we must make the usual transformation:

$$t = mx + b \qquad (12.5.3)$$

When $t = L$ (lower limit), $x = -1$; when $t = U$ (upper limit), $x = +1$. Then

$$\begin{cases} L = -m + b \\ U = m + b \end{cases} \qquad (12.5.4)$$

from which

$$t = \frac{(U-L)x + (U+L)}{2} \qquad (12.5.5)$$

$$dt = \frac{U-L}{2} dx \qquad (12.5.6)$$

The change of limits can easily be done within the computer program.

The thirteen integrals discussed in this chapter were approximated by Chebyshev polynomials with 25 coefficients, i.e., a 24th-degree polynomial of the form

$$p(x) = b_0 + b_1 x + \cdots + b_{24} x^{24} \qquad (12.5.7)$$

The coefficients found for $\Gamma(1.25)$ are shown in Figure 12.5-1.

	GAMMA FUNCTION	VARIABLE = 1.25
	CHEBYSHEV SUMMATION COEFFICIENTS	25 TERMS
K	FOR T(K)	FOR POWER SERIES
0	3.416096001641415D-01	7.357588823428847D-01
1	9.812621962406755D-02	-3.659608048082887D-01
2	-3.484404528165342D-01	-6.423591015186522D-01
3	2.159316051415930D-02	1.012891394693347D 00
4	7.252300818864032D-02	-1.401815795459549D 00
5	-8.310774284543321D-02	6.726965255860183D 00
6	6.769419819803032D-03	4.406179263778544D 00
7	1.039274571157033D-02	-8.394039537372296D 01
8	-2.606868081483937D-02	-5.816663384771957D 01
9	5.997445654242550D-03	6.351662239050528D 02
10	-4.250400252000910D-03	4.458602037895793D 02
11	-7.364283976172853D-03	-2.968849416399071D 03
12	1.663690413358362D-03	-1.996881380943581D 03
13	-4.664923561136928D-03	8.377384790337230D 03
14	-1.533216661253356D-03	5.705321009379389D 03
15	-9.570187414522357D-04	-1.716763176066942D 04
16	-2.376847960345312D-03	-1.058335392662328D 04
17	-5.058484276308826D-04	2.162531412171370D 04
18	-1.315329776394393D-03	1.270100958722425D 04
19	-8.478155524720598D-04	-1.697628850413086D 04
20	-5.173547833558964D-04	-9.549116745975299D 03
21	-7.203835418265454D-04	7.532530079607191D 03
22	-2.472999004817432D-04	4.102446513863840D 03
23	-3.436506094045848D-04	-1.441375125628087D 03
24	-9.181245157637059D-05	-7.701786657931550D 02

Fig. 12.5-1 Coefficients for the gamma function expressed as a Chebyshev summation.

ERROR TABLE

X	ERROR	X	ERROR
-1.000	-1.26D-04	0.0	-1.25D-16
-0.975	-1.08D-04	0.025	-4.58D-05
-0.950	1.28D-05	0.050	-7.57D-05
-0.925	1.48D-04	0.075	-7.76D-05
-0.900	-4.16D-05	0.100	-4.95D-05
-0.875	-1.34D-04	0.125	-7.14D-07
-0.850	-2.91D-05	0.150	5.07D-05
-0.825	6.47D-05	0.175	8.47D-05
-0.800	5.47D-05	0.200	8.66D-05
-0.775	6.45D-06	0.225	5.36D-05
-0.750	-1.29D-05	0.250	-3.25D-06
-0.725	-2.88D-06	0.275	-6.23D-05
-0.700	4.76D-06	0.300	-9.91D-05
-0.675	-6.13D-06	0.325	-9.59D-05
-0.650	-2.48D-05	0.350	-5.06D-05
-0.625	-3.07D-05	0.375	2.07D-05
-0.600	-1.43D-05	0.400	8.84D-05
-0.575	1.63D-05	0.425	1.21D-04
-0.550	4.21D-05	0.450	9.81D-05
-0.525	4.65D-05	0.475	2.50D-05
-0.500	2.57D-05	0.500	-6.83D-05
-0.475	-1.02D-05	0.525	-1.36D-04
-0.450	-4.32D-05	0.550	-1.37D-04
-0.425	-5.68D-05	0.575	-6.05D-05
-0.400	-4.42D-05	0.600	6.19D-05
-0.375	-1.10D-05	0.625	1.63D-04
-0.350	2.83D-05	0.650	1.73D-04
-0.325	5.64D-05	0.675	6.47D-05
-0.300	6.12D-05	0.700	-1.11D-04
-0.275	4.03D-05	0.725	-2.34D-04
-0.250	2.20D-06	0.750	-1.84D-04
-0.225	-3.78D-05	0.775	5.02D-05
-0.200	-6.37D-05	0.800	2.96D-04
-0.175	-6.48D-05	0.825	2.78D-04
-0.150	-4.04D-05	0.850	-1.10D-04
-0.125	5.91D-07	0.875	-4.94D-04
-0.100	4.26D-05	0.900	-1.68D-04
-0.075	6.94D-05	0.925	7.55D-04
-0.050	7.03D-05	0.950	9.75D-04
-0.025	4.41D-05	0.975	-1.63D-03
0.0	-1.25D-16	1.000	-7.53D-02

Fig. 12.5-2 Error table for the Chebyshev summation representation of the gamma function.

When the calculated values of the integrand were compared with those obtained by using the 25-term power series, the rather discouraging table in Figure 12.5-2 was obtained.

12.5.3 Step II. The Integration

The polynomial in (12.5.7) is integrated from -1 to $+1$:

$$\int_{-1}^{+1} p(x)\,dx = b_0 x + \tfrac{1}{2}b_1 x^2 + \cdots + \tfrac{1}{25}b_{24}x^{25}\Big|_{-1}^{+1} \qquad (12.5.8)$$

$$\int_{-1}^{+1} p(x)\,dx = 2\left(b_0 + \tfrac{1}{3}b_2 + \tfrac{1}{5}b_4 + \cdots + \tfrac{1}{25}b_{24}\right) \qquad (12.5.9)$$

Note that the actual integration results in a polynomial with only 13 terms. In fact, we did not need the odd-subscripted Chebyshev coefficients.

A partial flowchart for the computer program is shown in Figure 12.5-3, and the program is shown in Figure 12.5-4. The output for the gamma function is shown in Figure 12.5-5.

F(X)= $\cdots$ is a statement function for the integrand.

PART(X) is defined as 0.D0 except when the integration is done by parts.

ANG is a subroutine which supplies values for θ and $\cos\theta$.

A, B, D, P are, respectively, lower limit, upper limit, increment and parameter value. For variable B or P, the first value is read in.

NDEC is the number of decimal places in scientific notation, used to control output FORMAT.

JSTART is a code value used to control the printout of the error table, either from -1 to $+1$ or (for logarithms) from 0.025 to $+1$

CSUM is a subroutine to calculate Chebyshev coefficients and corresponding polynomial coefficients, and to print them and the error table.

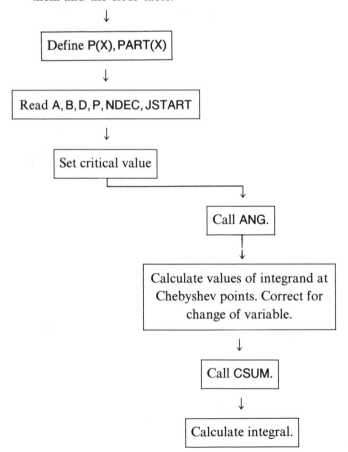

Fig. 12.5-3   A partial flowchart for the Chebyshev Integration procedure.

```
      DOUBLE PRECISION F(25),A(25),B(25),FCH(81),ANS(10),C(25),E(10),T1(    040
     125),TV(10),W(10),D,FI,P,PI,T,TL,TU,V,X,TU1,P1,CRIT,PART,R             050
      DIMENSION HEAD(12)                                                    060
      COMMON F,A,B,FCH/ANG/C,T1                                            070
C THE PURPOSE OF THIS PROGRAM IS TO PERFORM AN INTEGRATION BY AN APPROX-   080
C IMATION, USING A CHEBYSHEV SUMMATION WITH 25 COEFFICIENTS. THE INTE-     090
C GRAND IS DEFINED AS THE FOLLOWING STATEMENT FUNCTION. THIS ONE CARD IS   100
C CHANGED FOR EACH PROBLEM.                                                110
C CCCCCCCCCCCCCCCCCCCCCCCCCCCCCCCCCCCCCCCCCCCCCCCCCCCCCCCCCCCCCCCCCCC115
      FI(X)=(1.D0/X-1.D0)**(P-1.D0)*DEXP(1.D0-1.D0/X)/X**2              VAR
C CCCCCCCCCCCCCCCCCCCCCCCCCCCCCCCCCCCCCCCCCCCCCCCCCCCCCCCCCCCCCCCCCCC119
C FOR PROBLEMS REQUIRING INTEGRATION BY PARTS, A STATEMENT              120
C FUNCTION FOR PART(X) IS INSERTED, USED AFTER CARD 870.               121
C CCCCCCCCCCCCCCCCCCCCCCCCCCCCCCCCCCCCCCCCCCCCCCCCCCCCCCCCCCCCCCC
      PART(X) = 0.D0                                                    105
C THE NEXT CARD MAKES THE TRANSFORMATION TO STANDARD CHEBYSHEV LIMITS.   130
      T(X)=((TU-TL)*X+TU+TL)*5.D-1                                      140
C DATA CARDS ARE AS FOLLOWS.                                            150
C   CARD 1                                                              160
C NAME OF INTEGRAL CENTERED IN CC 1-48                                  170
C JSTART IN CC 50                                                       180
C ITYPE IN CC 51                                                        190
C NDEC IN CC 52-53                                                      200
C KOUNT IN CC 54                                                        205
C IF JSTART=0, THE ERROR TABLES START AT T=-1, OTHERWISE AT 0.025       210
C ITYPE=0 FOR VARIABLE PARAMETERS, 1 FOR VARIABLE UPPER LIMITS, 2 FOR   220
C     NO VARIABLES                                                      230
C KOUNT = NUMBER OF COPIES WANTED                                       240
C   CARD 2                                                              250
C CC1  LOWER LIMIT=TL                                                   260
C CC21 UPPER LIMIT=TU. IF VARIABLE, THE FIRST VALUE.                    270
C CC41 INCREMENT (IF ANY) = D                                           280
C CC61 PARAMETER (IF ANY) = P                                           290
C     CARDS 3,4 AND 5 HAVE TRUE VALUES.                                 300
C CCCCCCCCCCCCCCCCCCCCCCCCCCCCCCCCCCCCCCCCCCCCCCCCCCCCCCCCCCCCCCCCCCC305
C DEFINITIONS, INCLUDING CHEBYSHEV ANGLES AND COSINES                   310
      PI=3.141592653589793D0                                           320
      KT=10                                                            325
      CALL ANG25                                                       330
```

Fig. 12.5-4 (Continued on next page)

```
340  C INPUT.
350        READ(1,100)HEAD,JSTART,ITYPE,NDEC,KOUNT
361  100   FORMAT(12A4,1X,2I1,I2,I1)
362        CRIT=0.5D0*1.D1**(-NDEC)
370        NDEC=NDEC-4
380        READ(1,101)TL,TU,D,P
381  101   FORMAT(4D20.0)
382        TU1=TU
383        P1=P
384        IF(ITYPE.EQ.2)GO TO 21
385        READ(1,104)TV
386  104   FORMAT(3D20.0/3D20.0/3D20.0/D20.0)
390        GO TO 1
400  C IF NO VARIABLES.
410  21    READ(1,108)TV(1)
430  108   FORMAT(D20.0)
440        KT=1
450  C FOR PROBLEMS WITH VARIABLES. J IS THE PROBLEM NUMBER.
460  1     DO 9 J=1,KT
465        X=DFLOAT(J-1)
470        IF(ITYPE.EQ.2)GO TO 4
480  C VARIABLE PARAMETER.
490        IF(ITYPE.EQ.1)GO TO 2
500        P=P1+X*D
510        W(J)=P
520        GO TO 3
530  C VARIABLE UPPER LIMIT.
540  2     TU=TU1+X*D
550        W(J)=TU
560  C HEADING FOR PROBLEMS WITH VARIABLES.
570  3     WRITE(3,103)HEAD,W(J)
580  103   FORMAT(1H1,T32,12A4,'VARIABLE = ',G9.3//)
590  C CALCULATION OF INTEGRAND AT CHEBYSHEV POINTS.
600  4     DO 5 K=1,25
610        X=C(K)
620        X=T(X)
630  5     F(K)=FI(X)
     C CORRECTION FOR CHANGE FROM DX TO DT.
640        DO 6 K=1,25
650  6     F(K)=F(K)*(TU-TL)*5.D-1
```

```
                                                                      660
C   CALCULATION OF VALUES FOR CHECKING.                               670
      IX=-1025                                                        680
      IT=81                                                           690
      IF(JSTART.EQ.1)IX=0                                             700
      IF(JSTART.EQ.1)IT=40                                            710
C                                                                     720
      DO 7 I=1,IT                                                     730
      IX=IX+25                                                        740
      X=DFLOAT(IX)                                                    750
      X=X*1.D-3                                                       760
      X=T(X)                                                          770
7     FCH(I)=FI(X)*(TU-TL)*5.D-1                                      780
C   CALCULATION OF COEFFICIENTS.                                      790
      CALL CSUM25                                                     800
      CALL PR25(1,JSTART)                                             810
C   CALCULATION OF INTEGRAL.                                          820
      X=0.D0                                                          830
      DO 8 K=1,25,2                                                   840
      K1=26-K                                                         850
      V=DFLOAT(K1)                                                    860
      X=X+B(K1)/V                                                     870
      ANS(J)=X*2.D0                                                   880
8     ANS(J)=ANS(J)+PART(TU)                                          885
9     CONTINUE                                                        910
C   CALCULATION OF ERRORS, USING TRUE VALUES FROM TABLES.            940
      DO 10 J=1,KT                                                    950
      E(J)=TV(J)-ANS(J)                                               955
      IF(DABS(E(J)).LT.CRIT)E(J)=0.D0                                 956
10    CONTINUE                                                        960
C   PRINTOUT.                                                         970
      WRITE(3,102)HEAD                                                975
102   FORMAT(1H1,T32,12A4//)                                          980
C   SUBHEAD FOR VARIABLE PARAMETER.                                   990
      IF(ITYPE.NE.0)GO TO 11                                          1000
      WRITE(3,105)                                                    1010
105   FORMAT(1H0,T19,'PARAMETER',T42,'CALC. VALUE',T75,'TRUE VALUE',T93,1020
     1'ERROR'/)                                                       1030
      GO TO 12                                                        1040
C   SUBHEAD FOR VARIABLE UPPER LIMIT.                                 1050
11    WRITE(3,106)
```

Fig. 12.5-4 (Continued on next page)

```
106   FORMAT(1H0,T20,'UPPER LIMIT',T41,'CALC. VALUE',T75,'TRUE VALUE',T9    1060
     13,'ERROR'/)                                                           1070
C                                                                           1080
12    DO 13 J=1,KT                                                          1090
      GO TO(905,906,907,908,909,910,911,912,913,914,915),NDEC               1100
905   WRITE(3,925)W(J),ANS(J),TV(J),E(J)                                    1101
      GO TO 13                                                              1102
906   WRITE(3,926)W(J),ANS(J),TV(J),E(J)                                    1103
      GO TO 13                                                              1104
907   WRITE(3,927)W(J),ANS(J),TV(J),E(J)                                    1105
      GO TO 13                                                              1106
908   WRITE(3,928)W(J),ANS(J),TV(J),E(J)                                    1107
      GO TO 13                                                              1108
909   WRITE(3,929)W(J),ANS(J),TV(J),E(J)                                    1109
      GO TO 13                                                              1110
910   WRITE(3,930)W(J),ANS(J),TV(J),E(J)                                    1111
      GO TO 13                                                              1112
911   WRITE(3,931)W(J),ANS(J),TV(J),E(J)                                    1113
      GO TO 13                                                              1114
912   WRITE(3,932)W(J),ANS(J),TV(J),E(J)                                    1115
      GO TO 13                                                              1116
913   WRITE(3,933)W(J),ANS(J),TV(J),E(J)                                    1117
      GO TO 13                                                              1118
914   WRITE(3,934)W(J),ANS(J),TV(J),E(J)                                    1119
      GO TO 13                                                              1120
915   WRITE(3,935)W(J),ANS(J),TV(J),E(J)                                    1121
13    CONTINUE                                                              1122
925   FORMAT(1H0,T19,F9.5,T31,1PD22.5 ,T63,D22.5 ,T91,D9.2)                 1123
926   FORMAT(1H0,T19,F9.5,T31,1PD22.6 ,T63,D22.6 ,T91,D9.2)                 1124
927   FORMAT(1H0,T19,F9.5,T31,1PD22.7 ,T63,D22.7 ,T91,D9.2)                 1125
928   FORMAT(1H0,T19,F9.5,T31,1PD22.8 ,T63,D22.8 ,T91,D9.2)                 1126
929   FORMAT(1H0,T19,F9.5,T31,1PD22.9 ,T63,D22.9 ,T91,D9.2)                 1127
930   FORMAT(1H0,T19,F9.5,T31,1PD22.10,T63,D22.10,T91,D9.2)                 1128
931   FORMAT(1H0,T19,F9.5,T31,1PD22.11,T63,D22.11,T91,D9.2)                 1129
932   FORMAT(1H0,T19,F9.5,T31,1PD22.12,T63,D22.12,T91,D9.2)                 1130
933   FORMAT(1H0,T19,F9.5,T31,1PD22.13,T63,D22.13,T91,D9.2)                 1131
934   FORMAT(1H0,T19,F9.5,T31,1PD22.14,T63,D22.14,T91,D9.2)                 1132
935   FORMAT(1H0,T19,F9.5,T31,1PD22.15,T63,D22.15,T91,D9.2)                 1133
      GO TO 15                                                              1134
15    STOP                                                                  1200
      END                                                                   1210
```

Fig. 12.5-4   Computer program for the Chebyshev integration procedure.

GAMMA FUNCTION

PARAMETER	CALC. VALUE	TRUE VALUE	ERROR
1.25000	9.0639290090-01	9.0640247710-01	9.580-06
1.30000	8.9746214660-01	8.9747069630-01	8.550-06
1.35000	8.9114415290-01	8.9115144200-01	7.290-06
1.40000	8.8725781650-01	8.8726381750-01	6.000-06
1.45000	8.8565657090-01	8.8566138030-01	4.810-06
1.50000	8.8622314820-01	8.8622692550-01	3.780-06
1.55000	8.8886542110-01	8.8886834780-01	2.930-06
1.60000	8.9351309490-01	8.9351534930-01	2.250-06
1.65000	9.0011507320-01	9.0011681630-01	1.740-06
1.70000	9.0863736290-01	9.0863873290-01	1.370-06

Fig. 12.5-5   Output for the gamma function by the Chebyshev summation method.

12.5.4   Estimate of the Error

The Chebyshev summation results in

$$f(x) = p(x) + \epsilon(x) \qquad (12.5.10)$$

where $\epsilon(x)$ is the error at x. Integrating, we have

$$\int_{-1}^{+1} f(x)\,dx = \int_{-1}^{+1} p(x)\,dx + \int_{-1}^{+1} \epsilon(x)\,dx \qquad (12.5.11)$$

For n odd, the error curve seems to resemble a sequence of sine waves with different amplitudes which are the n extremals. We assume that

$$\int_{-1}^{+1} \epsilon(x)\,dx \sim \int_{-1}^{+1} A(x)\sin\theta_x\,dx \qquad (12.5.12)$$

Since the extremals are bounded and the error curve is continuous,

$$m\int_{-1}^{+1} \sin\theta_x\,dx \leqslant \int_{-1}^{+1} A(x)\sin\theta_x\,dx \leqslant M\int_{-1}^{+1} \sin\theta_x\,dx \qquad (12.5.13)$$

where m and M are the minimum and maximum extremals. (In the case of a Hastings approximation, they would be approximately equal.) Then there is some value of $x = c$ on the interval such that

$$\left| \int_{-1}^{+1} \epsilon(x)\,dx \right| \sim \left| A(c) \int_{-1}^{+1} \sin\theta_x\,dx \right| \tag{12.5.14}$$

It is obvious that

$$\left| \int_{-1}^{+1} \sin\theta_x\,dx \right| \leqslant \left| \int_{-1}^{+1} 1\,dx \right| = 2 \tag{12.5.15}$$

For n even, (12.5.12) becomes

$$\int_{-1}^{+1} \epsilon(x)\,dx \sim \int_{-1}^{+1} A(x)\cos\theta_x\,dx \tag{12.5.16}$$

with the same result.

In both cases,

$$\left| \int_{-1}^{+1} \epsilon(x)\,dx \right| \leqslant 2|A(c)| \tag{12.5.17}$$

It is now necessary to guess a value for $A(c)$. We make the reasonable guess that it is approximately equal to the average amplitude, i.e., the average extremal. This leads to the following rule of thumb, which worked very well for all the cases tested.

RULE OF THUMB: ERROR OF THE CHEBYSHEV-SUMMATION INTEGRATION
  The error of integration by use of a Chebyshev summation can be estimated by

$$E(n) = \frac{2}{n-1} \sum \text{extremals} \tag{12.5.18}$$

where n is the number of coefficients and each extremal is the absolute value of a maximum or minimum error.

It was found that there was very little difference when the maxima and

Table 12.5-1
Estimated and True Errors

FUNCTION	PRECISION	ESTIMATED	TRUE	EST/TRUE
Gamma	10^{-10}	6.6×10^{-5}	9.6×10^{-6}	7
Elliptic I	10^{-15}	1.4×10^{-9}	9.7×10^{-11}	14
Fresnel	10^{-7}	5.0×10^{-2}	2.2×10^{-3}	23
Bessel	10^{-15}	6.6×10^{-8}	2.8×10^{-9}	24
$\ln(1+x)\sin 10x$	10^{-8}	1.1×10^{-3}	4.4×10^{-5}	25

minima were read from an error table instead of being calculated by the extremal-point theorem. Table 12.5-1 shows the estimated errors and true errors for the worst results among the thirteen integrals displayed in this chapter. The column headed PRECISION refers to the precision of the true values given in published tables.

12.5.5 Reliability of the Chebyshev-Summation Method

We deliberately put our worst foot forward in presenting the Chebyshev-summation method. First of all, the 25-term summation reduced to a 13-term integration. That is probably too few. It would be safer to use a 50-term summation, which would reduce to 26 terms. Secondly, the power series for $\Gamma(x)$, for which the coefficients are shown in Figure 12.5-1, is unsatisfactory because the coefficients are increasing instead of decreasing. Thirdly, the error table, shown in Figure 12.5-2, displays errors which are large compared with the precision desired (10^{-10}).

In contrast, when the coefficients were calculated for the complete elliptic integral of the second kind, with parameter 0.5, they were small and decreasing. This is shown in Figure 12.5-6.

The corresponding error table (Figure 12.5-7) showed an excellent fit

```
COMPLETE ELLIPTIC INTEGRAL OF THE SECOND KIND    VARIABLE = 0.500


            CHEBYSHEV SUMMATION COEFFICIENTS      25 TERMS


  K                FOR T(K)                    FOR POWER SERIES

  0       7.754769782109207D-01              7.755185479351108D-01
  1      -1.127215859279776D-02             -1.561771392895180D-02
  2      -3.094607697473494D-05             -1.572579979624122D-04
  3       1.372703671884614D-03              6.419360495650850D-03
  4       9.651883749951515D-06              1.292597990521149D-04
  5      -4.456074444881607D-05             -7.884427824709124D-04
  6      -9.263575683271517D-07             -4.241993183866197D-05
  7       6.573165125112188D-07              4.446464089291167D-05
  8       4.413596487395920D-08              7.402339175159864D-06
  9      -4.435757899878822D-09             -9.559832616901068D-07
 10      -1.247839566120845D-09             -7.786132340470428D-07
 11      -4.557613064726239D-11             -8.541833039998886D-08
 12       2.192443226967100D-11              4.395447604110813D-08
 13       2.818482114363973D-12             -1.797106961021201D-09
 14      -1.726319087680394D-13              1.420401531504466D-08
 15      -7.679412661332205D-14              3.343306161696091D-08
 16      -4.310996004619482D-15             -2.916273388109402D-08
 17       1.887379141862766D-15             -4.233283107168971D-08
 18      -1.409983241273948D-16              3.577122697606681D-08
 19       6.494804694057164D-16              3.198336344212293D-08
 20      -4.740652315149417D-16             -2.728207618929445D-08
 21       8.115730310009893D-16             -1.366133801639080D-08
 22      -6.217248937900875D-16              1.172651536762714D-08
 23       6.017408793468347D-16              2.523884177207946D-09
 24      -2.588901315547786D-16             -2.171727828681468D-09
```

Fig. 12.5-6 Chebyshev summation coefficients for the complete elliptic integral of the second kind. Note that the coefficients are small and decreasing.

ERROR TABLE

X	ERROR	X	ERROR
-1.00	1.17D-14	0.0	2.78D-17
-0.975	-2.11D-15	0.025	1.53D-16
-0.950	1.57D-15	0.050	4.44D-16
-0.925	-5.00D-16	0.075	6.66D-16
-0.900	-1.64D-15	0.100	8.60D-16
-0.875	-4.16D-16	0.125	1.01D-15
-0.850	1.10D-15	0.150	1.19D-15
-0.825	1.35D-15	0.175	1.29D-15
-0.800	3.61D-16	0.200	1.44D-15
-0.775	-7.22D-16	0.225	1.48D-15
-0.750	-1.10D-15	0.250	1.50D-15
-0.725	-5.41D-16	0.275	1.36D-15
-0.700	5.69D-16	0.300	1.17D-15
-0.675	1.48D-15	0.325	9.30D-16
-0.650	2.00D-15	0.350	4.72D-16
-0.625	2.04D-15	0.375	-2.78D-17
-0.600	1.60D-15	0.400	-3.61D-16
-0.575	1.11D-15	0.425	-7.36D-16
-0.550	7.36D-16	0.450	-9.71D-16
-0.525	6.25D-16	0.475	-9.71D-16
-0.500	5.69D-16	0.500	-8.19D-16
-0.475	6.25D-16	0.525	-5.97D-16
-0.450	6.66D-16	0.550	-3.47D-16
-0.425	5.41D-16	0.575	-5.55D-17
-0.400	2.36D-16	0.600	2.91D-16
-0.375	-2.78D-17	0.625	6.38D-16
-0.350	-2.78D-16	0.650	1.01D-15
-0.325	-4.72D-16	0.675	1.35D-15
-0.300	-4.58D-16	0.700	1.40D-15
-0.275	-1.80D-16	0.725	1.39D-15
-0.250	4.16D-17	0.750	9.85D-16
-0.225	3.33D-16	0.775	3.89D-16
-0.200	5.97D-16	0.800	-9.71D-17
-0.175	7.22D-16	0.825	-2.91D-16
-0.150	6.25D-16	0.850	-1.39D-17
-0.125	4.02D-16	0.875	5.55D-16
-0.100	2.50D-16	0.900	9.71D-16
-0.075	1.39D-17	0.925	9.30D-16
-0.050	-8.33D-17	0.950	8.33D-16
-0.025	-6.94D-17	0.975	1.44D-15
0.0	2.78D-17	1.000	-1.36D-15

Fig. 12.5-7 Error table for the complete elliptic integral of the second kind. Note the excellent agreement.

COMPLETE ELLIPTIC INTEGRAL OF THE SECOND KIND

PARAMETER	CALC. VALUE	TRUE VALUE	ERROR
0.05000	1.550973352D 00	1.550973352D 00	0.0
0.10000	1.530757637D 00	1.530757637D 00	0.0
0.15000	1.510121832D 00	1.510121831D 00	-1.09D-09
0.20000	1.489035058D 00	1.489035058D 00	0.0
0.25000	1.467462209D 00	1.467462209D 00	0.0
0.30000	1.445363064D 00	1.445363064D 00	0.0
0.35000	1.422691133D 00	1.422691133D 00	0.0
0.40000	1.399392139D 00	1.399392139D 00	0.0
0.45000	1.375401972D 00	1.375401972D 00	0.0
0.50000	1.350643881D 00	1.350643881D 00	0.0

Fig. 12.5-8 Output of the Chebyshev-summation integration procedure for the complete elliptic integral of the second kind.

with the integrand. The rule-of-thumb estimate of the error was 1.7×10^{-16}. The published values were given correct to 10^{-9}. It might therefore be expected that the agreement would be perfect, or almost perfect.

The excellent results are shown in Figure 12.5-8.

PROBLEM SECTION 12.5

Use the problems for Section 12.3, but write a program involving either the Chebyshev summation or (much more difficult) a Hastings approximation. In the latter case, take the approximation through several cycles until an approximately equi-ripple effect is obtained.

Chapter 13

First-Order Ordinary
Differential Equations

In an abbreviated course, omit Section 9.1.

All computer programs are subject to at least five categories of error: input, conversion, truncation, roundoff and procedure. The saving grace is that we are often able to estimate the magnitude of the error and the relative error. In the case of the numerical solution of differential equations, there is, unfortunately, no really satisfactory method for this estimate. There are "rules of thumb" which we shall examine which work "almost always"; but one never knows, in a practical situation, whether or not they are right.

Under the circumstances, it seems best to avoid or reduce the computer procedure if at all possible, and also if possible to do the same problem by at least two different methods with different error characteristics.

Before proceeding to our discussion of computer methods, it is important to discuss the error (or relative error) due to input error and to the procedure. We define *conditioning* and *stability* as follows.

CONDITIONING   A problem is ill conditioned (relatively ill conditioned) if an input error (relative error) is increased in magnitude by the procedure. If it is not ill conditioned, it is well conditioned.

STABILITY   A step-by-step algorithm with interval $h > 0$ is unstable (relatively unstable) if the error (relative error) increases as h decreases. If it is not unstable, it is stable.

Note that under this definition, Simpson's rule for integration is unstable.

Conditioning

The following examples are intended to clarify the concept.

I.   Given: $y' = ay$, $y(0) = C$.
Then, by elementary methods,

$$y = Ce^{ax} \tag{13.1}$$

Suppose there is an error ΔC in the input C. Then the error in y is given by

$$\Delta y = e^{ax} \Delta C \tag{13.2}$$

If $ax > 0$, then Δy is magnified as follows:

ax	0.1	0.7	1.1	1.6	2.3	3.9	4.6
Factor	. 1.1	2	3	5	10	50	100

Therefore the problem is *absolutely ill-conditioned* for $ax > 0$. In the same problem,

$$\frac{\Delta y}{y} = \frac{e^{ax} \Delta C}{y} = \frac{e^{ax} \Delta C}{Ce^{ax}} = \frac{\Delta C}{C} \tag{13.3}$$

Assuming that $\Delta C \ll C$, the problem is *relatively well conditioned* for any ax.

In the same problem, suppose $ax < 0$. Then Δy in (13.2) is multiplied as follows:

ax	-0.7	-1.6	-2.3	-4.6	-6.9
Factor	0.5	0.2	0.1	0.01	0.001

Then the problem is *absolutely well-conditioned* for $ax < 0$. We have already pointed out that it is relatively well-conditioned for any value of ax.

II. Given: $y' = ay - ax + 1$, $y(0) = C$.
The method of solution is elementary. First, the homogeneous equation

$$y' - ay = 0 \tag{13.4}$$

is solved, leading to the complementary solution

$$y_c = Ae^{ax} \tag{13.5}$$

where A is an essential arbitrary constant (eac). Then, a

particular solution of the original is sought, leading to

$$y_p = x$$

Using the initial condition, we obtain

$$y = Ce^{ax} + x \tag{13.6}$$

consisting of a complementary solution plus a particular solution.

An error in input leads to

$$\Delta y = e^{ax}\Delta C \tag{13.7}$$

with results as in Example I. The relative error is

$$\frac{\Delta y}{y} = \frac{e^{ax}\Delta c}{Ce^{ax} + x} \tag{13.8}$$

If $y(0)$ is large and $ax > 0$, the Ce^{ax} *dominates* x in the denominator and the result is very much like (13.3). However, as $y(0)$ approaches 0, (13.8) is like

$$\frac{\Delta y}{y} \sim \frac{e^{ax}\Delta C}{x} \tag{13.9}$$

and small errors in $y(0)$ may very well cause large relative errors, depending upon the ratio e^{ax}/x.

III. Given: $y'' - k^2 y = 0$, $y(0)$, $y'(0)$.

Then, by elementary means,

$$y = A_1 e^{kx} + A_2 e^{-kx} \tag{13.10}$$

where

$$A_1 = \frac{1}{2}\left[y(0) + \frac{1}{k}y'(0) \right] \tag{13.11}$$

and

$$A_2 = \frac{1}{2}\left[y(0) - \frac{1}{k}y'(0) \right] \tag{13.12}$$

The following table is intended to give a general idea of the relationship between e^{kx} and e^{-kx} for various values of kx:

kx	0	0.1	0.5	1	5	10
e^{kx}	1	1.1	1.6	2.7	148.	22000.
e^{-kx}	1	0.9	0.6	0.4	0.007	0.00005

It is clear that for "reasonable" values of A_1 and A_2, the first term of (13.10) will dominate the second when $kx \gg 0$ and vice versa if $kx \ll 0$. Consider the case when $kx \gg 0$. Then

$$y \sim A_1 e^{kx} \tag{13.13}$$

and an error in $y(0)$ or $y'(0)$ leads to

$$\Delta y = e^{kx} \Delta A_1 \tag{13.14}$$

$$\frac{\Delta y}{y} = \frac{e^{kx} \Delta A_1}{e^{kx} A_1} = \frac{\Delta A_1}{A_1} \tag{13.15}$$

showing that for large kx, the problem is absolutely ill-conditioned but relatively well-conditioned. If $kx \ll 0$,

$$y \sim A_2 e^{-kx} \tag{13.16}$$

and we reach the same conclusion.

There is a further complication if $|A_1/A_2|$ is small, i.e., if

$$\frac{A_1}{A_2} = \frac{y_0 + (1/k)y'(0)}{y_0 - (1/k)y'(0)} = \frac{ky_0 + y'(0)}{ky_0 - y'(0)} \tag{13.17}$$

is close to 0. This is so if $y'(0) \sim -ky_0$. Then, regardless of the sign or size of kx, $y \sim A_2 e^{-kx}$, which may or may not resemble the physical situation. If it does not, the problem is said to be *physically ill-conditioned*. We cannot go into the detail or remedy at this time. For a detailed discussion to illuminate this rather nasty problem, read Fox and Mayers, pp. 194 ff.

We offer one more example to show that things are not always so horrible.

IV.   Given: $y'' + k^2 y = 0$, $y(0)$, $y'(0)$.

Then the solution is of the form

$$y = A_1 \cos kx + A_2 \sin kx \tag{13.18}$$

which is well-conditioned. Neither part of the solution dominates the other.

Stability

Consider the equation $y' = f(x,y)$ with solution $y = g(x)$. We shall explain an algorithm which operates step by step; for our present purposes, suffice it to say that the new value of y, namely y_{k+1}, is produced from the preceding value of y, namely y_k, plus some function of previous values of

y'. We indicate this by

$$y_{k+1} = y_k + \varphi(f_k, f_{k-1}, f_{k-2}, \dots) \qquad (13.19)$$

and, of course, we shall explain in detail in the body of the chapter. At any rate, the algorithm, however it works, produces an approximate solution which we indicate as $\tilde{y}$. The true value is y. The error is

$$\epsilon = y - \tilde{y} \qquad (13.20)$$

and the derivative with respect to x is

$$\epsilon' = y' - \tilde{y}' = f(x,y) - f(x,\tilde{y}) \qquad (13.21)$$

Using the mean-value theorem,

$$\epsilon' = (y - \tilde{y}) \frac{\partial f(x,\theta)}{\partial y}, \qquad \theta \in (y,\tilde{y}) \qquad (13.22)$$

$$\therefore \quad \epsilon' = \epsilon f_y(x,\theta) \qquad (13.23)$$

for which the solution is

$$\epsilon = A \exp\left(\int f_y(x,\theta)\,dx \right) \qquad (13.24)$$

where A is a constant.

Suppose $f_y < 0$. To make it definite, let $f_y = -1$. Then

$$\epsilon = Ae^{-x} \qquad (13.25)$$

As x increases, $|\epsilon|$ decreases. Suppose $f_y > 0$. In particular, suppose $f_y = +1$. Then

$$\epsilon = Ae^x \qquad (13.26)$$

and as x increases, $|\epsilon|$ increases. In either case, a great deal depends upon the size of A, the size of x and the specific algorithm used. If the algorithm is inefficient, the error produced in each value of y_k may produce a larger error (or relative error) in y_{k+1}, and the process is then said to be *unstable*. Many of the early methods, e.g. the Euler method and the Milne method for solving differential equations, proved to be unstable. The Adams method, which we shall discuss, is known to be relatively stable. A fine discussion of stability can be found in Chapter 19 of Scheid.

In this chapter, we shall introduce the computer procedure by a very brief discussion of the graphical method of isoclines, followed by the Runge-Kutta and Adams-Moulton algorithms for initial-value problems, probably the best combination for general use at the time of publication of this book.

13.1 GRAPHICAL APPROXIMATIONS FOR FIRST-ORDER DIFFERENTIAL EQUATIONS

13.1.1 Introduction

There are three reasons usually stated for making graphical approximation to the set of solutions of a first-order differential equation. *First*, the approximation may yield a graph satisfying the desired conditions and sufficiently accurate for the purpose. If the graph is good enough and it is desired to transform it into a formula, this may be done by a Fourier series or a Chebyshev-Hastings procedure, as explained previously. *Second*, the graph may point to methods of analytic approximation, or to regions where problems (such as singularities) may occur. *Third*, the graph may furnish a gross check on the solution reached by other methods.

The technique for first-order differential equations is demonstrated briefly.

13.1.2 First Order, First Degree

Consider the differential equation

$$y' = \frac{(x+1)y}{x} \tag{13.1.1}$$

We know that there is an infinity of solutions depending upon the given conditions such as *initial conditions* or *boundary values*. We solve for y, using $p = y'$:

$$y = \frac{xp}{x+1} \tag{13.1.2}$$

Substituting $p = -2.0, -1.5, -1.0, \ldots, +1.5, +2.0$ in (13.1.2), we obtain a *table of isoclines*. The word "isocline" means "equal slope." A small portion of the table is as follows:

$x \backslash p$	-2.00	-1.50	-1.00	-0.50	0.00	0.50	1.00	1.50	2.00
2	-1.33	-1.00	-0.67	-0.33	0.00	0.33	0.67	1.00	1.33
4	-1.60	-1.20	-0.80	-0.40	0.00	0.40	0.80	1.20	1.60
6	-1.71	-1.29	-0.86	-0.43	0.00	0.43	0.86	1.29	1.71
8	-1.78	-1.33	-0.89	-0.44	0.00	0.44	0.89	1.33	1.78
10	-1.82	-1.36	-0.91	-0.45	0.00	0.45	0.91	1.36	1.82

The graph is started, as shown in Figure 13.1-1, by finding the points in a single *column* of the table of isoclines. The numbers in the body of the table are the y-coordinates. For example, when $p = 2$, we have $(x, y) = (2, 1.33)$. At each value a short line segment of slope 2 is drawn. This is easy by using either a pair of transparent triangles or parallel rulers. On the

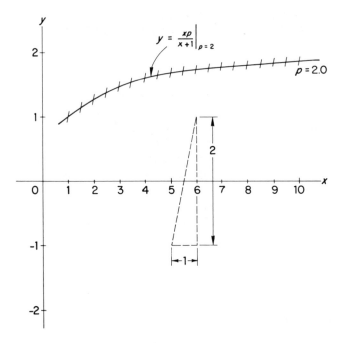

Fig. 13.1-1   Beginning the graph. This is the column for $p = 2$, $x = 1$ to 10.

dashed curve in Figure 13.1-1, every solution of the differential equation passes through with slope 2.

In Figure 13.1-2, the process has been continued for other columns of the table of isoclines. Now, if a solution of the differential equation through $(3.5, 0.6)$ is desired, it can be sketched as shown in the figure. Note that the graphical solution has the proper slope as it crosses the isocline curves.

13.1.3   First Order, Second Degree

The equation

$$y\left(\frac{dy}{dx}\right)^2 - 2x\frac{dy}{dx} + y = 0 \qquad (13.1.3)$$

can be written as

$$yp^2 - 2xp + y = 0 \qquad (13.1.4)$$

Solving for p,

$$p = \frac{2x \pm \sqrt{4x^2 - 4y^2}}{2y} \qquad (13.1.5)$$

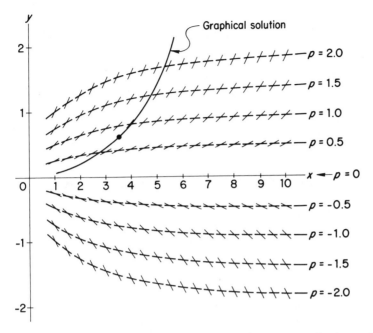

Fig. 13.1-2 Graphical solution of $y'=(x+1)y/x$. The particular solution through (3.5, 0.6) is shown.

leading to two first-order, first-degree equations:

$$p = \frac{x + \sqrt{x^2 - y^2}}{y} \qquad (13.1.6)$$

$$p = \frac{x - \sqrt{x^2 - y^2}}{y} \qquad (13.1.7)$$

each of which can be graphed as shown in Section 13.1.2.

PROBLEM SECTION 13.1

For practice, graph each of the following and compare a graphical solution with the exact solution.

A. First order, linear:

(1) $xyp = x^2 - y^2$. Find the graphical solution through $(1,2)$. The exact solution is $2x^2y - x^4 = 3$.

(2) $y + x(1 - x^2y^4)p = 0$. Find the graphical solution through $(2,1)$. The exact solution is $1 + x^2y^4 = 5x^2y^2/4$.

(3) $y - xp = 2(y^2 + p)$. Find the graphical solution through $(3, -1)$. The exact solution is $(2 + x)(2y - 1) = 15y$.

(4) $y(3x^2 + y) + x(x^2 - y)p = 0$. Find the graphical solution through $(-1, 3)$. The exact solution is $y(5x^2 - y)^4 = -48x^5$.

B. Non-linear:

(5) $4y^2 = p^2x^2$. Find the graphical solutions passing through $(1,3)$. The exact solutions are $y = 3x^2$ and $x^2y = 3$.

(6) $2x^2yp^2 - (2y - x^2)p - 1 = 0$. Find the graphical solutions passing through $(1,1)$. The exact solutions are $x + y^2 = 2$ and $xy + 1 = x$.

(7) $y^2p^2 + xyp - 2x^2 = 0$. Find the graphical solutions passing through $(0, -1)$. The exact solutions are $x^2 - y^2 = 1$, $2x^2 + y^2 = 1$.

(8) $p^2 + y^2 = 1$. Find the graphical solutions passing through $(1.05, 0.5)$. The exact solution is $y = \cos x$.

13.2 THE RUNGE-KUTTA ALGORITHM (FIRST ORDER)

13.2.1 Introduction

We are given a first-order differential equation

$$y' = f(x,y) \tag{13.2.1}$$

together with a value of y' at one point, (x_n, y_n):

$$y'_n = f(x_n, y_n) = f_n \tag{13.2.2}$$

Our immediate problem is to find the value of y at x_{n+1}, where

$$x_{n+1} = x_n + h \tag{13.2.3}$$

The first thought of a mathematician is to use a Taylor series:

$$y_{n+1} = y_n + hy'_n + \frac{h^2}{2!}y''_n + \cdots + \frac{h^m}{m!}y_n^{(m)} + R \tag{13.2.4}$$

where

$$R = \frac{h^{m+1}}{(m+1)!}y^{(m+1)}(\theta), \qquad \theta \in (x_n, x_{n+1}). \tag{13.2.5}$$

This procedure could be acceptable in a problem such as

$$y' = \sin x \tag{13.2.6}$$

since $y'', y''', \ldots, y^{(m)}$ are easily found. However, there are relatively few problems of this type which require computer solution. If the problem is slightly more complicated:

$$y' = \frac{-xy}{\sqrt{1 + x^2}} \tag{13.2.7}$$

the Taylor series becomes impractical. Furthermore, even if the derivatives are found, the number of approximations is so large that the roundoff error becomes unacceptable.

A first approximation to the Taylor series (13.2.4) would be

$$\tilde{y}_{n+1} = y_n + hy'_n = y_n + k_1 \qquad (13.2.8)$$

where $\tilde{y}_{n+1}$ is the extrapolated value of y_{n+1}. This situation is shown in Figure 13.2-1. It is apparent that (13.2.8) would bring about a gross error.

The Runge-Kutta algorithms depend upon the assumption that there exist values of a_i and k_i such that

$$\tilde{y}_{n+1} = y_n + a_1 k_1 + a_2 k_2 + \cdots + a_m k_m \qquad (13.2.9)$$

will yield an acceptable approximation and, furthermore, that values of the k_i can be found recursively from previous values of k_i starting with $k_1 = hy'_n$. (We shall explain and illustrate.) Then

$$k_2 = hf(x + b_2 h, y + c_2 k_1)$$
$$k_3 = hf(x + b_3 h, y + c_3 k_2),$$
$$\vdots \qquad (13.2.10)$$

The problem becomes that of finding suitable values of a_i and k_i in (13.2.9) using suitable values of b_i and c_i in (13.2.10) so that (13.2.9) will agree with the Taylor series (13.2.4) up to some specified number of terms without requiring any differentiation.

This sounds like an impossible task, but in fact there are only two difficulties: (1) the algebra is tedious, and (2) there are infinitely many

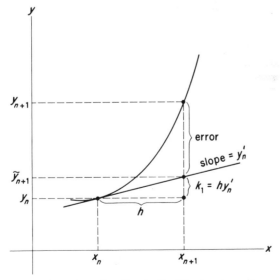

Fig. 13.2-1   A first approximation to the Maclaurin-series solution.

solutions. A new problem appears: that of selecting one that is suitable in terms of error and programming ease.

In the development, we shall need the two theorems presented in Sections 13.2.2 and 13.2.3.

13.2.2 Taylor Series for Many Variables

In Section 4.2.4 we proved the mean-value theorem for many variables. This can be written in the form

$$f(\mathbf{X}+\mathbf{H}) = f(\mathbf{X}) + (\mathbf{H}\cdot\nabla)f(\mathbf{X}_\theta) \tag{13.2.11}$$

where $\mathbf{X}$, $\mathbf{H}$ and ∇ are vectors defined by

$$\mathbf{X} = (x,y,z,\dots)$$
$$\mathbf{H} = (\Delta x,\Delta y,\Delta z,\dots) = (h,k,l,\dots)$$
$$\nabla = (\partial/\partial x,\partial/\partial y,\partial/\partial z,\dots)$$

and $\mathbf{X}_\theta$ marks a point on the segment joining $\mathbf{X}$ to $\mathbf{X}+\mathbf{H}$. Taylor's theorem for many variables is a generalization of the mean-value theorem, as follows:

TAYLOR'S THEOREM FOR MANY VARIABLES If f and all its partial derivatives are continuous through order n in neighborhood N, then

$$f(\mathbf{X}+\mathbf{H}) = f(\mathbf{X}) + (\mathbf{H}\cdot\nabla)f(\mathbf{X}) + \frac{1}{2!}(\mathbf{H}\cdot\nabla)^2 f(\mathbf{X})$$

$$+ \frac{1}{3!}(\mathbf{H}\cdot\nabla)f(\mathbf{X}) + \cdots + E_n \tag{13.2.12}$$

where

$$E_n = \frac{1}{n!}(\mathbf{H}\cdot\nabla)^n f(\mathbf{X}) \tag{13.2.13}$$

Proof As in Chapter 4, we define

$$\mathbf{R} = (x+th,y+th,z+th,\dots) = \mathbf{X}+t\mathbf{H} \tag{13.2.14}$$
$$g(t) = f(\mathbf{X}+t\mathbf{H}), \qquad t\in[0,1] \tag{13.2.15}$$

Then, by Taylor's theorem for one variable, namely t,

$$g(t) = g(0) + tg'(0) + \frac{1}{2!}t^2 g''(0) + \cdots + E_n \tag{13.2.16}$$

where

$$E_n = \frac{1}{n!} t^n g^{(n)}(\theta), \qquad \theta \in [0,1] \tag{13.2.17}$$

Setting $t=1$ and making substitutions as in Chapter 4, the theorem follows directly. □

Consider the calculation of $(\mathbf{H} \cdot \nabla)^n$, which is $(h\partial/\partial x + k\partial/\partial y + l\partial/\partial z + \cdots)^n$. If $(h,k,l,\ldots)$ is a constant vector, the expansion can be accomplished formally, treating the partials as a polynomial, as shown in the example which follows. The vertical line denotes, as usual, the value of the vector at which the function is evaluated.

ILLUSTRATIVE PROBLEM 13.2-1   Find $(\mathbf{H} \cdot \nabla)^3 f(x,y)|_{(1,\pi/2)}$, where $f(x,y)$ is $x^2 \sin y$ and $\mathbf{H} = (1,2)$.

Solution   Formally, we proceed as follows:

$$(\mathbf{H} \cdot \nabla)^3 = \left(h\frac{\partial}{\partial x} + k\frac{\partial}{\partial y} \right)^3$$

$$= h^3 \frac{\partial^3}{\partial x^3} + 3h^2 k \frac{\partial^3}{\partial x^2 \partial y} + 3hk^2 \frac{\partial^3}{\partial x \partial y^2} + k^3 \frac{\partial^3}{\partial y^3}$$

from which

$$(\mathbf{H} \cdot \nabla)^3 f = h^3 f_{xxx} + 3h^2 k f_{xxy} + 3hk^2 f_{xyy} + k^3 f_{yyy}$$

$$f_{xxx} = \frac{\partial^3}{\partial x^3}(x^2 \sin y) = \frac{\partial^2}{\partial x^2}(2x \sin y) = \frac{\partial}{\partial x}(2 \sin y) = 0$$

$$f_{xxy} = \frac{\partial^3}{\partial x^2 \partial y}(x^2 \sin y) = \frac{\partial^2}{\partial x^2}(x^2 \cos y) = \frac{\partial}{\partial x}(2x \cos y) = 2 \cos y$$

$$f_{xyy} = \frac{\partial^3}{\partial x \partial y^2}(x^2 \sin y) = \frac{\partial^2}{\partial x \partial y}(x^2 \cos y) = \frac{\partial}{\partial x}(-x^2 \sin y) = -2x \sin y$$

$$f_{yyy} = \frac{\partial^3}{\partial y^3}(x^2 \sin y) = \frac{\partial^2}{\partial y^2}(x^2 \cos y) = \frac{\partial}{\partial y}(-x^2 \sin y) = -x^2 \cos y$$

Substituting $x=1$, $y=\pi/2$, $h=1$ and $k=2$ from the given conditions, and using the above partial derivatives,

$$(\mathbf{H} \cdot \nabla)f = 3h^2 k(2 \cos y) + 3hk^2(-2x \sin y) + k^3(-x^2 \cos y)$$

$$= -24$$

ILLUSTRATIVE PROBLEM 13.2-2   Write Taylor's theorem for two variables to five sets of terms and indicate the error of truncation.

Solution   In the following, unless otherwise stated, f is evaluated at (x,y):

$$f(x+h, y+k) = f(x,y) + \left(h\frac{\partial}{\partial x} + k\frac{\partial}{\partial y} \right)f$$

$$+ \frac{1}{2!}\left(h\frac{\partial}{\partial x} + k\frac{\partial}{\partial y} \right)^2 f + \frac{1}{3!}\left(h\frac{\partial}{\partial x} + k\frac{\partial}{\partial y} \right)^3 f + \frac{1}{4!}\left(h\frac{\partial}{\partial x} + k\frac{\partial}{\partial y} \right)^4 f + E_4$$

This can be written as

$$f(x+h,y+k)=f+(hf_x+kf_y)+\frac{1}{2!}\left(h^2f_{xx}+2hkf_{xy}+k^2f_{yy}\right)$$
$$+\frac{1}{3!}\left(h^3f_{xxx}+3h^2f_{xxy}+3hk^2f_{xyy}+k^3f_{yyy}\right)$$
$$+\frac{1}{4!}\left(h^4f_{xxxx}+4h^3kf_{xxxy}+6h^2k^2f_{xxyy}+4hk^3f_{xyyy}+k^4f_{yyyy}\right)+E_4 \quad (13.2.28)$$

where

$$E_4=\frac{1}{5!}\left(h\frac{\partial}{\partial x}+k\frac{\partial}{\partial y}\right)^5 f(x+th,y+tk), \qquad t\in(0,1) \quad (13.2.19)$$

13.2.3 The P-Theorem

We are going to need to express the one-variable Taylor series (13.2.8) in terms of partial derivatives in order to compare it to the Runge-Kutta formulas. Assuming that the partial derivatives exist and are continuous,

$$y'=\frac{dy}{dx}=f(x,y) \quad (13.2.20)$$

$$dy'=\frac{\partial f}{\partial x}\,dx+\frac{\partial f}{\partial y}\,dy \quad (13.2.21)$$

$$\therefore\quad y''=\frac{dy'}{dx}=\frac{\partial f}{\partial x}+\frac{\partial f}{\partial y}\frac{dy}{dx} \quad (13.2.22)$$

$$y''=\left(\frac{\partial}{\partial x}+f\frac{\partial}{\partial y}\right)f=Pf \quad (13.2.23)$$

where P is defined as the operator in parentheses. Then

$$dy''=\frac{\partial y''}{\partial x}\,dx+\frac{\partial y''}{\partial y}\,dy \quad (13.2.24)$$

$$\frac{dy''}{dx}=\frac{\partial y''}{\partial x}+\frac{\partial y''}{\partial y}\frac{dy}{dx} \quad (13.2.25)$$

$$y'''=\left(\frac{\partial}{\partial x}+f\frac{\partial}{\partial y}\right)y'' \quad (13.2.26)$$

Using (13.2.23), we have

$$y'''=\left(\frac{\partial}{\partial x}+f\frac{\partial}{\partial y}\right)\left(\frac{\partial}{\partial x}+f\frac{\partial}{\partial y}\right)f \quad (13.2.27)$$

which we can represent as

$$y'''=P^2f \quad (13.2.28)$$

This is easily generalized by mathematical induction.

THE P-THEOREM   If $y' = f(x,y)$ has n partial derivatives continuous in the region of interest, then

$$y^{(n)} = P^{n-1}f \qquad (13.2.29)$$

where P is defined as the operator

$$P = \left(\frac{\partial}{\partial x} + f\frac{\partial}{\partial y} \right) \qquad (13.2.30)$$

The illustrative problem which follows shows that P^2, P^3,... *cannot* be expanded formally.

ILLUSTRATIVE PROBLEM 13.2-3   Express y''' in terms of its partials.

Solution

$$
\begin{aligned}
y''' = P^2 f &= \left(\frac{\partial}{\partial x} + f\frac{\partial}{\partial y} \right)\left(\frac{\partial}{\partial x} + f\frac{\partial}{\partial y} \right) f \\
&= \left[\frac{\partial^2}{\partial x^2} + \frac{\partial}{\partial x}\left(f\frac{\partial}{\partial y} \right) + f\frac{\partial^2}{\partial x\,\partial y} + f\frac{\partial}{\partial y}\left(f\frac{\partial}{\partial y} \right) \right] f \\
&= f_{xx} + ff_{xy} + f_x f_y + ff_{xy} + f\left[ff_{yy} + f_y^2 \right] \\
&= \left(f_{xx} + 2ff_{xy} + f^2 f_{yy} \right) + f_y\left(f_x + ff_y \right) \qquad (13.2.31)
\end{aligned}
$$

In the same way (left as an exercise),

$$
\begin{aligned}
y^{iv} = &\left(f_{xxx} + 3ff_{xxy} + 3f^2 f_{xyy} + f^3 f_{yyy} \right) + f_y\left(f_{xx} + 2ff_{xy} + f^2 f_{yy} \right) \\
&+ 3(f_x + ff_y)(f_{xy} + ff_{yy}) + f_y^2 (f_x + ff_y) \qquad (13.2.32)
\end{aligned}
$$

It is rather a misfortune that the results are so unwieldy. We define the following abbreviations:

$$Q = f_x + ff_y \qquad (13.2.33)$$

$$Q_2 = f_{xx} + 2ff_{xy} + f^2 f_{yy} \qquad (13.2.34)$$

$$Q_3 = f_{xxx} + 3ff_{xxy} + 3f^2 f_{xyy} + f^3 f_{yyy} \qquad (13.2.35)$$

$$Q_4 = f_{xxxx} + 4ff_{xxxy} + 6f^2 f_{xxyy} + 4f^3 f_{xyyy} + f^4 f_{yyyy} \qquad (13.2.36)$$

The subscripts on the Qs serve to point out that the coefficients are the same as those of a binomial expansion. Then we can abbreviate the derivatives as follows:

$$y' = f \tag{13.2.37}$$

$$y'' = Q \tag{13.2.38}$$

$$y''' = Q_2 + f_y Q \tag{13.2.39}$$

$$y^{\text{iv}} = Q_3 + f_y Q_2 + f_y^2 Q + 3(f_{xy} + ff_{yy})Q \tag{13.2.40}$$

and (13.2.8) can be rewritten as

$$y_{n+1} = y_n + hf + \frac{1}{2!}h^2 Q + \frac{1}{3!}h^3(Q_2 + f_y Q) + \frac{1}{4!}\left[Q_3 + f_y Q_2 + f_y^2 Q \right.$$
$$\left. + 3(f_{xy} + ff_{yy})Q \right] + O(h^5) \tag{13.2.41}$$

13.2.4   The Runge-Kutta Algorithm up to h^2

In order to clarify the slightly more general discussion which follows, we shall attempt to discover a Runge-Kutta algorithm of the form

$$y_{n+1} = y_n + a_1 k_1 + a_2 k_2 + a_3 k_3 + \cdots \tag{13.2.42}$$

where

$$k_1 = hf_n \tag{13.2.43}$$

$$k_2 = hf(x_n + b_2 h, y_n + b_2 k_1) \tag{13.2.44}$$

$$k_3 = hf(x_n + b_3 h, y_n + b_3 k_2) \tag{13.2.45}$$

$$\vdots$$

This has to agree with the Taylor series (13.2.41) up to the h^2 term. Expanding k_2 [from (13.2.11)] by Taylor's theorem for two variables (13.2.12), we have

$$k_2 = h\left[f_n + \left(b_2 h \frac{\partial}{\partial x} + b_2 k_1 \frac{\partial}{\partial y} \right) f_n + \frac{1}{2!}\left(b_2 h \frac{\partial}{\partial x} + b_2 k_1 \frac{\partial}{\partial y} \right)^2 f_n + \cdots \right]$$

$$\tag{13.2.46}$$

We note that $k_1 = hf_n$. Substituting, factoring and keeping terms up to h^2, we have

$$k_2 = hf_n + b_2 h^2 \left(\frac{\partial}{\partial x} + f \frac{\partial}{\partial y} \right) f_n \qquad (13.2.47)$$

$$= hf_n + b_2 h^2 P f_n \qquad (13.2.48)$$

$$= hf_n + b_2 y'' \qquad (13.2.49)$$

$$= hf_n + b_2 h^2 Q \qquad (13.2.50)$$

We now have enough information to write algorithms satisfying the conditions. Substituting from (13.2.44) and (13.2.50) into (13.2.42),

$$y_{n+1} \sim y_n + a_1 hf_n + a_2 hf_n + a_2 b_2 h^2 Q \qquad (13.2.51)$$

Comparing (13.2.51) and (13.2.41),

$$\begin{cases} a_1 + a_2 = 1 & (13.2.52) \\ \quad a_2 b_2 = \frac{1}{2} & (13.2.53) \end{cases}$$

There are three unknowns and two equations. The difference between the number of independent equations and the number of unknowns is often called the number of *degrees of freedom* (df) in the system. In this case, there is 1 df. We are free to choose one of the unknowns arbitrarily. For example, if we choose $b_2 = \frac{1}{2}$, then $a_2 = 1$ and $a_1 = 0$, leading to the algorithm

$$\tilde{y}_{n+1} = y_n + f \left(x + \tfrac{1}{2} h, y + \tfrac{1}{2} k_1 \right) \qquad (13.2.54)$$

where $k_1 = hf_n$.

This is not a very good approximation, but it is as good as the Taylor series up to the h^2 term.

13.2.5 The Runge-Kutta Algorithm up to h^4

Using precisely the same method with (13.2.12) and with (13.2.33) to (13.2.36), we obtain, by retaining terms up to and including h^4,

$$k_1 = hf_n \qquad (13.2.55)$$

$$k_2 = hf_n + b_2 h^2 Q + \frac{1}{2!} b_2^2 h^3 Q_2 + \frac{1}{3!} b_2^3 h^4 Q_3 \qquad (13.2.56)$$

$$k_3 = hf_n + b_3 h^2 Q + \frac{1}{2!} h^3 \left(b_3^2 Q_2 + 2 b_2 b_3 f_y Q \right)$$

$$+ \frac{1}{3!} h^4 \left(b_3^3 Q_3 + 3 b_3^2 b_2 f_y Q_2 + 6 b_3 b_2^2 \left[f_{xy} + f f_{yy} \right] Q \right) + \cdots \qquad (13.2.57)$$

$$k_4 = hf_n + b_4 h^2 Q + \frac{1}{2!} \left(b_4 Q_2 + 2 b_3 b_4 f_y Q \right) + \frac{1}{3!} \left(b_4 Q_3 + \cdots \right) + \cdots \qquad (13.2.58)$$

Comparing with (13.2.41), we obtain

$$
\begin{cases}
a_1 + a_2 + a_3 + a_4 = 1 & (13.2.59) \\[4pt]
a_2 b_2 + a_3 b_3 + a_4 b_4 = \tfrac{1}{2} & (13.2.60) \\[4pt]
a_2 (b_2)^2 + a_3 (b_3)^2 + a_4 (b_4)^2 = \tfrac{1}{3} & (13.2.61) \\[4pt]
a_2 (b_2)^3 + a_3 (b_3)^3 + a_4 (b_4)^3 = \tfrac{1}{4} & (13.2.62) \\[4pt]
a_3 b_2 b_3 + a_4 b_3 b_4 = \tfrac{1}{6} & (13.2.63) \\[4pt]
a_3 b_2 (b_3)^2 + a_4 b_3 (b_4)^2 = \tfrac{1}{8} & (13.2.64) \\[4pt]
a_3 (b_2)^2 b_3 + a_4 (b_3)^2 b_4 = \tfrac{1}{12} & (13.2.65) \\[4pt]
a_4 b_2 b_3 b_4 = \dfrac{1}{24} & (13.2.66)
\end{cases}
$$

There is one degree of freedom in this set of eight equations. There is evidently an infinity of choices, all of which agree with a Taylor series up to the h^4 term. Choosing $a_1 = 1$ arbitrarily, the result is

$$\tilde{y}(x_n + h) = y(x_n) + \tfrac{1}{6}(k_1 + 2k_2 + 2k_3 + k_4) \qquad (13.2.67)$$

where

$$k_1 = hf(x_n, y_n) \qquad (13.2.68)$$

$$k_2 = hf\left(x_n + \tfrac{1}{2}h, y_n + \tfrac{1}{2}k_1\right) \qquad (13.2.69)$$

$$k_3 = hf\left(x_n + \tfrac{1}{2}h, y_n + \tfrac{1}{2}k_2\right) \qquad (13.2.70)$$

$$k_4 = hf(x_n + h, y_n + k_3) \qquad (13.2.71)$$

The combination (13.2.67) to (13.2.71) is a very popular version of the Runge-Kutta algorithm. A more detailed discussion can be found in Ralston, pp. 191–200, and in Scheid, pp. 202–204. Fortunately or unfortunately, the real problems leading to differential equations do not seem to demand as extreme precision as other problems in applied mathematics.

13.2.6 Application of the Runge-Kutta Algorithm up to h^4

The procedure which we shall illustrate and explain is quite straightforward. We substitute our starting values of h, x_0 and y_0 in (13.2.68) and find k_1. Using this in (13.2.69), we find k_2, and then [from (13.2.70)] k_3 and [from (13.2.71)] k_4. Substituting in (13.2.67), we have a calculated value for y at $x_0 + h$. The values of x_0 and y_0 must, of course, be given.

An obvious question concerns the value of h to be used, and a very important question concerns the estimate of error. We shall attack these questions in the illustrative problem which follows.

ILLUSTRATIVE PROBLEM 13.2-4   For

$$y' = f(x,y) = -y\cos x \tag{13.2.72}$$

if $y(0) = 1$, find three additional points using the Runge-Kutta algorithm up to h^4. Find the actual error and relative error using the exact solution

$$y = e^{-\sin x} \tag{13.2.73}$$

Solution   We start, arbitrarily, with $h = 0.1$, $x_0 = 0$ and $y_0 = 1$. From (13.2.68),

$$k_1 = 0.1f(0,1) = 0.1(-1\cos 0) = -0.1$$

Using (13.2.69) with $x = 0 + \frac{1}{2}(0,1)$ and $y = 1 + \frac{1}{2}(-0.1)$,

$$k_2 = 0.1f(0.05, 0.95) = 0.1(-0.95\cos 0.05) = -0.09488\,12747$$

Using (13.2.70) with $x = 0 + \frac{1}{2}(0,1), y = 1 + \frac{1}{2}(-0.09488\,12747)$, we have

$$k_3 = 0.1f(0.05, 0.9525593626) = -0.0951368911$$

Using (13.2.71) with $x = 0 + 0.1$, $y = 1 + (-0.0951368911)$, we have

$$k_4 = 0.1f(0.1, 0.9048631089) = -0.0900342562$$

We define

$$k = k_1 + 2k_2 + 2k_3 + k_4 \tag{13.2.74}$$

Then

$$k = -0.57007\,05878$$

Substituting in (13.2.67),

$$y(0.1) = y(0) + \frac{1}{6}k = 0.90498\,82354$$

This is our *first* Runge-Kutta value. To find the second, we return to (13.2.68) with $h = 0.1$, $x = 0.1$, $y = 0.90498\,82354$:

$$k_1 = 0.1f(0.1, 0.90498\,82354) = -0.09004\,67063$$

Using (13.2.69) with $x = 0.1 + \frac{1}{2}(0.1)$, $y = 0.90498\,82354 + \frac{1}{2}(-0.09004\,67063)$, we have

$$k_2 = 0.1f(0.15, 0.8599648822) = -0.08503\,08403$$

Using (13.2.70) with $x = 0.1 + \frac{1}{2}(0.1)$, $y = 0.90498\,82354 + \frac{1}{2}(-0.08503\,08403)$, we have

$$k_3 = 0.1f(0.15, 0.8624728152) = -0.0852788175$$

Using (13.2.71) with $x = 0.1 + 0.1$, $y = 0.90498\ 82354 + (-0.08527\ 88175)$, we have

$$k_4 = 0.1 f(0.2, 0.81970\ 94179) = -0.08033\ 69804$$

Using (13.2.74),

$$k = -0.51100\ 30023$$

Using (13.2.67),

$$y(0.2) = y(0.1) + \tfrac{1}{6}k = 0.81982\ 10683$$

For later work, we need four *starting values*, i.e., the given value plus three Runge-Kutta values. Proceeding as before, we find that

$$y(0.3) = 0.74414\ 45473$$

Since we know the exact solution (13.2.73), we can find the actual error and relative error. These are as follows:

x	|ACTUAL ERROR|	|RELATIVE ERROR|
0.1	7.40×10^{-8}	8.17×10^{-8}
0.2	1.30×10^{-7}	1.59×10^{-7}
0.3	1.69×10^{-7}	2.28×10^{-7}

Let us consider the problem of estimating the error in a more realistic situation where we do not know the exact solution.

13.2.7 Estimating the Error

In actual practice, the first round of three calculations shown above is followed by successive rounds of *six* calculations, each round using the previous value of h. In Illustrative Problem 13.2-4, the second round starts with $x_0 = 0$, $y_0 = 1$, and then continues as follows:

FIRST ROUND	SECOND ROUND
$x = 0, y(0)$	$x = 0, y(0)$
	$x = 0.05, y(0.05) = R_2(1)$
$x = 0.1, y(0.1) = R_1(1)$	$x = 0.1, y(0.1) = R_2(2)$
	$x = 0.15, y(0.15) = R_2(3)$
$x = 0.2, y(0.2) = R_1(2)$	$x = 0.2, y(0.2) = R_2(4)$
	$x = 0.25, y(0.25) = R_2(5)$
$x = 0.3, y(0.3) = R_1(3)$	$x = 0.3, y(0.3) = R_2(6)$

We now compare $R_1(3)$ and $R_2(6)$. Our rule of thumb is as follows.

RULE OF FIFTEEN   The actual error of R is *usually* about

$$\frac{|R_1 - R_2|}{15}$$

There are variations due to closeness to singularities (where a Taylor series cannot apply) and to unusually high roundoff errors.

> *Discussion* Since the Runge-Kutta algorithm is essentially a Taylor series up to h^4, as developed in these pages, the first of every two rounds of calculations can be represented by

$$R_1 = y(x+h) = y(x) + hy'(x) + \cdots + \frac{1}{5!}h^5 y^v(\theta_1) \qquad (13.2.75)$$

To arrive at a corresponding value of y in the second of every two rounds of calculations, we must use two Taylor series, as follows:

$$y\left(x+\frac{h}{2}\right) = y(x) + \left(\frac{h}{2}\right)y'(x) + \cdots + \frac{1}{5!}\left(\frac{h}{2}\right)^5 y^v(\theta_2) \qquad (13.2.76)$$

$$R_2 = y(x+h) = y\left(x+\frac{h}{2}\right) + \left(\frac{h}{2}\right)y'\left(x+\frac{h}{2}\right) + \cdots + \frac{1}{5!}\left(\frac{h}{2}\right)^5 y^v(\theta_3) \qquad (13.2.77)$$

We can abbreviate (13.2.75) and (13.2.77), using T as the truncation error in (13.2.76). Assuming the fifth derivatives are equal,

$$R_1 = y + 32T \qquad (13.2.78)$$
$$R_2 = y + 2T \qquad (13.2.79)$$

Note that there are *two* truncation errors in (13.2.79), one from (13.2.76) and one from (13.2.77). Subtracting,

$$R_1 - R_2 = 30T \qquad (13.2.80)$$

But

$$E = R_2 - y = 2T \qquad (13.2.81)$$
$$\therefore \quad R_1 - R_2 = 15E \qquad (13.2.82)$$

According to this scheme, we may be able to estimate the error by continuing the Runge-Kutta algorithm. This is simple on a computer. It merely requires a redefinition of the R_2 values as R_1 values:

$$R_2(6) \text{ becomes } R_1(3)$$
$$R_2(4) \text{ becomes } R_1(2)$$
$$R_2(2) \text{ becomes } R_1(1)$$

followed by a division of h by 2 and the calculation of *six* values of (new) R_2. For every pair of calculations, the correspondences look like this:

$y(x_0)$	$y(x_0)$
	$y(x_0+h_3)$
$y(x_0+h_2)$	$y(x_0+2h_3)$
	$y(x_0+3h_3)$
$y(x_0+2h_2)$	$y(x_0+4h_3)$
	$y(x_0+5h_3)$
$y(x_0+3h_2)$	$y(x_0+6h_3)$

For the illustrative problem, the results of the operation are shown in Figure 13.2-2. N refers to the iteration, and H refers to the (halved) value for R_2. The computer program which produced these (and other) results is displayed in Figure 13.2-3.

Using the rule of thumb, the error for the last entry, $y(0.01875)$, should be approximately

$$\frac{2.11\times10^{-13}}{15} = 1.41\times10^{-14}$$

It is actually 1.40×10^{-14}. We hasten to state that this degree of agreement is not universal. Table 13.2-1 gives the ratio

$$\left|\frac{R_1-R_2}{\text{Actual error}}\right|$$

for the problems in the problem section.

Table 13.2-1
Application of the Rule of Fifteen

PROBLEM	RATIO
1	15
2	15
3	15
4	15
5	21
6	15
7	15
8	16
9	14
10	0.7

13.2.7 Exceptions to the Rule of Fifteen

Consider

$$y' = -\frac{xy}{\sqrt{1+x^2}} \tag{13.2.83}$$

which has the exact solution

$$y = 5e^{-\sqrt{1+x^2}} \tag{13.2.84}$$

FIRST THREE R-K VALUES FOR Y'=-Y COS X

N	H	X	R1	R2	R1-R2
2	0.0500000000	0.1000000000	9.0498823533325240D-01	9.0498816599150230D-01	6.93D-08
2	0.0500000000	0.2000000000	8.1982106825127590D-01	8.1982094608018680D-01	1.22D-07
2	0.0500000000	0.3000000000	7.4414454726320050D-01	7.4414438835763220D-01	1.59D-07
3	0.0250000000	0.0500000000	9.5124924191667750D-01	9.5124923965900820D-01	2.26D-09
3	0.0250000000	0.1000000000	9.0498816599150230D-01	9.0498816172457360D-01	4.27D-09
3	0.0250000000	0.1500000000	8.6119172259415070D-01	8.6119171657115250D-01	6.02D-09
4	0.0125000000	0.0250000000	9.7531245189847140D-01	9.7531245182660310D-01	7.19D-11
4	0.0125000000	0.0500000000	9.5124923965900820D-01	9.5124923951904930D-01	1.40D-10
4	0.0125000000	0.0750000000	9.2780870245823710D-01	9.2780870225403380D-01	2.04D-10
5	0.0062500000	0.0125000000	9.8757812197098720D-01	9.8757812196872150D-01	2.27D-12
5	0.0062500000	0.0250000000	9.7531245182660310D-01	9.7531245182213000D-01	4.47D-12
5	0.0062500000	0.0375000000	9.6302882745802240D-01	9.6302882739180510D-01	6.62D-12
6	0.0031250000	0.0062500000	9.9376953105997650D-01	9.9376953105990540D-01	7.11D-14
6	0.0031250000	0.0125000000	9.8757812196872150D-01	9.8757812196858020D-01	1.41D-13
6	0.0031250000	0.0187500000	9.8142576595500590D-01	9.8142576595479530D-01	2.11D-13

Fig. 13.2-2 The first three Runge-Kutta values for $y' = -y \cos x$.

```
C SOLUTION OF A FIRST-ORDER DIFFERENTIAL EQUATION BY A RUNGE-KUTTA           040
C FOURTH-ORDER ALGORITHM.  THE FIRST TRIAL VALUE OF H IS APPROXIMATELY       050
C ONE-TENTH OF X(FINAL) - X(INITIAL).  THE SECOND TRIAL USES H/2.  THE       060
C RESULT IS CONSIDERED TO BE CORRECT IF THE CORRESPONDING VALUES OF Y        070
C AGREE WITHIN THE SPECIFIED CRITICAL VALUE.  IF NOT, H IS HALVED AGAIN.     080
C THERE ARE TWO VARIABLE STATEMENT FUNCTION CARDS AFTER 210.                 090
C THERE IS ONE INPUT CARD -                                                  100
C       CC 1-30    DESCRIPTION OF FUNCTION, NAME                             110
C       CC 31-40   INITIAL VALUE OF X IN F-FORMAT, XO                        120
C       CC 41-50   INITIAL VALUE OF Y IN F-FORMAT, YO                        130
C       CC 51-60   STARTING VALUE FOR H, HO, IN F-FORMAT                     140
                                                                            160
      DOUBLE PRECISION R1(6),R2(320) ,A,B,CRIT,E,H,HO,H1,F,P               170
     11,R,S,TK,TK1,TK2,TK3,TK4,X,XO,X1,Y,YO,TV                            180
      DIMENSION NAME(7)                                                    200
C VARIABLE STATEMENT FUNCTIONS.                                            210
      F(X,Y)=-Y*DCOS(X)                                                   VAR21
      S(X)=DEXP(-DSIN(X))                                                 VAR22
C INPUT.                                                                   220
      READ(1,100)NAME,XO,YO,HO                                            230
100   FORMAT(7A4,2X,3F10.0)                                               240
C CONSTANTS.                                                               250
      PI=3.141592653589793D0                                             260
      B=2.718281828459045D0                                              270
      A=0.5D0                                                            280
      CRIT=5.D-17                                                         290
C SET BOX.                                                                350
      H=HO                                                               360
      X=XO-H                                                             370
      Y=YO                                                               380
      N=1                                                                390
C                                                                        391
      WRITE(3,108)NAME                                                   392
108   FORMAT(1H1,T30,'FIRST THREE R-K VALUES FOR ',7A4/)                 393
      WRITE(3,109)                                                       394
109   FORMAT(1H0,T7,'N',T21,'H',T40,'X',T64,'R1',T92,'R2',T111,'R1-R2'/) 395
C FIRST CALCULATION OF THREE R-K VALUES TO TEST H.                       400
      DO 2 K=2,6,2                                                       410
      X=X+H                                                              420
      TK1=H*F(X,Y)                                                       430
      TK2=H*F(X+A*H,Y+A*TK1)                                             440
```

```
      TK3=H*F(X+A*H,Y+A*TK2)                                          450
      TK4=H*F(X+H,Y+TK3)                                              460
      TK=(TK1+2.DO*TK2+2.DO*TK3+TK4)/6.DO                             470
      Y=Y+TK                                                          480
    2 R1(K)=Y                                                         490
C SECOND CALCULATION.                                                 500
    3 H=A*H                                                           510
      X=XO-H                                                          520
      Y=YO                                                            530
      N=N+1                                                           535
      DO 4 K=1,6                                                      540
      X=X+H                                                           550
      TK1=H*F(X,Y)                                                    560
      TK2=H*F(X+A*H,Y+A*TK1)                                          570
      TK3=H*F(X+A*H,Y+A*TK2)                                          580
      TK4=H*F(X+H,Y+TK3)                                              590
      TK=(TK1+2.DO*TK2+2.DO*TK3+TK4)/6.DO                             600
      Y=Y+TK                                                          610
    4 R2(K)=Y                                                         620
C COMPARISON OF R1 AND R2 VALUES.                                     621
      X1=XO                                                           622
      DO 41 K=2,6,2                                                   623
      X1=X1+H*2.DO                                                    624
      E=R1(K)-R2(K)                                                   625
   41 WRITE(3,110)N,H,X1,R1(K),R2(K),E                                626
  110 FORMAT(1H ,T6,I2,T15,F12.10,T33,F15.10,T53,1PD23.15,T81,D23.15,T10  627
     19,D9.2)                                                        628
      WRITE(3,111)                                                    629
  111 FORMAT(1H0)                                                     62A
C COMPARISON OF FOURTH VALUES, NAMELY R1(6) AND R2(6).                630
      IF(DABS(R2(6)-R1(6)).LE.CRIT)GO TO 5                            640
C LIMITING THE NUMBER OF TRIALS.                                      650
      IF(N.EQ.6)GO TO 5                                               660
C STEP, THEN RETURN TO A NEW SECOND CALCULATION.                      670
      R1(2)=R2(1)                                                     690
      R1(4)=R2(2)                                                     700
      R1(6)=R2(3)                                                     710
      GO TO 3                                                         730
C COMPLETED CALCULATION OF R-K VALUES (UP TO 320 VALUES)             740
```

Fig. 13.2-3 (Continued on next page)

```
 5      KF=10*2**(N-1)                                                      750
        DO 6 K=7,KF                                                         780
        X=X+H                                                               790
        TK1=H*F(X,Y)                                                        800
        TK2=H*F(X+A*H,Y+A*TK1)                                              810
        TK3=H*F(X+A*H,Y+A*TK2)                                              820
        TK4=H*F(X+H,Y+TK3)                                                  830
        TK=(TK1+2.DO*TK2+2.DO*TK3+TK4)/6.DO                                 840
        Y=Y+TK                                                              850
 6      R2(K)=Y                                                             860
C OUTPUT. INTERMEDIATE R-K VALUES.                                          870
        L=0                                                                 880
        WRITE(3,101)                                                        890
101     FORMAT(1H1,T38,'TABLE OF RUNGE-KUTTA VALUES'/)                      900
        WRITE(3,102)NAME,H,N                                                910
102     FORMAT(1HO,T5,7A4,      'H USED IN FINAL CALCULATION=',F12.10,5X,   920
       1'N=',I4/)                                                           921
        WRITE(3,103)                                                        930
103     FORMAT(1HO,T3 ,'X',T26,'CALC. Y',T51,'TRUE VALUE',T74,'ERROR',T85,  940
       1'REL. ERROR'/)                                                      950
        X=XO                                                                960
        DO 7 K=1,KF                                                         970
        L=L+1                                                               975
        X=X+H                                                               980
        TV=S(X)                                                             985
        E=TV-R2(K)                                                          990
        IF(DABS(E).LE.5.D-17)GO TO 71                                      1000
        R=E/R2(K)                                                          1010
        GO TO 72                                                           1020
71      E=0.DO                                                             1030
        R=0.DO                                                             1040
72      WRITE(3,104)X,R2(K),TV,E,R                                         1045
104     FORMAT(1H ,T2,F14.12,2X,1PD22.15,5X,D22.15,5X,2(D9.2,5X))          1050
        IF(L.NE.45) GO TO 7                                                1050
        WRITE(3,101)                                                       1051
        WRITE(3,102)                                                       1052
        WRITE(3,103)                                                       1053
        L=1                                                                1054
7       CONTINUE                                                           1055
        END                                                                1056
                                                                           1230
```

Fig. 13.2-3 Computer program for the Runge-Kutta algorithm (up to h^4) for a first-order differential equation.

FIRST THREE R-K VALUES FOR Y'=-XY/SQR(1+X**2)

N	H	X	R1	R2	R1-R2
2	0.0500000000	0.1000000000	1.830245924155814D 00	1.830245937398785D 00	-1.32D-08
2	0.0500000000	0.2000000000	1.803328242437608D 00	1.803328290796064D 00	-4.84D-08
2	0.0500000000	0.3000000000	1.760164384457524D 00	1.760164477960305D 00	-9.35D-08
3	0.0250000000	0.0500000000	1.837100829000066D 00	1.837100829209664D 00	-2.10D-10
3	0.0250000000	0.1000000000	1.830245937398785D 00	1.830245938217939D 00	-8.19D-10
3	0.0250000000	0.1500000000	1.818933777332999D 00	1.818933775100302D 00	-1.77D-09
4	0.0125000000	0.0250000000	1.838822573790048D 00	1.838822573793333D 00	-3.29D-12
4	0.0125000000	0.0500000000	1.837100829209664D 00	1.837100829222730D 00	-1.31D-11
4	0.0125000000	0.0750000000	1.834238403538799D 00	1.834238403567900D 00	-2.91D-11
5	0.0062500000	0.0125000000	1.839255141176215D 00	1.839255141176267D 00	-5.13D-14
5	0.0062500000	0.0250000000	1.838822573793333D 00	1.838822573793538D 00	-2.05D-13
5	0.0062500000	0.0375000000	1.838104788321033D 00	1.838104788321493D 00	-4.60D-13
6	0.0031250000	0.0062500000	1.839361280832192D 00	1.839361280832193D 00	-6.66D-16
6	0.0031250000	0.0125000000	1.839255141176267D 00	1.839255141176270D 00	-2.89D-15
6	0.0031250000	0.0187500000	1.839073931141081D 00	1.839073931141088D 00	-6.88D-15

Fig. 13.2-4 The results of the Runge-Kutta algorithm (up to h^4) for Equation (13.2.83).

We wish to find a solution through $x_0=0$, $y_0=5e^{-1}$. The results of the Runge-Kutta program are shown in Figure 13.2-4.

The Rule of Fifteen predicts that the actual error of the last entry will be in the neighborhood of

$$\frac{6.88\times10^{-15}}{15}=4.59\times10^{-16}$$

The actual error is 1.11×10^{-15}, about 2.5 times as large. Looking back at (13.2.83), we see that $y'(0)$ is 0. Other empirical evidence suggests that this situation may cause the rule of thumb to fail. The same observation holds for Problem (10) at the end of this section.

13.2.8 Extending the Runge-Kutta Algorithm

The results so far seem to be so good that one may wonder why we should not continue the algorithm indefinitely. It is instructive to continue at least to x_0+1 to see what happens to the error. For Illustrative Problem 13.2-4, the errors and relative errors are given in Table 13.2-2.

Table 13.2-2
Errors for $y' = -y\cos x$

x	ACTUAL ERROR	RELATIVE ERROR
0.1	-68.1×10^{-15}	-75.2×10^{-15}
0.2	-120×10^{-15}	-146×10^{-15}
0.3	-157×10^{-15}	-211×10^{-15}
0.4	-179×10^{-15}	-265×10^{-15}
0.5	-191×10^{-15}	-308×10^{-15}
0.6	-194×10^{-15}	-341×10^{-15}
0.7	-191×10^{-15}	-363×10^{-15}
0.8	-184×10^{-15}	-376×10^{-15}
0.9	-175×10^{-15}	-384×10^{-15}
1.0	-166×10^{-15}	-386×10^{-15}

For the following problems the actual error at x_0+1 is shown for those who wish to see what happens when the Runge-Kutta algorithm is extended that far. In the next section, we shall discuss the most common method actually used.

PROBLEM SECTION 13.2

A. The following problems, for which the exact solutions are known, were chosen for practice and experimentation. In each case, use the Runge-Kutta method up to h^4 to produce a table with headings X, CALC. Y, TRUE VALUE, ERROR and REL. ERROR. Start with $h=0.1$ and cut this in half until $h=0.003125$ (to make results comparable). Although only four values (i.e., the given initial condition and three Runge-Kutta values) are needed for use in later algorithms, it is instructive (if computer time is available) to continue the Runge-Kutta procedure, using $h=0.003125$, to examine the errors. In the following, the exact solutions are given as well as the exact value of $y(x_0)$; D

is the difference between $R_1(3)$ and $R_2(6)$; E_1 is the error of $R_2(6)$; and E_2 is the actual error of the Runge-Kutta value at x_0+1—all computed on an IBM 360 using double precision throughout.

(1) $y' = -(x^2+1)/y$, $y(0.5) = \sqrt{4.75+\ln 4}$. Exact solution: $y = \sqrt{5-x^2-2\ln x}$. $D = -2.71 \times 10^{-12}$; $E_1 = 9.44 \times 10^{-14}$; $E_2 = 3.25 \times 10^{-12}$

(2) $y' = -(\sin y + x^2 + 2x)/\cos y$. $y(0.1) = \arcsin(e^{-0.1} - 0.01)$. Exact solution: $y = \arcsin(e^{-x} - x^2)$. $D = -3.63 \times 10^{-10}$; $E_1 = 1.48 \times 10^{-11}$; $E_2 = 1.96 \times 10^{-9}$

(3) $y' = [\ln(x \sin y) - 1]/(x \cot y)$, $y(-2) = \arcsin[-1/(2e^2)]$. Exact solution: $y = \arcsin(e^x/x)$

(4) $y' = -y/(2x+1)$, $y(0) = 1$. Exact solution: $y = (2x+1)^{-1/2}$. $D = -3.0 \times 10^{-12}$; $E_1 = 1.06 \times 10^{-13}$; $E_2 = 8.53 \times 10^{-13}$

(5) $y' = (y-xy)/x^2$, $y(2) = 3/(2\sqrt{e}\,)$. Exact solution: $y = 3e^{-1/x}/x$. $D = 9.40 \times 10^{-15}$; $E_1 = -3.89 \times 10^{-16}$; $E_2 = 2.51 \times 10^{-15}$

(6) $y' = x(1+y^2)/(1+2x^2)$, $y(0) = 0$. Exact solution: $y = \tan[\frac{1}{4}\ln(1+2x^2)]$. $D = 4.70 \times 10^{-14}$; $E_1 = -8.25 \times 10^{-16}$; $E_2 = -5.44 \times 10^{-13}$

(7) $y' = (x+1)y/x$, $y(1) = 2e$. Exact solution: $y = 2xe^x$. $D = -3.78 \times 10^{-11}$; $E_1 = 1.25 \times 10^{-12}$; $E_2 = 3.41 \times 10^{-10}$

(8) $y' = (x+y)/x$, $y(1) = 3$. Exact solution: $y = x(3+\ln x)$. $D = -8.66 \times 10^{-13}$; $E_1 = 2.73 \times 10^{-14}$; $E_2 = 1.20 \times 10^{-12}$

(9) $y' = (x^2+y^2)/(xy)$, $y(1) = \sqrt{2}$. Exact solution: $y = x\sqrt{2(\ln x + 1)}$. $D = -2.00 \times 10^{-5}$, $E_1 = 4.10 \times 10^{-5}$, $E_2 = 1.58 \times 10^{-2}$

(10) $y' = 2xy$, $y(0) = 1$. Exact solution: $y = e^{x^2}$.

B.

(11) Find $(\mathbf{H} \cdot \nabla)^3 F(x,y)$, where $F(x,y) = e^x \ln y$, $y > 0$.

(12) Write Taylor's theorem for three variables to three sets of terms and indicate the error of truncation.

(13) Verify (13.2.32).

(14) Equations to derive a Runge-Kutta formula matching the Taylor series up to the h^3 term are

$$\begin{cases} a_1 + a_2 + a_3 = 1 \\ a_2 b_2 + a_3 b_3 = \frac{1}{2} \\ a_2(b_2)^2 + a_3(b_3)^2 = \frac{1}{3} \\ a_3 b_2 b_3 = \frac{1}{6} \end{cases}$$

Using $b_2 = \frac{1}{2}$ arbitrarily, find a_1, a_2, a_3 and b_3 to deduce a Runge-Kutta algorithm of the form

$$y_{n+1} = y_n + a_1 k_1 + a_2 k_2 + a_3 k_3$$

13.3 PREDICTOR-CORRECTOR METHODS: FIRST ORDER

13.3.1 Introduction

It is assumed that the problem requires a particular solution of

$$y' = f(x,y) \tag{13.3.1}$$

through (x_0,y_0) and that a sufficient number of starting values (x_1,y_1), $(x_2,y_2),\cdots$ have been found, e.g. by use of the Runge-Kutta algorithm. Using these values to substitute into (13.3.1), we have, in fact, (x_0,y_0,f_0), $(x_1,y_1,f_1),\ldots$. To find further values of (x,y,f), the most popular methods at this time are the many *predictor-corrector* methods.

13.3.2 Background

The fundamental thought behind the predictor-corrector methods is that an estimate of y_{k+1} can be found from previous values of y and y'. More specifically, we assume that there exist constants $a_0,a_1,\ldots,b_{-1},b_0,b_1,\ldots$ such that

$$\tilde{y}_{k+1} = a_0 y_k + a_1 y_{k-1} + a_2 y_{k-2} + \cdots + h(b_{-1}f_{k+1} + b_0 f_k + b_1 f_{k-1} + \cdots)$$
(13.3.2)

where f is given by (13.3.1).

An immediate difficulty presents itself. In (13.3.2), $f_{k+1} = f(x_{k+1}, y_{k+1})$, so that it is impossible to find f_{k+1} before finding y_{k+1}; yet we need f_{k+1} to find y_{k+1}. Therefore, for the first estimate, we must assume $b_{-1} = 0$. If $b_{-1} = 0$, we say that (13.3.2) is a *predictor*. We shall call the result pre y_{k+1}.

Once we have this estimate of y_{k+1}, we can use it in (13.3.2). This leads to another equation (with different values of a and b) called the *corrector*. We shall call the result cor y_{k+1}. By comparison of pre and cor, we can estimate the size and sign of the error, then adjust cor. This operation is called *mop-up*.

There are many different predictor-corrector systems. We shall illustrate the *Adams-Moulton system*, which requires *four* previous values. The methods of Milne, Hermite and Hamming are discussed in Ralston, pp. 202–210, and a detailed study of predictor-corrector methods is in Hamming, Chapter 15.

In the Adams-Moulton system, the predictor has the form

$$\text{pre}\, y_{k+1} = a_0 y_k + h(b_0 f_k + b_1 f_{k-1} + b_2 f_{k-2} + b_3 f_{k-3}) \quad (13.3.3)$$

and the corrector has the form

$$\text{cor}\, y_{k+1} = a_0 y_k + h(b_{-1} f_{k-1} + b_0 f_k + b_1 f_{k-1} + b_2 f_{k-2}) \quad (13.3.4)$$

The coefficients are *not* the same in these two formulas.

13.3.3 The Adams-Moulton Method: Predictor

We return to (13.3.3) to find the coefficients by the method of undetermined coefficients. We need five equations because there are five unknown coefficients.

> *Step I.* Make (13.3.3) exact for $y \equiv 1$. Then $y' = f = 0$. Substituting in (13.3.3),

$$1 = a_0 \quad (13.3.5)$$

Step II. Make (13.3.3) exact for $y \equiv x$. Then $y' = f = 1$. For ease of computation, we assume that f has been translated so that $x_k = 0$. Then $y_k = x_k = 0$, $y_{k+1} = x_k + h = h$, and so on. Substituting in (13.3.3),

$$h = h[b_0 + b_1 + b_2 + b_3] \tag{13.3.6}$$

$$1 = b_0 + b_1 + b_2 + b_3 \tag{13.3.7}$$

Step III. Make (13.3.3) exact for $y \equiv x^2$. Then $y' = f = 2x$. Also, $y_k = (x_k)^2 = 0$, $y_{k+1} = (x_k + h)^2 = h^2$, $y'_k = f_k = 2x_k = 0$, $y'_{k-1} = f_{k-1} = 2(x_k - h) = -2h$, and so on. Then

$$h^2 = h[-2hb_1 - 4hb_2 - 6hb_3] \tag{13.3.8}$$

$$1 = -2b_1 - 4b_2 - 6b_3 \tag{13.3.9}$$

Step IV. Make (13.3.3) exact for $y \equiv x^3$. Then $y' = f = 3x^2$. Also, $y_{k+1} = (x_k + h)^3 = h^3$, $y'_k = f_k = 0$, $y'_{k-1} = f_{k-1} = 3(x_k - h)^2 = 3h^2$, and so on. Then

$$h^3 = h[3h^2 b_1 + 12h^2 b_2 + 27h^2 b_3] \tag{13.3.10}$$

$$1 = 3b_1 + 12b_2 + 27b_3 \tag{13.3.11}$$

Step V. Make (13.3.3) exact for $y \equiv x^4$. The result is

$$1 = -4b_1 - 32b_2 - 108b_3 \tag{13.3.12}$$

In augmented matrix form, the five equations are

$$\begin{bmatrix} 1 & 0 & 0 & 0 & 0 & \vdots & 1 \\ 0 & 1 & 1 & 1 & 1 & \vdots & 1 \\ 0 & 0 & 1 & 2 & 3 & \vdots & -\frac{1}{2} \\ 0 & 0 & 1 & 4 & 9 & \vdots & \frac{1}{3} \\ 0 & 0 & 1 & 8 & 27 & \vdots & -\frac{1}{4} \end{bmatrix} \tag{13.3.13}$$

from which

$$a_0 = 1$$

$$b_0 = \frac{55}{24}$$

$$b_1 = -\frac{59}{24} \tag{13.3.14}$$

$$b_2 = \frac{37}{24}$$

$$b_3 = -\frac{3}{8}$$

leading to the predictor formula

$$\text{pre } y_{k+1} = y_k + \frac{h}{24} \left[55f_k - 59f_{k-1} + 37f_{k-2} - 9f_{k-3} \right] \quad (13.3.15)$$

The error can be calculated in several sophisticated ways (see Hamming, pp. 144–151, 196). For our purposes, we shall be content with a comparison using the Taylor series. If y_{k+1} is the true value, the error is given by

$$E = y_{k+1} - \left[y_k + \frac{h}{24} (55y_k' - 59y_{k-1}' + 37y_{k-2}' - 9y_{k-3}') \right] \quad (13.3.16)$$

$$E = (y_{k+1} - y_k) - \frac{h}{24} (55y_k' - 59y_{k-1}' + 37y_{k-2}' - 9y_{k-3}') \quad (13.3.17)$$

By Taylor's theorem, expanding about $x_k = 0$,

$$y_{k+1} - y_k = hy_k' + \frac{1}{2}h^2 y_k'' + \frac{1}{6}h^3 y_k''' + \frac{1}{24}h^4 y_k^{iv} + \frac{1}{120}h^5 y_k^v + \cdots \quad (13.3.18)$$

Also

$$y_{k-1}' = y_k' - hy_k'' + \frac{1}{2}h^2 y_k''' - \frac{1}{6}h^3 y_k^{iv} + \frac{1}{24}h^4 y_k^v + \cdots \quad (13.3.19)$$

$$y_{k-2}' = y_k' - (2h) y_k'' + \frac{1}{2}(2h)^2 y_5''' - \frac{1}{6}(2h)^3 y_k^{iv} + \frac{1}{24}(2h)^4 y_k^v - \cdots \quad (13.3.20)$$

$$y_{k-3}' = y_k' - (3h) y_k'' + \frac{1}{2}(3h)^2 y_k'' - \frac{1}{6}(3h)^3 y_k^{iv} + \frac{1}{24}(3h)^4 y_k^v - \cdots \quad (13.3.21)$$

We line up the coefficients as follows for addition:

hy_k'	$h^2 y_k''$	$h^3 y_k'''$	$h^4 y_k^{iv}$	$h^5 y_k^v$
1	$\dfrac{1}{2}$	$\dfrac{1}{6}$	$\dfrac{1}{24}$	$\dfrac{1}{120}$
$-\dfrac{55}{24}$	0	0	0	0
$\dfrac{59}{24}$	$-\dfrac{59}{24}$	$\dfrac{59}{24}$	$-\dfrac{59}{144}$	$\dfrac{59}{576}$
$-\dfrac{37}{24}$	$\dfrac{74}{24}$	$-\dfrac{74}{24}$	$\dfrac{296}{144}$	$-\dfrac{592}{576}$
$\dfrac{9}{24}$	$-\dfrac{27}{24}$	$\dfrac{31}{48}$	$-\dfrac{243}{144}$	$\dfrac{729}{576}$
0	0	0	0	$\dfrac{251}{720}$

We conclude that the Adams-Moulton predictor equation agrees with a Taylor series up to h^4, the error being

$$\frac{251}{720} h^5 y^v(\theta_1)$$

where θ_1 is between x_k and x_{k+1}. We now state the

ADAMS-MOULTON PREDICTOR EQUATION: $\text{pre } y_{k+1}$

$$= y_k + \frac{h}{24}(55f_k - 59f_{k-1} + 37f_{k-2} - 9f_{k-3}) + \frac{251}{720}h^5 y^v(\theta_1)$$

$$(13.3.22)$$

13.3.4 The Adams-Moulton Method: Corrector Equation

The corrector equation is derived from (13.3.4) in precisely the same way. We leave it as a useful exercise. The result is the

---

ADAMS-MOULTON CORRECTOR EQUATION: $\text{cor } y_{k+1}$

$$= y_k + \frac{h}{24}(9f_{k+1} + 19f_k - 5f_{k-1} + f_{k-2}) - \frac{19}{720}h^5 y^v(\theta_2). \quad (13.3.23)$$

13.3.5 The Adams-Moulton Method: Mop-Up

From (13.3.22) and (13.3.23), the errors are given by

$$E_p = y - \text{pre} = \frac{251}{720}h^5 y^v(\theta_1) \qquad (13.3.24)$$

$$E_c = y - \text{cor} = -\frac{19}{720}h^5 y^v(\theta_2) \qquad (13.3.25)$$

Assuming that $y^v(\theta_1)$ and $y^v(\theta_2)$ are equal,

$$\frac{E_p}{E_c} = -\frac{251}{19} \qquad (13.3.26)$$

$$E_p = -\frac{251}{19}E_c \qquad (13.3.27)$$

$$E_p - E_c = -\frac{251}{19}E_c - E_c = -\frac{270}{19}E_c \qquad (13.3.28)$$

$$E_c = -\frac{19}{270}(E_p - E_c) \qquad (13.3.29)$$

But

$$E_c = y - \text{cor} \qquad (13.3.30)$$

and

$$E_p - E_c = (y - \text{pre}) - (y - \text{cor}) = \text{cor} - \text{pre} \qquad (13.3.31)$$

$$\therefore \quad y = \text{cor} + E_c = \text{cor} + \frac{19}{270}(\text{pre} - \text{cor}) \qquad (13.3.32)$$

Equation (13.3.32) is the *mop-up equation* for the Adams-Moulton system. We summarize the system as follows:

ADAMS-MOULTON PREDICTOR-CORRECTOR SYSTEM

$$\text{pre } y_{k+1} = y_k + \frac{h}{24}(55f_k - 59f_{k-1} + 37f_{k-2} - 9f_{k-3}) \qquad (13.3.33)$$

$$\text{cor } y_{k+1} = y_k + \frac{h}{24}(9f_{k+1} + 19f_k - 5f_{k-1} + f_{k-1}) \qquad (13.3.34)$$

$$y_{k+1} = \text{cor} + \frac{19}{270}(\text{pre} - \text{cor}) \qquad (13.3.35)$$

13.3.6 An Example

We shall do a partial hand calculation to illustrate the application of (13.3.33)–(13.3.35). We need four starting values: the initial value (x_0, y_0), and three additional values, which are usually obtained from the Runge-Kutta procedure. We shall assume $h = 0.1$ merely for illustrative purposes, but to avoid clouding the issue, we shall use the true values for all four starters. We discuss the error estimate in the next section.

ILLUSTRATIVE PROBLEM 13.3-1   Given

$$y' = -y \cos x \qquad (13.3.36)$$

and four starting values from the exact solution

$$y = e^{-\sin x} \qquad (13.3.37)$$

find the fifth value, y_4, using $x_0 = 0$ and $h = 0.1$.

Solution   Substituting in (13.3.37), we have

$$y_0 = \exp(-\sin 0) = 1$$
$$y_1 = y(0.1) = \exp(-\sin 0.1) = 0.9049881614$$
$$y_2 = y(0.2) = \exp(-\sin 0.2) = 0.8198209381$$
$$y_3 = y(0.3) = \exp(-\sin 0.3) = 0.7441443779$$

For Equation (13.3.33), we need y_3, f_3, f_2, f_1 and f_0. Substituting in (13.3.36),

$$f_0 = -y_0 \cos x_0 = -1 \cos 0 = -1$$
$$f_1 = -y_1 \cos x_1 = -0.9049881614 \cos 0.1 = -0.9004669901$$
$$f_2 = -y_2 \cos x_2 = -0.8034791012$$
$$f_3 = -y_3 \cos x_3 = -0.7109082774$$

Substituting into the predictor formula (13.3.33), we have

$$\text{pre } y_4 = 0.7441443779 + \frac{0.1}{24}(55f_3 - 59f_2 + 37f_1 - 9f_0)$$

$$\text{pre } y_4 = 0.6774278490 \qquad (13.3.38)$$

We now calculate f_4 using the predicted value of y_4 in (13.3.38):

$$f_4 = -y_4 \cos x_4 = -0.6774278490 \cos 0.4 = -0.6239523680$$

and substitute in the corrector formula (13.3.34):

$$\text{cor } y_4 = 0.7441443779 + \frac{0.1}{24}(9f_4 + 19f_3 - 5f_2 + f_1)$$

$$\text{cor } y_4 = 0.6774531276 \tag{13.3.39}$$

We now use the mop-up formula (13.3.35):

$$y_4 = 0.67745\,1276 + \frac{19}{270}(0.6774278490 - 0.6774531276)$$

$$y_4 = 0.6774543487 \tag{13.3.40}$$

The result is correct to only five places, scarcely a welcome accuracy after working diligently with ten significant figures. Of course, h was miles too big. Figure 13.3-1 displays the result of a computer run with $h = 0.00625$, and Figure 13.3-2 displays the results with $h = 0.003125$. The computer program for the Adams-Moulton method is shown in Figure 13.3-3. We remind you that we used true values as starters only to avoid contaminating the Adams-Moulton algorithm. In a real problem, we would have started with Runge-Kutta results.

13.3.7 Estimates of Error

Testing the solutions of differential equations whose exact solutions are known shows that for each differential equation there is an optimum size for h. If h is too large, accuracy diminishes. If h is too small, roundoff errors drown the theoretically increased accuracy.

There is, however, no way to find this optimum value or, in fact, to estimate the error. A suggested rule of thumb for the error and the size of h is given by combining Equations (13.3.29) and (13.3.31):

$$E_c = -\frac{19}{270}(\text{cor} - \text{pre}) \tag{13.3.41}$$

If this relationship is really reliable, and if y is needed to a precision of 10^{-N}, then

$$|E_c| = \frac{19}{270}|\text{cor} - \text{pre}| < 10^{-N} \tag{13.3.42}$$

$$|\text{cor} - \text{pre}| < \frac{270}{19} \times 10^{-N} \sim 14 \times 10^{-N} \tag{13.3.43}$$

According to (13.3.43), h should be reduced until the difference between corrected and predicted values is no greater than 14 times the precision required. For example, if the precision required is 10^{-13}, h should be reduced until $|\text{cor} - \text{pre}| < 14 \times 10^{-13}$. Table 13.3-1 compares $|\text{cor} - \text{pre}|$, using two different values of h, with the actual error for Illustrative

ADAMS-MOULTON PREDICTOR-CORRECTOR METHOD, FIRST-ORDER

Y'=-Y COS X   XO = 0.0   H= 0.00625000

X	CALC.Y	TRUE VALUE	ERROR	REL.ERROR	C-P
0.10000000	9.0498816144430787D-01	9.0498816144245380D-01	-6.25D-13	-6.90D-13	2.70D-11
0.20000000	8.1982093805635558D-01	8.1982093805516340D-01	-1.19D-12	-1.45D-12	2.40D-11
0.30000000	7.4414437789931930D-01	7.4414437789977623D-01	-1.56D-12	-2.09D-12	2.02D-11
0.40000000	6.7745080439351950D-01	6.7745080439176750D-01	-1.75D-12	-2.59D-12	1.62D-11
0.50000000	6.1913896109954800D-01	6.1913896109773130D-01	-1.82D-12	-2.93D-12	1.24D-11
0.60000000	5.6856338696272340D-01	5.6856338696093290D-01	-1.79D-12	-3.15D-12	9.10D-12
0.70000000	5.2507315290298770D-01	5.2507315290127920D-01	-1.71D-12	-3.25D-12	6.32D-12
0.80000000	4.8040887442375550D-01	4.8040887440776600D-01	-1.60D-12	-3.28D-12	4.11D-12
0.90000000	4.5688347001736810D-01	4.5688347001588470D-01	-1.48D-12	-3.25D-12	2.44D-12
1.00000000	4.3107595064669683D-01	4.3107595064559240D-01	-1.38D-12	-3.19D-12	1.27D-12

Fig. 13.3-1 Application of the Adams-Moulton method with $h = 0.00625$.

ADAMS-MOULTON PREDICTOR-CORRECTOR METHOD, FIRST-ORDER

Y'=-Y COS X   XO = 0.0   H= 0.00312500

X	CALC.Y	TRUE VALUE	ERROR	REL.ERROR	C-P
0.10000000	9.0498816144247510D-01	9.0498816144245380D-01	-2.11D-14	-2.34D-14	8.39D-13
0.20000000	8.1982093805520170D-01	8.1982093805516340D-01	-3.79D-14	-4.63D-14	7.41D-13
0.30000000	7.4414437789781140D-01	7.4414437789977623D-01	-4.85D-14	-6.51D-14	6.22D-13
0.40000000	6.7745080439182210D-01	6.7745080439176750D-01	-5.38D-14	-7.95D-14	4.98D-13
0.50000000	6.1913896109778730D-01	6.1913896109773130D-01	-5.52D-14	-8.92D-14	3.81D-13
0.60000000	5.6856338696098770D-01	5.6856338696093290D-01	-5.39D-14	-9.48D-14	2.78D-13
0.70000000	5.2507315290133100D-01	5.2507315290127920D-01	-5.09D-14	-9.69D-14	1.92D-13
0.80000000	4.8040887440824550D-01	4.8040887440776600D-01	-4.70D-14	-9.64D-14	1.24D-13
0.90000000	4.5688347001592870D-01	4.5688347001588470D-01	-4.31D-14	-9.43D-14	7.36D-14
1.00000000	4.3107595064563260D-01	4.3107595064559240D-01	-3.94D-14	-9.14D-14	3.77D-14

Fig. 13.3-2 Application of the Adams-Moulton method with $h = 0.003125$.

```
C  IN THIS PROGRAM, IT IS ASSUMED THAT FOUR SATISFACTORY STARTING VALUES     040
C  OF Y ARE AVAILABLE, POSSIBLY FROM A RUNGE-KUTTA DETERMINATION. TRUE       050
C  VALUES ARE USED IN THIS DEMONSTRATION PROGRAM TO AVOID CONTAMINATING      060
C  THE ALGORITHM TO BE TESTED, BUT ARE EASILY ENTERED BY CARD ALONG WITH     070
C  OTHER INPUT. A TABLE OF VALUES IS GENERATED USING THE ADAMS-MOULTON       080
C  PREDICTOR-CORRECTOR ALGORITHM FOR A FIRST-ORDER EQUATION. THERE IS        090
C  ONE INPUT CARD, AS FOLLOWS.                                               100
C      CC  1-28   NAME OR DESCRIPTION OF THE EQUATION                        110
C      CC 31-40   INITIAL X, XO, IN F-FORMAT                                 120
C      CC 41-50   INITIAL H, HO, IN F-FORMAT                                 130
C      CC 61      NUMBER OF ITERATIONS, N1, FROM THE R-K PROGRAM             160
C  THERE ARE TWO VARIABLE STATEMENT FUNCTIONS. AFTER CARD 260. THE FIRST,    170
C  FN(X,A), DEFINES THE DIFFERENTIAL EQUATION. THE SECOND, S(X), GIVES       180
C  THE TRUE VALUES USED TO EVALUATE THE CALCULATED ANSWERS.                  190
C                                                                            200
      DOUBLE PRECISION Y(321,2),P(321),CP(321),ER(321),REL(321),A,B,C,      210
     1C1,C2,CY,D,DY,E,E1,F,FN,H,HO,HF,HP,PI,PY,RY,S,TV,X,X0,YC,YM,YP         220
      DIMENSION NAME(7)                                                      230
C  STATEMENT FUNCTIONS. IN THE FOLLOWING, A=Y(K),B=F(K+1),C=F(K),D=F(K-1)    240
C     E=F(K-2),PY=YP=PREDICTED Y(K+1),CY=YC=CORRECTED                        250
C     Y(K+1),YM=MOPPED-UP Y(K+1),HP=H/24,F=F(K-3)                            260
      FN(X,A)=-A*DCOS(X)                                                PC1  2
      S(X)=DEXP(-DSIN(X))                                               PC1  2
C  ALGORITHM.                                                               270
      YP(A)=A+HP*(55.D0*C-59.D0*D+37.D0*E-9.D0*F)                           280
      YC(A)=A+HP*(9.D0*B+19.D0*C-5.D0*D+E)                                  290
      YM(CY)=CY+19.D0*(PY-CY)/270.D0                                        300
C  CONSTANTS.                                                               310
      PI=3.141592653589793D0                                               320
      E1=2.718281828459045D0                                               330
      C1=0.5D0                                                             340
      C2=DSQRT(2.D0)                                                        350
C  INPUT.                                                                  360
      READ(1,100)NAME,X0,H0,N1                                             370
100   FORMAT(7A4,2X,2F10.0,10X,I1)                                         380
C  SET. H IS DOUBLED IN ORDER TO ALLOW COMPARISON.                         390
C  KD AND KF ARE USED IN DO LOOPS, LATER ON.                               395
      N=1                                                                  400
      H=H0*2.D0                                                            410
      KD=2**(N1-2)                                                         415
      KF=10*KD+1                                                           420
```

Fig. 13.3-3 (Continued on next page)

```
                                                                          430
C FIRST HEADING.
1      WRITE(3,101)                                                       440
101    FORMAT(1H1,T30,'ADAMS-MOULTON PREDICTOR-CORRECTOR METHOD, FIRST-OR 450
      1DER'/)                                                             460
       WRITE(3,102)NAME,XO,H                                              470
102    FORMAT(1H0,T20,7A4,5X,'XO = ',F11.8,T79,'H=',F11.8/)              480
       WRITE(3,103)                                                       490
103    FORMAT(1H0,T11,'X',T29,'CALC.Y' ,T55,'TRUE VALUE',T78,'ERROR',T91, 500
      1'REL.ERROR',T107,'C-P'/)                                           510
C CALCULATION OF STARTING VALUES, Y AND Y'=P.                             520
       X=XO-H                                                             530
       DO 2 K=1,4                                                         540
       X=X+H                                                              550
       Y(K,N)=S(X)                                                        570
       A=Y(K,N)                                                           570
2      P(K)=FN(X,A)                                                       580
C CALCULATION OF EXTRAPOLATED VALUES. L IS THE LINE COUNT.                590
       HP=H/24.DO                                                         600
       DO 3 K=5,KF                                                        620
       X=X+H                                                              640
       C=P(K-1)                                                           660
       D=P(K-2)                                                           670
       E=P(K-3)                                                           680
       F=P(K-4)                                                           690
       PY=YP(A)                                                           700
       B=FN(X,PY)                                                         710
       CY=YC(A)                                                           720
       Y(K,N)=YM(CY)                                                      730
       TV=S(X)                                                            740
       CP(K)=CY-PY                                                        750
       ER(K)=TV-Y(K,N)                                                    760
       IF(DABS(ER(K)).LE..5.D-17)GO TO 31                                 770
       REL(K)=ER(K)/Y(K,N)                                                780
       GO TO 32                                                           790
31     ER(K)=0.D0                                                         800
       REL(K)=0.D0                                                        810
32     A=Y(K,N)                                                           820
3      P(K)=FN(X,A)                                                       825
C FIRST TABLE.                                                            826
       K1=KD+1                                                            852
       X=XO                                                               860
       DO 33 K=K1,KF,KD                                                   865
       X=X+H*DFLOAT(KD)                                                   866
       TV=S(X)                                                            867
```

```
33      WRITE(3,107)X,Y(K,N),TV,ER(K),REL(K),CP(K)                       870
107     FORMAT(1H ,T6,F11.8,T22,1PD22.15,T50,D22.15,3(5X,D9.2))          875
C  RE-TRIAL AT ORIGINAL H.                                               880
        IF(N.EQ.2)GO TO 4                                                890
        N=N+1                                                            900
        H=HO                                                             910
        KF=2*KF-1                                                        920
        KD=2*KD                                                          925
        GC TO 1                                                          930
C  TABLE TO COMPARE Y1 AND Y2.                                           940
4       WRITE(3,101)                                                     950
        WRITE(3,104)NAME                                                 960
104     FORMAT(1HO,T42,7A4/)                                             970

        B=HO*2.DO                                                        980
        WRITE(3,105)B,HO                                                 990
105     FORMAT(1HO,T15,'X',T31,'Y1,H=',F9.7    ,T58,'Y2,H=',F9.7    ,T82,1000
       1'Y1-Y2',T94,'REL.ERROR'/)                                       1010
C                                                                       1030
        X=X0                                                            1031
        K1=(K1+3)/4                                                     1032
        K3=(KF+1)/2                                                     1033
        KD=K1-1                                                         1040
        DC 5 K=K1,K3,KD                                                 1050
        X=X+0.05D0                                                      1060
        J=2*K-1                                                         1070
        DY=Y(K,1)-Y(J,2)                                                1080
        IF(DABS(DY).LE.5.D-17)GO TO 51                                  1090
        RY=DY/Y(J,2)                                                    1100
        GO TO 5                                                         1110
51      DY=0.D0                                                         1120
        RY=0.D0                                                         1130
5       WRITE(3,106)X,Y(K,1),Y(J,2),DY,RY                               1140
106     FORMAT(1H ,T10,F11.8,T26,1PD22.15,T53,D22.15,2(5X,D9.2))        1160
C  FINAL TABLE, HIGHER VALUES OF X.                                     1170
        WRITE(3,101)                                                    1180
        WRITE(3,102)NAME,X0,H                                           1190
        WRITE(3,117)                                                    1200
117     FORMAT(1HO,T11,'X',T19,'CALC. Y',T48,'TRUE VALUE',T70,'ERROR',  1210
       1T80,'REL.ERROR',T95,'C-P'/)                                     1220
        KT=0                                                            1220
        K=321                                                           1225
```

Fig. 13.3-3 (Continued on next page)

```
C SET.
      C=P(K-1)                                                      1230
      D=P(K-2)                                                      1240
      E=P(K-3)                                                      1250
      F=P(K-4)                                                      1260
      A=Y(K,N)                                                      1270
C                                                                   1280
      K3=10*KF-9                                                    1290
      KD=(KF-1)/2                                                   1291
      DO 6 K=KF,K3                                                  1292
      KT=KT+1                                                       1300
      X=X+H                                                         1310
      PY=YP(A)                                                      1320
      B=FN(X,PY)                                                    1330
      CY=YC(A)                                                      1340
      A=YM(CY)                                                      1350
C                                                                   1360
      IF(KT.EQ.KD)GO TO 61                                          1370
      GO TO 62                                                      1380
61    TV=S(X)                                                       1390
      C1=TV-A                                                       1400
      IF(DABS(C1).LE.5.D-17)GO TO 611                               1410
      C2=C1/A                                                       1411
      GO TO 612                                                     1420
611   C1=0.D0                                                       1421
      C2=0.D0                                                       1422
612   DY=CY-PY                                                      1423
      KT=0                                                          1430
      WRITE(3,108)X,A,TV,C1,C2,DY                                   1440
108   FORMAT(1H ,T9 ,F4.1,3X,1PD23.15,3X,D23.15,3X,3(3X,D9.2))      1450
C                                                                   1460
62    F=E                                                           1470
      E=D                                                           1480
      D=C                                                           1490
6     C=FN(X,A)                                                     1500
      END                                                           1510
                                                                    1150
```

Fig. 13.3-3 Computer program for the Adams-Moulton algorithm for a first-order differential equation.

Table 13.3-1

Data for Evaluating Various Rules of Thumb for First-Order Differential Equations

PROB.	$dy/dx =$	$y =$	x_0	y_0	y_0'	$\left.\dfrac{\partial f}{\partial y}\right\|_{x_0}$	$\left\|\dfrac{\text{cor}-\text{pre}}{E}\right\|$ $h=0.00625$			$\left\|\dfrac{\text{cor}-\text{pre}}{E}\right\|$ $h=0.003125$			$\dfrac{\|Y_1 - Y_2\|}{E}$		
							$x_0+0.1$	$x_0+0.5$	x_0+1	$x_0+0.1$	$x_0+0.5$	x_0+1	$x_0+0.1$	$x_0+0.5$	x_0+1
13.3-1	$-y\cos x$	$e^{-\sin x}$	0	1	-1	-1	43	7	0.9	42	7	0.9	28	32	34
(1)[a]	$-\dfrac{x^2+1}{xy}$	$\sqrt{5-x^2-2\ln x}$	0.5	2.5	2.5	0.4	15	1	18	13	1	19	27	30	31
(2)	$-\dfrac{(\sin y+x^2+2x)}{\cos y}$	$\sin^{-1}(e^{-x}-x^2)$	0.1	1.1	-2.4	5.9	26	11	30	17	0.8	34	24	23	22
(3)	$\dfrac{\ln(x\sin y)-1}{x\cot y}$	$\sin^{-1}(e^x/x)$	-2	-0.1	-0.13	16	84	31	1761	68	29	1245	24	28	20
(4)	$\dfrac{-y}{2x+1}$	$\dfrac{1}{\sqrt{2x+1}}$	0	1	1	1	43	2	0.2	35	2	0.1	27	29	30
(5)	$\dfrac{y-xy}{x^2}$	$\dfrac{3}{x}e^{-1/x}$	2	0.9	-0.2	-0.25	125	11	2	47	3	0.4	5	9	6
(6)	$\dfrac{x(1+y^2)}{1+2x^2}$	$\tan[\dfrac{\ln(1+2x^2)}{4}]$	0	0	0	0	22	4	9	21	2	13	28	51	46
(7)	$\dfrac{(x+1)y}{x}$	$2xe^x$	1	5.4	11	2	21	3	2	25	5	2	15	44	42
(8)	$\dfrac{x+y}{x}$	$x(3+\ln x)$	1	3	4	1	20	2	0.3	11	0.7	0.1	8	14	11
(9)	$\dfrac{x^2+y^2}{xy}$	$x\sqrt{2(\ln x+1)}$	1	1.4	2	3.5	13	0.1	0.1	12	0.2	2	32	56	360

[a]Section 13.2.

Problem 13.3-1 and for some of problems for Section 13.2. It is perfectly clear that (13.3.43) does not hold up even at $x_0 + 1$.

Another suggested rule of thumb involves $\partial f / \partial y$. The table shows that this is also a failure.

Into this morass, we throw another empirical rule of thumb:

RULE OF THUMB FOR THE ADAMS-MOULTON SOLUTION OF A FIRST-ORDER DIFFERENTIAL EQUATION   The actual error is much less than $|Y_1 - Y_2|$, where Y_1 and Y_2 are two successive calculations determined by cutting h in two.

Consequently, if accuracy is required to 10^{-9}, we suggest halving h until $|Y_1 - Y_2| \leqslant 10^{-9}$. The actual error is usually less than one-twentieth of this difference.

13.3.8   Comparison of Runge-Kutta and Adams-Moulton Methods

It remains to compare the predictor-corrector method with the Runge-Kutta method when the latter is extended. In Table 13.3-2 the relative errors at $x_0 + 1$ are compared for the illustrative problem and some of the problems for Section 13.2. It can be seen that, in fact, the two methods are roughly comparable. This was to be expected, since both are derived from the same Taylor series; the differences are due to differences in roundoff depending upon the equations. Although the predictor-corrector method seems to have a built-in measure of error, namely $|\text{pre} - \text{cor}|$, we have already seen that this is an illusion. For certainty in extremely important calculation, it might be well to solve by both methods, retaining the digits which agree. Unfortunately, both are very slow methods and require a great deal of computer time.

Table 13.3-2
Comparison of Runge-Kutta and Adams-Moulton Results at $x_0 + 1$.

PROBLEM	RK RESULT	AM RESULT
13.3-1	3.9×10^{-13}	9.1×10^{-14}
(1)[a]	2.3×10^{-12}	1.4×10^{-12}
(2)	1.8×10^{-9}	1.2×10^{-8}
(3)	3.6×10^{-12}	8.9×10^{-14}
(4)	1.5×10^{-12}	2.3×10^{-12}
(5)	3.5×10^{-15}	2.3×10^{-14}
(6)	1.9×10^{-12}	1.2×10^{-13}
(7)	1.2×10^{-11}	1.7×10^{-13}
(8)	1.6×10^{-13}	5.1×10^{-14}
(9)	4.3×10^{-12}	1.8×10^{-15}

[a]Section 13.2.

PROBLEM SECTION 13.3

A. Using the method of undetermined coefficients, derive the following equations:

(1) Euler predictor equation:

$$y_{k+1} = y_k + \frac{1}{2}h(f_k + f_{k+1}) - \frac{1}{12}h^3 y'''(\theta)$$

(2) Milne predictor equation:

$$y_{k+1} = y_{k-3} + \frac{4}{3}h(2f_{k-2} - f_{k-1} + 2f_k) + \frac{14}{45}h^5 y^{\nu}(\theta)$$

(3) Milne corrector equation:

$$y_{k+1} = y_{k-1} + \frac{1}{3}h(f_{k+1} + 4f_k + f_{k-1}) - \frac{1}{90}h^5 y^{\nu}(\theta)$$

(4) Adams-Moulton corrector equation (see text)

B. Using the problems for Section 13.2, part A, calculate the first four true values from the exact answer; then apply the Adams-Moulton predictor-corrector system to prepare a table from x_0 to $x_0 + 1$ by intervals of 0.1 and, where possible, from $x_0 + 2$ to $x_0 + 10$ by intervals of 1. Use $h_1 = 0.00625$ and $h_2 = 0.003125$, and print both $|cor - pre|$ and $|Y_1 - Y_2|$ to compare with the actual errors.

Chapter 14

Second-Order Ordinary Differential Equations

In an abbreviated course there may not be enough time for this chapter, but it is important.

In an overwhelming number of practical problems, the result of scientific or engineering investigation involves the solution of a differential equation of the form

$$p(x)y'' + q(x)y' + r(x)y = s(x), \qquad ' = d/dx \qquad (14.1)$$

Apart from the usual standard methods, which we assume have been tried before resorting to approximations, there are the *series methods*, which usually require computer assistance in order to present a useful table of results. In this chapter, we tackle these first. Following the discussion of series methods, we explore the Runge-Kutta method and a predictor-corrector method, ending the book with a brief explanation of two special problems: the solution of a differential equation with the y' term missing, and the solution of a differential equation when boundary conditions are given instead of initial conditions.

14.1 SERIES METHODS FOR SECOND-ORDER DIFFERENTIAL EQUATIONS

14.1.1 Introduction

Our present concern is to investigate problems of the form

$$y'' + P(x)y' + Q(x)y = R(x), \qquad ' = d/dx \qquad (14.1.1)$$

The solution of (14.1.1) can often be approached through that of the homogeneous equation

$$y'' + P(x)y' + Q(x)y = 0 \qquad (14.1.2)$$

and we shall concentrate on this.

In the solution of these equations by series methods, there are three types of problems:

I. Replacement of y by a series leads immediately to the general solution.
II. Replacement of y by a series leads immediately to a set of particular solutions, but not to the general solution. Further work must be done to obtain the general solution.
III. Replacement of y by a series does not lead to a solution.

We shall limit ourselves to the simplest types, which fortunately are the most common in practice. Our main purpose is to display the kind of mathematical thinking required to solve practical problems by a series method.

A complete understanding of the problem involves a dip into the theory of complex variables. This is so because some of the results over the reals are incomprehensible unless the behavior over the complex region is explored. We therefore start with a definition of analytic functions, and then solve completely the Chebyshev equation

$$(1 - x^2)y'' - xy' + n^2 y = 0, \qquad n \neq 0 \qquad (14.1.3)$$

which is a type I problem. Then we solve the Bessel equation in part:

$$x^2 y'' + xy' + (x^2 - n^2)y = 0, \qquad n \neq 0 \qquad (14.1.4)$$

to illustrate type II. Finally, we discuss methods of improving our procedure by making use of the mathematical properties of the so-called *special functions*.

All calculations were done in double precision using nesting in the summation of the series produced.

14.1.2 Taylor Series for Analytic Functions

We omit the proof of the following important theorem:

> THEOREM: TAYLOR SERIES FOR ANALYTIC FUNCTIONS   For z complex, a function $f(z)$ is analytic at $z = z_0$ iff it can be expanded in a uniformly convergent power series about that point. Convergence extends at least up to a circle, with z_0 as center, passing through the nearest singular point, if any. If $f(z)$ is analytic at z_0,
>
> $$f(z) = f(z_0) + (z - z_0)f'(z_0) + \frac{(z - z_0)^2}{2!} f''(z_0) + \cdots \qquad (14.1.5)$$

14.1.3   The Series Solution about an Ordinary Point

The point z_0 is called an *ordinary point* of the differential equation

$$y'' + P(z)y' + Q(z)y = 0, \qquad ' = d/dz \qquad (14.1.6)$$

iff both $P(z)$ and $Q(z)$ are analytic at $z = z_0$. In practical situations, there is seldom a problem deciding whether or not $P(z)$ and $Q(z)$ are analytic. For example, in Bessel's equation of the second order,

$$x^2 y'' + xy' + (x^2 - 4)y = 0 \qquad (14.1.7)$$

we have, dividing by x^2,

$$y'' + \frac{1}{x}y' + \left(\frac{x^2 - 4}{x^2}\right)y = 0 \qquad (14.1.8)$$

and it is obvious that $P(x) = 1/x$ and $Q(x) = (x^2 - 4)/x^2$ have no derivatives at $x = 0$. Therefore, $x = 0$ is *not* an ordinary point; it is called a *singular point*. Every other point (e.g. $x = 5$) is an ordinary point for (14.1.8).

On the other hand, in Airy's function:

$$y'' - xy = 0 \qquad (14.1.9)$$

we have $P(x) = 0$ and $Q(x) = -x$, so that all points are ordinary.

The solution of a differential equation about an ordinary point is quite straightforward, explained in texts on differential equations. The justification of the procedure is the following theorem:

THEOREM: ORDINARY POINTS   If z_0 is an ordinary point of

$$y'' + P(z)y' + Q(z)y = 0$$

then there exists a unique function $f(z)$ which is analytic and satisfies the differential equation in the neighborhood of z_0 at least up to the singularity nearest to z_0.

ILLUSTRATIVE PROBLEM 14.1-1   Discuss the solution about $x = 0$ for Chebyshev's equation

$$(1 - x^2)y'' - xy' + n^2 y = 0, \qquad n \neq 0 \qquad (14.1.10)$$

Discussion   Using the usual substitution

$$y = \sum_{k=0}^{\infty} a_k x^k \qquad (14.1.11)$$

we obtain the *recursion relation*

$$a_k = \frac{k^2 - n^2}{(k+2)(k+1)} a_k \qquad (14.1.12)$$

Letting $a_0 = A$ and $a_1 = B$, we obtain a series which expresses the solution of (14.1.10) and converges uniformly for all x inside the circle of convergence, $|x| < 1$.

When (14.1.12) is used to compute the particular solution for $n = 2$, the calculated values are exact, as shown in Figure 14.1-1. The computer program will be displayed later in this section.

```
FROBENIUS METHOD   CHEBYSHEV EQUATION (N=2)      0(0.1)1      1(1)10
```

X	TRUE VALUE	CALC. VALUE.	ERROR	REL. ERROR
0.0	-1.000000000000000D 00	-1.000000000000000D 00	0.0	0.0
0.100	-9.800000000000000D-01	-9.800000000000000D-01	0.0	0.0
0.200	-9.200000000000000D-01	-9.200000000000000D-01	0.0	0.0
0.300	-8.200000000000000D-01	-8.200000000000000D-01	0.0	0.0
0.400	-6.800000000000000D-01	-6.800000000000000D-01	0.0	0.0
0.500	-5.000000000000000D-01	-5.000000000000000D-01	0.0	0.0
0.600	-2.800000000000000D-01	-2.800000000000000D-01	0.0	0.0
0.700	-2.000000000000003D-02	-2.000000000000002D-02	0.0	0.0
0.800	2.800000000000000D-01	2.800000000000000D-01	0.0	0.0
0.900	6.200000000000000D-01	6.200000000000000D-01	0.0	0.0
1.000	1.000000000000000D 00			
2.000	7.000000000000000D 00	7.000000000000000D 00	0.0	0.0
3.000	1.700000000000000D 01	1.700000000000000D 01	0.0	0.0
4.000	3.100000000000000D 01	3.100000000000000D 01	0.0	0.0
5.000	4.900000000000000D 01	4.900000000000000D 01	0.0	0.0
6.000	7.100000000000000D 01	7.100000000000000D 01	0.0	0.0
7.000	9.700000000000000D 01	9.700000000000000D 01	0.0	0.0
8.000	1.270000000000000D 02	1.270000000000000D 02	0.0	0.0
9.000	1.610000000000000D 02	1.610000000000000D 02	0.0	0.0
10.000	1.990000000000000D 02	1.990000000000000D 02	0.0	0.0

Fig. 14.1-1   Solution of the Chebyshev differential equation.

14.1.4   The Series Solution about a Regular Singularity

The point z_0 is called a *regular singular point* of the differential equation

$$y'' + P(z)y' + Q(z)y = 0 \qquad (14.1.13)$$

iff both $(z - z_0)P(z)$ and $(z - z_0)^2 Q(z)$ are analytic at $z = z_0$. If z_0 is neither ordinary nor regular singular, it is said to be *irregular singular*. It can be proved that power-series methods break down for irregular singularities.

THEOREM: REGULAR SINGULAR POINTS   If z_0 is a regular singular point of (14.1.13), there exists a unique function $f(z)$ which is analytic and which satisfies the differential equation in a neighborhood of z_0 at least up to the singularity nearest to z_0.

In solving such an equation, we assume the formal solution

$$y = z^c \sum_{k=0}^{\infty} a_k z^k \qquad (14.1.14)$$

This results in a quadratic:

$$c^2 + (p_0 - 1)c + q_0 = 0 \qquad (14.1.15)$$

called an *indicial equation*. It leads to two values of the index, c_1 and c_2. If $c_1 - c_2$ is not an integer, the general solution of (14.1.13) is always obtained. If $c_1 - c_2$ is an integer, but not 0, either the general solution is obtained by using the smaller index, or else a particular solution is obtained by using the larger. If $c_1 = c_2$, a particular solution is obtained. Details of the procedure may be found in texts on differential equations, sometimes under the headings *Frobenius methods* or *Fuchs-Frobenius methods*.

ILLUSTRATIVE PROBLEM 14.1-2   Discuss a solution for Bessel's equation

$$x^2 y'' + xy' + (x^2 - n^2)y = 0, \qquad n \neq 0 \qquad (14.1.16)$$

Discussion   Using the method mentioned, we find that $c_1 = -n$ and $c_2 = n$. If $c = -n$ is tried, we obtain the recursion relation

$$a_{k+2} = \frac{-a_k}{(k+2)(k+2-2n)} \qquad (14.1.17)$$

which satisfies (14.1.16) whenever n is not an integer. If n is an integer, however, (14.1.17) blows up whenever $k + 2 = 2n$, so that it cannot be used.

For integral n, we try $c = n$, which leads to

$$a_1 = 0$$
$$a_{k+2} = \frac{-a_k}{(k+2)(k+2+2n)} \qquad (14.1.18)$$

To continue the discussion, we now focus on the situation when $n = 2$. As an example, we use $a_0 = 1/8$, which provides the particular solution tabulated by scientists, represented by the symbol $J_2(x)$, and called the *Bessel function of order* 2.

When (14.1.18) is used with $n = 2$, the coefficients in Figure 14.1-2 are calculated. In the calculation, a_0 was taken as 1, and the computed answer was multiplied by the essential arbitrary constant $1/8$.

Then $J_2(x)$ was computed for two ranges: $0(0.1)1$ and $1(1)10$, with the results shown in Figure 14.1-3.

The computer program is shown in Figure 14.1-4.

It can be seen from Figure 14.1-3 that agreement is excellent up to $x = 5$. Then the roundoff error begins to build up rapidly.

We shall now digress to discuss several methods of dealing with functions for which there are recursion relations.

FROBENIUS METHOD   BESSEL FUNCTION, ORDER 2   O(0.1)1   1(1)10

K	COEFFICIENT
0	1.0000000000000000D 00
2	-8.333333333333333D-02
4	2.604166666666666D-03
6	-4.340277777777777D-05
8	4.521122685185183D-07
10	-3.229373346560846D-09
12	1.681965284667107D-11
14	-6.674465415345660D-14
16	2.085770442295519D-16
18	-5.267097076503835D-19
20	1.097311890938299D-21
22	-1.918377431710313D-24
24	2.854728320997489D-27
26	-3.659908103842935D-30
28	4.084718865896132D-33
30	-4.004626339113855D-36
32	3.476238141591887D-39
34	-2.690586796897746D-42
36	1.868463053401212D-45
38	-1.170716198872939D-48

THE ESSENTIAL ARBITRARY CONSTANT IS 1.249999999652964D-01

Fig. 14.1-2   Coefficients for $J_2(x)$.

FROBENIUS METHOD   BESSEL FUNCTION, ORDER 2   O(0.1)1   1(1)10

X	TRUE VALUE	CALC. VALUE.	ERROR	REL. ERROR
0.0	0.0	0.0	0.0	0.0
0.100	1.248958700D-03	1.248958658D-03	4.15D-11	3.33D-08
0.200	4.983354200D-03	4.983354151D-03	4.86D-11	9.75D-09
0.300	1.116586190D-02	1.116586195D-02	-4.60D-11	-4.12D-09
0.400	1.973466310D-02	1.973466311D-02	-1.16D-11	-5.85D-10
0.500	3.060402350D-02	3.060402345D-02	4.98D-11	1.63D-09
0.600	4.366509670D-02	4.366509670D-02	0.0	0.0
0.700	5.878694440D-02	5.878694435D-02	5.21D-11	8.87D-10
0.800	7.581776250D-02	7.581776246D-02	3.61D-11	4.76D-10
0.900	9.458630430D-02	9.458630425D-02	5.15D-11	5.44D-10
1.000	1.149034849D-01			
2.000	3.528340286D-01	3.528340285D-01	8.23D-11	2.33D-10
3.000	4.860912605D-01	4.860912605D-01	1.49D-10	3.07D-10
4.000	3.641281459D-01	3.641281458D-01	1.49D-10	4.09D-10
5.000	4.656511630D-02	4.656511626D-02	3.52D-11	7.55D-10
6.000	-2.428732100D-01	-2.428732099D-01	-1.07D-10	4.42D-10
7.000	-3.014172201D-01	-3.014172200D-01	-9.77D-11	3.24D-10
8.000	-1.129917204D-01	-1.129917204D-01	-7.29D-12	6.45D-11
9.000	1.448473415D-01	1.448473415D-01	8.68D-12	5.99D-11
10.000	2.546303137D-01	2.546303135D-01	1.65D-10	6.47D-10

Fig. 14.1-3   Results of a computer program for $J_2(x)$.

```
C THIS PROGRAM USES A RECURSION RELATION DERIVED BY THE FROBENIUS METHOD      040
C TO GENERATE THE FIRST FEW COEFFICIENTS OF AN INFINITE SERIES FOR ONE        050
C FAMILY OF PARTICULAR SOLUTIONS. THEN THE ESSENTIAL ARBITRARY CONSTANT       060
C (EAC) IS CALCULATED. FINALLY, A TABLE IS PRODUCED TO COMPARE TRUE AND       070
C CALCULATED VALUES. THE INPUT CARDS ARE AS FOLLOWS.                          080
C    CARD 1   CC 1-28 NAME OR DESCRIPTION OF EQUATION                         091
C             CC 29-40   RANGE1, FOR THE FIRST INCREMENT                      092
C             CC 41-52   RANGE2, FOR THE SECOND INCREMENT                     093
C    CARD 2   CC 1    XI, THE FIRST VALUE OF X                                100
C             CC 21   TV(1), THE FIRST TRUE VALUE GIVEN                       110
C             CC 61   H(1), THE FIRST INCREMENT IN X                          120
C             CC 71   ICODE EQUALS 1 IF THE TRUE VALUES ARE TO BE CALCULATED. 121
C    CARD 3   VARIABLE FORMAT, FORM1, FOR THE OUTPUT TABLE.                   130
C    CARDS 4-7 CONTAIN TV(2) THROUGH TV(11) IN CC 1,21,41 FOR THE            140
C             FIRST SET OF INCREMENTS.                                        150
C    CARD 8   CC 1-10  H(2), THE SECOND INCREMENT                            160
C             CC 11    KF, THE SUBSCRIPT OF THE FIRST NON-ZERO COEFFICIENT    170
C             CC 12    KS, THE SUBSCRIPT OF THE SECOND NON-ZERO COEFFICIENT   180
C             CC 13-14 NC, THE INDEX, FROM THE INDICIAL EQUATION              190
C             CC 15-16 NP, THE SUBSCRIPT OF THE TRUE VALUE CHOSEN AS PIVOT    200
C             CC 17    ITYPE=2 FOR A TWO-TERM RECURSION, 3 FOR THREE          230
C    CARDS 9-11 CONTAIN TV(12) THROUGH TV(20) FOR THE SECOND                  240
C             SET OF INCREMENTS.                                              250
C THE FIRST VARIABLE STATEMENT FUNCTION, C, IS THE COEFFICIENT IN A TWO-TERM  260
C RECURSION RELATION A(K+KD)=C*A(K). FOR A THREE-TERM RECURSION RELATION      270
C A(K+2*KD)=C*A(K)+D* A(K+KD). THE RELATION BETWEEN A(K) AND A(K+KD)          275
C MUST BE DEFINED.                                                            280
      DOUBLE PRECISION TV(20),FORM1(15),A(80),C,D,E,EAC,F,TK,X,XN,Y,H(2),Z,S,RE, 290
     1CRIT,XI                                                                 291
      DIMENSION NAME(7),FORM1(15),RANGE1(3),RANGE2(3)                         300
C VARIABLE STATEMENT FUNCTION. TK CORRESPONDS TO SUBSCRIPT, K.               310
C FOR TYPE 3, THE SECOND COEFFICIENT MUST BE GIVEN.                          311
      C(TK)=-1.DO/((TK+2.DO)*(TK+6.DO))                                       12B
      D(TK)=0.DO                                                             12C
      CRIT=5.D-12                                                            12E
C INPUT.                                                                      320
      READ(1,100)NAME ,RANGE1,RANGE2                                          330
100   FORMAT(7A4,3A4,3A4)                                                     340
      READ(1,101)XI,TV(1),H(1),ICODE                                          350
101   FORMAT(2F20.0,20X,F10.0,I1)                                             360
      READ(1,102)FORM1                                                        370
102   FORMAT(15A4)                                                            380
```

```
      IF(ICODE.EQ.1)GO TO 141                                              381
C
14    READ(1,103)(TV(J),J=2,11)                                            390
103   FORMAT(3F20.0)                                                       400
141   READ(1,104)H(2),KF,KS,NC,NP,ITYPE                                    410
104   FORMAT(F10.0,2I1,2I2,I1)                                             420
      IF(ICODE.EQ.1)GO TO 142                                              425
105   READ(1,105)(TV(J),J=12,20)                                           430
      FORMAT(3F20.0)                                                       440
      GO TO 15                                                             441
C A PROGRAM FOR CALCULATION OF TRUE VALUES MAY BE INSERTED AFTER T10       T01
142   KL1=-9                                                               T02
      KL2=0                                                                T02
      DO 140 L=1,2                                                         T03
      KL1=KL1+10                                                           T05
      KL2=KL2+10                                                           T06
      IF(L.EQ.1)X=XI-H(1)                                                  T07
      IF(L.EQ.2)X=X+H(1)-H(2)                                              T08
      DO 140 K=KL1,KL2                                                     T09
      X=X+H(L)                                                             T10
140   TV(K)=0.D0                                                           TEMP
C HEADING                                                                  450
15    WRITE(3,106)NAME,RANGE1,RANGE2                                       460
106   FORMAT(1H1,T10,'FROBENIUS METHOD  ',13A4/)                           470
C ECHO OF INPUT.                                                           480
500   WRITE(3,500)XI,H(1),H(2),ICODE,KF,KS,NC,NP,ITYPE,TV                  490
      FORMAT(1H0,' XI=',1PG22.15/' H(1)=',G22.15/' H(2)=',G22.15/          500
     1' ICODE=',I1/' KF=',I1/' KS=',I1/' NC=',I2/' NP=',I2/                510
     5' ITYPE=',I1/' TRUE VALUES ARE',20(/0PF22.15))                       520
      WRITE(3,501)FORM1                                                    521
501   FORMAT(15A4)                                                         522
C HEADING.                                                                 530
      WRITE(3,106)NAME,RANGE1,RANGE2                                       540
C CONSTANTS.   NOTE THE CHANGE IN KS AND KF BECAUSE OF FORTRAN.            550
      KF=KF+1                                                              560
      KS=KS+1                                                              570
      KD=KS-KF                                                             571
      Y=X**KD                                                              572
C THE FOLLOWING CARD MAKES IT POSSIBLE TO CHANGE N EASILY.                 573
      N=20                                                                 580
      KN=KF+(N-1)*KD                                                       590
```

Fig. 14.1-4 (Continued on next page)

```
      KT=KN-KD                                                              610
      A(KF)=1.D0                                                            615
C CALCULATION OF COEFFICIENTS.                                             620
      DO 1 K=KF,KN,KD                                                       630
      TK=DFLOAT(K-1)                                                        640
      IF(ITYPE.EQ.3)GO TO 11                                               650
C TWO-TERM RECURSION.                                                       660
      A(K+KD)=C(TK)*A(K)                                                    670
      GO TO 1                                                               680
C THREE-TERM RECURSION.                                                     690
   11 A(K+2*KD)=C(TK)*A(K)+D(TK)*A(K+KD)                                   700
    1 CONTINUE                                                              710
      WRITE(3,107)                                                         720
  107 FORMAT(1H0,T26,'K',T40,'COEFFICIENT'/)                              730
      DO 12 K=KF,KN,KD                                                     740
      J=K-1                                                                750
   12 WRITE(3,108)J,A(K)                                                   760
  108 FORMAT(1H ,T25,I2,T32,1PD22.15)                                     770
C USING T(NP) AS PIVOT, TO FIND THE EAC.                                  780
      X=XI+H(1)*DFLOAT(NP-1)                                              790
      IF(NC.EQ.0)XN=1.D0                                                  800
      IF(NC.EQ.0)GO TO 121                                                810
      XN=X**NC                                                            815
  121 IF(KD.EQ.1)Y=X                                                      820
      IF(KD.EQ.1)GO TO 122                                                825
      Y=X**KD                                                             830
  122 S=A(KN)*Y                                                           840
C                                                                         850
      KFP=KF+KD                                                           855
      DO 2 K=KFP,KT,KD                                                    860
      J=KFP+KT-K                                                          870
      S=Y*(S+A(J))                                                        880
    2 S=S+A(KF)                                                           885
      S=S*XN                                                              890
C                                                                         900
      EAC=TV(NP)/S                                                        910
C                                                                         920
  109 WRITE(3,109)EAC                                                     930
      FORMAT(1H0,T10,'THE ESSENTIAL ARBITRARY CONSTANT IS',1PD22.15)      940
      WRITE(3,106)NAME,RANGE1,RANGE2                                      950
C CALCULATION AND OUTPUT OF RESULTS.                                      1000
```

```
110   WRITE(3,110)                                              1001
      FORMAT(1H0,T9,'X',T25,'TRUE VALUE',T49,'CALC. VALUE.',T67,'ERROR',  1002
     1T75,'REL. ERROR'//)                                       1003
      KL1=-9                                                    1010
      KL2=0                                                     1020
      DO 9 L=1,2                                                1030
      KL1=KL1+10                                                1040
      KL2=KL2+10                                                1050
      IF(L.EQ.1)X=XI-H(1)                                       1060
      IF(L.EQ.2)X=X+H(1)-H(2)                                   1065
      DO 9 K=KL1,KL2                                            1070
      X=X+H(L)                                                  1080
      IF(K.NE.NP)GO TO 91                                       1090
      WRITE(3,FORM1)X,TV(K)                                     1100
      GOTO 9                                                    1110
91    IF(X.EQ.0.D0)GO TO 913                                    1120
      IF(NC.EQ.0)XN=1.D0                                        1121
      IF(NC.EQ.0)GO TO 914                                      1122
      XN=X**NC                                                  1130
914   Y=X**KD                                                   1135
      S=A(KN)*Y                                                 1140
C                                                               1150
      DO 92 I=KFP,KT,KD                                         1160
      J=KFP+KT-I                                                1170
92    S=Y*(S+A(J))                                              1180
      S=S+A(KF)                                                 1186
910   S=S*EAC*XN                                                1187
      E=TV(K)-S                                                 1190
      IF(DABS(E).LT.CRIT)E=0.D0                                 1188
      IF(S.NE.0.D0)GO TO 911                                    1191
915   E=0.D0                                                    1192
      RE=0.D0                                                   1193
      GO TO 912                                                 1194
913   IF(NC.EQ.0)S=EAC                                          1195
      IF(NC.NE.0)S=0.D0                                         1196
      IF(NC.NE.0)GO TO 915                                      1197
      E=TV(K)-S                                                 1198
      IF(DABS(E).LT.CRIT)E=0.D0                                 1199
911   RE=E/S                                                    1200
912   WRITE(3,FORM1)X,TV(K),S,E,RE                              1210
9     CONTINUE                                                  1220
      END                                                       1230
```

Fig. 14.1-4 Computer program for solution of a differential equation, given the recursion relation.

14.1.5   Transformation of the Equation

We originally expanded

$$x^2 y'' + xy' + (x^2 - n^2) y = 0, \qquad ' = d/dx \qquad (14.1.19)$$

about $x_0 = 0$. Theoretically, this is an excellent choice because $x_0 = 0$ is the only finite singular point, so that convergence is guaranteed for all finite values of x. In practice, however, a power series of the form

$$y = \sum_{k=0}^{\infty} a_k x^{k+c} \qquad (14.1.20)$$

has a large buildup of roundoff error as x becomes large. It would be advisable to shift the origin to cut down on roundoff error. We shall expand about $x_0 = m$ instead of $x_0 = 0$. We let $t = x - m$, $x = t + m$, $dt = dx$ in (14.1.19). Then

$$(t + m)^2 y'' + (t + m) y' + \left([t + m]^2 - n^2 \right) y = 0, \qquad ' = d/dt \quad (14.1.21)$$

We now let

$$y = \sum_{k=0}^{\infty} a_k t^k \qquad (14.1.22)$$

and proceed as before. Note that $x_0 = m$ is an ordinary point, so that $c = 0$. After manipulation, we obtain

$$\sum_{k=0}^{\infty} \Big[a_k + 2m a_{k+1} + \left(\{k+2\}^2 + m^2 - n^2 \right) a_{k+2} $$

$$+ m(k+3)(2k+5) a_{k+3} + m^2 (k+4)(k+3) a_{k+4} \Big] t^{k+2} \equiv 0 \quad (14.1.23)$$

from which

$$\begin{cases} 2m a_0 + (m^2 - n^2 + 1) a_1 + 6m a_2 + 6m^2 a_3 = 0 & (14.1.24) \\ (m^2 - n^2) a_0 + \qquad\qquad m a_1 + 2m^2 a_2 \qquad\quad = 0 & (14.1.25) \end{cases}$$

Writing $a_0 = A$ and $a_1 = B$,

$$\begin{cases} a_2 = \dfrac{-(m^2 - n^2)A - mB}{2m^2} \\[2mm] a_3 = \dfrac{(m^2 - 3n^2)A - m(m^2 - n^2 - 2)B}{6m^3} \\[2mm] a_{k+4} = -\dfrac{a_k + 2m a_{k+1} + \left([k+2]^2 + m^2 - n^2 \right) a_{k+2} + m(k+3)(2k+5) a_{k+3}}{m^2(k+4)(k+3)} \end{cases} \qquad (14.1.26)$$

from which the coefficients and hence the calculated values of $J_n(k)$ may theoretically be calculated for any ordinary point m. We shall illustrate with the simplest case, the calculation of $J_2(x)$ expanded about $m=2$, so that $m=n=2$. Then

$$
\left\{
\begin{aligned}
a_0 &= A \\
a_1 &= B \\
a_2 &= -\tfrac{1}{4}B \\
a_3 &= \tfrac{1}{12}(B-2A) \\
a_{k+4} &= -\frac{a_k+4a_{k+1}+(k+2)^2 a_{k+2}+2(k+3)(2k+5)a_{k+3}}{4(k+4)(k+3)}
\end{aligned}
\right.
\tag{14.1.27}
$$

An obvious problem is that of finding the values of a_0 and a_1. Note that

$$J_2(t)=a_0+a_1 t+a_2 t^2+\cdots \tag{14.1.28}$$

$$J_2'(t)=a_1+2a_2 t+\cdots \tag{14.1.29}$$

so that at $t=0$, $J_2(t)=a_0$ and $J_2'(t)=a_1$. In other words, for $J_2(x)$, $J_2(2)=a_0$ and $J_2'(2)=a_1$. We must, unfortunately, have this information or equivalent information. From previous calculations, $J_2(2)=0.3528340286$. To find $J_2'(2)$, we search through treatises on the special functions, of which J_2 is one, and find that

$$nJ_n(x)+xJ_n'(x)=xJ_{n-1}(x) \tag{14.1.30}$$

Equation (14.1.30) requires knowledge of $J_1(2)$ as well. By methods identical to those already discussed, we find that $J_1(2)=0.5767248078$. Then $J_2'(2)=0.2238907792$. When these values are used, the excellent results shown in Figure 14.1-5 are obtained. Figure 14.1-6 lists the coefficients, and Figure 14.1-7 displays the computer program.

The value of this procedure is that if values of $J_n(x)$ are known with great precision for $x=1,2,3,4,\ldots$, then the formulas (14.1.26) can be used to subtabulate the function quickly and precisely. In the following, other methods from the theory of special functions are mentioned briefly. The moral of this digression is that a computer programmer can do his job most efficiently if he is able to use the mathematical theory already worked out.

14.1.6 The Three-Term Recursion Relation

A great many special functions satisfy a three-term recursion relation. These may be found in handbooks, such as Abramowitz and Stegun. For the Bessel function,

$$J_{n+1}(x)=\frac{2n}{x}J_n(x)-J_{n-1}(x) \tag{14.1.31}$$

VALUES OF J2(X) ABOUT X=2

X	TRUE VALUES	CALC. VALUES	ERROR	REL. ERROR
1.0	.1149034849	.1149034847	2.02D-10	1.76D-09
1.1	.1365641540	.1365641538	1.75D-10	1.28D-09
1.2	.1593490183	.1593490183	0.0	0.0
1.3	.1830266988	.1830266987	1.03D-10	5.62D-10
1.4	.2073558995	.2073558995	0.0	0.0
1.5	.2320876721	.2320876721	0.0	0.0
1.6	.2569677514	.2569677514	0.0	0.0
1.7	.2817389424	.2817389423	8.21D-11	2.92D-10
1.8	.3061435353	.3061435353	0.0	0.0
1.9	.3299257277	.3299257277	0.0	0.0
2.0	.3528340286	.3528340286	0.0	0.0
2.1	.3746236252	.3746236251	5.90D-11	1.57D-10
2.2	.3950586875	.3950586875	0.0	0.0
2.3	.4139145917	.4139145917	0.0	0.0
2.4	.4309800402	.4309800402	0.0	0.0
2.5	.4460590584	.4460590585	-5.02D-11	-1.13D-10
2.6	.4589728517	.4589728517	0.0	0.0
2.7	.4695615027	.4695615027	0.0	0.0
2.8	.4776854954	.4776854954	0.0	0.0
2.9	.4832270505	.4832270505	0.0	0.0
3.0	.4860912606	.4860912604	1.58D-10	3.25D-10

Fig. 14.1-5   The expansion of $J_2(x)$ about $x = 2$.

COEFFICIENTS FOR J2(X) ABOUT X=2

K	COEFFICIENTS
0	3.528340286000000D-01
1	2.238907792000000D-01
2	-5.597269479999999D-02
3	-4.014810649999999D-02
4	3.748683933333334D-03
5	1.892583227916665D-03
6	-1.145526053819440D-04
7	-4.190634536210337D-05
8	1.927820556485717D-06
9	5.429207736027896D-07
10	-2.037857085328349D-08
11	-4.631051366858497D-09

Fig. 14.1-6   The coefficients for Figure 14.1-5.

```
C CALCULATION OF J2(X) ABOUT X=2
      DOUBLE PRECISION A(50),Y(21),TV(21),E(21),R(21),CRIT,C2,C3,DEN,T,S
C
C THE FOLLOWING CRITICAL VALUE MUST BE ADJUSTED FOR DIFFERENT N.
      CRIT=5.D-15
C INPUT. ALL SUBSCRIPTS INCREASED BY 1.
      A(1)=3.528340286D-1
      A(2)=2.238907792D-1
      A(3)=-A(2)/4.DO
      A(4)=(A(2)-2.DO*A(1))/12.DO
C
      TV(1)=0.1149034849DO
      TV(2)=0.1365641540DO
      TV(3)=0.1593490183DO
      TV(4)=0.1830266988DO
      TV(5)=0.2073558995DO
      TV(6)=0.2320876721DO
      TV(7)=0.2569677514DO
      TV(8)=0.2817389424DO
      TV(9)=0.3061435353DO
```

```
       TV(10)=0.3299257277D0
       TV(11)=A(1)
       TV(12)=0.3746236252D0
       TV(13)=0.3950586875D0
       TV(14)=0.4139145917D0
       TV(15)=0.4309800402D0
       TV(16)=0.4460590584D0
       TV(17)=0.4589728517D0
       TV(18)=0.4695615027D0
       TV(19)=0.4776854954D0
       TV(20)=0.4832270505D0
       TV(21)=0.4860912606D0
C
       K=0
C SUBSCRIPTS.
1      L0=K+1
       L1=K+2
       L2=K+3
       L3=K+4
       L4=K+5
C COEFFICIENTS.
       L=4+K*(4+K)
       C2=DFLOAT(L)
       L=15+K*(11+2*K)
       C3=DFLOAT(2*L)
       L=12+K*(7+K)
       DEN=DFLOAT(4*L)
C FORMULA FOR COEFFICIENTS.
       A(L4)=-(A(L0)+4.D0*A(L1)+C2*A(L2)+C3*A(L3))/DEN
C DECISION.
       IF(DABS(A(L4)).LE.CRIT)GO TO 2
       IF(K.EQ.50)GO TO 2
       K=K+1
       GO TO 1
C PRINT-OUT OF COEFFICIENTS.
2      DO 3 JCOPY=1,3
       WRITE(3,100)
100    FORMAT(1H1,T5 ,'COEFFICIENTS FOR J2(X) ABOUT X=2'//)
       WRITE(3,101)
101    FORMAT(1H0,T6,'K',T20,'COEFFICIENTS'//)
       DO 3 I=1,K
       L=I-1
3      WRITE(3,102)L,A(I)
102    FORMAT(1H ,T5,I2,T10,1PD22.15)
C CALCULATION OF TABLE OF VALUES.
       T=-1.1D0
       DO 4 I=1,21
       T=T+1.D-1
       S=A(K)*T
       L=K-1
       DO 41 J=2,L
       M=L-J+2
41     S=T*(S+A(M))
       S=S+A(1)
       Y(I)=S
       E(I)=TV(I)-S
       IF(DABS(E(I)).LT.5.D-11)E(I)=0.D0
4      R(I)=E(I)/S
C PRINT-OUT OF TABLE OF VALUES.
       DO 5 JCOPY=1,3
       WRITE(3,103)
103    FORMAT(1H1,T20,'VALUES OF J2(X) ABOUT X=2'//)
       WRITE(3,104)
104    FORMAT(1H0,T6,'X',T12,'TRUE VALUES',T27,'CALC. VALUES',T45,'ERROR'
      1,T60,'REL. ERROR'//)
       X=0.9D0
       DO 5 I=1,21
       X=X+0.1D0
5      WRITE(3,105)X,TV(I),Y(I),E(I),R(I)
105    FORMAT(1H ,T5,F3.1,T12,F11.10,T28,F11.10,T44,1PD9.2,T60,D9.2)
       END
```

Fig. 14.1-7 Computer program for the calculation of $J_2(x)$ about $x=2$.

CALCULATION OF J2(X) USING STANDARD FORMULA

X	J2(X), TRUE	J2(X), CALC.	ERROR	REL. ERROR
0.1	0.0012489587	0.0012489579	7.66D-10	6.13D-07
0.2	0.0049833542	0.0049833538	4.40D-10	8.82D-08
0.3	0.0111658619	0.0111658621	-2.28D-10	-2.05D-08
0.4	0.0197346631	0.0197346633	-2.40D-10	-1.22D-08
0.5	0.0306040235	0.0306040236	-5.92D-11	-1.93D-09
0.6	0.0436650967	0.0436650968	-1.36D-10	-3.12D-09
0.7	0.0587869444	0.0587869442	1.50D-10	2.56D-09
0.8	0.0758177625	0.0758177625	0.0	0.0
0.9	0.0945863043	0.0945863043	0.0	0.0
1.0	0.1149034849	0.1149034848	5.80D-11	5.04D-10
1.1	0.1365641540	0.1365641540	0.0	0.0
1.2	0.1593490183	0.1593490184	-1.02D-10	-6.42D-10
1.3	0.1830266988	0.1830266987	5.38D-11	2.94D-10
1.4	0.2073558995	0.2073558995	0.0	0.0
1.5	0.2320876721	0.2320876721	0.0	0.0
1.6	0.2569677514	0.2569677515	-8.56D-11	-3.33D-10
1.7	0.2817389424	0.2817389423	8.14D-11	2.89D-10
1.8	0.3061435353	0.3061435353	0.0	0.0
1.9	0.3299257277	0.3299257277	0.0	0.0
2.0	0.3528340286	0.3528340287	-5.88D-11	-1.67D-10
2.5	0.4460590584	0.4460590585	-6.82D-11	-1.53D-10
3.0	0.4860912606	0.4860912606	0.0	0.0
3.5	0.4586291842	0.4586291842	0.0	0.0
4.0	0.3641281459	0.3641281459	0.0	0.0
4.5	0.2178489837	0.2178489837	0.0	0.0
5.0	0.0465651163	0.0465651163	0.0	0.0

Fig. 14.1-8   Calculation of J_2 from J_0 and J_1.

For example, when $n = 1$,

$$J_2(x) = \frac{2}{x} J_1(x) - J_0(x) \tag{14.1.32}$$

The tables in Abramowitz and Stegun were calculated in this manner. Since $J_0(x)$ and $J_1(x)$ were calculated with great precision over a century ago, (14.1.31) is easy to use. There is, however, the ever-present problem of propagated error, as shown in Figure 14.1-8, which was computed according to the three-term formula, using double precision. The computer program is shown in Figure 14.1-9.

14.1.7   A Two-Term Recursion Relation

From the recursion relations, it can be shown that

$$J_0(x) = \sum_{k=0}^{\infty} \left(-\frac{1}{4} x^2 \right)^k \frac{1}{(k!)^2} \tag{14.1.33}$$

and also

$$J_n(x) = \left(\frac{1}{2} x \right)^n \sum_{k=0}^{\infty} \left(-\frac{1}{4} x^2 \right)^k \frac{1}{k!(k+n)!} \tag{14.1.34}$$

```
C THE PURPOSE OF THIS PROGRAM IS TO EXPLORE THE STANDARD THREE-TERM
     FORMULA
C FOR FINDING JN(X).
     DOUBLE PRECISION ZJ0(26),ZJ1(26),ZJ2(26),CJ2(26),E(26),RE(26),X
C INPUT.
     READ(1,100)ZJ0
100  FORMAT(3F20.0)
     READ(1,100)ZJ1
     READ(1,100)ZJ2
C FIRST CALCULATION.
     X=0.D0
     DO 1 I=1,20
     X=X+0.1D0
     CJ2(I)=2.D0*ZJ1(I)/X-ZJ0(I)
     E(I)=ZJ2(I)-CJ2(I)
     IF(DABS(E(I)).LT.5.D-11)E(I)=0.D0
1    RE(I)=E(I)/ZJ2(I)
C SECOND CALCULATION.
     X=2.D0
     DO 2 I=21,26
     X=X+0.5D0
     CJ2(I)=2.D0*ZJ1(I)/X-ZJ0(I)
     E(I)=ZJ2(I)-CJ2(I)
     IF(DABS(E(I)).LT.5.D-11)E(I)=0.D0
2    RE(I)=E(I)/ZJ2(I)
C OUTPUT.
     WRITE(3,101)
101  FORMAT(1H1,T15,'CALCULATION OF J2(X) USING STANDARD FORMULA'//)
     WRITE(3,102)
102  FORMAT(1H0,T8,'X',T16,'J2(X), TRUE',T33,'J2(X), CALC.'T54,'ERROR',
    1T63,'REL. ERROR'//)
     X=0.D0
     DO 3 I=1,20
     X=X+0.1D0
3    WRITE(3,103)X,ZJ2(I),CJ2(I),E(I),RE(I)
103  FORMAT(1H ,T5,F4.1,T14,F13.10,T32,F13.10,T50,1PD9.2,T63,D9.2)
     WRITE(3,104)
104  FORMAT(1H )
     X=2.D0
     DO 4 I=21,26
     X=X+0.5D0
4    WRITE(3,103)X,ZJ2(I),CJ2(I),E(I),RE(I)
     END
```

Fig. 14.1-9 Computer program for the standard formula.

Knowing that $J_0(x)$ has been computed to a large number of places, it is natural to wonder whether the other $J_n(x)$ might not be calculated from those of $J_0(x)$ using a different approach based upon computing, rather than upon the pure mathematics of the special functions. We search for a computer algorithm as follows:

$$J_n(x) - J_0(x) = \sum_{k=0}^{\infty} \left(-\frac{1}{4}x^2\right)^k \frac{1}{k!}\left[\left(\frac{1}{2}x\right)^n \frac{1}{(k+n)!} - \frac{1}{k!}\right] \quad (14.1.35)$$

$$= \sum_{k=0}^{\infty} b_k y^k \quad (14.1.36)$$

where

$$b_k = \frac{x^n - (k+n)^{(n)}2^n}{2^n k!(k+n)!}$$

$$y = -\tfrac{1}{4}x^2 \quad (14.1.37)$$

We now examine the numerators and denominators of b_k *separately* in order to develop a computer procedure:

k	NUMERATOR	DENOMINATOR
0	$x^n - n!2^n$	$2^n 0! n!$
1	$x^n - (n+1)^{(n)}2^n$	$2^n 1!(n+1)!$
2	$x^n - (n+2)^{(n)}2^n$	$2^n 2!(n+2)!$
3	$x^n - (n+3)^{(n)}2^n$	$2^n 3!(n+3)!$

From this, the first numerator (at $k=0$) is defined as $x^n - n!2!$, and the first denominator as $2^n n!$. The only change in the numerator is in one factor, which we shall call F_k. Then $F_0 = n!$, $F_1 = (n+1)^{(n)}$, $F_2 = (n+2)^{(n)}, \ldots$. Each F_k is obtained from the previous one by multiplying by $n + k$ and dividing by k. In computer language,

$$F = F*(N+K)/K \qquad (14.1.38)$$

(Of course, mixed mode is avoided in the actual program.)

The denominators are even easier. Each denominator is formed by multiplying the previous one by $k(n+k)$. Then b_k is calculated from (14.1.37).

The results are excellent, as shown in Figures 14.1-10 and 14.1-11. The computer program for getting J_2 from J_0 is shown in Figure 14.1-12.

CALCULATION OF J2(X) FROM J0(X)

X	J1(X), TRUE	J1(X), CALC.	ERROR	REL. ERROR
0.1	0.0499375260	0.0499375260	0.0	0.0
0.2	0.0995008326	0.0995008326	0.0	0.0
0.3	0.1483188163	0.1483188163	0.0	0.0
0.4	0.1960265780	0.1960265780	0.0	0.0
0.5	0.2422684577	0.2422684577	0.0	0.0
0.6	0.2867009881	0.2867009881	0.0	0.0
0.7	0.3289957415	0.3289957415	0.0	0.0
0.8	0.3688420461	0.3688420461	0.0	0.0
0.9	0.4059495461	0.4059495461	0.0	0.0
1.0	0.4400505857	0.4400505857	0.0	0.0
1.1	0.4709023949	0.4709023949	0.0	0.0
1.2	0.4982890576	0.4982890576	0.0	0.0
1.3	0.5220232474	0.5220232474	0.0	0.0
1.4	0.5419477139	0.5419477139	0.0	0.0
1.5	0.5579365079	0.5579365079	0.0	0.0
1.6	0.5698959353	0.5698959353	0.0	0.0
1.7	0.5777652315	0.5777652315	0.0	0.0
1.8	0.5815169517	0.5815169517	0.0	0.0
1.9	0.5811570727	0.5811570727	0.0	0.0
2.0	0.5767248078	0.5767248078	0.0	0.0
2.5	0.4970941025	0.4970941025	0.0	0.0
3.0	0.3390589585	0.3390589585	0.0	0.0
3.5	0.1373775274	0.1373775274	0.0	0.0
4.0	-.0660433280	-.0660433280	0.0	0.0
4.5	-.2310604319	-.2310604319	0.0	0.0
5.0	-.3275791376	-.3275791376	0.0	0.0

Fig. 14.1-10 Calculation of $J_1(x)$ from $J_0(x)$.

X	J2(X), TRUE	J2(X), CALC.	ERROR	REL. ERROR
0.1	0.0012489587	0.0012489587	0.0	0.0
0.2	0.0049833542	0.0049833542	0.0	0.0
0.3	0.0111658619	0.0111658619	0.0	0.0
0.4	0.0197346631	0.0197346631	0.0	0.0
0.5	0.0306040235	0.0306040235	0.0	0.0
0.6	0.0436650967	0.0436650967	0.0	0.0
0.7	0.0587869444	0.0587869444	0.0	0.0
0.8	0.0758177625	0.0758177625	0.0	0.0
0.9	0.0945863043	0.0945863043	0.0	0.0
1.0	0.1149034849	0.1149034849	0.0	0.0
1.1	0.1365641540	0.1365641540	0.0	0.0
1.2	0.1593490183	0.1593490183	0.0	0.0
1.3	0.1830266988	0.1830266988	0.0	0.0
1.4	0.2073558995	0.2073558995	0.0	0.0
1.5	0.2320876721	0.2320876721	0.0	0.0
1.6	0.2569677514	0.2569677514	0.0	0.0
1.7	0.2817389424	0.2817389424	0.0	0.0
1.8	0.3061435353	0.3061435353	0.0	0.0
1.9	0.3299257277	0.3299257277	0.0	0.0
2.0	0.3528340286	0.3528340286	0.0	0.0
2.5	0.4460590584	0.4460590584	0.0	0.0
3.0	0.4860912606	0.4860912606	0.0	0.0
3.5	0.4586291842	0.4586291842	0.0	0.0
4.0	0.3641281459	0.3641281459	0.0	0.0
4.5	0.2178489837	0.2178489837	0.0	0.0
5.0	0.0465651163	0.0465651163	0.0	0.0

Fig. 14.1-11   Calculation of $J_2(x)$ from $J_0(x)$.

```
C IN THIS PROGRAM , JN(X) IS CALCULATED FROM KNOWN VALUES OF  JO(X)
      DOUBLE PRECISION ZJO(26),ZJ(26),B(16),E(26),RE(26),CJ(26),DX,F,G,S
     1,TK,X,Y,ZDEN,ZNUM,A,ZM,C
C TRUE VALUES OF JO(X) FROM A&S PAGE 390.
      READ(1,100)ZJO
100   FORMAT(3F20.0)
C TRUE VALUES OF THE PREDICTED VARIABLE.  N IS THE SUBSCRIPT OF J.
      READ(1,106)N
106   FORMAT(I2)
      READ(1,100)ZJ
C INITIAL VALUES.
      JX=1
      X=0.1D0
      DX=0.1D0
      G=DFLOAT(2**N)
C OUTSIDE LOOP, FOR STEPPED VALUES OF X.SUBSCRIPTS INCREASED BY 1.
1     K=1
      M=1
      DO 11 I=1,N
11    M=M*I
      F=DFLOAT(M)
        C=X**N
        ZNUM=C-F*G
      ZDEN=F*G
      B(1)=ZNUM/ZDEN
      K=2
C INSIDE LOOP, FOR COEFFICIENTS.
2     TK=DFLOAT(K-1)
      F=F*DFLOAT(N+K-1)/TK
        ZNUM=C-F*G
      ZM=DFLOAT((K-1)*(N+K-1))
      ZDEN=ZDEN*ZM
      B(K)=ZNUM/ZDEN
```

Fig. 14.1-12   (Continued on next page)

```
C
        IF(K.EQ.16) GO TO 3
        K=K+1
        GC TO 2
C
3       Y=-X*X/4.DO
        S=B(16)*Y
        DO 4 K=2,15
        L=17-K
4       S=Y*(S+B(L))
        CJ(JX)=S+B(1)
C
        IF(JX.EQ.26)GO TO 5
        IF(JX.GE.20)DX=0.5DO
        X=X+DX
        JX=JX+1
        GO TO 1
C CALCULATIONS FOR OUTPUT.
5       DO 6 JX=1,26
        CJ(JX)=ZJO(JX)+CJ(JX)
        E(JX)=ZJ(JX)-CJ(JX)
        IF(DABS(E(JX)).LT.5.D-11)E(JX)=0.DO
        IF(DABS(E(JX)).LT.5.D-11)RE(JX)=0.DO
        IF(DABS(E(JX)).LT.5.D-11)GO TO 6
        RE(JX)=E(JX)/ZJ(JX)
6       CONTINUE
C PRINTOUT.
        WRITE(3,102)N
102     FORMAT(1H1,T19,'CALCULATION OF J',I1,'(X) FROM J0(X)'//)
        WRITE(3,103)N,N
103     FORMAT(1H0,T8,'X',T15,'J',I1,'(X), TRUE',T31,'J',I1,'(X), CALC.',T
        149,'ERROR',T61,'REL. ERROR'//)
C FIRST TWENTY RESULTS.
        DX=0.1DO
        DO 7 JX=1,20
        Y=DFLOAT(JX-1)
        X=0.1DO+DX*Y
7       WRITE(3,104)X,ZJ(JX),CJ(JX),E(JX),RE(JX)
104     FORMAT(1H ,T5,F4.1,T14,F12.10,T31,F12.10,T48,1PD9.2,T62,D9.2)
C OTHER RESULTS.
        WRITE(3,105)
105     FORMAT(1H )
        DX=0.5DO
        DO 8 JX=21,26
        Y=DFLOAT(JX-21)
        X=2.5DO+DX*Y
8       WRITE(3,104)X,ZJ(JX),CJ(JX),E(JX),RE(JX)
        END
```

Fig. 14.1-12   Computer program for calculating $J_n(x)$ from $J_0(x)$.

PROBLEM SECTION 14.1

Using series methods, as demonstrated in this section, find solutions for the following equations. Figure 14.1-13 gives the "true values" for certain particular solutions. In each case use $x_0 = 0$.

A. The following have particular solutions in closed form.
  (1) *Laguerre equation:*

$$xy'' + (1 - \omega - x)y' + ny = 0$$

  (a) Show that $a_{k+1} = \dfrac{k-n}{(k+1)(k+1+\omega)} a_k$.

(b) With $\omega=0$ and $n=6$, form a table of values for the particular solution using $a_0=1.0$.
The exact solution is given by

$$L_6^{(0)} = \frac{1}{720}[x^6 - 36x^5 + 450x^4 - 2400x^3 + 5400x^2 - 4320x + 720]$$

(2) *Legendre equation:*

$$(1-x^2)y'' - 2xy' + n(n+1)y = 0$$

(a) Show that $a_{k+2} = \dfrac{k(k+1) - n(n+1)}{(k+2)(k+1)} a_k$.

(b) With $n=2$, form a table for the particular solution using $a_0 = -0.5$. The table should be for $x=0(0.01)0.1$ and $0.1(0.1)1$. The exact solution is given by $P_2(x) = \frac{1}{2}(3x^2 - 1)$.

B. Form a table for the particular solution indicated. The "true values" are given in Figure 14.1-13.

(3) *Bessel equation, order* 0; $0(0.1)1$, $1(1)10$:

$$xy'' + y' + xy = 0$$

(a) Show that $a_{k+2} = [-1/(k+2)^2]a_k$.
(b) Form a table for the particular solution using $a_0 = 1.0$, $a_1 = 0$.
(c) From the recursion relation, derive

$$J_0(x) = \sum_{k=0}^{\infty} \frac{(-1)^k x^{2k}}{4^k (k!)^2}$$

(4) *Modified Bessel equation, order* 1; $0(0.1)1$, $1(0.5)5.5$:

$$x^2 y'' + xy' - (x^2 + n^2)y = 0$$

(a) Show that $a_{k+2} = [1/(k+2)(k+2+2n)]a_k$.
(b) Form a table for the particular solution with $n=1$, using $a_0 = 0.5$, $a_1 = 0$.
(c) From the recursion relation, derive

$$I_n(x) = \frac{x^n}{2^n} \sum_{k=0}^{\infty} \frac{x^{2k}}{4^k k!(k+n)!}$$

(5) *Kummer equation*; $0(0.1)1$, $1(1)10$:

$$xy'' + (b-x)y' - ay = 0$$

(a) Show that $a_{k+1} = (a+k)/[(k+1)(k+b)]a_k$.
(b) With $a = -0.1$, $b = 0.2$, form a table for the particular solution using $a_0 = 1$.
(c) From the recursion relation, derive

$$M(a,b,x) = \sum_{k=0}^{\infty} \frac{(a+k-1)^{(k)}}{(b+k-1)^{(k)}} \frac{x^k}{k!}$$

BESSEL ORDER 0 1.0(0.1)2.0   INITIAL SLOPE= -4.40050585700000D-01

X	Y	X	Y
1.000	7.65197686557967D-01	1.500	5.11827671735918D-01
1.100	7.19622018527511D-01	1.600	4.55402167639381D-01
1.200	6.71132744264363D-01	1.700	3.97984859446109D-01
1.300	6.20085989561509D-01	1.800	3.39986411042558D-01
1.400	5.66851203742289D-01	1.900	2.81815593743850D-01
1.500	5.11827671735918D-01	2.000	2.23890779141236D-01

BESSEL ORDER 2 1.0(0.1)2.0   INITIAL SLOPE= 2.102436159D-01

X	Y	X	Y
1.000	1.149034849D-01	1.500	2.320876721D-01
1.100	1.365641540D-01	1.600	2.569677514D-01
1.200	1.593490183D-01	1.700	2.817389424D-01
1.300	1.830266988D-01	1.800	3.061435353D-01
1.400	2.073558995D-01	1.900	3.299252770D-01
1.500	2.320876721D-01	2.000	3.528340286D-01

MODIFIED BESSEL ORDER 1 5.0(0.1)6.0   INITIAL SLOPE= 2.237274339D 01

X	Y	X	Y
5.000	2.433564213D 01	5.500	3.858816461D 01
5.100	2.668043568D 01	5.600	4.232828803D 01
5.200	2.925430987D 01	5.700	4.643550394D 01
5.300	3.207989157D 01	5.800	5.094618498D 01
5.400	3.518205851D 01	5.900	5.590031750D 01
5.500	3.858816461D 01	6.000	6.134193677D 01

KUMMER'S EQUATION (A=-0.1,B=+0.2) 0.0(0.1)1.0   INITIAL SLOPE= -5.0000000D-01

x	Y	x	Y
0.0	1.0000000D 00	0.500	6.9553657D-01
0.100	9.4806978D-01	0.600	6.19C6129D-01
0.200	8.9204786D-01	0.700	5.3624653D-01
0.300	8.3156277D-01	0.800	4.4650560D-01
0.400	7.6620759D-01	0.900	3.4919537D-01
0.500	6.9553657D-01	1.000	2.4361069D-01

COULOMB WAVE FUNCTION, ORDER 0, ETA=5, L=0 1.0(1.0)11.0 INITIAL SLOPE= 6.7615D-05

x	Y	x	Y
1.000	2.0413D-05	6.000	7.5384D-02
2.000	2.8622D-04	7.000	1.7351D-01
3.000	1.8829D-03	8.000	3.4502D-01
4.000	8.2690D-03	9.000	6.0014D-01
5.000	2.7673D-02	10.000	9.1794D-01
6.000	7.5384D-02	11.000	1.2318D 00

AIRY FUNCTION 0.01(0.02)0.20   INITIAL SLOPE= -2.5881940D-01

x	Y	x	Y
0.0	3.5502805D-01	0.100	3.2910313D-01
0.020	3.4985214D-01	0.120	3.2406751D-01
0.040	3.4467901D-01	0.140	3.1894743D-01
0.060	3.3951139D-01	0.160	3.1384521D-01
0.080	3.3435191D-01	0.180	3.0876307D-01
0.100	3.2910313D-01	0.200	3.0370315D-01

Fig. 14.1-13 (Continued on next page)

AIRY FUNCTION 0.0(0.01)0.10

INITIAL SLOPE= -2.5881940D-01

X	Y		X	Y
0.0	3.5502805D-01		0.050	3.4209435D-01
0.010	3.5243992D-01		0.060	3.3951139D-01
0.020	3.4985214D-01		0.070	3.3693047D-01
0.030	3.4726505D-01		0.080	3.3435191D-01
0.040	3.4467901D-01		0.090	3.3177603D-01
0.050	3.4209435D-01		0.100	3.2920313D-01

PARABOLIC CYLINDER FUNCTION 0.1(0.2)2.0

INITIAL SLOPE= 1.0000D 00

X	Y		X	Y
0.0	0.0		1.000	7.7880D-01
0.200	1.9801D-01		1.200	8.3721D-01
0.400	3.8432D-01		1.400	8.5768D-01
0.600	5.4836D-01		1.600	8.4367D-01
0.800	7.3502D-01		1.800	8.0074D-01
1.000	7.7880D-01		2.000	7.3576D-01

CHEBYSHEV EQUATION (N=2) 0.0(0.1)1.0

INITIAL SLOPE= 0.0

X	Y		X	Y
0.0	-1.0000000000000000D 00		0.500	-5.0000000000000000D-01
0.100	-9.8000000000000000D-01		0.600	-2.8000000000000000D-01
0.200	-9.2000000000000000D-01		0.700	-2.0000000000000000D-02
0.300	-8.2000000000000000D-01		0.800	2.8000000000000000D-01
0.400	-6.8000000000000000D-01		0.900	6.2000000000000000D-01
0.500	-5.0000000000000000D-01		1.000	1.0000000000000000D 00

LEGENDRE EQUATION (N=2) 0.20(0.01)0.30   INITIAL SLOPE=   6.0000D-01

x	Y	x	Y
0.200	-4.400000000000000D-01	0.250	-4.062500000000000D-01
0.210	-4.338500000000000D-01	0.260	-3.986000000000000D-01
0.220	-4.274000000000000D-01	0.270	-3.906500000000000D-01
0.230	-4.206500000000000D-01	0.280	-3.824000000000000D-01
0.240	-4.136000000000000D-01	0.290	-3.738500000000000D-01
0.250	-4.062500000000000D-01	0.300	-3.650000000000000D-01

HERMITE EQUATION (N=5) 1.5(0.1)2.5   INITIAL SLOPE=   -1.50000000D 02

x	Y	x	Y
1.500	-1.170000000000000D 02	2.000	-1.600000000000000D 01
1.600	-1.278156800000000D 02	2.100	7.715232000000000D 01
1.700	-1.277257600000000D 02	2.200	2.094822400000000D 02
1.800	-1.124582400000000D 02	2.300	3.889097599999999D 02
1.900	-7.708832000000000D 01	2.400	6.241968000000000D 02
2.000	-1.600000000000000D 01	2.500	9.250000000000000D 02

Fig. 14.1-13 "True values" for selected differential equations. The initial slopes are needed for a later problem section.

(6) *Coulomb equation*; 1(1)20:

$$y'' + \left[1 - \frac{2E}{x} - \frac{L(L+1)}{x^2} \right] y = 0$$

(a) Show that there are two indicial values, $c = L + 1$ and $c = -L$.

(b) For $E = 5$, $L = 0$, show that $a_{k+2} = \dfrac{-a_k + 10a_{k+1}}{(k+3)(k+2)}$.

(c) Form a table for the particular value which has $a_0 = 8.4468135740 \times 10^{-7}$ and $a_1 = 5a_0$.

(7) *Parabolic cylinder function*; 0(0.1)1, 1(0.5)5:

$$y'' - \left(\tfrac{1}{4}x^2 + b \right) y = 0$$

(a) Show that $a_{k+4} = (0.25a_k + ba_{k+2})/(k+4)(k+2)$.

(b) With $b = -1.5$, form a table for the particular solution using $a_0 = 1$, $a_1 = 0$. [*Hint:* Show that $a_2 = (b/2)a_0$ and $a_3 = (b/6)a_1$.]

(8) *Airy equation*; 0(0.01)0.1, 0.10(0.05)0.55:

$$y'' - xy = 0$$

(a) Show that $a_{k+3} = 1/[(k+3)(k+2)]$.

(b) One solution starts as follows:

$$\text{Ai}(x) = 0.3550280538\,87817 \left(1 + \frac{1}{6}x^3 + \cdots \right)$$

$$- 0.2588194037\,92807 \left(x + \frac{2}{24}x^4 + \cdots \right)$$

Using this information, form a table for Ai(x).

14.2 HIGHER-ORDER DIFFERENTIAL EQUATIONS: THE RUNGE-KUTTA ALGORITHM

14.2.1 Introduction

Consider the system of simultaneous differential equations

$$\frac{dy}{dx} = f_1(x,y,p)$$

$$\frac{dp}{dx} = f_2(x,y,p) \tag{14.2.1}$$

where $p = dy/dx$. If $y(x_0)$ and $p(x_0)$ are given, this system can be solved using a *double* set of Runge-Kutta equations followed by a *double* set of predictor-corrector equations. The Runge-Kutta system used in the previous chapter would be as follows:

$$\begin{cases} k_{11} = hf_1(x_n, y_n, p_n) \\ k_{12} = hf_2(x_n, y_n, p_n) \end{cases} \tag{14.2.2}$$

$$\begin{cases} k_{21} = hf_1\left(x_n + \tfrac{1}{2}h, y_n + \tfrac{1}{2}k_{11}, p_n + \tfrac{1}{2}k_{12}\right) \\ k_{22} = hf_2\left(x_n + \tfrac{1}{2}h, y_n + \tfrac{1}{2}k_{11}, p_n + \tfrac{1}{2}k_{12}\right) \end{cases} \tag{14.2.3}$$

$$\begin{cases} k_{31} = hf_1\left(x_n + \tfrac{1}{2}h, y_n + \tfrac{1}{2}k_{21}, p_n + \tfrac{1}{2}k_{22}\right) \\ k_{32} = hf_2\left(x_n + \tfrac{1}{2}h, y_n + \tfrac{1}{2}k_{21}, p_n + \tfrac{1}{2}k_{22}\right) \end{cases} \tag{14.2.4}$$

$$\begin{cases} k_{41} = hf_1(x_n + h, y_n + k_{31}, p_n + k_{32}) \\ k_{42} = hf_2(x_n + h, y_n + k_{31}, p_n + k_{32}) \end{cases} \tag{14.2.5}$$

$$\begin{cases} k_1 = \tfrac{1}{6}\left[k_{11} + 2k_{21} + 2k_{31} + k_{41}\right] \\ k_2 = \tfrac{1}{6}\left[k_{12} + 2k_{22} + 2k_{32} + k_{42}\right] \end{cases} \tag{14.2.6}$$

$$\begin{cases} y_{n+1} = y_n + k_1 \\ p_{n+1} = p_n + k_2 \end{cases} \tag{14.2.7}$$

Now, consider the second-order differential equation

$$y'' + g_1(x,y)y' + g_0(x,y)y = g(x,y) \tag{14.2.8}$$

If we define

$$\begin{cases} p = y' = f_1(x,y,p) \\ q = y'' = f_2(x,y,p) = g(x,y) - g_0(x,y)y - g_1(x,y)p \end{cases} \tag{14.2.9}$$

it becomes clear that the simultaneous system of two equations (14.2.9) is equivalent to the second-order differential equation (14.2.8) and that a Runge-Kutta algorithm like equations (14.2.2) to (14.2.7) can be used.

The extension to higher-order differential equations is immediate. For

$$y''' + g_2(x,y)y'' + g_1(x,y)y' + g_0(x,y)y = g(x,y) \tag{14.2.10}$$

we start with

$$\begin{cases} p = y' = f_1(x,y,p,q) \\ q = y'' = f_2(x,y,p,q) \\ r = y''' = f_3(x,y,p,q) = g(x,y) - g_0(x,y)y - g_1(x,y)p - g_2(x,y)q \end{cases} \tag{14.2.11}$$

and then develop the algorithm. Corresponding to the equations (14.2.3) would be

$$\begin{cases} k_{21} = hf_1\left(x_n + \tfrac{1}{2}h, y_n + \tfrac{1}{2}k_{11}, p_n + \tfrac{1}{2}k_{12}, q_n + \tfrac{1}{2}k_{13}\right) \\ k_{22} = hf_2\left(x_n + \tfrac{1}{2}h, y_n + \tfrac{1}{2}k_{11}, p_n + \tfrac{1}{2}k_{12}, q_n + \tfrac{1}{2}k_{13}\right) \quad (14.2.12) \\ k_{23} = hf_3\left(x_n + \tfrac{1}{2}h, y_n + \tfrac{1}{2}k_{11}, p_n + \tfrac{1}{2}k_{12}, q_n + \tfrac{1}{2}k_{13}\right) \end{cases}$$

with the others done similarly.

14.2.2 A Runge-Kutta Algorithm for Second Order

We shall illustrate the algorithm for second-order differential equations. Because x and y are not involved explicitly in k_{11}, k_{21}, k_{31} and k_{41}, the equations can be simplified somewhat. In the following, $p = y'$ and $q = y''$.

Step I. $x = x_n, y = y_n, p = p_n$:

$$\begin{cases} k_{11} = hp \\ k_{12} = hq \end{cases} \quad (14.2.13)$$

Step II. $x = x_n + \tfrac{1}{2}h, y = y_n + \tfrac{1}{2}k_{11}, p = p_n + \tfrac{1}{2}k_{12}$:

$$\begin{cases} k_{21} = hp \\ k_{22} = hq \end{cases} \quad (14.2.14)$$

Step III. x is unchanged, $y = y_n + \tfrac{1}{2}k_{21}, p = p_n + \tfrac{1}{2}k_{22}$:

$$\begin{cases} k_{31} = hp \\ k_{32} = hq \end{cases} \quad (14.2.15)$$

Step IV. $x = x_n + h, y = y_n + k_{31}, p = p_n + k_{32}$:

$$\begin{cases} k_{41} = hp \\ k_{42} = hq \end{cases} \quad (14.2.16)$$

Step V.

$$\begin{cases} k_1 = \tfrac{1}{6}(k_{11} + 2k_{21} + 2k_{31} + k_{41}) \\ k_2 = \tfrac{1}{6}(k_{12} + 2k_{22} + 2k_{32} + k_{42}) \end{cases} \quad (14.2.17)$$

Step VI.

$$\begin{cases} x_{n+1} = x_n + h \\ y_{n+1} = y_n + k_1 \\ p_{n+1} = p_n + k_2 \end{cases} \quad (14.2.18)$$

We illustrate with a problem for which we know the exact solution.

ILLUSTRATIVE PROBLEM 14.2-1   Find three Runge-Kutta starters for

$$xy'' - 3y' + \frac{3}{x}y = 2x^2 - x \tag{14.2.19}$$

given the initial conditions $x_0 = 1$, $y_0 = 1$, $p_0 = y_0' = 10$. For checking, the exact solution is

$$y = 3x^3 - 2x + x^2(1 + x\ln x) \tag{14.2.20}$$

Solution   We start by transforming (14.2.19) into a simultaneous system:

$$\begin{cases} p = y' \\ q = y'' = \dfrac{3p}{x} - \dfrac{3y}{x^2} + 2x - 1 \end{cases} \tag{14.2.21}$$

Then

$$\begin{cases} p_0 = 10 \\ q_0 = \dfrac{3(10)}{1} - \dfrac{3(2)}{1^2} + 2(1) - 1 = 25 \end{cases} \tag{14.2.22}$$

Using the system (14.2.13) with $h = 0.1$,

$$\begin{cases} k_{11} = 0.1p_0 = 1 \\ k_{12} = 0.1q_1 = 2.5 \end{cases} \tag{14.2.23}$$

For the system (14.2.14), we recalculate x, y and p:

$$\begin{cases} x = x_0 + \frac{1}{2}h = 1 + \frac{1}{2}(0.1) = 1.05 \\ y = y_0 + \frac{1}{2}k_{11} = 2 + \frac{1}{2}(1) = 2.5 \\ p = p_0 + \frac{1}{2}k_{12} = 10 + \frac{1}{2}(2.5) = 11.25 \end{cases} \tag{14.2.24}$$

Substituting, we have

$$\begin{cases} k_{21} = hp = 0.1(11.25) = 1.125 \\ k_{22} = hq = 0.1\left[\dfrac{3(11.25)}{1.05} - \dfrac{3(2.5)}{(1.05)^2} + 2(1.05) - 1 \right] = 2.64401\,3605 \end{cases} \tag{14.2.25}$$

For system (14.2.15), we recalculate y and p; x is unchanged:

$$\begin{cases} y = y_0 + \frac{1}{2}k_{21} = 2 + \frac{1}{2}(1.125) = 2.5625 \\ p = p_0 + \frac{1}{2}k_{22} = 10 + \frac{1}{2}(2.64401\,3605) = 11.32200\,680 \end{cases} \tag{14.2.26}$$

Substituting, we have

$$k_{31} = hp = 0.1(11.32200680) = 1.1322000680$$

$$k_{32} = hq = 0.1 \left[\frac{3(11.32200680)}{1.05} - \frac{3(2.5625)}{(1.05)^2} + 2(1.05) - 1 \right] = 2.647580175 \quad (14.2.27)$$

Recalculating x, y and p for the system (14.2.16), we have

$$\begin{cases} x = x_0 + h = 1 + 0.1 = 1.1 \\ y = y_0 + k_{31} = 2 + 1.1322000680 = 3.132200680 \\ p = p_0 + k_{32} = 10 + 2.647580175 = 12.647580175 \end{cases} \quad (14.2.28)$$

Substituting, we have

$$\begin{cases} k_{41} = hp = 0.1(12.647580175) = 1.2647580175 \\ k_{42} = hq = 0.1 \left[\frac{3(12.647580175)}{1.1} - \frac{3(3.132200680)}{(1.1)^2} + 2(1.1) - 1 \right] \end{cases}$$

$$= 2.792761365 \quad (14.2.29)$$

Now we apply (14.2.17) to obtain

$$\begin{cases} k_1 = \frac{1}{6} [1 + 2(1.125) + 2(1.132200680) + 1.2647580175] \\ \quad = 1.129859896 \\ k_2 = \frac{1}{6} [2.5 + 2(2.644013605) + 2(2.647580175) + 2.792761265] \\ \quad = 2.646024821 \end{cases} \quad (14.2.30)$$

The result of the first set of steps is as follows:

$$\begin{cases} x_1 = x_0 + h = 1.1 \\ y_1 = y_0 + k_1 = 3.129859896 \\ p_1 = p_0 + k_2 = 12.646024821 \end{cases} \quad (14.2.31)$$

We can now return to step I using these values for x, y and p. When this is done, we find that

$$\begin{cases} x_2 = 1.2 \\ y_2 = 4.539060954 \\ p_2 = 15.58770344 \end{cases} \quad (14.2.32)$$

Returning to step I with these values,

$$\begin{cases} x_3 = 1.3 \\ y_3 = 6.257433335 \\ p_3 = 18.83028802 \end{cases} \tag{14.2.33}$$

This gives us four starting values at $h = 0.1$, namely $y(0)$, $y(0.1)$, $y(0.2)$ and $y(0.3)$. We designate these as $R_1(0)$, $R_1(2)$, $R_1(4)$ and $R_1(6)$. The next step is to calculate *six* sets of Runge-Kutta values for y and p, comparing them as follows:

FIRST ROUND	SECOND ROUND
$x = 1, y(1), p(1)$	$x = 1, y(1), p(1)$
	$x = 1.05, R_2(1) = y(1.05), p(1.05)$
$x = 1.1, R_1(2) = y(1.1), p(1.1)$	$x = 1.1, R_2(2) = y(1.1), p(1.1)$
	$x = 1.15, R_2(3) = y(1.15), p(1.15)$
$x = 1.2, R_1(4) = y(1.2), p(1.2)$	$x = 1.2, R_2(4) = y(1.2), p(1.2)$
	$x = 1.25, R_2(5) = y(1.25), p(1.25)$
$x = 1.3, R_1(6) = y(1.3), p(1.3)$	$x = 1.3, R_2(6) = y(1.3), p(1.3)$

We compare $R_1(6)$ with $R_2(6)$ to estimate the error. To illustrate the procedure explicitly, we now display one set of calculations. We start by copying (14.2.22):

$$\begin{cases} p_0 = 10 \\ q_0 = 25 \end{cases} \tag{14.2.34}$$

Using the system (14.2.13) with $h = 0.05$,

$$\begin{cases} k_{11} = 0.05 p_0 = 0.5 \\ k_{12} = 0.05 q_0 = 1.25 \end{cases} \tag{14.2.35}$$

For the system (14.2.14), we recalculate x, y and p:

$$\begin{cases} x = x_0 + \frac{1}{2}h = 1 + \frac{1}{2}(0.05) = 1.025 \\ y = y_0 + \frac{1}{2}k_{11} = 2 + \frac{1}{2}(0.5) = 2.25 \\ p = p_0 + \frac{1}{2}k_{12} = 10 + \frac{1}{2}(1.25) = 10.625 \end{cases} \tag{14.2.36}$$

Then

$$\begin{cases} k_{21} = 0.05p = 0.53125 \\ k_{22} = 0.05q = 0.05(25.72281380) = 1.286140690 \end{cases} \tag{14.2.37}$$

For the system (14.2.15), x is unchanged; we recalculate y and p:

$$\begin{cases} y = y_0 + \frac{1}{2}k_{21} = 2.2656255 \\ p = p_0 + \frac{1}{2}k_{22} = 10.64307035 \end{cases} \tag{14.2.38}$$

Then

$$\begin{cases} k_{31} = 0.05p = 0.5321535173 \\ k_{32} = 0.05q = 1.2865543316 \end{cases} \tag{14.2.39}$$

For the system (14.2.16),

$$\begin{cases} x = x_0 + h = 1 + 0.05 = 1.05 \\ y = y_0 + k_{31} = 2.5321535173 \\ p = p_0 + k_{32} = 11.28655432 \end{cases} \tag{14.2.40}$$

Then

$$\begin{cases} k_{41} = 0.05p = 0.5643277158 \\ k_{42} = 0.05q = 1.3228544220 \end{cases} \tag{14.2.41}$$

Using (14.2.17),

$$\begin{cases} k_1 = 0.5318557917 \\ k_2 = 1.2863740439 \end{cases} \tag{14.2.42}$$

and therefore

$$\begin{cases} x_1 = 1.05 \\ y_1 = y_0 + k_1 = 2.531855792 \\ p_1 = p_0 + k_2 = 11.2863740439 \end{cases} \tag{14.2.43}$$

The next set of calculations leads to

$$\begin{cases} x_2 = 1.1 \\ y_2 = 3.1298558028 \\ p_2 = 12.64537709 \end{cases} \tag{14.2.44}$$

The computer results are shown in Figure 14.2-1, and the computer program in Figure 14.2-2. On each line of Figure 14.2-1, the corresponding values for $2h$ and h are shown.

14.2.3 Estimating the Error

In Table 14.2-1 we display the ratio $|(R_1 - R_2)/E|$, where E is the actual error. In the sample, comprising Illustrative Problem 14.2-1 and the ten problems at the end of the section, the Rule of Fifteen is not as persuasive as it was for the first-order equations, but it is a great deal better than nothing.

FIRST THREE R-K VALUES FOR XY''-3Y'+(3/X)Y

N	H FOR R2	X	R1(Y)	R2(Y)	R1-R2
2	0.0500000	1.1000000000	3.1298598963394880 00	3.1298580280620770 00	1.87D-06
2	0.0500000	1.2000000000	4.5390571532812950 00	4.5390521025880650 00	5.05D-06
2	0.0500000	1.3000000000	6.2574247428804030 00	6.2574151169912810 00	9.63D-06
3	0.0250000	1.0500000000	2.5318557917156350 00	2.5318557195074490 00	7.22D-08
3	0.0250000	1.1000000000	3.1298580280620770 00	3.1298578622187680 00	1.66D-07
3	0.0250000	1.1500000000	3.7976857474127690 00	3.7976854657350070 00	2.82D-07
4	0.0125000	1.0250000000	2.2578881206797880 00	2.2578881181860010 00	2.49D-09
4	0.0125000	1.0500000000	2.5318557195074490 00	2.5318557141809210 00	5.33D-09
4	0.0125000	1.0750000000	2.8223593659948640 00	2.8223593574896360 00	8.51D-09
5	0.0062500	1.0125000000	2.1269625712799880 00	2.1269625711981700 00	8.18D-11
5	0.0062500	1.0250000000	2.2578881181860010 00	2.2578881180170500 00	1.69D-10
5	0.0062500	1.0375000000	2.3928334988145760 00	2.3928334985312000 00	2.61D-10
6	0.0031250	1.0062500000	2.0629894616446812D 00	2.0629894616441930 00	2.62D-12
6	0.0031250	1.0125000000	2.1269625711981700 00	2.1269625711928480 00	5.32D-12
6	0.0031250	1.0187500000	2.1919264223941150 00	2.1919264223860080 00	8.11D-12
7	0.0015625	1.0031250000	2.0313722178380430 00	2.0313722178379600 00	8.30D-14
7	0.0015625	1.0062500000	2.0629894616441930 00	2.0629894616440250 00	1.67D-13
7	0.0015625	1.0093750000	2.0948526172842950 00	2.0948526172840420 00	2.52D-13

Fig. 14.2-1 Computer results for Illustrative Problem 14.2-1.

```
C THE PURPOSE OF THIS PROGRAM IS TO FIND THREE SATISFACTORY RUNGE-KUTTA
C VALUES OF Y TO USE AS STARTERS FOR A PREDICTOR-CORRECTOR ALGORITHM.
C THERE ARE TWO VARIABLE STATEMENT FUNCTION CARDS, ONE TO DEFINE THE
C DIFFENTIAL EQUATION, THE OTHER TO CALCULATE TRUE VALUES. THERE IS ONE
C INPUT CARD AS FOLLOWS.
C          CC  1-28  DESCRIPTION OF THE D. E.
C          CC 31-40  INITIAL X, XO, IN F-FORMAT
C          CC 41-50  INITIAL Y, YO, IN F-FORMAT
C          CC 51-60  INITIAL P, PO, IN F-FORMAT
C          CC 61-70  INITIAL H, HO, IN F-FORMAT
C WHERE X, Y AND/OR P IS IRRATIONAL, THE VALUE IS RECALCULATED AFTER
C THE INPUT CARD HAS BEEN READ IN WITH A DUMMY VALUE.
C
       DOUBLE PRECISION RY1(6),RY2(640),RP1(6),RP2(640),A,B,CRIT,E,E1,F,
      1HO,P,PI,PO,R,S,TKP,TKP1,TKP2,TKP3,TKP4,TKY,TKY1,TKY2,TKY3,TKY4,TV,
      2X,XO,Y,YO,X1,D,H
       DIMENSION NAME(7)
C VARIABLE STATEMENT FUNCTIONS.
       F(X,Y,P)=3.DO*P/X-3.DO*Y/X**2+2.DO*X-1.DO
       S(X)=X**2*(3.DO*X+1.DO+X*DLOG(X))-2.DO*X
C INPUT.
       READ(1,100)NAME,XO,YO,PO,HO
100    FORMAT(7A4,2X,4F10.0)
C CONSTANTS.
       PI=3.141592653589793DO
       E1=2.718281828459045DO
       A=0.5DO
       B=DSQRT(3.DO)
       D=DSQRT(2.DO)
       CRIT=5.D-17
C SET BOX.
       H=HO
       X=XO-H
       Y=YO
       P=PO
       N=1
C FIRST CALCULATION OF R-K VALUES.
       DO 1 K=2,6,2
       X=X+H
       TKY1=H*P
       TKP1=H*F(X,Y,P)
C
       TKY2=H*(P+A*TKP1)
       TKP2=H*F(X+A*H,Y+A*TKY1,P+A*TKP1)
C
       TKY3=H*(P+A*TKP2)
       TKP3=H*F(X+A*H,Y+A*TKY2,P+A*TKP2)
C
       TKY4=H*(P+TKP3)
       TKP4=H*F(X+H,Y+TKY3,P+TKP3)
C
       TKY=(TKY1+TKY4+2.DO*(TKY2+TKY3))/6.DO
       TKP=(TKP1+TKP4+2.DO*(TKP2+TKP3))/6.DO
C
       Y=Y+TKY
       P=P+TKP
       RY1(K)=Y
1      RP1(K)=P
C FIRST OUTPUT HEADING.
       WRITE(3,101)NAME
101    FORMAT(1H1,T30,'FIRST THREE R-K VALUES FOR ',7A4/)
       WRITE(3,102)
102    FORMAT(T7,'N',T10,'H FOR R2',T30,'X',T49,'R1(Y)',T74,'R2(Y)',T95,'
      1R1-R2'/)
C SECOND .CALCULATION.
2      H=A*H
       X=XO-H
```

```
      Y=Y0
      P=P0
      N=N+1
      DO 3 K=1,6
      X=X+H
      TKY1=H*P
      TKP1=H*F(X,Y,P)
C
      TKY2=H*(P+A*TKP1)
      TKP2=H*F(X+A*H,Y+A*TKY1,P+A*TKP1)
C
      TKY3=H*(P+A*TKP2)
      TKP3=H*F(X+A*H,Y+A*TKY2,P+A*TKP2)
C
      TKY4=H*(P+TKP3)
      TKP4=H*F(X+H,Y+TKY3,P+TKP3)
C
      TKY=(TKY1+TKY4+2.D0*(TKY2+TKY3))/6.D0
      TKP=(TKP1+TKP4+2.D0*(TKP2+TKP3))/6.D0
C
      Y=Y+TKY
      P=P+TKP
      RY2(K)=Y
3     RP2(K)=P
C FIRST OUTPUT TABLE.
      X1=X0
      DO 4 K=2,6,2
      X1=X1+H*2.D0
      E=RY1(K)-RY2(K)
4     WRITE(3,103)N,H,X1,RY1(K),RY2(K),E
103   FORMAT(1H ,T6,I2,T10,F10.7 ,T23,F15.10,T40,1PD23.15,T65,D23.15,T93
     1,D9.2)
      WRITE(3,104)
104   FORMAT(1H )
C COMPARISON OF FOURTH VALUES, RY1(6) AND RY2(6).
      IF(DABS(RY2(6)-RY1(6)).LE.CRIT)GO TO 5
C LIMITING THE NUMBER OF TRIALS.
      IF(N.EQ.7) GO TO 5
C STEP, THEN RETURN TO A NEW SECOND CALCULATICN.
      RY1(2)=RY2(1)
      RY1(4)=RY2(2)
      RY1(6)=RY2(3)
      GO TO 2
C COMPLETED TABLE OF UP TO 640 VALUES.
5     KF=10*2**(N-1)
      DO 6 K=7,KF
      X=X+H
      TKY1=H*P
      TKP1=H*F(X,Y,P)
C
      TKY2=H*(P+A*TKP1)
      TKP2=H*F(X+A*H,Y+A*TKY1,P+A*TKP1)
C
      TKY3=H*(P+A*TKP2)
      TKP3=H*F(X+A*H,Y+A*TKY2,P+A*TKP2)
C
      TKY4=H*(P+TKP3)
      TKP4=H*F(X+H,Y+TKY3,P+TKP3)
C
      TKY=(TKY1+TKY4+2.D0*(TKY2+TKY3))/6.D0
      TKP=(TKP1+TKP4+2.D0*(TKP2+TKP3))/6.D0
C
      Y=Y+TKY
      P=P+TKP
      RY2(K)=Y
6     RP2(K)=P
```

Fig. 14.2-2 **(Continued on next page)**

```
C HEADING FOR SECOND OUTPUT.
      L=0
      WRITE(3,105)NAME
105   FORMAT(1H1,T18,'TABLE OF RUNGE-KUTTA VALUES FOR ',7A4/)
      WRITE(3,106)N,H
106   FORMAT(1H0,T5,'FINAL N = ',I2,5X,'FINAL H = ',F10.7/)
      WRITE(3,107)
107   FORMAT(1H0,T19,'X',T31,'CALC. Y',T67,'TRUE VALUE',T91,'ERROR',T102
     1,'REL. ERROR'/)
      X=X0
      DO 7 K=1,KF
      L=L+1
      X=X+H
      TV=S(X)
      E=TV-RY2(K)
      IF(DABS(E).LE.5.D-17)GO TO 71
      R=E/RY2(K)
      GO TO 72
71    E=0.D0
      R=0.D0
72    WRITE(3,108)X,RY2(K),TV,E,R
108   FORMAT(1H ,T11,F17.15,T23,1PD23.15,T61,D23.15,T89,D9.2,T103,D9.2)
      IF(L.NE.45)GO TO 7
      WRITE(3,105)NAME
      WRITE(3,106)N,H
      WRITE(3,107)
      L=0
7     CONTINUE
      END
```

Fig. 14.2-2 Computer program for determination of Runge-Kutta starting values for a second-order differential equation.

Table 14.2-1
Rule of Thumb for Second-Order Runge-Kutta Determination

PROBLEM	$|(R_1 - R_2)/(\text{ACTUAL ERROR})|$
I. P. 14.2-1	11
(1)	9
(2)	16
(3)	16
(4)	15
(5)	13
(6)	7
(7)	15
(8)	15
(9)	12
(10)	15

14.2.4 Extending the Algorithm

We continued the Runge-Kutta algorithm to $x_0 + 1$ to see what would happen. Table 14.2-2 traces the errors and relative errors for Illustrative Problem 14.2-1. This suggests that a more suitable rule of thumb may be based upon the relative error than upon the absolute error.

Table 14.2-2
Actual errors for the Runge-Kutta Results Applied to Illustrative Problem 14.2-1

x	ACTUAL ERROR	REL. ERROR
1.1	29×10^{-14}	9×10^{-14}
1.2	71×10^{-14}	16×10^{-14}
1.3	126×10^{-14}	20×10^{-14}
1.4	197×10^{-14}	24×10^{-14}
1.5	283×10^{-14}	26×10^{-14}
1.6	387×10^{-14}	28×10^{-14}
1.7	507×10^{-14}	30×10^{-14}
1.8	638×10^{-14}	31×10^{-14}
1.9	790×10^{-14}	32×10^{-14}
2.0	966×10^{-14}	33×10^{-14}

PROBLEM SECTION 14.2

In each problem, start with $h=0.1$ and cut h in half until there have been seven iterations, resulting in a table like Figure 14.2-1. If time permits, continue the algorithm to x_0+1, calculating the actual and relative errors. In the following, D is the difference between $R_1(6)$ and $R_2(6)$ in the seventh iteration, E_1 is the actual error of R_2 in this iteration, and E_2 is the actual error at x_0+1. These values were obtained using double precision on an IBM 360 and are supplied for checking. You are reminded that different computers, or even the same computers with different kinds or levels of compiler, will display different errors, but not very different.

(1) $x^2y'' - 3xy' + 3y = 2x^3 - x^2$, $y(1)=6$, $p(1)=12$
Exact solution: $y = 2x^3 + 3x + x^3\ln x + x^2$

$D = 1.48 \times 10^{-13}$, $E_1 = 1.58 \times 10^{-14}$, $E_2 = 6.63 \times 10^{-12}$

(2) $x^2(x+1)y'' - x(x^2+4x+2)y' + (x^2+4x+2)y = -x^4 - 2x^3$, $y(0)=10$, $p(0)=5$
Exact solution: $y = 4x^2e^x + 5x + x^2$.

$D = 1.98 \times 10^{-12}$, $E_1 = 1.21 \times 10^{-13}$, $E_2 = 6.01 \times 10^{-11}$

(3) $xy'' - (2x+1)y' + (x+1)y = (x^2+x-1)e^{2x}$, $y(0)=7$, $p(0)=8$
Exact solution: $y = 6x^2e^x + 7e^x + xe^{2x}$

$D = 1.07 \times 10^{-11}$, $E_1 = 6.82 \times 10^{-13}$, $E_2 = 3.63 \times 10^{-10}$

(4) $(x-1)y'' - (4x-7)y' + (4x-6)y = 0$, $y(0)=41$, $p(0)=50$
Exact solution: $y = 8e^{2x}(x-2)^2 + 9e^{2x}$

$D = 1.34 \times 10^{-12}$, $E_1 = 8.88 \times 10^{-14}$, $E_2 = 3.80 \times 10^{-11}$

(5) $y'' - 2(\tan x)y' + 3y = 2\sec x$, $y(0)=1$, $p(0)=3$
Exact solution: $y = \cos x + 3\sin x$

$D = 5.55 \times 10^{-14}$, $E_1 = 4.22 \times 10^{-15}$, $E_2 = 7.97 \times 10^{-14}$

(6) $x^2y'' - 2xy' + (x^2 + 2)y = x^3 e^x$, $y(\pi/2) = (\pi/4)(8 + e^{\pi/2})$, $p(\pi/2) = 4 - \pi + \frac{1}{4}e^{\pi/2}(\pi + 2)$

Exact solution: $y = 2x\cos x + 4x\sin x + \frac{1}{2}xe^x$

$D = 2.35 \times 10^{-14}$, $E_1 = 3.33 \times 10^{-15}$, $E_2 = 1.94 \times 10^{-12}$

(7) $y'' - 2xy' + (x^2 + 2)y = \exp[\frac{1}{2}(x^2 + 2x)]$, $y(0) = -\frac{3}{4}$, $p(0) = 5\sqrt{3} + \frac{1}{4}$

Exact solution: $y = (\exp[x^2/2])(5\sin x\sqrt{3} - \cos x\sqrt{3}) + \frac{1}{4}\exp[\frac{1}{2}(x^2 + 2x)]$

$D = 6.46 \times 10^{-14}$, $E_1 = 4.22 \times 10^{-15}$, $E_2 = 2.46 \times 10^{-13}$

(8) $y'' - 4xy' + 4x^2y = x\exp(x^2/2)$, $y(0) = -1$, $p(0) = 4\sqrt{2} + \frac{1}{2}e^4$

Exact solution: $y = \exp(x^2)(4\sin x\sqrt{2} - 2\cos x\sqrt{2}) + \frac{1}{2}x\exp(x^2)$

$D = 1.94 \times 10^{-13}$, $E_1 = 1.31 \times 10^{-14}$, $E_2 = 2.38 \times 10^{-12}$

(9) $y'' - (\cot x)y' - (\sin^2 x)y = \cos x\sin^2 x$, $y(\pi/4) = e^a - 3e^{-a} - a$, $p(\pi/4) = 3ae^{-a} - ae^{-a} + a$, $a = \sqrt{2}/2$
Exact solution: $y = 3\exp(-\cos x) + \exp(\cos x) - \cos x$

$D = 4.97 \times 10^{-14}$, $E_1 = 4.00 \times 10^{-15}$, $E_2 = 1.10 \times 10^{-13}$

(10) $y'' + (2/x)y' + y/x^4 = (2x^2 + 1)/x^6$, $y(2/\pi) = \pi^2/4 - 1$, $p(2/\pi) = (\pi^2/4)(2 - \pi)$
Exact solution: $y = 2\cos(1/x) - \sin(1/x) + 1/x^2$

$D = 1.60 \times 10^{-11}$, $E_1 = 1.06 \times 10^{-12}$, $E_2 = 3.31 \times 10^{-12}$

14.3 PREDICTOR-CORRECTOR METHOD: SECOND ORDER

14.3.1 Introduction

The thought behind the second-order predictor-corrector method is similar to that of the Runge-Kutta second-order method. There is no need to elaborate. We shall merely illustrate the method.

The Adams-Moulton equations are as follows, where p is, as usual, dy/dx, and F is y'':

$$\text{pre } y_{k+1} = y_k + \frac{h}{24}\big[55p(x_k, y_k, p_k) - 59p(x_{k-1}, y_{k-1}, p_{k-1})$$

$$+ 37p(x_{k-2}, y_{k-2}, p_{k-2}) - 9p(x_{k-3}, y_{k-3}, p_{k-3})\big] \quad (14.3.1)$$

$$\text{pre } p_{k+1} = p_k + \frac{h}{24}\big[55F(x_k, y_k, p_k) - 59F(x_{k-1}, y_{k-1}, p_{k-1})$$

$$+ 37F(x_{k-2}, y_{k-2}, p_{k-2}) - 9F(x_{k-3}, y_{k-3}, p_{k-3})\big] \quad (14.3.2)$$

followed by

$$\begin{cases} \operatorname{cor} y_{k+1} = y_k + \dfrac{h}{24} \big[9p(x_{k+1}, y_{k+1}, p_{k+1}) + 19p(x_k, y_k, p_k) \\ \qquad\qquad - 5p(x_{k-1}, y_{k-1}, p_{k-1}) + p(x_{k-2}, y_{k-2}, p_{k-2}) \big] \quad (14.3.3) \\[2mm] \operatorname{cor} p_{k+1} = p_k + \dfrac{h}{24} \big[9F(x_{k+1}, y_{k+1}, p_{k+1}) + 19F(x_k, y_k, p_k) \\ \qquad\qquad - 5F(x_{k-1}, y_{k-1}, p_{k-1}) + F(x_{k-2}, y_{k-2}, p_{k-2}) \big] \quad (14.3.4) \end{cases}$$

and completed by the mop-up:

$$\begin{cases} y = \operatorname{cor} y + \dfrac{19}{270}(\operatorname{pre} y - \operatorname{cor} y) & (14.3.5) \\[2mm] p = \operatorname{cor} p + \dfrac{19}{270}(\operatorname{pre} p - \operatorname{cor} p) & (14.3.6) \end{cases}$$

14.3.2 An Example

As usual, we illustrate by a hand calculation.

ILLUSTRATIVE PROBLEM 14.3-1   Given

$$xy'' - 3y' + \frac{3}{x}y = 2x^2 - x \qquad (14.3.7)$$

calculate values for the particular solution passing through

$$x_0 = 1, \qquad y_0 = 2, \qquad p_0 = 10 \qquad (14.3.8)$$

Compare the results with the true values calculated from

$$y = 3x^3 - 2x + x^2(1 + x\ln x) \qquad (14.3.9)$$

Solution   For purposes of illustration, we use (14.3.9) and its derivative

$$y' = p = x^2(10 + 3\ln x) + 2x - 2 \qquad (14.3.10)$$

to obtain starting values with $h = 0.0015625 = 0.1/64$:

k	x	y	$y' = p$	$y'' = F$
1	1.0	2.0	10.	25.
2	1.00156250	2.0156 5536	10.0390 9790	25.0453 1947
3	1.00312500	2.0313 72218	10.0782 6663	25.0906 5281
4	1.00468750	2.047150156	10.1175 0621	25.1360 0278

We apply (14.3.1) and (14.3.2) to find the predictor values of y_{k+1} and p_{k+1}:

$$\operatorname{pre} y_5 = y_4 + \frac{h}{24}(55p_4 - 59p_3 + 37p_2 - 9p_1) \qquad (14.3.11)$$

$$= 2.0629 89676 \qquad (14.3.12)$$

$$\operatorname{pre} p_5 = p_4 + \frac{h}{24}(55F_4 - 59F_3 + 37F_2 - 9F_1) \qquad (14.3.13)$$

$$= 10.1568 1665 \qquad (14.3.14)$$

We now calculate the corrector values of y_5 using (14.3.3). The value of p_5 is the one calculated in (14.3.13) and (14.3.14):

$$\operatorname{cor} y_5 = y_4 + \frac{h}{24}(9p_5 + 19p_4 - 5p_3 + p_2) \tag{14.3.15}$$

$$= 2.06298\,9461 \tag{14.3.16}$$

To use (14.3.4), we must calculate F_5. We use (14.3.7) for this, entering with $x_5 = x_4 + 0.0015625 = 1.00625$, y_5 from (14.3.16) and p_5 from (14.3.14). Using (14.3.7) and solving for $y'' = F$, we have

$$y_5'' = F_5 = \frac{3p}{x} - \frac{3y}{x^2} + 2x - 1 \tag{14.3.17}$$

$$= 25.18136\,694 \tag{14.3.18}$$

We now can use (14.3.4):

$$\operatorname{cor} p_5 = p_4 + \frac{h}{24}[9F_5 + 19F_4 - 5F_3 + F_2] \tag{14.3.19}$$

$$= 10.15681\,665 \tag{14.3.20}$$

The mop-up uses Equations (14.3.5) and (14.3.6):

$$y_5 = \operatorname{cor} y + \frac{19}{270}[\operatorname{pre} y - \operatorname{cor} y] \tag{14.3.21}$$

$$= 2.06298\,9476 \tag{14.3.22}$$

$$p_5 = \operatorname{cor} p + \frac{19}{270}[\operatorname{pre} p - \operatorname{cor} p] \tag{14.3.26}$$

$$= 10.15681\,665 \tag{14.3.24}$$

The computer output is shown in Figures 14.3-1 to 14.3-4, and the computer program in Figure 14.3-5.

As in the first-order situation, $|Y_2 - Y_1|$ showed a better relationship to the true error than $|\operatorname{cor} - \operatorname{pre}|$. This is shown in Figure 14.3-6.

FIRST FOUR VALUES

```
        EQUATION
        XY''-3Y'+(3/X)Y                    X0= 1.00000000        H = 0.00156250


    X                   Y                          P                      F

1.00000000   2.0000000000000000D 00   1.0000000000000000D 01   2.5000002434497070 01
1.00156250   2.0156555360173160 00   1.0039097904203830 01   2.5045321896897210 01
1.00312500   2.0313722178379530 00   1.0078266632056260 01   2.5090655233108830 01
1.00468750   2.0471501561417010 00   1.0117506206391970 01   2.5136005203348200 01
```

Fig. 14.3-1   The first four (starting) values, calculated from the exact solution. In actual practice, they would probably be the result of a Runge-Kutta calculation. These are for $h = 0.00156250$.

CALCULATED VALUES OF Y, ADAMS-MOULTON ALGORITHM

EQUATION XY''-3Y'+(3/X)Y   X0= 1.00000000   H = 0.00156250

X	Y(CALC)	TRUE VALUE	C-P	TV-Y	REL.ERROR
1.1000	3.1298578495208760 00	3.1298578493194830 00	-1.75D-14	-2.01D-10	-6.43D-11
1.2000	4.5390516506113960 00	4.5390516501397760 00	-1.47D-14	-4.72D-10	-1.04D-10
1.3000	6.2574142898423330 00	6.2574142890347560 00	-1.24D-14	-8.08D-10	-1.29D-10
1.4000	8.3152798185028160 00	8.3152798172880990 00	-1.07D-14	-1.21D-09	-1.46D-10
1.5000	1.0743444741562830 01	1.0743444473986431 01	-9.33D-15	-1.70D-09	-1.58D-10
1.6000	1.3573134867653950 01	1.3573134865389490 01	-8.22D-15	-2.26D-09	-1.67D-10
1.7000	1.6835976600385030 01	1.6835976597467050 01	-7.11D-15	-2.92D-09	-1.73D-10
1.8000	2.0563971833371830 01	2.0563971829707350 01	-7.11D-15	-3.66D-09	-1.78D-10
1.9000	2.4789475809763700 01	2.4789475805254160 01	-3.55D-15	-4.51D-09	-1.82D-10
2.0000	2.9545177449935340 01	2.9545177444476700 01	-7.11D-15	-5.46D-09	-1.85D-10
3.0000	1.3662531815901D 02	1.3662531794024900 02	0.0	-2.19D-08	-1.92D-10
4.0000	2.8872283916635600 02	2.8872283911163330 02	0.0	-5.47D-08	-1.90D-10
5.0000	5.9117973916363520 02	5.9117973905417630 02	0.0	-1.09D-07	-1.85D-10
6.0000	1.0590200455446620 03	1.0590200453530990 03	0.0	-1.92D-07	-1.81D-10
7.0000	1.7314471814322130 03	1.7314471811257030 03	0.0	-3.07D-07	-1.77D-10
8.0000	2.6486740697994270 03	2.6486740693396570 03	0.0	-4.60D-07	-1.74D-10
9.0000	3.8517767175343050 03	3.8517767168774850 03	0.0	-6.57D-07	-1.71D-10
10.0000	5.3825850938960830 03	5.3825850929931700 03	0.0	-9.03D-07	-1.68D-10
11.0000	7.2835986092971200 03	7.2835986080934360 03	0.0	-1.20D-06	-1.65D-10

Fig. 14.3-2 When the numbers in Figure 14.3-1 are used as starters, these are the results. Note that C − P = cor − pre does not correspond to the true error, TV − Y.

FIRST FOUR VALUES

EQUATION XY''-3Y'+(3/X)Y

X0= 1.00000000   H = 0.00312500

X	Y	P	F
1.00000000	2.0000000000000000D 00	1.000000000000000D 01	2.500000000000004D 01
1.00312500	2.031372217837953D 00	1.007826663205626D 01	2.509065280619591D 01
1.00625000	2.062989461644011D 00	1.015681665001010D 01	2.518136444841530D 01
1.00937500	2.094852617284020D 00	1.023565023611311D 01	2.527213735920790D 01

Fig. 14.3-3 These are the starting values calculated using $h=0.003125$ using the exact value as source. In actual practice, the starting values would probably be obtained from the Runge-Kutta algorithm.

CALCULATED VALUES OF Y, ADAMS-MOULTON ALGORITHM

EQUATION XY''-3Y'+(3/X)Y   X0= 1.00000000   H = 0.00312500

X	Y(CALC)	TRUE VALUE	C-P	TV-Y	REL.ERROR
1.1000	3.1298578491902300D 00	3.1298578493194830D 00	-5.62D-13	1.29D-10	4.13D-11
1.2000	4.5390516498331260D 00	4.5390516501397760D 00	-4.71D-13	3.07D-10	6.76D-11
1.3000	6.2574142885075150D 00	6.2574142890347560D 00	-4.01D-13	5.27D-10	8.43D-11
1.4000	8.3152798164934640D 00	8.3152798172880099D 00	-3.46D-13	7.95D-10	9.56D-11
1.5000	1.0743444738751880D 01	1.0743444739864310D 01	-3.01D-13	1.11D-09	1.04D-10
1.6000	1.3573134863905280D 01	1.3573134865389490D 01	-2.64D-13	1.48D-09	1.09D-10
1.7000	1.6835976595553420D 01	1.6835976597467050D 01	-2.34D-13	1.91D-09	1.14D-10
1.8000	2.0563971827303040D 01	2.0563971829707350D 01	-2.10D-13	2.40D-09	1.17D-10
1.9000	2.4789475802294360D 01	2.4789475805254160D 01	-1.85D-13	2.96D-09	1.19D-10
2.0000	2.9545177440893040D 01	2.9545177444447670D 01	-1.67D-13	3.58D-09	1.21D-10
3.0000	1.1366253177965010D 02	1.1366253179402490D 02	-7.46D-14	1.44D-08	1.26D-10
4.0000	2.8872283907567250D 02	2.8872283911163330D 02	-5.68D-14	3.60D-08	1.25D-10
5.0000	5.9117938982229980D 02	5.9117939054176300D 02	-5.68D-14	7.19D-08	1.22D-10
6.0000	1.0590200452271690D 03	1.0590200453530990D 03	0.0	1.26D-07	1.19D-10
7.0000	1.7314471809241950D 03	1.7314471811257030D 03	-5.68D-14	2.02D-07	1.16D-10
8.0000	2.6486740690373780D 03	2.6486740693396570D 03	0.0	3.02D-07	1.14D-10
9.0000	3.8517767168774850D 03	3.8517767168774850D 03	0.0	4.32D-07	1.12D-10
10.0000	5.3825850923992770D 03	5.3825850929931700D 03	0.0	5.94D-07	1.10D-10
11.0000	7.2835986073014870D 03	7.2835986080934360D 03	0.0	7.92D-07	1.09D-10

Fig. 14.3-4 The results of the Adams-Moulton predictor-corrector system when the starting values of Figure 14.3-3 were used. This should be compared with Figure 14.3-2. Once again, there is little or no relationship between cor−pre and TV−Y.

```
C THIS PROGRAM DISPLAYS THE ADAMS-MOULTON ALGCRITHM FOR A SECOND-ORDER
C DIFFERENTIAL EQUATION. IT IS ASSUMED THAT FOUR STARTING VALUES ARE
C AVAILABLE, FOR EXAMPLE FROM A RUNGE-KUTTA PROCESS. HOWEVER, TO TEST
C THE ALGORITHM, THE FOUR STARTING VALUES ARE CALCULATED FROM THE EXACT
C ANSWERS. IN AN ACTUAL PROBLEM, THEY WOULD BE INPUT. THERE IS ONE
C DATA CARD AS FOLLOWS.
C          CC  1-28   NAME OR DESCRIPTION OF THE EQUATION
C          CC 31-40   INITIAL X, XO, IN F-FORMAT
C          CC 61-70   INITIAL VALUE OF H, HO, IN F-FORMAT, FROM THE
C                     RUNGE-KUTTA PROGRAM
C IN THIS PROGRAM, HO IS DIVIDED BY 2**(N1-1) WHERE N1 REFERS TO THE
C NUMBER OF ITERATIONS IN THE R-K PROGRAM. THERE ARE THREE VARIABLE
C STATEMENT FUNCTIONS AFTER CARD 310. THE FIRST, F(X,Y,P), DEFINES Y''.
C THE SECOND, FP(X), DEFINES Y'. THE THIRD, S(X), IS THE EXACT SOLUTION.
C IN AN ACTUAL PROBLEM, NEITHER FP NOR S WOULD BE KNOWN. THEY ARE
C USED IN THIS PROGRAM TO FIND THE ACTUAL ERRORS.
      REAL*8 MU
      DOUBLE PRECISION DIFF(20),ER(20,2),FF(5),P(5,2),PC(5,2),REL(20,2),
     1XT(20)   ,YT(20,2),Y(5,2),YP(5,2),YC(5,2),A,A1,A2,B1,B2,C1,C2,CO,
     2CON3,CON5,COR,D1,D2,DY,E1,E2,EX,F,F1,F2,FP,H,HO,HF,HP,P1,
     3PPR,PR,PRE,R,S,TV,X,XO,Z
      DIMENSION NAME(7)
C STATEMENT FUNCTIONS. THE DUMMIES IN THESE ARE AS FOLLOWS. A=Y(K) OR
C     P(K), B=Y'(K+1) OR Y''(K+1), C=Y'(K) OR Y''(K) D=Y'(K-1) OR
C     Y''(K-1), E=Y'(K-2) OR Y''(K-2), F=Y'(K-3) OR Y''(K-3), PREDICTED
C     VALUES ARE YP AND PPR. CORRECTED VALUES ARE YC AND PC. MOP-UP
C     VALUES ARE Y AND P. IN THE ALGORITHM, PR IS PREDICTED AND CO
C     IS CORRECTED. MU IS FOR MOP-UP. HP=H/24.
C
      F(X,A,P1)=3.D0*P1/X-3.D0*A/X**2+2.D0*X-1.D0
        FP(X)=X*X*(10.D0+3.D0*DLOG(X))+2.D0*X-2.D0
      S(X)=X**2*(3.D0*X+1.D0+X*DLOG(X))-2.D0*X
C ADAMS-MOULTON ALGORITHM.
      PR(A1,C1,D1,E1,F1)=A1+HP*(55.D0*C1-59.D0*D1+37.D0*E1-9.D0*F1)
      CO(A1,B1,C1,D1,E1)=A1+HP*(9.D0*B1+19.D0*C1-5.D0*D1+E1)
      MU(N)=COR+19.D0*(PRE-COR)/270.D0
C CONSTANTS.
      PI=3.141592653589793D0
      EX=2.718281828459045D0
      CON2=DSQRT(2.D0)
      CON3=DSQRT(3.D0)
      CON5=DSQRT(5.D0)
      HF=0.1D0
C INPUT.
      READ(1,100)NAME,XO,HO
100   FORMAT(7A4,2X,F10.0,20X,F10.0)
      N1=7
      H=HO/DFLOAT(2**(N1-1))
      N=1
C FIRST HEADING.
1     WRITE(3,101)
101   FORMAT (1H1,T58,'FIRST FOUR VALUES'/)
      WRITE(3,102)NAME,XO,H
102   FORMAT(1HO,T16,'EQUATION  ',7A4,5X,'XO=',F11.8,9X,'H =',F11.8/)
      WRITE(3,103)
103   FORMAT(1HO,T30,'X',T49,'Y',T72,'P',T98,'F'/)
C CALCULATION OF STARTERS.
      X=XO-H
      DO 21 K=1,4
      X=X+H
      Y(K,N)=S(X)
      P(K,N)=FP(X)
      A=Y(K,N)
      C=P(K,N)
21    FF(K)=F(X,A,C)
C FIRST OUTPUT.
      X=XO-H
```

```
        DO 22 K=1,4
        X=X+H
22      WRITE(3,104)X,Y (K,N),P (K,N),FF(K)
104     FORMAT(1H ,T25,F11.8,T38,1PD22.15,2(2X,D22.15))
C SECOND HEADING.
        WRITE(3,105)
105     FORMAT(1H1,T41,'CALCULATED VALUES OF Y, ADAMS-MOULTON ALGORITHM'/)
        WRITE(3,102)NAME,XO,H
        WRITE(3,106)
106     FORMAT(1H0,T12,'X',T25,'Y(CALC)',T51,'TRUE VALUE',T73,'C-P',
       1T84,'TV-Y ',T94,'REL.ERROR'/)
C SET. L IS THE NUMBER OF LINES OF OUTPUT. K COUNTS THE TRIALS.
        HP=H/24.DO
        LF1=IDINT(0.1DO/H)
        LF2=10*LF1
        L=1
        K=4
C CALCULATIONS.
4       X=X+H
C PREDICTED VALUES.
        A1=Y(4,N)
        C1=P(4,N)
        D1=P(3,N)
        E1=P(2,N)
        F1=P(1,N)
        YP(5,N)=PR(A1,C1,D1,E1,F1)
C
        A2=P(4,N)
        C2=FF(4)
        D2=FF(3)
        E2=FF(2)
        F2=FF(1)
        PPR=PR(A2,C2,D2,E2,F2)
C CORRECTED VALUES.
        B1=PPR
        YC(5,N)=CO(A1,B1,C1,D1,E1)
        Z=YC(5,N)
        B2=F(X,Z,PPR)
        PC(5,N)=CO(A2,B2,C2,D2,E2)
C MOP-UP.
        PRE=YP(5,N)
        COR=YC(5,N)
        DY=COR-PRE
        Y(5,N)=MU(N)
        A=Y(5,N)
C
        PRE=PPR
        COR=PC(5,N)
        P(5,N)=MU(N)
        P1=P(5,N)
C
        FF(5)=F(X,A,P1)
C RE-SHUFFLING.
        DO 11 KK=2,5
        J=KK-1
        Y(J,N)=Y(KK,N)
        P(J,N)=P(KK,N)
11      FF(J)=FF(KK)
        IF(L.LE.10)GO TO 41
        IF(K.EQ.LF2)GO TO 5
        GO TO 42
41      IF(K.EQ.LF1)GO TO 5
42      K=K+1
        GO TO 4
C OUTPUT.
5       XT(L)=X
        YT(L,N)=Y(5,N)
```

Fig. 14.3-5   (Continued on next page)

```
         DIFF(L)=DY
         TV=S(X)
         ER(L,N)=TV-YT(L,N)
         IF(DABS(ER(L,N)).LE.5.D-17)GO TO 51
         REL(L,N)=ER(L,N)/YT(L,N)
         GO TO 52
51       ER(L,N)=0.D0
         REL(L,N)=0.D0
52       WRITE(3,107)XT(L),YT(L,N),TV,DY,ER(L,N),REL(L,N)
107      FORMAT(1H ,T7 ,F8.4,3X,1PD23.15,3X,D23.15,3(3X,D9.2))
C RESET.
         L=L+1
         K=1
C
         IF(L.NE.20)GO TO 4
         IF(N.EQ.2)GO TO 6
         N=2
         H=2.D0*H
         GO TO 1
C FINAL TABLE.
6        WRITE(3,108)
108      FORMAT(1H1,T42,'COMPARISON OF SUCCESSIVE VALUES OF Y'/)
         WRITE(3,102)NAME,X0,H
         Z=2.D0*H
         WRITE(3,109)Z,H
109      FORMAT(1H0,T12,'X',T23,'Y1(H=',F8.7,')',T49,'Y2(H=',F8.7,')',
        1T72,'Y2-Y1',T83,'C-P(Y2)',T94,'ERROR(Y2)'/)
         DO 60 L=1,19
         Z=YT(L,2)-YT(L,1)
60       WRITE(3,110)XT(L),YT(L,1),YT(L,2),Z,DIFF(L),ER(L,2)
110      FORMAT(1H ,T10,F5.1,3X,1PD23.15,3X,D23.15,3(3X,D9.2))
         END
```

Fig. 14.3-5   Computer program for the Adams-Moulton algorithm, second order.

COMPARISON OF SUCCESSIVE VALUES OF Y

EQUATION XY''-3Y'+(3/X)Y   XO= 1.00000000   H = 0.00312500

x	Y1(H=.0062500)	Y2(H=.0031250)	Y2-Y1	C-P(Y2)	ERROR(Y2)
1.1	3.1298578495208760D 00	3.1298578491902230D 00	-3.31D-10	-5.62D-13	1.29D-10
1.2	4.5390516506113960D 00	4.5390516498331260D 00	-7.78D-10	-4.71D-13	3.07D-10
1.3	6.2574142898423330D 00	6.2574142885075150D 00	-1.33D-09	-4.01D-13	5.27D-10
1.4	8.3152798185028160D 00	8.3152798164934640D 00	-2.01D-09	-3.46D-13	7.95D-10
1.5	1.0743444741562830D 01	1.0743444738751880D 01	-2.81D-09	-3.01D-13	1.11D-09
1.6	1.3573134867653950D 01	1.3573134863905280D 01	-3.75D-09	-2.64D-13	1.48D-09
1.7	1.6835976600385030D 01	1.6835976595553420D 01	-4.83D-09	-2.34D-13	1.91D-09
1.8	2.0563971833371830D 01	2.0563971827303040D 01	-6.07D-09	-2.10D-13	2.40D-09
1.9	2.4789475809763700D 01	2.4789475802294360D 01	-7.47D-09	-1.85D-13	2.96D-09
2.0	2.9545177449935340D 01	2.9545177440893040D 01	-9.04D-09	-1.67D-13	3.58D-09
3.0	1.1366253181590170D 02	1.1366253177965010D 02	-3.63D-08	-7.46D-14	1.44D-08
4.0	2.8872283916635600D 02	2.8872283907567250D 02	-9.07D-08	-5.68D-14	3.60D-08
5.0	5.9117973916363520D 02	5.9117973898222980D 02	-1.81D-07	-5.68D-14	7.19D-08
6.0	1.0590200455446620D 03	1.0590200452271690D 03	-3.17D-07	0.0	1.26D-07
7.0	1.7314471814322130D 03	1.7314471809241950D 03	-5.08D-07	-5.68D-14	2.02D-07
8.0	2.6486740697994270D 03	2.6486740690373780D 03	-7.62D-07	0.0	3.02D-07
9.0	3.8517767175343050D 03	3.8517767164456470D 03	-1.09D-06	0.0	4.32D-07
10.0	5.3825850938960830D 03	5.3825850923992770D 03	-1.50D-06	0.0	5.94D-07
11.0	7.2835986092971200D 03	7.2835986073014870D 03	-2.00D-06	0.0	7.92D-07

Fig. 14.3-6 The quantity $|Y_2 - Y_1|$ was a fairly good indicator of the true error.

PROBLEM SECTION 14.3

Using the equations in the problems for Section 14.2, calculate the first four values from the exact answer, then apply the Adams-Moulton predictor-corrector algorithm to prepare a table of values from x_0 to $x_0 + 1$ by intervals of 0.1 and, where possible, from $x_0 + 1$ to $x_0 + 10$ by intervals of 1. This is, unfortunately, a slow program. Use $h_1 = 0.0031 25$ and $h_2 = 0.00156 25$ to compare your values with ours, then print both cor $-$ pre and $Y_1 - Y_2$ to compare with the actual error.

14.4 SECOND-ORDER DIFFERENTIAL EQUATIONS LACKING THE y' TERM

14.4.1 The Numerov Method

Many practical problems are of the form

$$y'' = F(x,y) = f(x) \cdot y^n \tag{14.4.1}$$

We have already discussed the series method for dealing with these when $n = 1$—e.g., the Airy, Coulomb and parabolic-cylinder differential equations. The Numerov method is an alternative method which we shall illustrate. The basic assumption is that

$$y_{k+1} = A y_k + B y_{k-1} + h^2 (C F_{k+1} + D F_k + E F_{k-1}) \tag{14.4.2}$$

where $F = y''$. Using Taylor series to express y_{k+1}, F_{k+1}, y_{k-1} and F_{k-1}, we obtain

$$y_{k+1} = y_k + h y_k' + \frac{h^2}{2!} y_k'' + \frac{h^3}{3!} y_k''' + \cdots + \frac{h^6}{6!} y^{vi}(\theta_1) \tag{14.4.3}$$

$$y_{k-1} = y_k - h y_k' + \frac{h^2}{2!} y_k'' - \frac{h^3}{3!} y_k''' + \cdots + \frac{h^6}{6!} y^{vi}(\theta_2) \tag{14.4.4}$$

$$F_{k+1} = y_k'' + h y_k''' + \cdots + \frac{h^6}{4!} y^{vi}(\theta_3) \tag{14.4.5}$$

$$F_{k-1} = y_k'' - h y_k''' + \cdots + \frac{h^6}{4!} y^{vi}(\theta_4) \tag{14.4.6}$$

Substituting in (14.4.2) and matching coefficients of like powers of h, we now find that if y^{vi} can be considered equal for θ_1, θ_2, θ_3 and θ_4, then

$$
\begin{cases}
y_k = A y_k + B y_k \quad \Rightarrow \quad A + B = 1 & (14.4.7) \\[2mm]
y_k' = - B y_k' \quad \Rightarrow \quad B = -1 & (14.4.8) \\[2mm]
\frac{1}{2!} y_k'' = \frac{1}{2!} B y_k'' + C y_k'' + D y_k'' + E y_k'' \quad \Rightarrow \quad \frac{1}{2} B + C + D + E = \frac{1}{2} & (14.4.9) \\[2mm]
\frac{1}{3!} y_k''' = -\frac{1}{3!} B y_k''' + C y_k''' - E y_k''' \quad \Rightarrow \quad \frac{1}{6} = -\frac{1}{6} B + C - E & (14.4.10) \\[2mm]
\frac{1}{4!} y_k^{iv} = \frac{1}{4!} B y_k^{iv} + \frac{1}{2!} C y_k^{iv} + \frac{1}{2!} E y_k^{iv} \quad \Rightarrow \quad \frac{1}{24} = \frac{1}{24} B + \frac{1}{2} C + \frac{1}{2} E
\end{cases}
$$

$$\tag{14.4.11}$$

from which $A = 2$, $B = -1$, $C = \frac{1}{12}$, $D = \frac{5}{6}$, and $E = \frac{1}{12}$.

To obtain an estimate of the error in (14.4.2), we assume that f^{vi} is the same at the four values of θ, and add the coefficients of $h^6 y^{vi}(\theta)$:

$$-\frac{1}{6!} + B \cdot \frac{1}{6!} + C \cdot \frac{1}{4!} + E \cdot \frac{1}{4!} = \frac{1}{240} \qquad (14.4.12)$$

NUMEROV'S FORMULA   If $y'' = F(x, y)$, then

$$y_{k+1} = 2y_k - y_{k-1} + \frac{1}{12} h^2 [F_{k+1} + 10F_k + F_{k-1}] + R \qquad (14.4.13)$$

where R is approximately $-\dfrac{1}{240} h^6 y^{vi}(\theta)$, $\theta \in (x_{k-1}, x_{k+1})$.

14.4.2 Application

We shall illustrate by solving an equation for which the exact solution is known, in order to illuminate the method and its accuracy.

ILLUSTRATIVE PROBLEM 14.4-1   Given the equation

$$y'' = e^{2y} \qquad (14.4.14)$$

with starting values for y and y', make a table of values 0.1(0.1)1.0. The exact solution, for checking, is

$$y = -\ln \cos x \qquad (14.4.15)$$

Solution   For a second-order differential of this kind, it is possible to find y_1 when y_0 and y_0' are given. We may use the Taylor series

$$y_1 = y(x + h) = y_0 + y_0' h + \frac{1}{2!} y_0'' h^2 + \frac{1}{3!} y_0''' h^3 + \cdots \qquad (14.4.16)$$

The values of y_0 and y_0' are given. The value of y_0'' is calculated from (14.4.14). Then

$$y''' = e^{2y} \cdot 2y_0' \qquad (14.4.17)$$

$$y^{iv} = 2e^{2y} y_0'' + 4e^{2y} (y_0')^2 \qquad (14.4.18)$$

and so on.

In this example, we wish to avoid contaminating the algorithm by the error of the Taylor series. Therefore, we shall use (14.4.15), merely for illustrative purposes, to obtain the starting values. We start with $h = 0.1$, $x_0 = 0$, $y_0 = 0$, so that

$$h = 0.1$$

$$y(0) = 0 \qquad (14.4.19)$$

$$y(0.1) = -\ln \cos 0.1 = 5.00835\,5621 \times 10^{-3}$$

We now divide h by 2 and start with

$$h = 0.05$$

$$y(0) = 0 \tag{14.4.20}$$

$$y(0.05) = -\ln\cos 0.05 = 1.250521184 \times 10^{-3}$$

We wish to find $y(0.1)$ to compare with the value in (14.4.19). First we use a *portion* of (14.4.13) as a predictor:

$$\text{pre}\, y_2 = 2y_1 - y_0 + \frac{1}{12}[10F_1 + F_0] \tag{14.4.21}$$

$$\text{pre}\, y(0.1) = 2y(0.05) - y(0) + \frac{1}{12}(0.05)^2[10y''(0.05) + y''(0)] \tag{14.4.22}$$

y'' is, of course, calculated from the original equation, (14.4.14):

$$\text{pre}\, y(0.1) = 2(1.250521184 \times 10^{-3}) - 0 + \frac{1}{12}(0.05)^2[10(1.002504172) + 1]$$

$$\tag{14.4.23}$$

$$= 4.797926061 \times 10^{-3} \tag{14.4.24}$$

To correct this, we must add on $\frac{1}{12}F_{k+1}$. Then

$$y(0.1) = \text{pre}\, y + \frac{1}{12}(0.05)^2 y''(0.1) \tag{14.4.25}$$

$$= 4.797926061 \times 10^{-3} + \frac{1}{12}(0.05)^2(1.009642039) \tag{14.4.26}$$

$$= 5.008568152 \times 10^{-3} \tag{14.4.27}$$

If the difference between (14.4.19) and (14.4.27) is too large, we continue as follows. We divide h by 2 again and start with

$$h = 0.025$$

$$y(0) = 0 \tag{14.4.28}$$

$$y(0.025) = -\ln\cos 0.025 = 3.1253255563 \times 10^{-4}$$

We wish to find $y(0.05)$ to compare with the value in (14.4.20). Using the predictor (14.4.21) again,

$$\text{pre}\, y(0.05) = 2y(0.025) - y(0) + \frac{(0.025)^2}{12}[10y''(0.025) + y''(0)] \tag{14.4.29}$$

$$= 3(3.1253255563 \times 10^{-4}) - 0 + \frac{1}{12}(0.025)^2[10(1.000625260) + 1] \tag{14.4.30}$$

$$= 1.198307414 \times 10^{-3} \tag{14.4.31}$$

Adding $\frac{1}{12} h^2 F_{k+1}$ to obtain the corrected value, we have

$$y(0.05) = 1.198307414 \times 10^{-3} + \frac{(0.025)^2}{12} [1.002399489] \qquad (14.4.32)$$

$$= 1.250515721 \times 10^{-3} \qquad (14.4.33)$$

This procedure eventually sets h. Then the table can be generated. For example, if the results above are sufficiently accurate, we can use $y(0.025)$ and $y(0.05)$ to find $y(0.075)$, and so on. The errors in a computer run are given in Table 14.4-1. It turns out that the error is quite close to the difference between successive calculations.

Table 14.4-1

x	ERROR	RELATIVE ERROR
0.1	6.61×10^{-10}	1.32×10^{-7}
0.2	2.73×10^{-9}	1.35×10^{-7}
0.3	6.33×10^{-9}	1.39×10^{-7}
0.4	1.17×10^{-8}	1.43×10^{-7}
0.5	1.93×10^{-8}	1.48×10^{-7}
0.6	2.97×10^{-8}	1.55×10^{-7}
0.7	4.40×10^{-8}	1.64×10^{-7}
0.8	6.39×10^{-8}	1.77×10^{-7}
0.9	9.25×10^{-8}	1.95×10^{-7}
1.0	1.36×10^{-7}	2.20×10^{-7}

The computer program is shown in Figure 14.4-1.

```
C THE PURPOSE OF THIS PROGRAM IS TO FIND A PARTICULAR SOLUTION OF A
C SECOND-ORDER DIFFERENTIAL EQUATION LACKING THE Y-PRIME TERM. THE
C NUMEROV ALGORITHM IS USED. THERE IS ONE INPUT CARD, AS FOLLOWS.
C               CC  1-28   NAME OR DESCRIPTION OF THE EQUATION
C               CC 31-40   INITIAL X IN F-FORMAT, XO
C               CC 41-50   INITIAL H IN F-FORMAT, HO
C THERE ARE TWO VARIABLE STATEMENT FUNCTIONS AFTER CARD 200, ONE TO
C DEFINE THE EQUATION, THE OTHER FOR TRUE VALUES.
C
      REAL*8 N1,N2(320)
      DOUBLE PRECISION A,B,C,C1,C2,C3,C5,CRIT,D,E,E1,F,H,HO,PI,Q,R,S,TV,
     1X,XO,X1,X2,Y,YDP,YC,YP,Z
      DIMENSION NAME(7)
C
C STATEMENT FUNCTIONS. A=Y(K),B=Y(K+1), C=CORRECTED Y(K+2),D=YDP(K),
C                      E=YDP(K+1),F=YDP(K+2),Z=H*H/12,Q=YP=PREDICTED
C                      VALUE OF Y(K+2)
      YDP(X,Y)=DEXP(2.D0*Y)
      S(X)=-DLOG(DCOS(X))
      YP(Z)=2.D0*B-A+Z*(10.D0*E+D)
      YC(Z)=Q+Z*F
C CONSTANTS.
      PI=3.14159265358979300
      E1=2.71828182845904500
      C1=0.5D0
      C2=DSQRT(2.D0)
      C3=DSQRT(3.D0)
      C5=DSQRT(5.D0)
      CRIT=5.D-17
```

Fig. 14.4-1 (Continued on next page)

```
C INPUT.
      READ(1,100)NAME,X0,H0
100   FORMAT(7A4,2X,2F10.0)
C FIRST HEADING.
      WRITE(3,101)
101   FORMAT(1H1,T35,'COMPARISON OF SUCCESSIVE VALUES'/)
      WRITE(3,102)NAME
102   FORMAT(1H0,T16,'EQUATION ',7A4/)
      WRITE(3,103)
103   FORMAT(1H0,T11,'X',T25,'H',T34,'FIRST CALCULATION,N1',T61,'SECOND
     1 CALCULATION,N2',T90,'N1-N2'/)
C INITIAL CONDITIONS. N COUNTS THE ITERATIONS.
      N=1
      H=H0
      A=S(X0)
      D=YDP(X0,A)
      N1=S(X0+H)
C LOOP TO DETERMINE AN ACCEPTABLE H.
1     N=N+1
      H=C1*H
      X=X0+H
      Z=H*H/12.D0
      B=S(X0+H)
C PREDICTED Y(2).
      E=YDP(X0+H,B)
      Q=YP(Z)
C CORRECTED Y(2).
      F=YDP(X0+H*2.D0,Q)
      C=YC(Z)
C OUTPUT.
      R=N1-C
      X2=X+H
      WRITE(3,104)X2,H,N1,C,R
104   FORMAT(1H ,T6,F11.8,T20,F11.8,T32,1PD23.15,T60,D23.15,T88,D9.2)
C DECISIONS.
      IF(DABS(N1-C).LE.5.D-17)GO TO 2
      IF(N.EQ.6)GO TO 2
C STEP.
      N1=B
      GO TO 1
C CALCULATION OF A TABLE OF VALUES.
2     KF=10*2**(N-1)
      N2(1)=B
      N2(2)=C
      X=X0
      DO 3 K=3,KF
C PREDICTED VALUES.
      A=N2(K-2)
      B=N2(K-1)
      X=X+H
      D=YDP(X,A)
      X1=X+H
      E=YDP(X1,B)
      Q=YP(Z)
C CORRECTED VALUES.
      X2=X1+H
      F=YDP(X2,Q)
3     N2(K)=YC(Z)
C OUTPUT.
      L=0
      WRITE(3,105)
105   FORMAT(1H1,T38,'TABLE OF NUMEROV VALUES'/)
      WRITE(3,106)NAME,H,N
106   FORMAT(1H0,T5,7A4,T35,'H USED IN CALCULATIONS= ',F10.8,10X,'N=',I1
     1/)
      WRITE(3,107)
107   FORMAT(1H0,T3,'X',T23,'CALC. Y',T49,'TRUE VALUE',T72,'ERROR',T83,
     1'REL. ERROR'/)
```

```
C
      X=XO
      DO 4 K=1,KF
      L=L+1
      X=X+H
      TV=S(X)
      E=TV-N2(K)
      IF(DABS(E).LE.CRIT)GO TO 41
      R=E/N2(K)
      GO TO 42
41    E=0.DO
      R=0.DO
42    WRITE(3,108)X,N2(K),TV,E,R
108   FORMAT(1H ,T2,F11.8, 2X,1PD22.15,5X,D22.15,5X,2(D9.2,5X))
      IF(L.NE.45)GO TO 4
      WRITE(3,105)
      WRITE(3,106)NAME,H,N
      WRITE(3,107)
      L=1
4     CONTINUE
      END
```

Fig. 14.4-1   Computer program for the Numerov method.

PROBLEM SECTION 14.4

In each of the following, use the Numerov method to develop, where possible, a table of ten values with increment 0.1. Use the exact solution to obtain starters and to check results. The numbers given in parentheses are our results for the relative error at $x_0 + 0.1$ and at the last calculation, usually $x_0 + 1.0$.

(1) $y'' - y = e^x$, $x_0 = 0$. Exact solution: $y = xe^x/2$
$(6.03 \times 10^{-9}, 4.59 \times 10^{-8})$.

(2) $y'' + 4y = \sin 2x$, $x_0 = 0$. Exact solution: $y = -(x \cos 2x)/4$
$(6.00 \times 10^{-9}, 7.26 \times 10^{-7})$.

(3) $y'' + 5y = \cos x\sqrt{5}$, $x_0 = 0$. Exact solution: $y = (\sqrt{5}\, x \sin x\sqrt{5})/10$
$(3.27 \times 10^{-7}, 9.96 \times 10^{-8})$.

(4) $y'' + 2y = e^{2x}$, $x_0 = 0$. Exact solution: $y = e^{2x}/6$
$(2.32 \times 10^{-9}, 7.18 \times 10^{-8})$.

(5) $y'' + 2y = x^3 + x^2 + e^{-2x} + \cos 3x$, $x_0 = 0$. Exact solution: $y = (x^3 + x^2 - 3x - 1)/2 + e^{-2x}/6 - (\cos 3x)/7$
$(3.06 \times 10^{-9}, 2.00 \times 10^{-7})$.

(6) $y'' - y = xe^{3x}$, $x_0 = 0$. Exact solution: $y = e^{3x}(4x - 3)/32$
$(1.70 \times 10^{-10}, 9.86 \times 10^{-8})$.

(7) $y'' = y + 2/(1 + e^x)$, $x_0 = 0$. Exact solution: $y = e^x - 1 + \ln(1 - e^{-x}) - e^{-x}\ln(1 + e^x)$
$(2.36 \times 10^{-9}, 2.45 \times 10^{-8})$.

(8) $y'' = -y/(1 - x^2)^2$, $x_0 = 0$. Exact solution: $y = \sqrt{1 - x^2}\,\ln[(1 + x)(1 - x)]$
$(1.25 \times 10^{-10}, 3.50 \times 10^{-3})$.

(9) $y'' = [(2/x^2) - 1]y + x^2$, $x_0 = \pi/2$. Exact solution: $y = x^2 + \sin x + (\cos x)/x$
$(3.54 \times 10^{-11}, 3.44 \times 10^{-9})$.

(10) $y'' = 2y \csc^2 x$, $x_0 = \pi/4$. Exact solution: $y = 3 + (1 - 3x)\cot x$
$(4.84 \times 10^{-9}, 1.48 \times 10^{-7})$.

14.5   THE BOUNDARY-VALUE PROBLEM

14.5.1   Introduction

In this last section of the book, we introduce the problem of solving a second-order differential equation in which two endpoint values (x_0, y_0) and (x_n, y_n) are given, instead of (x_0, y_0, p_0). These problems are called *boundary-value problems*.

As an example, we choose the Bessel equation of the first kind, of order 2, with solutions defined by

$$x^2 y'' + xy' + (x^2 - 4) y = 0 \tag{14.5.1}$$

We are given boundary values $J_2(0) = 0$ and $J_2(1) = 0.1149034849$ and required to furnish a table for 0(0.1)1.

14.5.2   The Difference Equations

We start with

$$P(x) y'' + Q(x) y' + R(x) y = 0 \tag{14.5.2}$$

and we are given that

$$f_0 = f(x_0) \tag{14.5.3}$$

We now write two Taylor series:

$$f_1 = f(x_0 + h) = f_0 + h f'(x_0) + \frac{h^2}{2!} f''(x_0) + \frac{h^3}{3!} f'''(\theta_1) \tag{14.5.4}$$

$$f_{-1} = f(x_0 - h) = f_0 - h f'(x_0) + \frac{h^2}{2!} f''(x_0) - \frac{h^3}{3!} f'''(\theta_2) \tag{14.5.5}$$

Writing a and b for the error terms in (14.5.4) and (14.5.5), we use (14.5.3), (14.5.4) and (14.5.5) as a simultaneous system which can be written as the following augmented matrix:

$$\begin{bmatrix} 1 & 0 & 0 & \vdots & f_0 \\ 1 & 1 & 1 & \vdots & f_1 - a \\ 1 & -1 & 1 & \vdots & f_{-1} + b \end{bmatrix} \tag{14.5.6}$$

Solving this system, the results are

$$-2hf'(x_0) = -f_1 + f_{-1} + a + b \qquad (14.5.7)$$

$$\frac{2h^2 f''(x_0)}{2!} = f_1 - 2f_0 + f_{-1} - a - b \qquad (14.5.8)$$

from which

$$f'(x_0) = \frac{1}{2h}(f_1 - f_{-1}) - \frac{1}{2h}(a+b) \qquad (14.5.9)$$

$$f''(x_0) = \frac{1}{h^2}(f_1 - 2f_0 + f_{-1}) - \frac{1}{h^2}(a-b) \qquad (14.5.10)$$

If the intervals are small, it is possible that a and b are approximately equal. Call their value E. Then

$$f' = \frac{1}{2h}(f_1 - f_{-1}) - \frac{E}{h} \qquad (14.5.11)$$

$$f'' = \frac{1}{h^2}(f_1 - 2f_0 + f_{-1}) \qquad (14.5.12)$$

As usual, we will suppose that E/h is negligible, and substitute into (14.5.2):

$$\frac{f_1 - 2f_0 + f_{-1}}{h^2} P(x) + \frac{f_1 - f_{-1}}{2h} Q(x) + f_0 R = 0 \qquad (14.5.13)$$

Multiplying by $2h^2$ and simplifying, we obtain

$$(2P + hQ)f_1 + (2h^2 R - 4P)f_0 + (2P - hQ)f_{-1} = 0 \qquad (14.5.14)$$

Changing subscripts for convenience, we arrive at the operational equation which we shall use:

$$(2P - hQ)f_k + (2h^2 R - 4P)f_{k+1} + (2P + hQ)f_{k+2} = 0 \qquad (14.5.15)$$

14.5.3   Setting Up the Equations

In (14.5.1), $P(x) = x^2$, $Q(x) = x$, and $R(x) = x^2 - 4$. We are using $h = 0.1$. The first equation, at $k = 0$, uses P, Q and R at $x_1 = x_0 + h = 0.1$. Then

$$\begin{aligned}
P &= 0.01 \\
Q &= 0.1 \qquad\qquad\qquad (14.5.16) \\
R &= 0.01 - 4 = -3.99
\end{aligned}$$

Therefore, the first equation is

$$(0.01)f_0 + (-0.1198)f_1 + (0.03)f_2 = 0 \qquad (14.5.17)$$

In this case, we know that $f_0 = J_2(0) = 0$, which is the first boundary value. The second equation, at $k = 1$, uses P, Q and R at $x = x_0 + 2h = 0.2$. Then

$$\begin{aligned} P &= 0.04 \\ Q &= 0.2 \\ R &= -3.96 \end{aligned} \qquad (14.5.18)$$

so that the second equation is

$$(0.06)f_1 + (-0.2392)f_2 + (0.1000)f_3 = 0 \qquad (14.5.19)$$

We skip to the last equation, at $k = 9$, which uses P, Q and R at $x = x_0 + 9h = 0.9$. Then

$$\begin{aligned} P &= 0.81 \\ Q &= 0.9 \\ R &= -3.19 \end{aligned} \qquad (14.5.20)$$

from which the equation is

$$(1.53)f_9 + (-3.3038)f_{10} + (1.71)f_{11} = 0 \qquad (14.5.21)$$

In this illustrative problem, $f_{11} = J_2(1) = 0.1149034849$. Equation (14.5.21) becomes

$$(1.53)f_9 + (-3.3038)f_{10} = -0.1964849592 \qquad (14.5.22)$$

We obtain a set of *nine* simultaneous equations, for which the augmented matrix is shown in Figure 14.5-1. In the figure, the coefficients have been abbreviated for printing convenience, but will enable the reader to check a program. The tenth column is, of course, the column of constants.

```
ECHO MATRIX      BESSEL FUNCTION, ORDER 2      0(0.1)1

-0.120   0.030   0.0     0.0      0.0      0.0      0.0      0.0      0.0      0.0
 0.060  -0.239   0.100   0.0      0.0      0.0      0.0      0.0      0.0      0.0
 0.0     0.150  -0.438   0.210    0.0      0.0      0.0      0.0      0.0      0.0
 0.0     0.0     0.280  -0.717    0.360    0.0      0.0      0.0      0.0      0.0
 0.0     0.0     0.0     0.450   -1.075    0.550    0.0      0.0      0.0      0.0
 0.0     0.0     0.0     0.0      0.660   -1.513    0.780    0.0      0.0      0.0
 0.0     0.0     0.0     0.0      0.0      0.910   -2.030    1.050    0.0      0.0
 0.0     0.0     0.0     0.0      0.0      0.0      1.200   -2.627    1.360    0.0
 0.0     0.0     0.0     0.0      0.0      0.0      0.0      1.530   -3.304   -0.196
```

Fig. 14.5-1 Augmented matrix for the illustrative problem.

Because there are so many zero values in each row of this augmented matrix, it is called a *sparse matrix*. It is easy to solve, even by calculator.

The results are shown in Figure 14.5-2, and the computer program in Figure 14.5-3.

SOLUTION OF BOUNDARY-VALUE D.E. BY DIFFERENCE EQUATIONS

BESSEL FUNCTION, ORDER 2   0(0.1)1

x	TRUE VALUE	CALC. VALUE	PCT.ERROR
0.0	0.0		
0.100	1.248958700D-03	1.245469341D-03	2.80D-01
0.200	4.983354200D-03	4.973574234D-03	1.97D-01
0.300	1.116586190D-02	1.114950796D-02	1.47D-01
0.400	1.973466310D-02	1.971275359D-02	1.11D-01
0.500	3.060402350D-02	3.057844318D-02	8.37D-02
0.600	4.366509670D-02	4.363834055D-02	6.13D-02
0.700	5.878694440D-02	5.876193473D-02	4.26D-02
0.800	7.581776250D-02	7.579770474D-02	2.65D-02
0.900	9.458630430D-02	9.457456488D-02	1.24D-02
1.000	1.149034849D-01		

Fig. 14.5-2   Solutions for the boundary-value problem. The errors are large, but the method can be extended by using more terms of the Taylor series, and by using smaller h.

```
C THE PURPOSE OF THIS PROGRAM IS TO FIND A PARTICULAR SOLUTION OF A
C SECOND-ORDER DIFFERENTIAL EQUATION LACKING THE Y-PRIME TERM. THE
C NUMEROV ALGORITHM IS USED. THERE IS ONE INPUT CARD, AS FOLLOWS.
C              CC  1-28  NAME OR DESCRIPTION OF THE EQUATION
C              CC 31-40  INITIAL X IN F-FORMAT, XO
C              CC 41-50  INITIAL H IN F-FORMAT, HO
C THERE ARE TWO VARIABLE STATEMENT FUNCTIONS AFTER CARD 200, ONE TO
C DEFINE THE EQUATION, THE OTHER FOR TRUE VALUES.
C
      REAL*8 N1,N2(320)
      DOUBLE PRECISION A,B,C,C1,C2,C3,C5,CRIT,D,E,E1,F,H,HO,PI,Q,R,S,TV,
     1X,XO,X1,X2,Y,YDP,YC,YP,Z
      DIMENSION NAME(7)
C
C STATEMENT FUNCTIONS. A=Y(K),B=Y(K+1), C=CORRECTED Y(K+2),D=YDP(K),
C                 E=YDP(K+1),F=YDP(K+2),Z=H*H/12,Q=YP=PREDICTED
C                 VALUE OF Y(K+2)
      YDP(X,Y)=DEXP(2.D0*Y)
      S(X)=-DLOG(DCOS(X))
      YP(Z)=2.D0*B-A+Z*(10.D0*E+D)
      YC(Z)=Q+Z*F
C CONSTANTS.
      PI=3.141592653589793D0
      E1=2.718281828459045D0
      C1=0.5D0
      C2=DSQRT(2.D0)
      C3=DSQRT(3.D0)
      C5=DSQRT(5.D0)
      CRIT=5.D-17
C INPUT.
      READ(1,100)NAME,XO,HO
100   FORMAT(7A4,2X,2F10.0)
C FIRST HEADING.
      WRITE(3,101)
101   FORMAT(1H1,T35,'COMPARISON OF SUCCESSIVE VALUES'/)
      WRITE(3,102)NAME
102   FORMAT(1H0,T16,'EQUATION ',7A4/)
      WRITE(3,103)
```

Fig. 14.5-3   (Continued on next page)

```
103    FORMAT(1H0,T11,'X',T25,'H',T34,'FIRST CALCULATION,N1',T61,'SECOND
      1 CALCULATION,N2',T90,'N1-N2'/)
C INITIAL CONDITIONS. N COUNTS THE ITERATIONS.
       N=1
       H=H0
       A=S(X0)
       D=YDP(X0,A)
       N1=S(X0+H)
C LOOP TO DETERMINE AN ACCEPTABLE H.
1      N=N+1
       H=C1*H
       X=X0+H
       Z=H*H/12.D0
       B=S(X0+H)
C PREDICTED Y(2).
       E=YDP(X0+H,B)
       Q=YP(Z)
C CORRECTED Y(2).
       F=YDP(X0+H*2.D0,Q)
       C=YC(Z)
C OUTPUT.
       R=N1-C
       X2=X+H
       WRITE(3,104)X2,H,N1,C,R
104    FORMAT(1H ,T6,F11.8,T20,F11.8,T32,1PD23.15,T60,D23.15,T88,D9.2)
C DECISIONS.
       IF(DABS(N1-C).LE.5.D-17)GO TO 2
       IF(N.EQ.6)GO TO 2
C STEP.
       N1=B
       GO TO 1
C CALCULATION OF A TABLE OF VALUES.
2      KF=10*2**(N-1)
       N2(1)=B
       N2(2)=C
       X=X0
       DO 3 K=3,KF
C PREDICTED VALUES.
       A=N2(K-2)
       B=N2(K-1)
       X=X+H
       D=YDP(X,A)
       X1=X+H
       E=YDP(X1,B)
       Q=YP(Z)
C CORRECTED VALUES.
       X2=X1+H
       F=YDP(X2,Q)
3      N2(K)=YC(Z)
C OUTPUT.
       L=0
       WRITE(3,105)
105    FORMAT(1H1,T38,'TABLE OF NUMEROV VALUES'/)
       WRITE(3,106)NAME,H,N
106    FORMAT(1H0,T5,7A4,T35,'H USED IN CALCULATIONS= ',F10.8,10X,'N=',I1
      1/)
       WRITE(3,107)
107    FORMAT(1H0,T3,'X',T23,'CALC. Y',T49,'TRUE VALUE',T72,'ERROR',T83,
      1'REL. ERROR'/)
C
       X=X0
       DO 4 K=1,KF
       L=L+1
       X=X+H
       TV=S(X)
       E=TV-N2(K)
       IF(DABS(E).LE.CRIT)GO TO 41
```

```
        R=E/N2(K)
        GO TO 42
41      E=0.D0
        R=0.D0
42      WRITE(3,108)X,N2(K),TV,E,R
108     FORMAT(1H ,T2,F11.8, 2X,1PD22.15,5X,D22.15,5X,2(D9.2,5X))
        IF(L.NE.45)GO TO 4
        WRITE(3,105)
        WRITE(3,106)NAME,H,N
        WRITE(3,107)
        L=1
4       CONTINUE
        END
```

Fig. 14.5-3   Computer program for the boundary-value problem.

PROBLEM SECTION 14.5

Using the true values in Figure 14.1-13, select the two end values as boundary conditions and calculate the in-between values.
(1) Bessel equation, order 0
(2) Modified Bessel equation, order 1
(3) Kummer's equation
(4) Coulomb wave function
(5) Airy function, 0.01(0.02)0.20
(6) Airy function, 0.00(0.01)0.10
(7) Parabolic-cylinder function
(8) Chebyshev equation
(9) Legendre equation
(10) Hermite equation

Appendix—Tables

LIST OF TABLES

Table A.1
Selected Constants, 16 sf

$\sqrt{2}$	1.41421 35623 73095
$\sqrt{3}$	1.73205 08075 68877
e	2.71828 18284 59045
$\ln 2$	0.69314 71805 599453
π	3.14159 26535 89793
$\pi/2$	1.57079 63267 94897
$\pi/3$	1.04719 75511 96598
$\pi/4$	0.78539 81633 974483
$\sqrt{\pi}$	1.77245 38509 05516
2π	6.28318 53071 79586
1 rad	57.29577 95130 8232°
1°	0.01745 32925 19943 30 rad
$\sin 15°$	0.25881 90451 02520 8
$\sin 45°$	0.70710 67811 86547 5
$\sin 60°$	0.86602 54037 84438 6
$\sin 75°$	0.96592 58262 89068 3

Table A.2
Maximum Relative Errors, Except for Roundoff

OPERATION	RELATIVE ERROR[a]
$x+y$	$\dfrac{x}{x+y}i_x + \dfrac{y}{x+y}i_y$
$x-y$	$\dfrac{x}{x-y}i_x - \dfrac{y}{x-y}i_y$
$x \times y$	$i_x + i_y$
x/y	$i_x - i_y$
$\ln x$	$\dfrac{1}{\ln x}i_x + T_{\ln}$
$\exp x = e^x$	$xi_x + T_{\exp}$
x^n, n in fl. pt.	$ni_x + (n \ln x)(T_{\ln} + R) + T_{\exp}$
$\sqrt{x}$	$\frac{1}{2}i_x + T_{\sqrt{x}}$
$\sin x$	$(x \cot x)i_x + T_{\sin}$
$\cos x$	$(-x \tan x)i_x + T_{\cos}$
$\tan x$	$(x \sec x \csc x)i_x + T_{\tan}$

[a]T is the value of the relative truncation error as supplied by the manufacturer.

Table A.3
Stirling Numbers
Stirling Numbers of the First Kind

FACTORIAL FUNCTION	FUNCTION
$x^{(1)}$	x
$x^{(2)}$	$x^2 - x$
$x^{(3)}$	$x^3 - 3x^2 + 2x$
$x^{(4)}$	$x^4 - 6x^3 + 11x^2 - 6x$
$x^{(5)}$	$x^5 - 10x^4 + 35x^3 - 50x^2 + 24x$

Stirling Numbers of the Second Kind

FUNCTION	FACTORIAL FUNCTION
x	$x^{(1)}$
x^2	$x^{(2)} + x^{(1)}$
x^3	$x^{(3)} + 3x^{(2)} + x^{(1)}$
x^4	$x^{(4)} + 6x^{(3)} + 7x^{(2)} + x^{(1)}$
x^5	$x^{(5)} + 10x^{(4)} + 25x^{(3)} + 15x^{(2)} + x^{(1)}$

Table A.4
First Forward Differences, $\Delta x = h$

	f	Δf
1	c	0
2	$u + v$	$\Delta u + \Delta v$
3	$u - v$	$\Delta u - \Delta v$
4	uv	$u(x+h)\Delta v + v\Delta u$
5	uv	$v(x+h)\Delta u + u\Delta v$
6	u/v	$\dfrac{v\Delta u - u\Delta v}{v(x)v(x+h)}$
7	x^n	$\displaystyle\sum_{k=1}^{n}\binom{n}{k}x^{n-k}h^k$
8	$\sin(ax+b)$	$2\cos\left[a\left(x+\dfrac{h}{2}\right)+b\right]\sin\dfrac{ah}{2}$
9	$\cos(ax+b)$	$-2\sin\left[a\left(x+\dfrac{h}{2}\right)+b\right]\sin\dfrac{ah}{2}$
10	$\tan(ax+b)$	$\dfrac{\sin ah}{\cos[a(x+h)+b]\cos(ax+b)}$
11	$\csc(ax+b)$	$\dfrac{-2\cos[a(x+h/2)+b]\sin(ah/2)}{\sin[a(x+h)+b]\sin(ax+b)}$
12	$\sec(ax+b)$	$\dfrac{2\sin[a(x+h/2)+b]\sin(ah/2)}{\cos[a(x+h)+b]\cos(ax+b)}$
13	$\cot(ax+b)$	$\dfrac{-\sin ah}{\sin[a(x+h)+b]\sin(ax+b)}$
14	a^{cx+d}	$a^{cx+d}(a^{ch}-1)$
15	$\ln(ax+b)$	$\ln\left(1+\dfrac{ah}{ax+b}\right)$
16	$\sinh(ax+b)$	$\cosh\left[a\left(x+\dfrac{h}{2}\right)+b\right]\sinh\dfrac{ah}{2}$
17	$\cosh(ax+b)$	$2\sinh\left[a\left(x+\dfrac{h}{2}\right)+b\right]\sinh\dfrac{ah}{2}$
18	$\tanh(ax+b)$	$\dfrac{\sinh ah}{\cosh(ax+ah+b)\cosh(ax+b)}$

Table A.5
First Forward Differences, $h = 1$

	f	Δf
1	$\sin ax$	$2\sin(a/2)\cos[a(x+\tfrac{1}{2})]$
2	$\cos ax$	$-2\cos(a/2)\sin[a(x+\tfrac{1}{2})]$
3	$\tan ax$	$\dfrac{\sin a}{\cos[a(x+1)]\cos ax}$
4	$\csc ax$	$\dfrac{-2\sin(a/2)\cos\left[a\left(x+\tfrac{1}{2}\right)\right]}{\sin[a(x+1)]\sin ax}$
5	$\sec ax$	$\dfrac{2\sin(a/2)\sin\left[a\left(x+\tfrac{1}{2}\right)\right]}{\cos[a(x+1)]\cos ax}$
6	$\cot ax$	$\dfrac{-\sin a}{\sin[a(x+1)]\sin ax}$
7	a^{cx}	$a^{cx}(a^c-1)$
8	2^x	2^x
9	$x^{(n)}$	$nx^{(n-1)}$
10	$(x+c)!$	$(x+c)\cdot(x+c)!$
11	$\dfrac{1}{(x+c)!}$	$\dfrac{1-(x+c)}{(x+c+1)!}$
12	$\binom{x}{n}$	$\binom{x}{n-1}$
13	$\binom{n}{x}$	$\dfrac{n-2x-1}{x+1}\binom{n}{x}$
14	$\arctan ax$	$\arctan\dfrac{a}{1+a^2x+a^2x^2}$
15	$B_n(x)$	nx^{n-1} for $n=1,2,3,$

Table A.6
Sum Theorems

(1) $\displaystyle\sum_{k=a}^{b-1} \Delta f(k) = f(b) - f(a)$

(2)^a $\displaystyle\sum_{k=a}^{b-1} k^{(n)} = \frac{1}{n+1}(b^{(n+1)} - a^{(n+1)})$

(3) $\displaystyle\sum_{k=a}^{b-1} c^{tk} = \frac{1}{c^t - 1}(c^{tb} - c^{ta})$

(4) $\displaystyle\sum_{k=a}^{b-1} \cos(pk + q) = \frac{\cos\frac{1}{2}[p(a+b-1)+2q]\sin\frac{1}{2}p(b-a)}{\sin\frac{1}{2}p}$, den $\neq 0$

(5) $\displaystyle\sum_{k=a}^{b-1} \sin(pk + q) = \frac{\sin\frac{1}{2}[p(a+b-1)+2q]\sin\frac{1}{2}p(b-a)}{\sin\frac{1}{2}p}$, den $\neq 0$

(6) $\displaystyle\sum_{k=a}^{b-1} \cos^2(pk + q) = \frac{1}{2}\left[b - a + \frac{\cos[p(a+b-1)+2q]\sin p(b-a)}{\sin p}\right]$, den $\neq 0$

(7) $\displaystyle\sum_{k=a}^{b-1} \sin^2(pk + q) = \frac{1}{2}\left[b - a - \frac{\cos[p(a+b-1)+2q]\sin p(b-a)}{\sin p}\right]$, den $\neq 0$

(8) $\displaystyle\sum_{k=a}^{b-1} \sinh(pk + q) = \frac{\cosh\left[p\left(b-\frac{1}{2}\right)+q\right] - \cosh\left[p\left(a-\frac{1}{2}\right)+q\right]}{2\sinh(p/2)}$, den $\neq 0$

(9) $\displaystyle\sum_{k=a}^{b-1} \cosh(pk + q) = \frac{\sinh\left[p\left(b-\frac{1}{2}\right)+q\right] - \sinh\left[p\left(a-\frac{1}{2}\right)+q\right]}{2\sinh(p/2)}$

^{a}Note that $k^{(-n)} = 1/(k+n)^{(n)}$ and $1/k^{(n)} = (k-n)^{(-n)}$ for $n > 0$.

Table A.7
Indefinite Summands, $h = 1$

	IF Δv IS	THEN v MAY BE
(1)	$ax + b$	$\dfrac{a}{2}x^2 + \left(b - \dfrac{a}{2}\right)x$
(2)	$ax^2 + b$	$\dfrac{a}{3}x^3 - \dfrac{a}{2}x^2 + \left(b + \dfrac{a}{6}\right)x$
(3)	$ax^{(n)} + b$	$\dfrac{a}{n+1}x^{(n+1)} + bx$
(4)	$\sin(ax + b)$	$-\dfrac{\cos\left[a\left(x - \frac{1}{2}\right) + b\right]}{2\sin\frac{1}{2}a}$
(5)	$\cos(ax + b)$	$\dfrac{\sin\left[a\left(x - \frac{1}{2}\right) + b\right]}{2\sin\frac{1}{2}a}$
(6)	a^{cx}	$\dfrac{1}{a-1}a^{cx}$
(7)	$\dbinom{x}{n}$	$\dbinom{x}{n+1}$
(8)	$\sinh(ax + b)$	$\dfrac{\cosh\left[a\left(x - \frac{1}{2}\right) + b\right]}{2\sinh\frac{1}{2}a}$
(9)	$\cosh(ax + b)$	$\dfrac{\sinh\left[a\left(x - \frac{1}{2}\right) + b\right]}{2\sinh\frac{1}{2}a}$
(10)	$x \cdot x!$	$x!$
(11)	x^n	$\dfrac{1}{n+1}B_{n+1}(x)$
(12)	$\dfrac{-x}{(x+1)!}$	$\dfrac{1}{x!}$

Table A.8
Miscellaneous Trigonometric Formulas

(1)	$\sin 2A = 2\sin A \cos A$
(2)	$\sin 3A = 3\sin A - 4\sin^3 A$
(3)	$\sin 4A = 4\sin A \cos A - 8\sin^3 A \cos A$
(4)	$\sin 5A = 5\sin A - 20\sin^3 A + 16\sin^5 A$
(5)	$\sin 6A = 6\sin A \cos A - 32\sin^3 A \cos A + 32\sin^5 A \cos A$
(6)	$\cos 2A = 2\cos^2 A - 1 = 1 - 2\sin^2 A$
(7)	$\cos 3A = 4\cos^3 A - 3\cos A$
(8)	$\cos 4A = 8\cos^4 A - 8\cos^2 A + 1$
(9)	$\cos 5A = 16\cos^5 A - 20\cos^3 A + 5\cos A$
(10)	$\cos 6A = 32\cos^6 A - 48\cos^4 A + 18\cos^2 A - 1$
(11)	$\tan 2A = (2\tan A)/(1 - \tan^2 A)$
(12)	$\tan 3A = (3\tan A - \tan^3 A)/(1 - \tan^2 A)$
(13)[a]	$\sin A + \sin B = 2\sin HS \cos HD$
(14)[a]	$\sin A - \sin B = 2\cos HS \sin HD$
(15)[a]	$\cos A + \cos B = 2\cos HS \cos HD$
(16)[a]	$\cos A - \cos B = -2\sin HS \sin HD$
(17)[b]	$2\sin A \sin B = \cos D - \cos S$
(18)[b]	$2\cos A \cos B = \cos D + \cos S$
(19)[b]	$2\sin A \cos B = \sin S + \sin D$

[a] $HS = \frac{1}{2}(A + B)$ and $HD = \frac{1}{2}(A - B)$.
[b] $S = A + B$ and $D = A - B$.

Table A.9
$$\sum_{k=1}^{n} k^c$$

c	Sum
1	$n(n+1)/2$
2	$n(n+1)(2n+1)/6$
3	$n^2(n+1)^2/4$
4	$n(n+1)(2n+1)(3n^2+3n-1)/30$
5	$n^2(n+1)^2(2n^2+2n-1)/12$
6	$n(n+1)(2n+1)(3n^4+6n^3-3n+1)/42$
7	$n^2(n+1)^2(3n^4+6n^3-n^2-4n+2)/24$
8	$n(n+1)(2n+1)(5n^6+15n^5+5n^4-15n^3-n^2+9n+3)/90$
9	$n^2(n+1)^2(2n^6+6n^5+n^4-8n^3+n^2+6n-3)/20$
10	$n(n+1)(2n+1)(3n^8+12n^7+8n^6-18n^5-10n^4+24n^3+2n^2-15n+5)/66$

Table A.10

$\sum_{k=1}^{n} 1/k^p$ **from** $p=2$ **to** $13, n=3$ **to** 15

THE SUM OF RECIPROCALS TO A POWER

EXPONENT

UPPER LIMIT	2	3	4	5	6	7
3	1.3611111111	1.1620370370	1.0748456790	1.0353652263	1.0169967421	1.0082697474
4	1.4236111111	1.1776620370	1.0787519290	1.0363417888	1.0172408827	1.0083307825
5	1.4636111111	1.1856620370	1.0803519290	1.0366617888	1.0173048827	1.0083435825
6	1.4913888889	1.1902916667	1.0811235340	1.0367903897	1.0173263162	1.0083471548
7	1.5117970522	1.1932071186	1.0815400271	1.0368498887	1.0173348161	1.0083483690
8	1.5274220522	1.1951602436	1.0817841677	1.0368804063	1.0173386308	1.0083488459
9	1.5397677312	1.1965319857	1.0819365835	1.0368973413	1.0173405124	1.0083490550
10	1.5497677312	1.1975319857	1.0820365835	1.0369073413	1.0173415124	1.0083491550
11	1.5580321940	1.1982833005	1.0821048848	1.0369135506	1.0173420769	1.0083492063
12	1.5649766384	1.1988620042	1.0821531101	1.0369175693	1.0173424118	1.0083492342
13	1.5708937982	1.1993171703	1.0821881229	1.0369202626	1.0173426190	1.0083492501
14	1.5759958390	1.1996816018	1.0822141537	1.0369221220	1.0173427518	1.0083492596
15	1.5804402834	1.1999778981	1.0822339068	1.0369234388	1.0173428396	1.0083492654

THE SUM OF RECIPROCALS TO A POWER

EXPONENT

UPPER LIMIT	8	9	10	11	12	13
3	1.0040586658	1.0020039303	1.0009934976	1.0004939263	1.0002460223	1.0001226975
4	1.0040739246	1.0020077450	1.0009944513	1.0004941647	1.0002460819	1.0001227124
5	1.0040764846	1.0020082570	1.0009945537	1.0004941852	1.0002460860	1.0001227133
6	1.0040770800	1.0020083562	1.0009945702	1.0004941879	1.0002460865	1.0001227133
7	1.0040772534	1.0020083810	1.0009945737	1.0004941884	1.0002460865	1.0001227133
8	1.0040773130	1.0020083884	1.0009945747	1.0004941886	1.0002460866	1.0001227133
9	1.0040773363	1.0020083910	1.0009945750	1.0004941886	1.0002460866	1.0001227133
10	1.0040773463	1.0020083920	1.0009945751	1.0004941886	1.0002460866	1.0001227133
11	1.0040773509	1.0020083924	1.0009945751	1.0004941886	1.0002460866	1.0001227133
12	1.0040773532	1.0020083926	1.0009945751	1.0004941886	1.0002460866	1.0001227133
13	1.0040773545	1.0020083927	1.0009945751	1.0004941886	1.0002460866	1.0001227133
14	1.0040773551	1.0020083928	1.0009945751	1.0004941886	1.0002460866	1.0001227133
15	1.0040773555	1.0020083928	1.0009945751	1.0004941886	1.0002460866	1.0001227133

Table A.11[a]

$$\int x^n \sin kx\, dx \text{ and } \int x^n \cos kx\, dx$$

(1) $\int x \sin kx\, dx = \dfrac{1}{k^2}\sin kx - \dfrac{x}{k}\cos kx$ [2]

(2) $\int x^2 \sin kx\, dx = \dfrac{2x}{k^2}\sin kx - \dfrac{k^2x^2-2}{k^3}\cos kx$ [0]

(3) $\int x^3 \sin kx\, dx = \dfrac{3k^2x^2-6}{k^4}\sin kx - \dfrac{k^2x^3-6x}{k^3}\cos kx$ $\left[\dfrac{3}{2}\pi^2-12\right]$

(4) $\int x^4 \sin kx\, dx = \dfrac{4k^2x^3-24x}{k^4}\sin kx - \dfrac{k^4x^4-12k^2x^2+24}{k^5}\cos kx$ [0]

(5) $\int x^5 \sin kx\, dx = \dfrac{5k^4x^4-60k^2x^2+120}{k^6}\sin kx - \dfrac{k^4x^5-20k^2x^3+120x}{k^5}\cos kx$ $\left[\dfrac{5}{8}\pi^4-30\pi^2+240\right]$

(6) $\int x^6 \sin kx\, dx = \dfrac{6k^4x^5-120k^2x^3+720x}{k^6}\sin kx - \dfrac{k^6x^6-30k^4x^4+360k^2x^2-720}{k^7}\cos kx$ [0]

(7) $\int x \cos kx\, dx = \dfrac{1}{k^2}\cos kx + \dfrac{x}{k}\sin kx$ [0]

(8) $\int x^2 \cos kx\, dx = \dfrac{2x}{k^2}\cos kx + \dfrac{k^2x^2-2}{k^3}\sin kx$ $\left[\dfrac{1}{2}\pi^2-4\right]$

(9) $\int x^3 \cos kx\, dx = \dfrac{3k^2x^2-6}{k^4}\cos kx + \dfrac{k^2x^3-6x}{k^3}\sin kx$ [0]

(10) $\int x^4 \cos kx\, dx = \dfrac{4k^2x^3-24x}{k^4}\cos kx + \dfrac{k^4x^4-12k^2x^2+24}{k^5}\sin kx$ $\left[\dfrac{1}{8}\pi^4-6\pi^2+48\right]$

(11) $\int x^5 \cos kx\, dx = \dfrac{5k^4x^4-60k^2x^2+120}{k^6}\cos kx + \dfrac{k^4x^5-20k^2x^3+120x}{k^5}\sin kx$ [0]

(12) $\int x^6 \cos kx\, dx = \dfrac{6k^4x^5-120k^2x^3+720}{k^6}\cos kx + \dfrac{k^6x^6-30k^4x^4+360k^2x^2-720}{k^7}\sin kx$ [0]

[a]The constants of integration for the indefinite integrals have been omitted. The values in brackets are the definite integrals from $-\pi/2$ to $+\pi/2$, for $k=1$.

Table A.12
The Riemann Zeta-Function $\zeta(n) = \sum\limits_{k=1}^{\infty} \dfrac{1}{k^n}$

n	$\zeta(n)$ Rounded to 16 sf
1	[diverges]
2	$\pi^2/6 \sim 1.64493\ 40668\ 48226$
3	$1.20205\ 69031\ 59594$
4	$\pi^4/90 \sim 1.08232\ 32337\ 11138$
5	$1.03692\ 77551\ 43370$
6	$\pi^6/945 \sim 1.01734\ 30619\ 84449$
7	$1.00834\ 92773\ 81923$
8	$\pi^8/9450 \sim 1.00407\ 73561\ 97944$
9	$1.00200\ 83928\ 26082$
10	$1.00099\ 45751\ 27818$
11	$1.00049\ 41886\ 04119$
12	$1.00024\ 60865\ 53308$

Table A.13
Selected Comparison Series[a]

Series[b]	Lower Limit	Restriction	Value
$\sum x^k/k!$	0		e^x
$\sum (-1)^{k+1} \dfrac{x^{2k-1}}{(2k-1)!}$	1		$\sin x$
$\sum (-1)^k \dfrac{x^{2k}}{(2k)!}$	0		$\cos x$
$\sum \dfrac{2x}{x^2 - k^2\pi^2}$	0	$x \neq k\pi$	$\cot x$
$\sum \dfrac{2x}{x^2 - k^2\pi^2}(-1)^k$	0	$x \neq k\pi$	$\csc x$
$\sum \dfrac{x^{2k-1}}{(2k-1)!}$	1		$\sinh x$
$\sum \dfrac{x^{2k}}{(2k)!}$	0		$\cosh x$
$\sum \dfrac{x^{2k-1}}{2k-1}$	1	$|x| < 1$	$\operatorname{arctanh} x$
$\sum \dfrac{1}{(2k-1)x^{2k-1}}$	1	$|x| < 1$	$\operatorname{arccoth} x$
$\sum \dfrac{\cos kx}{k}$	1	$0 < x < 2\pi$	$-\ln\left(2\sin\dfrac{x}{2}\right)$
$\sum \dfrac{\cos kx}{k^2}$	1	$0 < x < 2\pi$	$\dfrac{\pi^2}{6} - \dfrac{x\pi}{2} + \dfrac{x^2}{4}$
$\sum \dfrac{\cos kx}{k^4}$	1	$0 < x < 2\pi$	$\dfrac{\pi^4}{90} - \dfrac{\pi^2 x^2}{12} + \dfrac{\pi x^3}{12} - \dfrac{x^4}{48}$
$\sum \dfrac{\sin kx}{k}$	1	$0 < x < 2\pi$	$\frac{1}{2}(\pi - x)$
$\sum \dfrac{\sin kx}{k^3}$	1	$0 < x < 2\pi$	$\dfrac{\pi^2 x}{6} - \dfrac{\pi x^2}{4} + \dfrac{x^3}{12}$

Table A.13 (Continued)

Series[b]	Lower Limit	Restriction	Value
$\displaystyle\sum \frac{(-1)^k x^{2k+1}}{(2k+1)(2k+1)!}$	0		Sine-integral
$\displaystyle\frac{2}{\sqrt{\pi}}\sum \frac{(-1)^k x^{2k+1}}{(2k+1)k!}$	0		Error function
$\displaystyle\sum (-1)^{k-1} k^{-n}$	1	$n > 1$	Riemann eta-function
$\displaystyle\sum (2k+1)^{-n}$	0	$n > 1$	Riemann lambda-function
$\displaystyle\sum (-1)^{k-1}(2k+1)^{-n}$	0	$n > 1$	Riemann beta-function
$\displaystyle\sum (-1)^{k+1}\frac{x^k}{k}$	1	$-1 < x \leqslant 1$	$\ln(1+x)$

[a]The above are either known in closed form or are available to many places in handbooks such as Abramowitz and Stegun. Other comparison series can be found in that handbook or in A. D. Wheelon, *Tables of Summable Series*, Holden-Day, 1968.
[b]The upper limit is ∞.

Table A.14
Exact Sines and Cosines

Angle (°)	Sine[a]	Cosine[a]
0, 360, 720, 1080	0	1
15, 375, 735, 1095	A	D
30, 390, 750, 1110	$\frac{1}{2}$	C
45, 405, 765, 1125	B	B
60, 420, 780, 1140	C	$\frac{1}{2}$
75, 435, 795, 1155	D	A
90, 450, 810, 1170	1	0
105, 465, 825, 1185	D	$-A$
120, 480, 840, 1200	C	$-\frac{1}{2}$
135, 495, 855, 1215	B	$-B$
150, 510, 870, 1230	$\frac{1}{2}$	$-C$
165, 525, 885, 1245	A	$-D$
180, 540, 900, 1260	0	-1
195, 555, 915, 1275	$-A$	$-D$
210, 570, 930, 1290	$-\frac{1}{2}$	$-C$
225, 585, 945, 1305	$-B$	$-B$
240, 600, 960, 1320	$-C$	$-\frac{1}{2}$
255, 615, 975, 1335	$-D$	$-A$
270, 630, 990, 1350	-1	0
285, 645, 1005, 1365	$-D$	A
300, 660, 1020, 1380	$-C$	$\frac{1}{2}$
315, 675, 1035, 1395	$-B$	B
330, 690, 1050, 1410	$-\frac{1}{2}$	C
345, 705, 1065, 1425	$-A$	D

[a]$A = \frac{1}{2}\sqrt{2-\sqrt{3}}$, $B = \sqrt{2}/2$, $C = \sqrt{3}/2$,
$D = \frac{1}{2}\sqrt{2+\sqrt{3}}$.

Table A. 15
Values of $\sin k\theta$ for Harmonic Analysis

θ (°)	$k=1$	2	3	4	5	6	7	8	9	10	11
0	0	0	0	0	0	0	0	0	0	0	0
15	A	0.5	B	C	D	1	D	C	B	0.5	A
30	0.5	C	1	C	0.5	0	-0.5	$-C$	-1	$-C$	-0.5
45	B	1	B	0	$-B$	-1	$-B$	0	B	1	B
60	C	C	0	$-C$	$-C$	0	C	C	0	$-C$	$-C$
75	D	0.5	$-B$	$-C$	A	1	A	$-C$	$-B$	0.5	D
90	1	0	-1	0	1	0	-1	0	1	0	-1
105	D	-0.5	$-B$	C	A	-1	A	C	$-B$	-0.5	D
120	C	$-C$	0	C	$-C$	0	C	$-C$	0	C	$-C$
135	B	-1	B	0	$-B$	1	$-B$	0	B	-1	B
150	0.5	$-C$	1	$-C$	0.5	0	-0.5	C	-1	C	-0.5
165	A	-0.5	B	$-C$	D	-1	D	$-C$	B	-0.5	A

Table A.16.
Values of $\cos k\theta$ for Harmonic Analysis

θ (°)	$k=1$	2	3	4	5	6	7	8	9	10	11	12
0	1	1	1	1	1	1	1	1	1	1	1	1
15	D	C	B	0.5	A	0	$-A$	-0.5	$-B$	$-C$	$-D$	-1
30	C	0.5	0	-0.5	$-C$	-1	$-C$	-0.5	0	0.5	C	1
45	B	0	$-B$	-1	$-B$	0	B	1	B	0	$-B$	-1
60	0.5	-0.5	-1	-0.5	0.5	1	0.5	-0.5	-1	-0.5	0.5	1
75	A	$-C$	$-B$	0.5	D	0	$-D$	-0.5	B	C	$-A$	-1
90	0	-1	0	1	0	-1	0	1	0	-1	0	1
105	$-A$	$-C$	B	0.5	$-D$	0	D	-0.5	$-B$	C	A	-1
120	-0.5	-0.5	1	-0.5	-0.5	1	-0.5	-0.5	1	-0.5	-0.5	1
135	$-B$	0	B	-1	B	0	$-B$	0	B	0	$-B$	-1
150	$-C$	0.5	0	-0.5	C	-1	C	-0.5	0	0.5	$-C$	1
165	$-D$	C	$-B$	$.5$	$-A$	0	A	-0.5	B	$-C$	D	-1

Table A.17
Chebyshev Polynomials $(-1, +1)$

$T_0 = 1$

$T_1 = x$

$T_2 = 2x^2 - 1$

$T_3 = 4x^3 - 3x$

$T_4 = 8x^4 - 8x^2 + 1$

$T_5 = 16x^5 - 20x^3 + 5x$

$T_6 = 32x^6 - 48x^4 + 18x^2 - 1$

$T_7 = 64x^7 - 112x^5 + 56x^3 - 7x$

$T_8 = 128x^8 - 256x^6 + 160x^4 - 32x^2 + 1$

$T_9 = 256x^9 - 576x^7 + 432x^5 - 120x^3 + 9x$

$T_{10} = 512x^{10} - 1280x^8 + 1120x^6 - 400x^4 + 50x^2 - 1$

$T_{11} = 1024x^{11} - 2816x^9 + 2816x^7 - 1232x^5 + 220x^3 - 11x$

$T_{12} = 2048x^{12} - 6144x^{10} + 6912x^8 - 3584x^6 + 840x^4 - 72x^2 + 1$

$1 = T_0$

$x = T_1$

$x^2 = \frac{1}{2}(T_0 + T_2)$

$x^3 = \frac{1}{4}(3T_1 + T_3)$

$x^4 = \frac{1}{8}(3T_0 + 4T_2 + T_4)$

$x^5 = \frac{1}{16}(10T_1 + 5T_3 + T_5)$

$x^6 = \frac{1}{32}(10T_0 + 15T_2 + 6T_4 + T_6)$

$x^7 = \frac{1}{64}(35T_1 + 21T_3 + 7T_5 + T_7)$

$x^8 = \frac{1}{128}(35T_0 + 56T_2 + 28T_4 + 8T_6 + T_8)$

$x^9 = \frac{1}{256}(126T_1 + 84T_3 + 36T_5 + 9T_7 + T_9)$

$x^{10} = \frac{1}{512}(126T_0 + 210T_2 + 120T_4 + 45T_6 + 10T_8 + T_{10})$

Table A.18
Properties of the Hyperbolic Functions

(1)	$\cosh u = \frac{1}{2}(e^u + e^{-u})$	Range $[0, \infty)$
(2)	$\sinh u = \frac{1}{2}(e^u - e^{-u})$	Range $(-\infty, +\infty)$
(3)	$\tanh u = \sinh u / \cosh u$	Range $[0, 1)$
(4)	$\operatorname{sech} u = 1/\cosh u$	Range $[1, \infty)$
(5)	$\cosh^2 u - \sinh^2 u = 1$	
(6)	$1 - \tanh^2 u = \operatorname{sech}^2 u$	
(7)	$D \sinh u = \cosh u$	
(8)	$D \cosh u = \sinh u$	
(9)	$D \tanh u = \operatorname{sech}^2 u$	
(10)	$D \operatorname{sech} u = -\operatorname{sech} u \tanh u$	
(11)	$\int \sinh u \, du = \cosh u + C$	
(12)	$\int \cosh u \, du = \sinh u + C$	
(13)	$\int \tanh u \, du = \ln \cosh u + C$	
(14)	$\int \operatorname{sech} u \, du = 2 \tan^{-1} e^u + C$	
(15)	$D \sinh^{-1} u = 1/(1 + u^2)^{1/2}$	
(16)	$D \cosh^{-1} u = 1/(u^2 - 1)^{1/2}$	
(17)	$D \tanh^{-1} u = 1/(1 - u^2), \; |u| < 1$	
(18)	$D \coth^{-1} u = 1/(1 - u^2), \; |u| > 1$	
(19)	$D \operatorname{sech}^{-1} u = 1/[u(1 - u^2)^{1/2}]$	
(20)	$D \operatorname{csch}^{-1} u = -1/[|u|(1 + u^2)^{1/2}]$	
(21)	$\displaystyle \int \frac{du}{\sqrt{1 + u^2}} = \sinh^{-1} u + C$	
(22)	$\displaystyle \int \frac{du}{\sqrt{u^2 - 1}} = \cosh^{-1} u + C$	
(23)	$\displaystyle \int \frac{du}{1 - u^2} = \begin{cases} \tanh^{-1} u + C \text{ if } |u| < 1 \\ \coth^{-1} u + C \text{ if } |u| > 1 \end{cases}$	
(24)	$\displaystyle \int \frac{du}{u\sqrt{1 - u^2}} = -\operatorname{sech}^{-1}|u| + C$	

Table A. 19
First-Order Bernoulli Polynomials

$B_0(k) = 1$

$B_1(k) = k - \frac{1}{2}$

$B_2(k) = k^2 - k + \frac{1}{6}$

$B_3(k) = k^3 - \frac{3}{2}k^2 + \frac{1}{2}k$

$B_4(k) = k^4 - 2k^3 + k^2 - \frac{1}{30}$

$B_5(k) = k^5 - \frac{5}{2}k^4 + \frac{5}{3}k^3 - \frac{1}{6}k$

$B_6(k) = k^6 - 3k^5 + \frac{5}{2}k^4 - \frac{1}{2}k^2 + \frac{1}{42}$

$B_7(k) = k^7 - \frac{7}{2}k^6 + \frac{7}{2}k^5 - \frac{7}{6}k^3 + \frac{1}{6}k$

$B_8(k) = k^8 - 4k^7 + \frac{14}{3}k^6 - \frac{7}{3}k^4 + \frac{2}{3}k^2 - \frac{1}{30}$

Selected Bibliography

ABRAMOWITZ, A. AND STEGUN, I. A. EDS., *Handbook of Mathematical Functions,* Appl. Math. Ser 55, Nat. Bur. of Stand., June 1964

ACTON, F. S. DOVER., *Numerical Methods That Work,* Harper and Row, 1970

BROMWICH, T. J., *An Introduction to the Theory of Infinite Series,* Macmillan, 1959

DORN, W. S. and McCracken, D. D., *Numerical Methods with Fortran IV Case Studies,* Wiley, 1972

FOX, L. AND MAYERS, D. F., *Computing Methods for Scientists and Engineers,* Clarendon Press, Oxford, 1968

FRÖBERG, C-E., *Introduction to Numerical Analysis,* Addison-Wesley, 1965

HAMMING, R. W., *Numerical Methods for Scientists and Engineers,* McGraw-Hill, 1962

HASTINGS, C., JR., *Approximations for Digital Computers,* Princeton U. P., 1955

HILDEBRAND, F. B., *Introduction to Numerical Analysis,* McGraw-Hill, 1956

LEVY, H. AND LESSMAN, F., *Finite Difference Equations,* Macmillan, 1961

MILNE, W. E., *Numerical Calculus,* Princeton, 1949

MILNE-THOMSON, L. M., *The Calculus of Finite Differences,* Macmillan, 1960

RALSTON, A., *A First Course in Numerical Analysis,* McGraw-Hill, 1965

SCARBOROUGH, J. B., *Numerical Mathematical Analysis,* Oxford, 1955

SCHEID, F., *Numerical Analysis,* Schaum's Outline Ser., McGraw-Hill, 1968

SOKOLNIKOFF, I. S. AND REDHEFFER, R. M., *Mathematics of Physics and Modern Engineering,* McGraw-Hill, 1966

WHITTAKER, E. T. AND WATSON, G. N., *A Course of Modern Analysis,* Cambridge, 1952

WYLIE, C. R., *Advanced Engineering Mathematics,* McGraw-Hill, 1960

Answers to Numerical Problems

CHAPTER 1

Section 1.1

(1) $0.F46666 \times 10_x^1$, 15.274994, 3.37×10^{-7}
(2) $0.156B02 \times 10_x^2$, 21.417999, 3.42×10^{-8}
(3) $0.166EFD \times 10_x^3$, 358.893676, 6.48×10^{-7}
(4) $0.199D37 \times 10_x^3$, 409.82592, 1.76×10^{-7}
(5) $0.14719E \times 10_x^4$, 5233.6171, 1.55×10^{-7}
(6) $0.1B0F69 \times 10_x^4$, 6927.4101, 1.30×10^{-7}
(7) 0.2654×10^2 (8) 0.8100×10^2
(9) 0.3141×10^1 (10) 0.4332×10^{-3}
(11) 0.2718×10^1 (12) 0.6118×10^{-3}
(13) $(0.8124 + 0.877 \times 10^{-4}) \times 10^1$
(14) $(0.9023 + 0.506 \times 10^{-4}) \times 10^1$
(15) $(0.1582 + 0.6228 \times 10^{-4}) \times 10^3$
(16) $(0.5816 + 0.3572 \times 10^{-4}) \times 10^3$
(17) $(0.1588 + 0.946 \times 10^{-4}) \times 10^{-2}$
(18) $(0.1973 + 0.926 \times 10^{-4}) \times 10^{-2}$

Section 1.2

(1) 85.24, 0.009, 1.06×10^{-4}
(2) 83.68, 0.003, 3.59×10^{-5}
(3) 6.260, 0.0000281, 4.49×10^{-5}
(4) 3.584, 0.000856, 2.39×10^{-4}
(5) 0.02506, 0.0000023, 9.18×10^{-5}
(6) 0.09427, 0.0000044, 4.67×10^{-5}

Section 1.3

(1) 59.364266, 59, 0.364266, 6.17×10^{-3}
(2) 59.350152, 59, 0.350152, 5.93×10^{-3}
(3) $1.71680484 \times 10^{-4}$, 1.716×10^{-4}, 8.0484×10^{-8}, 4.69×10^{-4}
(4) 1588.899..., 1588, 0.899, 5.66×10^{-4}
(5) 11.31891..., 11.31, 0.00891, 7.87×10^{-4}
(6) 7.93078..., 7.930, 0.00078, 9.84×10^{-5}
(7) 103.6675..., 103.6, 0.0675, 6.52×10^{-4}
(8) 244.9516..., 244.9, 0.0516, 2.11×10^{-4}
(9) 421.8272..., 421.5, 0.3272, 7.76×10^{-4}
(10) 1368.2832..., 1369, 0.7169, 5.24×10^{-4}

Section 1.4

The following intermediate results are probably more useful than the numerical answers.

(5) $i_u^0 = T_{\sqrt{x}}$; $i_v^0 = T_{\sqrt{x}} + r$; $i_y^I = \frac{1}{2} i_x^I$

(6) $i_u^0 = 0$; $i_v^0 = T_{\exp(0.41)} + T_{\ln 1.5}$; $i_y^I = i_x^I$

(7) $i_u^0 = T_{\exp(8.7)} + T_{\ln 32} + r_1$; $i_v^0 = T_{\sqrt{32}} + r_2 + r_3$; $i_y^I = 2.5 i_x^I$

(8) $i_u^0 = (\frac{1}{60} \ln 16)[T_{\ln 16} + 1.5 r_1 - 0.5 r_2] + r_3$; $i_v^0 = T_{\sqrt{16}} + r_4 + r_5$; $i_y^I = 1.57 i_x^I$

(9) $i_u^0 = T_{\cos x} + T_{\tan x} + r$; $i_v^0 = T_{\sin x}$; $i_y^I = (x \cot x) i_x^I$

(10) $i_u^0 = T_{\sin 1} + r_1$; $i_v^0 = T_{\sin 0.5} + T_{\cos 0.5} + r_2 + r_3$; $i_y^I = (2x \cot 2x) i_x^I$

CHAPTER 2

Section 2.1

(1)	−4	(7)	−2	(13)	0
(2)	−2	(8)	1	(14)	2
(3)	3	(9)	1	(15)	1
(4)	3	(10)	−1	(16)	0
(5)	3	(11)	−5	(17)	−2
(6)	4	(12)	−3	(18)	−5

Section 2.2

(1)	−3.141818207	(10)	−0.539835277
(2)	−1.058543221	(11)	−4.390638426
(3)	3.752237076	(12)	−2.124520201
(4)	3.735167876	(13)	0.912208941
(5)	3.180899877	(14)	2.705740726
(6)	4.090014812	(15)	1.615021480
(7)	−1.728466319	(16)	0.882013620
(8)	1.285666579	(17)	−1.548722534
(9)	1.188900906	(18)	−4.926657575

Section 2.3

(1)	4.333526	(10)	3.735168
(2)	12.206556	(11)	−0.976371
(3)	6.588532	(12)	0.860440
(4)	8.168309	(13)	−2.379135229
(5)	4.400559	(14)	−0.740669176
(6)	4.945102	(15)	4.856553205
(7)	−3.141818	(16)	1.024026944
(8)	−1.058543	(17)	−1.636986500
(9)	3.752237	(18)	−0.324991971

Section 2.4

(1)	3.180900	(4)	1.285667
(2)	4.090015	(5)	1.188901
(3)	−1.728466	(6)	−0.539835

(7) − 4.390638
(8) − 2.124520
(9)   0.912409
(10) 2.705741

(11) 1.615021
(12) 0.882014
(13) − 1.548723
(14) − 4.926658

CHAPTER 3

Section 3.1

(1) $x(1) =$ − 4.353712
     $x(2) =$ 5.120932
     $x(3) =$ − 0.980187
     $x(4) =$ − 0.530119
(2) $x(1) =$ − 5.923429
     $x(2) =$ − 4.113363
     $x(3) =$ − 1.029629
     $x(4) =$ 18.461975
(3) $x(1) =$ 0.160055
     $x(2) =$ 0.765538
     $x(3) =$ − 0.915442
     $x(4) =$ 1.103409
(4) $x(1) =$ 1.226849
     $x(2) =$ 7.521628
     $x(3) =$ − 6.350254
     $x(4) =$ − 1.170100
(5) $x(1) =$ − 0.304083
     $x(2) =$ − 0.103485
     $x(3) =$ 0.354872
     $x(4) =$ 1.193169

(6) $x(1) =$ − 1.782385
     $x(2) =$ 3.336623
     $x(3) =$ 2.248588
     $x(4) =$ − 1.701545
(7) $x(1) =$ − 0.748751
     $x(2) =$ 1.486626
     $x(3) =$ 0.184672
     $x(4) =$ − 0.981853
(8) $x(1) =$ 0.329281
     $x(2) =$ 1.117162
     $x(3) =$ − 0.219182
     $x(4) =$ 0.89992
(9) $x(1) =$ 0.240763
     $x(2) =$ − 4.412409
     $x(3) =$ 4.136817
     $x(4) =$ − 1.436729
(10) $x(1) =$ 0.380273
     $x(2) =$ − 0.871830
     $x(3) =$ 1.939354
     $x(4) =$ − 0.694339

Section 3.3

(13) $\begin{pmatrix} 0.262794 & -0.137771 & 0.041703 \\ -0.209004 & 0.163081 & 0.030581 \\ 0.061243 & -0.097063 & 0.055139 \end{pmatrix}$

(14) $\begin{pmatrix} -0.035216 & 0.097910 & -0.049775 \\ 0.059959 & 0.102359 & 0.064300 \\ 0.149203 & 0.010825 & 0.018745 \end{pmatrix}$

(15) $\begin{pmatrix} 0.430973 & 0.090504 & -0.029182 \\ 0.255768 & 0.132968 & -0.102909 \\ -0.058816 & -0.105215 & -0.047445 \end{pmatrix}$

(16) $\begin{pmatrix} -0.536043 & 0.879790 & 0.558417 \\ 0.027821 & 0.135882 & 0.075809 \\ -0.456366 & 0.812808 & 0.621706 \end{pmatrix}$

(17) $\begin{pmatrix} 0.050602 & 0.121435 & -0.100473 & 0.016054 \\ 0.177952 & -0.215930 & 0.112914 & 0.057694 \\ 0.184749 & -0.122687 & 0.229996 & -0.217530 \\ -0.319711 & 0.204368 & -0.100972 & 0.169242 \end{pmatrix}$

$$(18) \begin{bmatrix} 0.181403 & 0.063252 & -0.030726 & -0.032211 \\ -0.067713 & 0.004562 & 0.208298 & -0.183369 \\ 0.223498 & -0.091844 & -0.134541 & 0.180594 \\ -0.313077 & 0.077575 & 0.047206 & 0.050931 \end{bmatrix}$$

$$(19) \begin{bmatrix} 0.236533 & -0.102805 & -0.222786 & 0.099496 \\ -0.190304 & 0.264748 & -0.038786 & -0.068226 \\ 0.131639 & -0.167981 & 0.201823 & -0.053118 \\ -0.073679 & 0.071160 & -0.064571 & 0.139651 \end{bmatrix}$$

$$(20) \begin{bmatrix} 0.071084 & -0.244975 & 0.138191 & 0.009221 \\ -0.034554 & 0.157690 & -0.114844 & 0.108840 \\ 0.204740 & -0.011572 & -0.006023 & -0.085171 \\ -0.314176 & 0.091732 & 0.132318 & 0.025610 \end{bmatrix}$$

Section 3.4

(The numbers correspond to those for Section 3.3.)

(13) 579.0797 87   (17) −729.439222
(14) 660.457961   (18) 1291.411725
(15) −192.971055   (19) 982.2169 08
(16) −93.867763   (20) 1157.435893

Section 3.5

$$(1) \quad \{1,2,3\}, \quad V = c_1 \begin{bmatrix} -1 \\ 1 \\ 2 \end{bmatrix} + c_2 \begin{bmatrix} -2 \\ 1 \\ 4 \end{bmatrix} + c_3 \begin{bmatrix} -1 \\ 1 \\ 4 \end{bmatrix}$$

$$(2) \quad \{1,-2,3\}, \quad V = c_1 \begin{bmatrix} 1 \\ -1 \\ 1 \end{bmatrix} + c_2 \begin{bmatrix} 11 \\ 1 \\ -14 \end{bmatrix} + c_3 \begin{bmatrix} 1 \\ 1 \\ 1 \end{bmatrix}$$

$$(3) \quad \{1,2\}, \quad V = c_1 \begin{bmatrix} 1 \\ 0 \\ 0 \end{bmatrix} + c_2 \begin{bmatrix} 1 \\ 1 \\ -1 \end{bmatrix}$$

$$(4) \quad \{0, \pm 2\sqrt{2}\}, \quad V = c_1 \begin{bmatrix} 2 \\ -3 \\ 8 \end{bmatrix} + c_2 \begin{bmatrix} 2+2\sqrt{2} \\ 1 \\ 0 \end{bmatrix} + c_3 \begin{bmatrix} 2-2\sqrt{2} \\ 1 \\ 0 \end{bmatrix}$$

In problems (5) through (8), our answers have been rounded to the nearest hundredth.

(5) $\{8.98, 4.55, 1.23\}$ (7) $\{2.54, -0.0166, 1.480\}$
(6) $\{14.39, -1.34, -0.052\}$ (8) $\{1, 5\}$

CHAPTER 4

Section 4.2

(1) (2, 2)   (5) (−7, 13)
(2) (1, 1)   (6) (−8, 16)
(3) (5, 11)   (7) (−1, −2)
(4) (8, −15)   (8) (−3, 1)

Section 4.3

(1)	2.512020, 4.182999	(7)	-3.516000, -2.468141
(2)	3.179980, -2.461986	(8)	-1.403005, -0.926981
(3)	-2.536582, 0.556488	(9)	1.401992, 1.372014, -1.815004
(4)	-2.536582, -2.585105	(10)	0.527997, 0.317009, 0.441988
(5)	-2.536582, 3.698081	(11)	1.063369, 2.614218, -3.662252
(6)	3.562059, 1.414979	(12)	3.499676, 2.700028, -3.700001

CHAPTER 5

Section 5.1

(1) $2x^3 - 5x^2 + 8x - 3$ (2) $-2x^2 + 5x + 7$

(3) $3x^3 - x^2 + 7x + 4$ (4) $3.4x - 5.8$

(6)   (a) $p[\dfrac{-69.27509}{x} + \dfrac{363.76921}{x-0.18} - \dfrac{772.11211}{x-0.36}$
$$+ \dfrac{833.31657}{x-0.54} - \dfrac{464.38084}{x-0.72} + \dfrac{113.69839}{x-0.90}]$$

(7)

x	2.5	7.5	12.5	17.5	22.5
$10^9 E$	9.03	3.02	2.16	3.02	9.06

(8)

(a)

x	1.5	2.5	3.5	4.5	5.5	6.5
$10^5 E$	68000	520	27	4.7	2.1	2.4

(b)

x	10.5	11.5	12.5	13.5	14.5	15.5
$10^7 E$	8.2	1.2	0.37	0.22	0.23	0.54

(9)

```
                PROBLEM 13

                                        PROBLEM 13

ERRORS AT MIDVALUES

                                   COEFFICIENTS

MIDVALUE       ERROR

                                   A(0) =   0.0

.105           3.39D-04            A(1) =   9.841190488989335D-01

.290          -5.50D-05            A(2) =   1.968453088030873D-01

.440           2.76D-05            A(3) =  -5.862193281319805D-01

.600          -5.07D-05            A(4) =   2.061121300131508D 00

.780           1.16D-04            A(5) =  -2.134661216136917D 00

.935          -2.77D-04            A(6) =   1.036202611090270D 00
```

(10)   (11)   PROBLEM 15
           PROBLEM 14

                                                          ERRORS AT MIDVALUES

ERRORS AT MIDVALUES

                                                     MIDVALUE   ERROR

MIDVALUE   ERROR

                                                       .042   5.05D-06
  .090   6.00D-04
                                                       .188   -8.20D-06
  .210   -3.12D-05
                                                       .380   4.27D-06
  .300   8.68D-05
                                                       .550   -3.83D-06
  .450   -3.90D-04
                                                       .770   1.77D-05
  .660   3.16D-03
                                                       .955   -8.38D-06
  .890   -3.96D-02

           PROBLEM 14   PROBLEM 15

COEFFICIENTS   COEFFICIENTS

A(0) = 0.0   A(0) = 0.0

A(1) = 9.612147738400021D-01   A(1) = 9.996480374898214D-01

A(2) = 6.081362218001096D-01   A(2) = 7.213161064536113D-03

A(3) = -3.416246821030550D 00   A(3) = -3.788386432800417D-01

A(4) = 9.881847356288168D 00   A(4) = 1.204128840091598D-01

A(5) = -1.268826574923512D 01   A(5) = 8.151952268674343D-02

A(6) = 6.224110545132291D 00   A(6) = -4.455679857277103D-02

(12)

PROBLEM 16

ERRORS AT MIDVALUES

MIDVALUE	ERROR
.034	3.91D-09
.104	-2.54D-09
.260	2.15D-08
.470	-2.40D-08
.720	1.47D-07
.940	-1.13D-07

PROBLEM 16

COEFFICIENTS

A(0) = 0.0
A(1) = 9.9999961768743380-01
A(2) = 1.0843325155850800-05
A(3) = 1.6656635435822590-01
A(4) = 3.9264075619962670-04
A(5) = 7.6066511480407680-03
A(6) = 6.2508636874503540-04

(13)

ERRORS AT MIDVALUES

MIDVALUE	ERROR
.039	6.66D-07
.145	-8.10D-07
.290	1.33D-06
.540	-8.93D-06
.820	8.32D-06
.965	-2.80D-06

PROBLEM 17-

COEFFICIENTS

A(0) = 0.0
A(1) = 9.9994966091088010-01
A(2) = 1.1055834662341740-03
A(3) = -3.4049399532387680-01
A(4) = 1.5690423064429180-02
A(5) = 1.3573850862392960-01
A(6) = -5.0396024785831690-02

(14)   (15)

ERRORS AT MIDVALUES   ERRORS AT MIDVALUES

MIDVALUE	ERROR
.110	2.95D-07
.310	-5.29D-08
.490	2.96D-08
.700	-4.06D-08
.890	1.79D-08
.980	-4.11D-09

MIDVALUE	ERROR
.030	2.50D-05
.196	-1.19D-04
.440	5.15D-05
.630	-2.76D-05
.800	5.22D-05
.945	-7.37D-05

COEFFICIENTS   PROBLEM 20

A(0) = 1.000000000000000D 00

A(1) = 9.999867834454911D-01   COEFFICIENTS

A(2) = 5.001574426668179D-01

A(3) = 1.659499094985164D-01   A(0) = 0.0

A(4) = 4.328010638777396D-02   A(1) = -9.676528937043742D-01

A(5) = 6.436040170680779D-03   A(2) = -1.278720019944953D 00

A(6) = 2.471546289764565D-03   A(3) = 5.906773631920310 00

                                                      A(4) = -2.146698344879946D 01

                                                      A(5) = 3.137089528941973D 01

                                                      A(6) = -1.731172065112032D 01

(16)

PROBLEM 20

PROBLEM 19

COEFFICIENTS

ERRORS AT MIDVALUES

MIDVALUE	ERROR
.041	-3.88D-04
.127	2.18D-04
.250	-6.94D-04
.460	3.65D-03
.680	-7.02D-03
.855	4.05D-02

$A(0) = 1.000000000000000D\ 00$

$A(1) = -2.103263328860820D-03$

$A(2) = 1.051129669598412D\ 00$

$A(3) = -3.214030310255913D-01$

$A(4) = 1.357396621251663D\ 00$

$A(5) = -1.066662324306545D\ 00$

$A(6) = 6.999241562699671D-01$

Section 5.2

(1) 2; (2) -2; (3) 12; (4) -8; (5) 15;
(6) -15; (7) 10; (8) 66; (9) -12; (10) 284;
(11) 100; (12) $385+10i$; (13) 745; (14) 390; (15) 72;
(16) 2880; (17) $\frac{5}{6}$; (18) 20; (19) $\frac{1}{320}$; (20) $-\frac{1}{2}$;
(21) 1, 12,66,220; (22) 1, 13,78,286

Section 5.3

(6) 15, 24,2; (7) 8, 36,30,3; (8) 5, 40,75,26, -1;
(9) 6, 75,260,270,62,1; (10) 7, 122,700,1400,903,126,1;
(11) $2\cos[(u_+ + u)/2]\sin(\Delta u/2)$;
(12) $a2^{-x-1}$;
(13) $-2^{x+2}\cos(3a/2^{x+2})\sin(a/2^{x+1})+2^x\sin(a/2^x)$;
(14) $\sin(\Delta u)/(\cos u_+ \cos u)$;
(15) $-[\sin(a/2^{x+1})]/[\cos(a/2^{x+1})\cos(a/2^x)]$

CHAPTER 6

Section 6.1

(9) $\frac{1}{4}(n-2)(n-1)n(n+1)$

(10) $n(2n^3+16n^2+43n+44)$

(11) $\frac{1}{4}n(n+1)(n^2+n+14)$

(12) $\frac{1}{24} - \frac{1}{6(3n+1)(3n+4)}$

Section 6.2

(1) 2.59752; (2) 2.30953; (3) 0.061968; (4) -0.33677;

(5) 2.01501; (6) 8.40098;

(7) $\dfrac{\sin\frac{1}{2}n(b+\pi)\cos\left[a+\frac{1}{2}(n-1)(b+\pi)\right]}{\sin\frac{1}{2}(b+\pi)}$

(8) $\dfrac{\sin\frac{1}{4}n(2b+\pi)\sin\left[a+\frac{1}{4}(n-1)(2b+\pi)\right]}{\sin\frac{1}{4}(2b+\pi)}$

Section 6.3

(1) $\dfrac{1}{4}ax^4-\dfrac{1}{2}ax^3+\dfrac{1}{4}ax^2+bx$

(2) $\dfrac{1}{5}ax^5-\dfrac{1}{2}ax^4+\dfrac{1}{3}ax^3+(b-\dfrac{1}{3})x$

(3) $\dfrac{n}{2}-\dfrac{\sin nr\cos[2q+(n-1)r]}{2\sin r}$

(4) $\dfrac{3\sin 2nt}{8\sin t}+\dfrac{\sin 6nt}{8\sin 3t}$

(5) $\tan a-\tan(a/2^n)$

(6) $\csc 2p\,[\cot p-\cot(2n+1)p]$

(7) $\sin ns$

(8) $\cot(\pi/2n)$

(9) $[(n-1)a^n-na^{n-1}+1]/(a-1)^2$

(10) $(n^2+n+2)2^n-2$

(11) $\dfrac{2x\sinh\frac{1}{2}(n-1)\sinh\frac{1}{2}n-\cosh n+1}{2\sinh\frac{1}{2}}$

(12) $\dfrac{2^n\sin(n-1)-2^{n-1}\sin n+\sin 1}{10-8\cos 1}$

(13) $(n^2-n+4)2^n-4$

(14) $(n-1)3^{n+1}+3$

(15) $\dfrac{1}{60}[n(n+1)(12n^3+33n^2+37n+8)]$

(16) $\dfrac{1-x^n}{(1-x)^3}-\dfrac{nx^n}{(1-x)^2}-\dfrac{n(n+1)x^n}{2(1-x)}$

Section 6.4

(1) $Ac^k a^{k^2-k}$

(2) $Ac^{k(k-1)/2}a^{k(k+1)/2}$

(3) $A[(k-1)!]^2$

(4) A/k

(5) $A\pi^{k^2}$

(6) Ax^{-k^2}

(7) $A\,2^k+B(-1)^k$

(8) $A\,4^k + B$
(9) $A\cos(\pi k/2) + B\sin(\pi k/2)$
(10) $A(-1)^k + B\cos(\pi k/3) + C\sin(\pi k/3)$
(11) A
(12) $(A + Bk + Ck^2)^{1/2}$
(13) $Ae^{Bk + Ck^2 + Dk^3}$
(14) $Ae^{Bk + Ck^2}$

Section 6.5

(1) $A(3+\sqrt{5}\,)^k + B(3-\sqrt{5}\,)^k - 10$
(2) $A\,3^k + B(-2)^k - \dfrac{1}{6}k - \dfrac{1}{36}$
(3) $A\,2^k + B\,3^k + \dfrac{1}{6}5^k$
(4) $-3^k/4 + A\,4^k + B(-1)^k$
(5) $A + 2^k(B + Ck - \dfrac{3}{2}k^2 + \dfrac{1}{2}k^3)$

(6) $n^k(A\cos\dfrac{\pi k}{2} + B\sin\dfrac{\pi k}{2}) + \dfrac{n^2\cos mk + \cos m(k-2)}{n^4 + 2n^2\cos 2m + 1}$

(7) $A/k! + 1/k$
(8) $k + Ae^{k!}$
(9) $k(k-1) + Ak^2$
(10) $\dfrac{1}{2}e^{k(k-1)}[A + k(k-1)(2k-1)]$
(11) $Ak + \dfrac{1}{2}k^2 + (B+k)k^2$
(13) $A^{1/k}e^{(k-1)/2}$

(14) $(k-1)!\left(A\sum_{n=1}^{k-1}\dfrac{1}{n} + B\right)$

CHAPTER 8

Section 8.1

(7) 5525, (8) 44,100, (9) 65,666,665

Section 8.2

(5) $\zeta(2) - \zeta(4) + \zeta(6) - \displaystyle\sum_{k=1}^{\infty}\dfrac{1}{k^6(k^2+1)}$

(6) $\zeta(3) - \displaystyle\sum_{k=1}^{\infty}\dfrac{1}{k^3(k^4+1)}$

(7) $\zeta(4) - \zeta(6) + \displaystyle\sum_{k=1}^{\infty}\dfrac{1}{k^6(k^2+1)}$

(8) $\zeta(3) - 2\zeta(5) + 3\displaystyle\sum_{k=1}^{\infty}\dfrac{1}{k^3(k^2+1)^2} + 2\displaystyle\sum_{k=1}^{\infty}\dfrac{1}{k^5(k^2+1)^2}$

(9) (a) $\dfrac{1}{4} + \dfrac{4}{67} + \dfrac{1}{18} - \displaystyle\sum_{k=3}^{\infty} \dfrac{2k^5 + k^4 - 2k^3 + 3}{(k^6 + 3)k(k^2 - 1)(k - 2)}$, $E_T \leqslant 2 \times 10^{-5}$

(b) $\zeta(4) - \displaystyle\sum_{k=1}^{\infty} \dfrac{3}{k^4(k^6 + 3)}$, $E_T \leqslant 6 \times 10^{-9}$

(10) (a)

$$-\dfrac{6}{7} + \dfrac{9}{112} + \dfrac{14}{2163} + \dfrac{4}{16352} + \dfrac{1}{1440} - \sum_{k=3}^{\infty} \dfrac{21k^2 - 12}{k(k^6 - 8)(k^2 - 4)(k^2 - 1)},$$

$E_T \leqslant 1.9 \times 10^{-8}$

(b) $\zeta(5) + \displaystyle\sum_{k=1}^{\infty} \dfrac{5k^4 - 8}{k^5(k^6 - 8)}$, $E_T \leqslant 6.1 \times 10^{-7}$

CHAPTER 9

Section 9.1

(12) $\dfrac{4\pi^2}{3} + 4 \displaystyle\sum_{k=1}^{\infty} \left(\dfrac{\cos kx}{k^2} - \dfrac{\pi \sin kx}{k} \right)$

(13) $-\dfrac{\pi}{2} - \dfrac{2}{\pi} \displaystyle\sum_{k=1}^{\infty} \dfrac{\cos(2k-1)x}{(2k-1)^2} + \displaystyle\sum_{k=1}^{\infty} \dfrac{3 \sin(2k-1)x}{2k-1} - \dfrac{\sin kx}{k}$

(14) $\sin x = \dfrac{2}{\pi} - \dfrac{4}{\pi} \displaystyle\sum_{k=1}^{\infty} \dfrac{\cos 2kx}{4k^2 - 1}$

(15) $g(x) = \dfrac{\pi}{4} - \dfrac{2}{\pi} \displaystyle\sum_{k=1}^{\infty} \dfrac{\cos(2k-1)x}{(2k-1)^2} - \displaystyle\sum_{k=1}^{\infty} (-1)^k \dfrac{\sin kx}{k}$

Section 9.2

	PROB. (3)	(4)	(5)	(6)	(7)	(8)
A_0	18.4333	14.8333	2.4083	27.1133	334.1667	9.0500
A_1	−6.9028	88.2076	0.0841	−0.8914	−20.2256	28.0530
A_2	3.4500	2.3333	−0.0625	−0.7958	−3.5000	−2.3667
A_3	3.3000	−34.8333	−0.0117	−0.6450	5.1667	0.9000
A_4	−0.6167	−25.1667	−0.0092	−0.6592	−0.8333	4.8500
A_5	0.6028	3.6258	−0.0025	−0.6836	0.5590	1.6970
A_6	0.5000	6.8333	0.0150	−0.6633	1.5000	1.4833
B_1	21.0896	41.1909	0.1647	4.7097	−12.1395	−0.1317
B_2	−2.8001	5.1962	0.0014	1.6411	−16.1658	−12.4996
B_3	1.9667	21.3333	0.0033	0.9617	−1.0000	−3.1500
B_4	1.0681	4.3301	−0.0072	0.5528	1.7321	3.4064
B_5	−1.0229	1.6424	−0.0114	0.2670	1.1395	−0.2183

(18) (a) $\dfrac{2}{3} - \dfrac{2}{3} \cos(2\pi x/3)$

(b) $\dfrac{1}{2} - \dfrac{1}{2} \cos(\pi x/2) + \dfrac{1}{2} \sin(\pi x/2)$

CHAPTER 10

Section 10.1

For the purpose of checking, we offer a sample of the numerical values of the errors at selected values of x.

PROB.	-0.75	-0.50	-0.25	$+0.25$	$+0.50$	$+0.75$
(1)	2.1D$-$2	8.3D$-$2	2.1D$-$2	2.1D$-$2	8.3D$-$2	2.1D$-$2
(2)	1.9D$-$2	4.4D$-$2	2.8D$-$2	$-$2.2D$-$2	$-$4.4D$-$2	$-$3.1D$-$2
(3)	—	—	—	1.7D$-$5	$-$4.7D$-$5	2.2D$-$5
(4)	2.1D$-$6	2.9D$-$6	$-$2.3D$-$6	2.3D$-$6	$-$2.9D$-$6	$-$2.1D$-$6
(5)	—	—	—	$-$1.2D$-$4	1.3D$-$4	1.9D$-$5

(6) At $x \sim 0.6$, the error is about -4×10^{-4}; at $x \sim 1.6$, it is about 4.9×10^{-4}.

Section 10.2

(9) $3T_0 - 2T_1 + T_2 - T_3$; $2x^2 - 2x + 2$; 1

(10) $2T_0 + T_1 - 4T_2 + 3T_3$; $-8x^2 + x + 6$; 3

(13) 2^{2n-1}

(14) $\pi/2 - (4/\pi)(T_1 + \frac{1}{9}T_3 + \frac{1}{25}T_5 + \cdots)$

(15) $(4/\pi)(\frac{1}{2} - \frac{1}{3}T_2 - \frac{1}{15}T_4 - \frac{1}{35}T_6 - \frac{1}{63}T_8 - \cdots)$

Section 10.3

(1) $\dfrac{46079}{46080}x - \dfrac{959}{5760}x^3 + \dfrac{23}{2880}x^5$; 3.1×10^{-6}

(2) $\dfrac{23041}{23040} + \dfrac{639}{1280}x^2 + \dfrac{7}{160}x^4$; 4.3×10^{-5}

(3) $\dfrac{2897}{2880}x + \dfrac{103}{360}x^3 + \dfrac{41}{180}x^5$; 8.4×10^{-4}

(4) $\dfrac{63}{64}x - \dfrac{5}{24}x^3 - \dfrac{1}{20}x^5$; 2.2×10^{-3}

(5) $\dfrac{65}{64} + x + \dfrac{3}{8}x^2$; 1.6×10^{-2}

(6) $e\left[\dfrac{23009}{23040} - \dfrac{609}{1280}x^2 - \dfrac{49}{480}x^4\right]$; 3.7×10^{-3}

(7) $x\left(1 - \dfrac{\pi^6}{2949120}\right) + x^3\left(-\dfrac{1}{6} + \dfrac{\pi^4}{92160}\right) + x^5\left(\dfrac{1}{120} - \dfrac{\pi^2}{11520}\right)$; 7.3×10^{-5}

(8) $x\left(1 + \dfrac{17\pi^6}{184320}\right) + x^3\left(\dfrac{1}{3} - \dfrac{17\pi^4}{5760}\right) + x^5\left(\dfrac{2}{15} + \dfrac{17\pi^2}{720}\right)$; 2.0×10^{-2}

Index